Lecture Notes in Bioinformatics   5251

Edited by S. Istrail, P. Pevzner, and M. Waterman

Editorial Board: A. Apostolico   S. Brunak   M. Gelfand
T. Lengauer   S. Miyano   G. Myers   M.-F. Sagot   D. Sankoff
R. Shamir   T. Speed   M. Vingron   W. Wong

Subseries of Lecture Notes in Computer Science

Keith A. Crandall   Jens Lagergren (Eds.)

Algorithms in Bioinformatics

8th International Workshop, WABI 2008
Karlsruhe, Germany, September 15-19, 2008
Proceedings

 Springer

Series Editors

Sorin Istrail, Brown University, Providence, RI, USA
Pavel Pevzner, University of California, San Diego, CA, USA
Michael Waterman, University of Southern California, Los Angeles, CA, USA

Volume Editors

Keith A. Crandall
Brigham Young University
Department of Biology
Provo, UT, USA
E-mail: keith_crandall@byu.edu

Jens Lagergren
KTH, Royal Institute of Technology
Computational Biology Department
Stockholm, Sweden
E-mail: jensl@kth.se

Library of Congress Control Number: 2008934600

CR Subject Classification (1998): F.1, F.2.2, E.1, G.1-3, J.3

LNCS Sublibrary: SL 8 – Bioinformatics

ISSN   0302-9743
ISBN-10   3-540-87360-0 Springer Berlin Heidelberg New York
ISBN-13   978-3-540-87360-0 Springer Berlin Heidelberg New York

This work is subject to copyright. All rights are reserved, whether the whole or part of the material is concerned, specifically the rights of translation, reprinting, re-use of illustrations, recitation, broadcasting, reproduction on microfilms or in any other way, and storage in data banks. Duplication of this publication or parts thereof is permitted only under the provisions of the German Copyright Law of September 9, 1965, in its current version, and permission for use must always be obtained from Springer. Violations are liable to prosecution under the German Copyright Law.

Springer is a part of Springer Science+Business Media

springer.com

© Springer-Verlag Berlin Heidelberg 2008
Printed in Germany

Typesetting: Camera-ready by author, data conversion by Scientific Publishing Services, Chennai, India
Printed on acid-free paper   SPIN: 12515653   06/3180   5 4 3 2 1 0

Preface

We are very pleased to present the proceedings of the *8th Workshop on Algorithms in Bioinformatics* (*WABI 2008*), held as part of the *ALGO 2008* meeting at the University of Karlsruhe, Germany on September 15–19, 2008. *WABI 2008* covered all research on algorithmic work in bioinformatics and systems biology with an emphasis on discrete algorithms and machine-learning methods that address important problems in molecular biology. Such algorithms are founded on sound models, are computationally efficient, and have been implemented and tested in simulations and on real datasets. The goal of these proceedings is to present recent research results, including significant work-in-progress, and to identify and explore directions for future research. Original research papers were solicited in all aspects of algorithms in bioinformatics, targeting the following areas in particular:

- Exact, approximate, and machine-learning algorithms for sequence analysis, gene and signal recognition, alignment and assembly, molecular evolution, structure determination or prediction, gene expression, pathways, gene networks, proteomics, functional genomics, and drug design;
- High-performance computing approaches to computationally hard learning and optimization problems in bioinformatics;
- Methods, software, and dataset repositories for development and testing of such algorithms and their underlying models.

The goal of the workshop is to bring together an interdisciplinary group of individuals with research interests in bioinformatic algorithms. For this reason, this year's workshop was held in conjunction with the *16th Annual European Symposium on Algorithms (ESA)*, the *6th Workshop on Approximation and Online Algorithms (WAOA)*, the *8th Workshop on Algorithmic Approaches for Transportation Modeling, Optimization, and Systems (ATMOS)*, and the *ALGO 2008 Graduate Students Meeting*. We believe this interdisciplinary group, especially with the specific inclusion of graduate students and young scientists, is essential for moving the field of bioinformatics forward in exciting and novel ways. We hope this volume attests to that fact with the assortment of exceptional papers contributed to the volume.

We received a total of 81 submissions in response to our call for papers for *WABI 2008* from which 32 were selected to be contributions to this volume. The topics range in biological applicability from genome mapping, to sequence assembly, to microarray quality, to phylogenetic inference, to molecular modeling.

This volume was made possible by the excellent submissions to *WABI 2008*. We thank all the authors for their submissions and especially those of papers selected for inclusion in this volume for their help in revising their work according to peer-review comments. We also thank our distinguished Program Committee for their great help in making difficult decisions on which papers to include in this volume. Anyone who has served on a program committee knows that this involves a great deal of concentrated effort over a short period of time. We are truly grateful for the excellent reviews and recommendations received by our Program Committee. These individuals are listed on the next page.

We were very fortunate to attract Eytan Ruppin of Tel Aviv University to present the Keynote Address for *WABI 2008* on the topic of large scale in *silico* studies of human metabolic diseases. We appreciate his willingness to deliver this address and establish the quality of the meetings. Finally, we would like to thank Peter Sanders and Dorothea Wagner and their local organizing committee (Veit Batz, Reinhard Bauer, Lilian Beckert, Anja Blancani, Daniel Delling, Dennis Luxen, Elke Sauer, and Dominik Schultes) for their excellent job in organizing these joint conferences.

We hope that you will consider contributing to future *WABI* events, through a submission or by participating in the workshops.

September 2008   Keith Crandall
                                                          Jens Lagergren

Conference Organization

Program Chairs

Keith A. Crandall, BYU, USA
Jens Lagergren, KTH, Sweden

Program Committee

Vineet Bafna, UCSD, USA
Dan Brown U. Waterloo, Canada
David Bryant, U. Auckland, New Zealand
Benny Chor, Tel Aviv, Israel
Mark Clement, BYU, USA
Keith A. Crandall (Co-chair), BYU, USA
Miklos Csuros, U. Montreal, Canada
Aaron Darling, U. Queensland, Australia
Matt Dyer, Virginia Tech, USA
Nadia El-Mabrouk, U. Montreal, Canada
Eleazar Eskin, UCLA, USA
Olivier Gascuel, LIRMM-CNRS, France
Roderico Guigo, U. Barcelona, Spain
Eran Halperin, Berkeley, USA
Arndt Von Haeseler, U. Vienna, Austria
Daniel Huson, U. Tübingen, Germany
John Kececioglu, U. Arizona, USA
Jens Lagergren (Co-chair), KTH, Sweden
Jim Leebens-Mack, U. Georgia, USA
Stefano Lonardi, UC Riverside, USA
Istvan Miklos, UPPS, Hungary
Bernard Moret, EPFL, Switzerland
Luay Nakhleh, Rice U., USA
Uwe Ohler, Duke U., USA
Laxmi Parida, IBM TJ Watson Research, USA
Ron Pinter, Technion, Israel
David Posada, U. Vigo, Spain
Ben Raphael, Brown, USA
Knut Reinert, Freie U. Berlin, Germany
Allen Rodrigo, U. Auckland, New Zealand
David Sankoff, U. Ottawa, Canada
Adam Siepel, Cornell U., USA
Mona Singh, Princeton U., USA

Saurabh Sinha, UIUC, USA
Jens Stoye, U. Bielefeld, Germany
Glenn Tesler, UCSD, USA
Esko Ukkonen, Helsinki, Finland
Steve Woolley, Washington U., USA
Christopher Workman, DTU, Denmark

Local Organization

Peter Sanders (Co-chair)
Dorothea Wagner (Co-chair)
Veit Batz
Reinhard Bauer
Lilian Beckert
Anja Blancani
Daniel Delling
Dennis Luxen
Elke Sauer
Dominik Schultes

External Reviewers

Max Alekseyev
José Augusto Amgarten Quitzau
Sandro Andreotti
Alexander Auch
Serdar Bozdag
Robert Beiko
Vincent Berry
Brona Brejova
Michael Brudno
Minh Bui Quang
Robert Castelo
Mark Chaisson
Cedric Chauve
Benny Chor
Matteo Comin
Alexis Criscuolo
Tahir Ejaz
Olof Emanuelsson
N. Fernandez-Fuentes
Philippe Gambette
Jennifer Gardy
Claudio Garruti
Ilan Gronau

Clemens Gröpl
Hugo Gutierrez de Teran
Eran Halperin
Elena Harris
Katharina Jahn
Crystal Kahn
Eun Yong Kang
Miriam Ruth Kantorovitz
Steven Kelk
Gad Kimmel
Dmitry Kondrashov
Carolin Kosiol
Anne Kupczok
Mathieu Lajoie
Guy Lapalme
Sébastien Lemieux
Qiyuan Li
William Majoros
Bob Mau
Paul Medvedev
Martin Milanic
Julia Mixtacki
John Moriarty

Cedric Notredame
Sean O'Rourke
Sebastian Oehm
Stephan Ossowski
Simon Puglisi
Tobias Rausch
Anna Ritz
Regula Rupp
Michael Sammeth
Heiko Schmidt
Sagi Snir
Todd Treangen

Vladimir Vacic
Tomas Vinar
Wei-Bung Wang
David Weese
Sebastian Will
Roland Wittler
Haim Wolfson
Christopher Workman
Nir Yosef
Matthias Zytnicki
Yonghui Wu

Table of Contents

Multichromosomal Genome Median and Halving Problems

Eric Tannier[1], Chunfang Zheng[2], and David Sankoff[2]

[1] INRIA Rhône-Alpes, Université de Lyon 1, Villeurbanne, France
[2] University of Ottawa, Canada

Abstract. Genome median and halving problems aim at reconstructing ancestral genomes and the evolutionary events leading from the ancestor to extant species. The NP-hardness of the breakpoint distance and reversal distance median problems for unichromosomal genomes do not easily imply the same for the multichromosomal case. Here we find the complexity of several genome median and halving problems, including a surprising polynomial result for the breakpoint median and guided halving problems in genomes with circular and linear chromosomes; the multichromosomal problem is thus easier than the unichromosomal one.

1   Introduction

The gene order or syntenic arrangement of ancestral genomes may be reconstructed based on comparative evidence from present-day genomes — the phylogenetic approach — or on internal evidence in the case of genomes descended from an ancestral polyploidisation event, or from a combination of the two. The computational problem at the heart of phylogenetic analysis is the *median problem*, while internal reconstruction inspires the *halving problem*, and the combined approach gives rise to *guided halving*. How these problems are formulated depends first on the karyotypic framework: the number of chromosomes in a genome and whether they are constrained to be linear, and second on the objective function used to evaluate solutions. This function is based on some notion of genomic distance, either the number of *breakpoints*, adjacent elements on a chromosome in one genome that are disrupted in another, or the number of evolutionary operations necessary to transform one genome to another.

While the karyotypes allowed in an ancestor vary only according to the dimensions of single versus multiple chromosome, and linear versus circular versus mixed, the genomic distances of interest have proliferated according to the kinds of evolutionary operations considered, from the classic, relatively constrained, reversals/transocations distance to the more inclusive *double cut and join* measure, and many others.

The complexity of each of the problems is known for one or more distances, in one or more specific karyotypic contexts, and it is sometimes taken for granted that these results carry over to other combinations of context and distance. This is not necessarily the case. In this paper, we survey the known results and

K.A. Crandall and J. Lagergren (Eds.): WABI 2008, LNBI 5251, pp. 1–13, 2008.
© Springer-Verlag Berlin Heidelberg 2008

unsolved cases for three distance measures in three kinds of karyotype, including several results presented here for the first time, including both new polynomial-time algorithms and NP-hardness proofs.

2   Genomes, Breakpoints and Rearrangements

Multichromosomal Genomes. We follow the general formulation of a genome in [3]. A *gene A* is an oriented sequence of DNA, identified by its *tail* A_t and its *head* A_h. Tails and heads are the *extremities* of the genes. An *adjacency* is an unordered pair of gene extremities; a *genome* is a set of adjacencies on a set of genes. Each adjacency in a genome means that two gene extremities are consecutive on the DNA molecule. In a genome, each gene extremity is adjacent to zero or one other extremity. An extremity x that is not adjacent to any other extremity is called a *telomere*, and can be written as an adjacency $x\circ$ with a null symbol $\circ$. Consider the graph G_Π whose vertices are all the extremities of the genes, and the edges include all the adjacencies in a genome Π as well as an edge joining the head and the tail of each gene. This graph is a set of disjoint paths and cycles. Every connected component is called a *chromosome* of Π. A chromosome is *linear* if it is a path, and *circular* if it is a cycle.

A genome with only one chromosome is called *unichromosomal*. These are *signed permutations* (linear or circular). A genome with only linear chromosomes is called a *linear* genome.

Genomes can be represented as a set of strings, by writing the genes for each chromosome in the order in which they appear in the paths and cycles of the graph G_Π, with a bar over the gene if the gene is read from the head to the tail (we say it has *negative* sign), and none if it is read from the tail to the head (it has *positive* sign). For each linear chromosome, there are two possible equivalent strings, according to the arbitrary chosen starting point. One is obtained from the other by reversing the order and switching the signs of all the genes. For circular chromosomes, there are also two possible circular string representations, according to the direction in which the cycle is traversed.

For example, if a genome Π is defined as the set of adjacencies on the set of genes $\{1, 2, 3, 4, 5, 6, 7, 8, 9, 10\}$

$$\{\circ 2_h, 2_t 1_h, 1_t 9_h, 9_t T, T 10_t, 10_h 6_h, 6_t 4_t, 4_h 3_h, 3_t T, T 8_t, 8_h 5_t, 5_h 7_t, 7_h \circ\},$$

we write it as the linear genome with 3 chromosomes:

$$\Pi = (\ \overline{2}\ \overline{1}\ \overline{9}\ ,\quad 10\ \overline{6}\ 4\ \overline{3}\ ,\quad 8\ 5\ 7\ )$$

A *duplicated gene A* is a couple of homologous oriented sequences of DNA, identified by two tails $A1_t$ and $A2_t$, and two heads $A1_h$ and $A2_h$. An *all-duplicates genome* Δ is a set of adjacencies on a set of duplicated genes.

For example, the following genome Δ is an all-duplicates genome on the set of genes $\{1, 2, 3, 4, 5\}$.

$$\Delta = (\ \overline{2}\ \overline{1}\ 2\ \overline{5}\ ,\quad 4\ \overline{3}\ 4\ \overline{1}\ ,\quad 3\ 5\ )$$

For a genome Π on a gene set $\mathcal{G}$, a *doubled genome* $\Pi \oplus \Pi$ is an all-duplicates genome on the set of duplicated genes from $\mathcal{G}$ such that if $A_x B_y$ is an adjacency of Π ($x, y \in \{t, h\}$), either $A1_x B1_y$ and $A2_x B2_y$, or $A2_x B1_y$ and $A1_x B2_y$ are adjacencies of $\Pi \oplus \Pi$. This includes telomeric adjacencies, so that a telomere in Π should yield two telomeres in $\Pi \oplus \Pi$.

Note the difference between a general all-duplicates genome and the special case of a doubled genome: the former has two copies of each gene, while in the latter these copies are organised in such a way that there are two identical copies of each chromosome (when we ignore the 1's and 2's in the $A1_x$'s and $A2_x$'s): it has two linear copies of each linear chromosome, and for each circular chromosome, either two circular copies or one circular chromosome containing the two successive copies. Note also that for a genome Π, there is an exponential number of possible doubled genomes $\Pi \oplus \Pi$ (exactly two to the power of the number of non telomeric adjacencies).

In discussing all-duplicates genomes, we will sometimes contrast them with *ordinary genomes* which have a single copy of each gene.

The Breakpoint Distance. We construct a breakpoint distance on multichromosomal genomes that depends on common adjacencies, or rather their absence, and also on common telomeres (or lack thereof) in two genomes. For two genomes Π and Γ on n genes, suppose Π has N_Π chromosomes, and Γ has N_Γ chromosomes. Let a be the number of common adjacencies, e be the number of common telomeres of Π and Γ. Then insofar as it should depend additively on these components, we may suppose the breakpoint distance has form

$$d_{BP}(\Pi, \Gamma) = n - a\beta - e\theta + (N_\Pi + N_\Gamma)\gamma + (|N_\Pi - N_\Gamma|)\psi,$$

where β, θ and γ are positive parameters, while ψ may have either sign. Taking $\Pi = \Gamma$ and imposing $d_{BP}(\Pi, \Pi) = 0$ yields the relations $\beta = 1$ and $1 - 2\theta + 2\gamma = 0$, so $\theta = \gamma + 1/2$. Now it is most plausible to count a total of 1 breakpoint for a fusion or fission of linear chromosomes, which implies $\gamma = \psi = 0$, so the most natural choice of *breakpoint distance* between Π and Γ is

$$d_{BP}(\Pi, \Gamma) = n - a - \frac{e}{2}.$$

For an all-duplicates genome Δ and an ordinary genome Π, the *breakpoint distance* between Π and Δ is $d_{BP}(\Pi, \Delta) = \min_{\Pi \oplus \Pi} d_{BP}(\Pi \oplus \Pi, \Delta)$.

The Double-Cut-and-Join Distance. Given a genome Π, which is defined as a set of adjacencies, a double-cut-and-join (DCJ) is an operation ρ acting on two adjacencies pq and rs (possibly some of p, q, r, s are $\circ$ symbols and even an adjacency may be composed of two $\circ$ symbols). The DCJ operation replaces pq and rs either by pr and qs, or ps and qr.

A DCJ can reverse an interval of a genome (DCJs include reversals), and may also fission one chromosome into two, fusion two chromosomes into a single one, or achieve a reciprocal translocation between two chromosomes. Two consecutive

DCJ operations may result in a block interchange: two segments of a genome exchange their positions, which results in a transposition if the two intervals are contiguous in the permutation. DCJ is thus a very general framework. It was introduced by Yancopoulos *et al.* [22], as well as by Lin *et al.* in a special case [14], and has since been adopted by Bergeron *et al.* [3] and many others, and has also been called "2-break rearrangement" [2].

If Π and Γ are two genomes on n genes, the *DCJ distance* $d_{DCJ}(\Pi, \Gamma)$ is the minimum number of DCJ operations needed to transform Π into Γ.

For an all-duplicates genome Δ and an ordinary genome Π, the *DCJ distance* between Π and Δ is $d_{DCJ}(\Pi, \Delta) = \min_{\Pi \oplus \Pi} d_{DCJ}(\Pi \oplus \Pi, \Delta)$.

The Reversal/Translocation Distance. The reversal/translocation distance was introduced by Hannenhalli and Pevzner [11], and is equivalent to the DCJ distance constrained to linear genomes.

If Π is a linear genome, a *linear* DCJ operation is a DCJ operation on Π that results in a linear genome. This allows reversals, reciprocal translocations, and chromosome fusions, fissions, which are special cases of translocations. Other DCJs, that create temporary circular chromosomes, are not allowed. If Π and Γ are linear genomes, the *RT* distance between Π and Γ is the minimum number of linear DCJ operations that transform Π into Γ, and is noted $d_{RT}(\Pi, \Gamma)$.

3   Computational Problems

The classical literature on genome rearrangements aims at reconstructing the evolutionary events and ancestral configurations that explain the differences between extant genomes. The focus has been on the genomic distance, median and halving problems. More recently the doubled distance and guiding halving problems have also emerged as important. In each of the ensuing sections of this paper, these five problems are examined for a specific combination of distance d (breakpoint, DCJ or RT) and kind of multichromosomal karyotype.

1- Distance. Given two genomes Π, Γ, compute $d(\Pi, \Gamma)$. Once the distance is calculated, an additional problem in the cases of DCJ and RT is to reconstruct the rearrangement scenario, i.e., the events that differentiate the genomes.

2- Double Distance. Given an all-duplicates genome Δ and an ordinary genome Π, compute $d(\Delta, \Pi)$. Because the assignment of labels "1" or "2" to the two identical (for our purposes) copies of a duplicated gene in Δ is arbitrary, the double distance problem is equivalent to finding such an assignment that minimises the distance between Δ and a genome $\Pi \oplus \Pi$ considered as ordinary genomes, where all the genes on any one chromosome in $\Pi \oplus \Pi$ are uniformly labeled "1" or "2" [2,26]. The double distance function is not symmetric because Δ is an all-duplicates genome and Π is an ordinary one, thus capturing the presumed asymmetric temporal and evolutionary relationship between the ancestor Π and the present-day genome Δ.

3- Median. Given three genomes Π_1, Π_2, Π_3, find a genome M which minimises $d(\Pi_1, M) + d(\Pi_2, M) + d(\Pi_3, M)$. The median problem estimates the common ancestor of two genomes, given a third one (not necessarely specifies) as an outgroup. It may be used as a hint for phylogenetic studies.

4- Halving. Given an all-duplicates genome Δ, find an ordinary genome Π which minimises $d(\Delta, \Pi)$. The goal of a halving analysis is to reconstruct the ancestor of an all-duplicates genome at the time of the doubling event.

5- Guided Halving. Given an all-duplicates genome Δ and an ordinary genome Π, find an ordinary genome M which minimises $d(\Delta, M) + d(M, \Pi)$. The guided halving problem is similar to the genome halving problem for Δ, but it takes into account the ordinary genome Π of an organism presumed to share a common ancestor with M, the reconstructed undoubled ancestor of Δ.

We will survey these five computational problems for the three distances that we have introduced, in the cases of multichromosomal genomes containing all linear chromosomes or permitting circular chromosomes.

4   Breakpoint Distance, General Case

In this section, $d = d_{BP}$, and genomes are considered in their most general definition, that is, multichromosomal with both circular and linear chromosomes allowed. As the nuclear genome of a eukaryotic species, such a configuration would be rare and unstable. Nevertheless this case is of great theoretical interest, as it is the only combination of distance and karyotype where all five problems mentioned in Section 3 prove to be polynomially solvable, including the only genomic median problem that is polynomially solvable to date. Furthermore, the solutions in this context may suggest approaches for other variants or the problems, as well as providing a rapid bound for other distances, through the Watterson *et al.* bound [21].

Distance and Double Distance. The distance computation follows directly from the definition, and is easily achievable in linear time.

The double distance computation is also easy: let Π be a genome and Δ be an all-duplicates genome. Let ab be an adjacency in Π (a or b may be $\circ$ symbols). If $a1b1$ or $a2b2$ is an adjacency in Δ, choose $a1b1$ and $a2b2$ for adjacencies in $\Pi \oplus \Pi$. If $a1b2$ or $a2b1$ is an adjacency in Δ, choose $a1b2$ and $a2b1$ for adjacencies in $\Pi \oplus \Pi$. The two cases are mutually exclusive, so the assignment is made without ambiguity. Assign all remaining adjacencies arbitrarily.

It is easy to see that this procedure minimises $d(\Pi \oplus \Pi, \Delta)$, as every possible common adjacency or telomere in Δ and Π is a common adjacency or telomere in $\Pi \oplus \Pi$ and Δ.

Median. The following result contrasts with the NP-completeness proofs of all genome median problems in the literature [6,7,17]. The problem is NP-complete for unichromosomal genomes, whether they are linear or circular [6,17], but the multichromosomal case happens to be easier.

Theorem 1. *There is a polynomial time algorithm for the multichromosomal genome median problem.*

Proof. For this extended abstract, we show only the principle of the algorithm, and omit the details of the proofs. Let Π_1, Π_2, Π_3 be three genomes on a gene set $\mathcal{G}$ of size n. Draw a complete graph G on the vertex set containing the union of all the extremities of the genes in $\mathcal{G}$ and a set containing one supplementary vertex t_x for every gene extremity x. For any pair of gene extremities x, y, weight the edge xy by the number of genomes, among Π_1, Π_2, Π_3, for which xy is an edge. Then each edge in G joining two gene extremities is weighted by 0, 1, 2 or 3. Now for any vertex x, weight the edge xt_x by half the number of genomes, among Π_1, Π_2, Π_3, having x as a telomere. Each edge xt_x is then weighted by 0, $\frac{1}{2}$, 1, or $\frac{3}{2}$. To every other edge of the complete graph G, assign the weight 0.

Let M be a perfect matching in G. Clearly, the edges between gene extremities in G define the adjacencies of a genome, that we also call M. The relation between the weight of the perfect matching M and the median score of the genome M is easy to state:

Claim. The weight of the perfect matching M in G is $3n-(d(\Pi_1, M)+d(\Pi_2, M)+d(\Pi_3, M))$.

This implies that a maximum weight perfect matching M is a minimum score median genome. As the maximum weight perfect matching problem is polynomial, so is the breakpoint median problem. $\square$

Note that this algorithm remains valid if the median of more than three genomes is to be computed.

Halving. To our knowledge, the genome halving with breakpoint distance has not yet been studied. In this framework, it has an easy solution, using a combination of elements from the maximum weight perfect matching technique in Theorem 1 and the double distance computation: let Δ be an all-duplicates genome on a gene set $\mathcal{G}$, and G be the complete graph on the vertex set containing all the extremities of the genes in $\mathcal{G}$, plus one supplementary vertex t_x for every gene extremity x. For any pair of gene extremities x, y, weight the edge xy by zero, one or two, according to the number of times an xy adjacency is present in Δ. Now for any vertex x, weight the edge xt_x by half the number of times x is a telomere in Δ. Weight the remaining edges between t vertices by zero.

Claim. A maximum weight perfect matching M in G defines the adjacencies of a genome M minimising $d(\Delta, M)$.

Guided Halving. Again, this will be the only polynomial result for the guided genome halving problem. The solution combines elements of the three algorithms (double distance, median, halving) previously discussed in this section.

Let Δ be an all-duplicates genome on a gene set $\mathcal{G}$, and Π be a genome on $\mathcal{G}$. Let G be the complete graph on the vertex set containing all the extremities of the genes in $\mathcal{G}$, plus one supplementary vertex t_x for every gene extremity x. For any pair of gene extremities x, y, weight the edge xy by the number of times x is adjacent to y in Δ and Π, and weight the edge xt_x by half the number of times x is a telomere in Δ and Π. Weight the remaining edges between t vertices by zero.

Claim. A maximum weight perfect matching M in G defines the adjacencies of a genome M minimising $d(\Delta, M) + d(M, \Pi)$.

5 Breakpoint Distance, Linear Case

In this section, $d = d_{BP}$ and all genomes must be linear, as is most appropriate for modeling for the eukaryotic nuclear genome. The solutions to the distance and double distance problems are the same as in the previous section, where circularity was allowed. But in contrast to the model of Section 4, all the problems concerning at least three genomes are NP-complete.

The Median Problem

Theorem 2. *The breakpoint median problem for multichromosomal linear genomes is NP-hard.*

Proof. For this extended abstract, we show only the principle of the reduction, and omit the details. We use a reduction from the circular permutation median (CPM) problem, which asks: Given three circular genomes Π_1, Π_2, and Π_3 with only one chromosome, find a circular genome M with only one chromosome, which minimises $d(\Pi_1, M) + d(\Pi_2, M) + d(\Pi_3, M)$. This problem is NP-hard [6,17].

Let Π_1, Π_2, Π_3, be an instance of the CPM problem, on the gene set $\{1, \ldots, n\}$. Let $n + 1$ be a new gene, and Π_i' be the genome constructed from Π_i ($1 \leq i \leq 3$) by deleting the adjacency $x1_t$ (x is the extremity of a gene in $\{2, \ldots, n\}$), and adding the adjacency $x(n + 1)_t$. Genomes Π_1', Π_2' and Π_3' are linear. Let k be a positive integer.

Claim. There exists a unichromosomal and circular genome M on $\{1, \ldots, n\}$ with $d(\Pi_1, M) + d(\Pi_2, M) + d(\Pi_3, M) \leq k$ if and only if there exists a multichromosomal and linear genome M' on $\{1, \ldots, n + 1\}$ with $d(\Pi_1', M') + d(\Pi_2', M') + d(\Pi_3', M') \leq k$. (This claim implies the theorem). $\qquad\square$

Halving and Guided Halving. Surprisingly, the halving problem has not been treated in the literature. We conjecture it has a polynomial solution, because the halving problem for all other rearrangement distances is polynomial. Constructing a solution is beyond the scope of this paper, and the problem remains open.

This guided halving problem is NP-hard, as proved in [24], using the NP-completeness result for the median proved above.

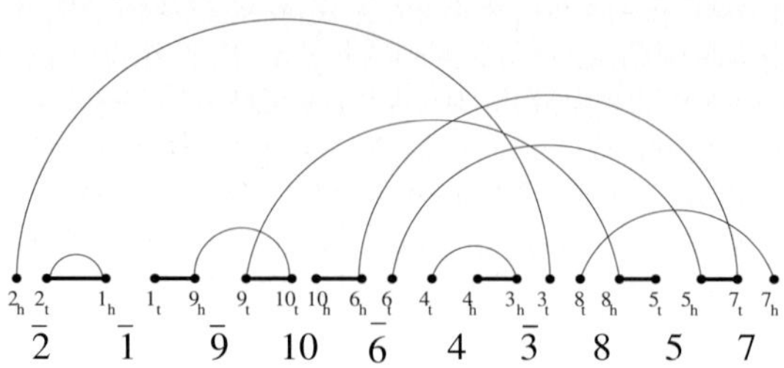

Fig. 1. The breakpoint graph of the genomes $\Pi = (\overline{2}\ \overline{1}\ \overline{9}\ 10\ \overline{6}\ ;\ 4\ \overline{3}\ ;\ 8\ 5\ 7)$ and $\Gamma = (1\ 2\ 3\ 4\ ;\ 5\ 6\ 7\ 8\ 9\ 10)$. Π-edges are drawn with bold segments, and Γ-edges are the thin circle arcs.

6   DCJ Distance, General Case

In this section, $d = d_{DCJ}$. Genomes can have several chromosomes, circular or linear. This is the most general context in which the DCJ distance has been explicitly formulated [3].

The complexity of the genome median probem is not established by the work of Caprara [7], who proved the unichromosomal result only. We show its NP-hardness here. The double distance problem was proposed by Alekseyev and Pevzner [2], and we will show its NP-hardness as well.

Distance. There is an easy linear solution, both for the distance and the scenario computation [3,22].

The *breakpoint graph* of two genomes Π and Γ, denoted by $BP(\Pi, \Gamma)$, is the bipartite graph whose vertex set is the set of extremities of the genes, and there is an edge between two vertices x and y if xy is an adjacency in either Π (these are Π-edges) or Γ (Γ-edges). Note that we do not invoke any $\circ$ symbols. Vertices in this graph have degree zero, one or two, so that the graph is a set of paths (possibly including some with no edges) and cycles. It is also the line-graph of the *adjacency graph*, an alternate representation in [3], and is commonly used in genome rearrangement studies. Fig. 1 shows an example of a breakpoint graph. Theorem 3 shows how to obtain the distance directly from the graph.

Theorem 3. $[3]^1$ *For two genomes Π and Γ, let $c(\Pi, \Gamma)$ be the number of cycles of the breakpoint graph $BP(\Pi, \Gamma)$, and $p(\Pi, \Gamma)$ be the number of paths with an even number of edges. Then*

$$d(\Pi, \Gamma) = n - c(\Pi, \Gamma) - \frac{p(\Pi, \Gamma)}{2}.$$

Note the similarity to the breakpoint distance formula in Section 2. The number of genes n is the same in both formulae, the parameter c is related to parameter

---

[1] The formula is presented in [3] with the cycles and odd paths of the adjacency graph. This corresponds to cycles and even paths of the breakpoint graph, as it is the line-graph of the adjacency graph.

a in the breakpoint formula in that each common adjacency is a cycle of the breakpoint graph (with two parallel edges), and parameter p is related to parameter e, as each shared telomere is an even path (with no edge) in the breakpoint graph. Although these two measures of genomic distance were derived in different contexts and through different reasoning, their formulae show a remarkably similar form. They differ in that the DCJ formula counts non trivial cycles and paths, but for distant genomes, they tend to give the same values.

Double Distance

Theorem 4. *The DCJ double distance problem is NP-hard for multichromosomal genomes.*

Proof. For this extended abstract, we show only the principle of the reduction, and omit the details. Reduction is from the *breakpoint graph decomposition* (BGD) problem (see [7]). A graph G is *bicoloured* if all its edges are coloured in either red or blue; it is *balanced* if it has only degree 2 or degree 4 vertices, every vertex is incident to the same number of red and blue edges, and there is no cycle formed by only red or only blue edges.

Given a balanced bicoloured graph G, the breakpoint graph decomposition problem is to find a decomposition of the edges of G into a maximum number of edge-disjoint cycles, each alternating between red and blue edges. Berman and Karpinski [4] proved APX-hardness of this problem.

Let G be a balanced bicoloured graph on n vertices, defining an instance of the BGD problem. Define the gene set $\mathcal{G}$ as the vertex set of G. Construct an all-duplicates genome Δ and a genome Π on $\mathcal{G}$ in the following way. First, for each vertex X of G, let X_t and X_h be its extremities; let $X_t X_h$ be an adjacency in Π. Then, for every vertex X of G, let $X1_t$, $X1_h$, $X2_t$ and $X2_h$, be the extremities of the duplicated gene X. For each blue edge XY in G, construct an adjacency in Δ joining the heads of genes $X1$ or $X2$, and $Y1$ or $Y2$: if vertex X has degree 4, one of the two adjacencies defined by the two blue edges involves $X1_h$, and the other $X2_h$ (arbitrarily). If vertex X has degree 2, define the adjacency with $X1_h$ and add another adjacency $X1_t X2_h$ in Δ. For each red edge, add an adjacency in Δ according to the same principle, but joining tails of genes.

We then have an all-duplicates genome Δ, and a genome Π. Note that Π is composed of n circular chromosomes, one for each gene, and that neither Π nor Δ have telomeres.

Claim. The maximum number of edge-disjoint alternating cycles in G is equal to $2n - d(\Delta, \Pi)$. (This claim implies the theorem). □

Median. Though effective heuristics are available [1], we have:

Theorem 5. *The DCJ median problem for multichromosomal genomes is NP-hard.*

Proof. We use a reduction from the breakpoint graph decomposition defined in the proof of Theorem 4, in a way very similar to part of Caprara's proof [7] for the unichromosomal case.

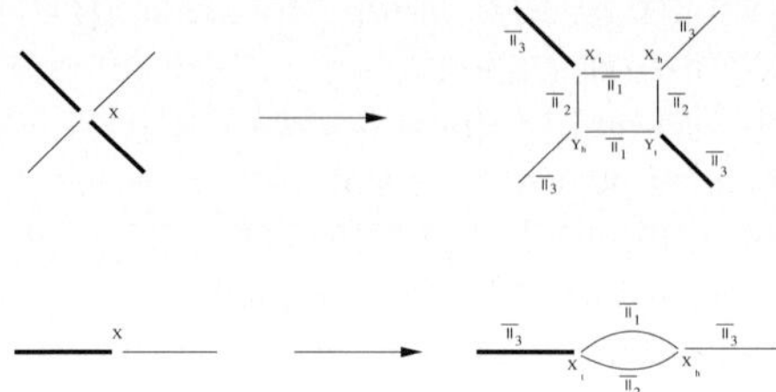

Fig. 2. Strategy for reducing the breakpoint graph decomposition to the DCJ median problem. Red edges are represented by thick lines, while blue edges are thin.

Let G be a balanced bicoloured graph on n vertices. Define the gene set $\mathcal{G}$ as a set containing one gene X for every degree 2 vertex of G, and two genes X and Y for every degree 4 vertex of G.

Apply the following transformation to G, which is similar to the transformation in [7], as illustrated in Figure 2.

Let v be a vertex of degree 2 in G. Replace v by two vertices labeled by the two extremities of the associated gene X, namely X_t and X_h. The blue edge incident to v becomes incident to X_h and the red edge to X_t. Add one Π_1-edge and one Π_2-edge between X_h and X_t. Now let v be a vertex of degree 4 in G. Replace v by four vertices labeled by the four extremities of X and Y, X_t, X_h, Y_h, and Y_t. The blue edges incident to v become incident to X_h and Y_h, while the red edges become incident to X_t and Y_t. Add two Π_1-edges X_tX_h and Y_tY_h, and two Π_2-edges X_tY_h and Y_tX_h. Red and blue edges are the Π_3-edges. Call the final graph G'.

It is easy to see that Π_1, Π_2, and Π_3 define genomes on the set of genes $\mathcal{G}$, and they have no telomeres. Let $w2$ be the number of degree 2 vertices of Γ, and $w4$ be the number of degree 4 vertices of Γ.

Claim. There exists a genome M such that $d(M, \Pi_1) + d(M, \Pi_2) + d(M, \Pi_3) \leq w2 + 3w4 - k$ if and only if there exists at least k edge-disjoint alternating cycles in G. (This claim implies the theorem.) □

Halving and Guided Halving. This problem has a polynomial solution, as recently stated for unichromosomal genomes by [2] and in the general case by [15,20]. All these algorithms are simplified versions of the algorithm by El-Mabrouk and Sankoff [9], developed for the RT rearrangement distance.

Theorem 6. *Guided halving is NP-complete for multichromosomal genomes.*

We omit here the proof of this theorem, based on a reduction of the same problem and similar ideas than in the previous one.

7   DCJ and Reversal/Translocation, Linear Chromosomes

In the original formulation of the DCJ distance [22], it was shown that there is a solution where each excision of a circular intermediate could be followed directly by its reinsertion. Thus the median and halving problems can be stated in terms of exclusively linear chromosomes in both the data genomes and the reconstructed ancestor. They all remain open.

Hannenhalli and Pevzner proposed a polynomial-time algorithm for calculating $d_{RT}(\Pi, \Gamma)$ for two genomes Π and Γ [11]. This was reformulated in [19] and minor corrections were added by [16] and [12]. A polynomial time genome halving algorithm was given in [9]. Though the constrained DCJ distance in the preceding paragraph is arguably just as realistic, because of the long history of d_{RT}, effective heuristics have been developed and applied for the double distance [23,26], median [5,13] and guided halving problems [23,25,26], but their complexities remain open. Note that Chen *et al.* give an NP-completeness result on a problem which slightly generalizes the RT double-distance probem.

8   Conclusions

Table 1. Current knowledge of the status of complexity questions for five problems related to ancestral genome reconstruction, for eight genomic distances in the unichromosomal and multichromosomal contexts, including the new results in this paper. Other versions of the halving problem are less restrictive [2,9,20]. P and NP stand for polynomial and NP-hard, respectively; when followed by ?, represent our conjectures.

problem context	distance	halving	double distance	median	guided halving
breakpoint uni	P	open	P	NP [6,17]	open
breakpoint general multi	P new	P new	P new	P new	P new
breakpoint linear multi	P new	open P?	P new	NP new	NP [24]
DCJ uni	P [3,22]	P [2]	open	NP [7]	open
DCJ general multi	P [3,22]	P [15,20]	NP new	NP new	NP new
DCJ linear multi	P [22]	open	open	open NP?	open NP?
RT uni	P [10]	open	open	NP [7]	open
RT multi	P [11,12,16,19]	P [9]	open NP?	open NP?	open NP?

Acknowledgements

Research supported in part by a grant to DS and a doctoral fellowship to CZ from the Natural Sciences and Engineering Research Council of Canada (NSERC). ET is funded by the Agence Nationale pour la Recherche (GIP ANR JC05_49162 and NT05-3_45205) and the Centre National de la Recherche Scientifique (CNRS). DS holds the Canada Research Chair in Mathematical Genomics.

References

1. Adam, Z., Sankoff, D.: The ABCs of MGR with DCJ. Evol. Bioinform. 4, 69–74 (2008)
2. Alekseyev, M., Pevzner, P.: Colored de Bruijn graphs and the genome halving problem. TCBB 4, 98–107 (2008)

3. Bergeron, A., Mixtacki, J., Stoye, J.: A unifying view of genome rearrangements. In: Bücher, P., Moret, B.M.E. (eds.) WABI 2006. LNCS (LNBI), vol. 4175, pp. 163–173. Springer, Heidelberg (2006)
4. Berman, P., Karpinski, M.: On some tighter inapproximability results: Further improvements. ECCC Report 65, Univ. of Trier. (1998), http://www.eccc.uni-trier.de/
5. Bourque, G., Pevzner, P.A.: Genome-scale evolution: Reconstructing gene orders in the ancestral species. Genome Res. 12, 26–36 (2002)
6. Bryant, D.: The complexity of the breakpoint median problem. TR CRM-2579. Centre de recherches mathématiques, Université de Montréal (1998)
7. Caprara, A.: The reversal median problem. INFORMS J. Comput. 15, 93–113 (2003)
8. Chen, X., Zheng, J., Fu, Z., Nan, P., Zhong, Y., Lonardi, S., Jiang, T.: Assignement of orthologous genes via genome rearrangement. IEEE/ACM Transactions on Computational Biology and Bioinformatics 2, 302–315 (2005)
9. El-Mabrouk, N., Sankoff, D.: The reconstruction of doubled genomes. SIAM J. Comput. 32, 754–792 (2003)
10. Hannenhalli, S., Pevzner, P.A.: Transforming cabbage into turnip: Polynomial algorithm for sorting signed permutations by reversals. JACM 46, 1–27 (1999)
11. Hannenhalli, S., Pevzner, P.A.: Transforming men into mice (polynomial algorithm for genomic distance problem). In: FOCS 1995, pp. 581–592 (1995)
12. Jean, G., Nikolski, M.: Genome rearrangements: a correct algorithm for optimal capping. Inf. Process Lett. 104, 14–20 (2007)
13. Lenne, R., Solnon, C., Stützle, T., Tannier, E., Birattari, M.: Reactive stochastic local search algorithms for the genomic median problem. In: van Hemert, J., Cotta, C. (eds.) EvoCOP 2008. LNCS, vol. 4972, pp. 266–276. Springer, Heidelberg (2008)
14. Lin, Y.C., Lu, C.L., Chang, H.Y., Tang, C.: An efficient algorithm for sorting by block-interchange and its application to the evolution of vibrio species. JCB 12, 102–112 (2005)
15. Mixtacki, J.: Genome halving under DCJ revisited. In: Hu, X., Wang, J. (eds.) COCOON 2008. LNCS, vol. 5092. Springer, Heidelberg (2008)
16. Ozery-Flato, M., Shamir, R.: Two notes on genome rearrangement. JBCB 1, 71–94 (2003)
17. Pe'er, I., Shamir, R.: The median problems for breakpoints are NP-complete. In: ECCC (1998)
18. Sankoff, D., Blanchette, M.: The median problem for breakpoints in comparative genomics. In: Jiang, T., Lee, D.T. (eds.) COCOON 1997. LNCS, vol. 1276, pp. 251–263. Springer, Heidelberg (1997)
19. Tesler, G.: Efficient algorithms for multichromosomal genome rearrangements. JCSS 65, 587–609 (2002)
20. Warren, R., Sankoff, D.: Genome halving with general operations. APBC 2008, Adv. Bioinform. Comput. Biol. 6, 231–240 (2008)
21. Watterson, G., Ewens, W., Hall, T., Morgan, A.: The chromosome inversion problem. J. Theoret. Biol. 99, 1–7 (1982)
22. Yancopoulos, S., Attie, O., Friedberg, R.: Efficient sorting of genomic permutations by translocation, inversion and block interchange. Bioinform. 21, 3340–3346 (2005)
23. Zheng, C., Wall, P.K., Leebens-Mack, J., de Pamphilis, C., Albert, V.A., Sankoff, D.: The effect of massive gene loss following whole genome duplication on the algorithmic reconstruction of the ancestral Populus diploid. In: CSB 2008 (in press, 2008)

24. Zheng, Z., Zhu, Q., Adam, Z., Sankoff, D.: Guided genome halving: hardness, heuristics and the history of the Hemiascomycetes. In: ISMB 2008 (in press, 2008)
25. Zheng, Z., Zhu, Q., Sankoff, D.: Genome halving with an outgroup. Evol. Bioinform. 2, 319–326 (2006)
26. Zheng, Z., Zhu, Q., Sankoff, D.: Descendants of whole genome duplication within gene order phylogeny. In: JCB (in press, 2008)

A Branch-and-Bound Method for the Multichromosomal Reversal Median Problem

Meng Zhang[1], William Arndt[2], and Jijun Tang[2]

[1] College of Computer Science and Technology
Jilin University, China
`zhangmeng@jlu.edu.cn`
[2] Department of Computer Science and Engineering
University of South Carolina, USA
`{arndtw, jtang}@cse.sc.edu`

Abstract. The ordering of genes in a genome can be changed through rearrangement events such as reversals, transpositions and translocations. Since these rearrangements are "rare events", they can be used to infer deep evolutionary histories. One important problem in rearrangement analysis is to find the median genome of three given genomes that minimizes the sum of the pairwise genomic distance between it and the three others. To date, MGR is the most commonly used tool for multichromosomal genomes. However, experimental evidence indicates that it leads to worse trees than an optimal median-solver, at least on unichromosomal genomes. In this paper, we present a new branch-and-bound method that provides an exact solution to the multichromosomal reversal median problem. We develop tight lower bounds and improve the enumeration procedure such that the search can be performed efficiently. Our extensive experiments on simulated datasets show that this median solver is efficient, has speed comparable to MGR, and is more accurate when genomes become distant.

1 Introduction

Annotation of genomes with computational pipelines can yield the ordering and strandedness of genes for genomes; each chromosome can then be represented by an ordering of signed genes, where the sign indicates the strand. Rearrangement of genes under reversal (also known as inversion), transposition, and other operations such as translocations, fissions and fusions, is an important evolutionary mechanism [7]. Since genome rearrangement events are "rare", these changes of gene orders enable biologists to reconstruct evolutionary histories far back in time.

One important problem in genome rearrangement analysis is to find the median of three genomes, that is, finding a fourth genome that minimizes the sum of the pairwise genomic distances between it and the three given genomes. This problem is important since it provides a maximum parsimony solution to the smallest binary tree and thus can be used as the basis for more complex methods. However,

K.A. Crandall and J. Lagergren (Eds.): WABI 2008, LNBI 5251, pp. 14–24, 2008.

© Springer-Verlag Berlin Heidelberg 2008

the median problem is NP-hard for genome rearrangement data [5,9] even under the simplest distance definition. To date, MGR (Multiple Genome Rearrangements) [4] is the only widely used tool that is able to handle multichromosomal genomes. However, experimental evidence indicates that MGR leads to worse trees than an optimal median solver [10], at least on small unichromosomal genomes. With more and more whole genome information available, it becomes very important to develop accurate median solvers for these multichromosomal genomes.

In this paper, we present an efficient branch-and-bound method to find the exact median for three multichromosomal genomes. We use an easy-to-compute and tight lower bound to prune bad branches and introduce a method that enumerates each genome at most once. Such median solver can be easily integrated with the existing methods such that datasets with more than three genomes can be analyzed.

2   Background and Notions

2.1   Genome Rearrangements

We assume a reference set of n genes $\{1, 2, \cdots, n\}$, and a genome is represented by an *ordering* of these genes. A gene g is assigned with an orientation that is either positive, written g, or negative, written $-g$. Specifically, we regard a multichromosomal genome as a set $A = A(1), \ldots, A(N_c)$ of N_c chromosomes partitioning genes $1, \ldots, n$; where $A(i) = \langle A(i)_1, \ldots, A(i)_{n_i} \rangle$ is the sequence of signed genes in the ith chromosome. In this paper, we also assume that each gene occurs exactly once in the genome.

In this study, we only consider *undirected* chromosomes [12], i.e. the flip of chromosomes is regarded as equivalent. We consider the following four operations on a genome: reversal, translocation, fission and fusion. Let $a = \langle a_1, \ldots, a_k \rangle$ and $b = \langle b_1, \ldots, b_m \rangle$ be two chromosomes. A *reversal* on the indices i and j ($i \leq j$) of chromosome a produces the chromosome with linear ordering $a_1, a_2, \cdots, a_{i-1}, -a_j, -a_{j-1}, \cdots, -a_i, a_{j+1}, \cdots, a_k$. A *translocation* transforms $a = \langle E, F \rangle$ and $b = \langle X, Y \rangle$ into $\langle E, Y \rangle$ and $\langle X, F \rangle$, where E, F, X, Y are gene segments. The *fusion* of a and b results in a chromosome $c = \langle a_1, \ldots, a_k, b_1, \ldots, b_m \rangle$. A *fission* of a results in two new chromosomes $\pi = \langle a_1, \ldots, a_{i-1} \rangle$ and $\sigma = \langle a_i, \ldots, a_k \rangle$.

An important concept in genome rearrangement analysis is the number of breakpoints between two genomes. Given genomes A and B, a *breakpoint* is defined as an ordered pair of genes (i, j) such that i and j are adjacent in A but not in B.

2.2   Genomic Distance for Multichromosomal Genomes

We define the *edit distance* as the minimum number of operations required to transform one genome into the other. Hannenhalli and Pevzner (HP) [8] provided a polynomial-time algorithm to compute the distance (HP distance) for reversals, translocations, fissions and fusions, as well as the corresponding sequence of events. Tesler [12] corrected the HP algorithm, which was later improved by Bergeron et al. [3].

Yancopoulos et al. [13] proposed a "universal" double-cut-and-join (DCJ) operation, resulting in a new genomic distance that can be computed in linear time. Although there is no direct biological evidence for DCJ operations, these operations are very attractive because they provide a unifying model for genome rearrangement [2]. Given two genomes A and B, computing the DCJ distance between these two genomes (denote $d_{DCJ}(A, B)$) is much easier to implement than computing the HP distance (denote $d_{HP}(A, B)$).

2.3   Reversal Median Problem

The median problem on three genomes is to find a single genome that minimizes the sum of the pairwise distances between itself and each of the three given genomes[1]. It has been proven that this problem is NP-hard [5] for unichromosomal genomes using reversal distance. Specifically the *reversal median problem* (RMP) is to find a median genome that minimizes the summation of the multichromosomal HP distances on the three edges.

Several solvers have been proposed for the unichromosomal reversal median problem (including MGR), among them, the one developed by Caprara [6] is the most accurate. Caprara's median solver is exact and treats the problem in a graph model, where each permutation corresponds to a matching of a point set. As a branch-and-bound algorithm, it enumerates all possible solutions and tests them edge by edge. At first, a lower bound is computed from the graph of the given genome's matchings. In each step of testing, the graph is reduced to a smaller one according to each edge of the solution being tested. A new bound is then computed from the new graph for bound testing. If the test failed, all solutions containing the edges tested so far in the current solution are excluded.

On the other hand, when used for three genomes, MGR attempts to find a longest sequence of reversals from one of the three given genomes that, at each step in the sequence, moves closer to the other two genomes. Since it is limited to a small subset of possible paths, MGR is less accurate than Caprara's median solver. Our method presented in this paper is inspired by Caprara's solver, and to our knowledge, is the first exact solver for the multichromosomal reversal median problem.

3   Graph Model for Undirected Genome

In this section, we introduce the graph model and a lower bound on the HP distance, which will be used in our new median solver.

3.1   Capless Breakpoint Graph

We modify the breakpoint graph to deal with genomes consisting of undirected chromosomes. In [8,12], *caps* are introduced to transform the multichromosomal genomes problem to unichromosomal problem. Caps play an important role in

---

[1] The median problem can be generalized for q ($q \geq 3$) genomes.

deriving the rearrangement scenario; however, since they are not necessary to compute the HP distance, capping nodes are not used here. We also do not distinguish between undirected unichromosomal and multichromosomal genomes, and treat both in a uniform way. This model is equivalent to the graph model of Bergeron et al. [1].

Given a node set V, we call an edge set $M \subset \{(i,j) : i,j \in V, i \neq j\}$ a *matching* of V if each node in V is incident to at most one edge in M. If each node in V is incident to exactly one edge in M, the matching is called *perfect*, otherwise *partial*. A genome A on genes $1, \ldots, n$ can be transformed to an unsigned genome $\mathcal{A}$ on $1, \ldots, 2n$, by replacing each positive entry g with $2g - 1, 2g$ and each negative entry g with $2|g|, 2|g| - 1$.

Consider the node set $\mathcal{V} := 1, \ldots, 2n$, and the associated perfect matching $\mathcal{H} := \{(2i - 1, 2i) : i = 1, \ldots, n\}$ (the base matching of $\mathcal{V}$). There is a correspondence between genomes composed of linear chromosomes and matchings $\mathcal{M}$ of $\mathcal{V}$ such that there are no cycles in $\mathcal{M} \cup \mathcal{H}$. These matchings are called *genome matchings*. In particular, the genome matching $\mathcal{M}(A)$ associated with a genome A is defined by

$$\mathcal{M}(A) := \big\{(\mathcal{A}(i)_k, \mathcal{A}(i)_{k+1}) : k \in \{2, 4, \ldots, 2n_i - 2\}, i \in \{1, \ldots, N_c\}\big\}.$$

The nodes in $\{\mathcal{A}(i)_1 : 1 \leq i \leq n_c\} \cup \{\mathcal{A}(i)_{2n_i} : 1 \leq i \leq n_c\}$ are called *end nodes* of A, denoted by `A-ends`. The genome matching has no capping node appended and all end nodes are not incident to any edge, thus the defined genome matchings are partial matchings. The absence of caps is a crucial step to reduce the complexity of the median problem for multichromosomal genomes.

Given two genomes A and B, the capless breakpoint graph $\mathcal{G}(A, B) = (\mathcal{V}, \mathcal{M}(A) \cup \mathcal{M}(B))$ defines a set of cycles and paths whose edges are alternate in $\mathcal{M}(A)$ and in $\mathcal{M}(B)$. An example of $\mathcal{G}$ can be found in Fig. 1. $c(A, B)$ denotes the number of cycles. Paths start from an end node and terminate at another end node. According to the type of ends, all paths in the graph can be classified into three groups: AA-paths, BB-paths, and AB-paths(called *odd paths* in [2]). A node which is an end of both A and B forms an AB-path of length 0. Denote the number of AB-paths by $|AB|$, and the number of AA-paths by $|AA|$.

3.2   Lower Bound of the HP Distance

We derive the following lower bound on the HP distance for undirected genomes, using only the parameters of the number of cycles and paths.

$$d_{HP} \geq n - \big(c(A, B) + |AB|/2\big) \tag{1}$$

This bound is indeed the same as the double-cut-and-join distance formula [2]. It can be directly derived from the HP distance formula for two multichromosomal genomes [8,12] or simply from the result that $d_{DCJ} \leq d_{HP}$ [3], where $d_{DCJ} = n - \big(c(A, B) + |AB|/2\big)$ [2].

For convenience, call $c(A, B) + |AB|/2$ the *pseudo-cycle* of $\mathcal{G}(A, B)$, denoted by $\tilde{c}(A, B)$. By the aid of DCJ distance, many useful results are easy to prove. Since the DCJ distance satisfies the triangle inequality, we have

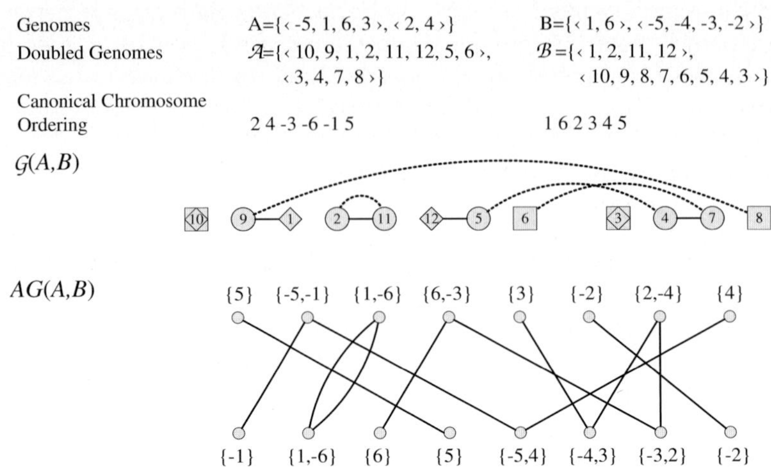

Fig. 1. The $\mathcal{G}(A, B)$ is the capless breakpoint graph of genome A and B. In $\mathcal{G}(A, B)$ diamonds represent B-ends, squares represent A-ends. Squares with a diamond inside indicate nodes are both A-ends and B-ends. In this figure, $n = 6$, $c(A, B) = 1$, $|AB| = 4$, and pseudo-cycle is 3. The graph $AG(A, B)$ is defined in [2] and the concept of canonical chromosome ordering is introduced in 4.3.

Lemma 1. *Given three genomes A,B,C, $n - \tilde{c}(A, C) + n - \tilde{c}(C, B) \geq n - \tilde{c}(A, B)$.*

3.3   Contraction Operation

We define the Multi-Breakpoint (MB) graph associated with q genomes $G_1, G_2, \ldots, G_q$ as the graph $\mathcal{G}(G_1, G_2, \ldots, G_q)$ with node set $\mathcal{V}$ and edge multiset $\mathcal{M}(G_1) \cup \mathcal{M}(G_2), \ldots, \cup \mathcal{M}(G_q)$. Note that for two matchings $\mathcal{M}(G_j)$ and $\mathcal{M}(G_k)$, $j \neq k$, some edges may be common in both matchings, but they are considered distinct parallel edges in the MB graph $\mathcal{G}(G_1, G_2, \ldots, G_q)$. In this paper, q always equals three.

Let $Q := \{1, \ldots, q\}$ and let τ be a genome, $\gamma(\tau) = \sum_{k \in Q} \tilde{c}(\tau, G_k)$. By Equation 1, for any genome σ, $\delta(\sigma) = \sum_{k \in Q} d_{HP}(\sigma, G_k) \geq qn - \gamma(\sigma)$, where n is the number of genes. We introduce the *Pseudo-Cycle Median Problem* (PMP): given undirected genomes $G_1, G_2, \ldots, G_q$, find a genome τ such that $qn - \gamma(\tau)$ is minimized. We have the following theorem:

Theorem 1. *Given an RMP instance and the associate PMP instance on q undirected genomes. Let δ^* and $qn - \gamma^*$ denote the optimal solution values of RMP and PMP, then $\delta^* \geq qn - \gamma^*$.*

Proof. If σ is an optimal solution of RMP and τ an optimal solution of PMP, $\delta^* = \delta(\sigma) \geq qn - \gamma(\sigma) \geq qn - \gamma(\tau) = qn - \gamma^*$.   □
Theorem 1 implies that the solution of PMP yields a lower bound on the optimal solution of RMP.

In Caprara's solver [6], the *contraction* on breakpoint graphs for perfect matchings is introduced. We modify it to fit the MB graphs for partial matchings which are dealt with here. Given a partial matching $\mathcal{M}$ on node set $\mathcal{V}$ and an edge $e = (i,j) \in E = \{(i,j) : i,j \in \mathcal{V}, i \neq j\}$, we define $\mathcal{M}/e$, the *contraction* of e on $\mathcal{M}$ (or on the genome corresponding to $\mathcal{M}$) as follows. If $e \in \mathcal{M}$, let $\mathcal{M}/e := \mathcal{M} \setminus \{e\}$. If one of i and j is an end, say i, and (b,j) is the edge incident to j, then $\mathcal{M}/e := \mathcal{M} \setminus \{(b,j)\}$. If both i and j are ends, $\mathcal{M}/e := \mathcal{M}$. Otherwise, let (a,i), (j,b) be the two edges in $\mathcal{M}$ incident to i and j, $\mathcal{M}/e := \mathcal{M} \setminus \{(a,i),(j,b)\} \cup \{(a,b)\}$. Given an MB graph $\mathcal{G}(G_1, G_2, \ldots, G_q)$, the *contraction* of an edge e is the operation that modifies the MB graph $\mathcal{G}(G_1, G_2, \ldots, G_q)$ as follows. Edge (i,j) is removed along with nodes i and j. For $k = 1, \ldots, q$, $\mathcal{M}(G_k)$ is replaced by $\mathcal{M}(G_k)/e$, and the base matching $\mathcal{H}$ is replaced by $\mathcal{H}/e$.

4 Branch-and-Bound Algorithm

In this section, we describe a branch-and-bound algorithm for the multichromosomal reversal median problem.

4.1 A Basic Branch-and-Bound Algorithm for PMP

The following lower bound on reversal median is based on Lemma 1.

Lemma 2. *Given a PMP instance associated with genomes* $G_1, \ldots, G_q$,

$$\gamma^* \leq \frac{qn}{2} + \sum_{k=1}^{q-1} \sum_{l=k+1}^{q} \frac{\tilde{c}(G_k, G_l)}{q - 1}. \tag{2}$$

The lower bound on the optimal PMP solution given by qn minus the right-hand side of (2), called LD, can be computed in $O(nq^2)$ time. For any PMP solution T, the cycles and paths in $T \cup \mathcal{M}(G_k)$ has one-to-one correspond to that in $(T/e) \cup (\mathcal{M}(G_k)/e)$ except for cycle of two copies of e. Therefore, for partial matchings and the new contraction defined on it, the same lemma in [6] holds.

Lemma 3. *Given a PMP instance and an edge* $e \in E$, *the best PMP solution containing* e *is given by* $\tilde{T} \cup \{e\}$, *where* $\tilde{T}$ *is the optimal solution of the PMP instance obtained by contracting edge* e.

According to Lemma 3, if we fix an edge e in the matching of PMP solution, an upper bound on the pseudo-cycle is given by $|\{k : e \in \mathcal{M}(G_k)\}|$ plus the upper bound (2) computed after the contraction of e. A branch-and-bound algorithm can be designed by these Lemmas. It enumerates all genomes gene by gene. In each step, it either selects a gene as an end or an inner gene in a chromosome of the solution (median) genome. In term of matching, the former operation fixes an end node in the matching of the solution and the latter one fixes an edge e in the matching. The former is called *end fixing* and the later *edge fixing*.

In edge fixing, the algorithm applies a contraction of edge e on the input genomes and computes a lower bound from these altered genomes. The number of newly generated cycles of length 0 is added to a counter. Based on these two values, the lower bound of all PMP solutions containing all edges added so far up to e can be derived. If it is greater than the current lower bound of the median problem, then e is not an edge in the matching of the best solution containing current fixed edges, thus the algorithm enumerates another gene or mark the last gene in the current solution as an end. Otherwise, the edge is fixed in the current solution.

In end fixing, no contraction operation is applied on the intermediate genomes and the lower bound is not changed. The number of newly generated paths of length 0 is added to a counter. When a complete solution is available, the lower bound that takes all the ends into account is computed by the aid of the counters and compared with the current lower bound. If they are equal the algorithm stops and the current solution is optimal.

4.2   Genome Enumeration

In the beginning, the algorithm enumerates all partial matchings by fixing, in turn, either 1, or 2, ..., or $2n$ as an end in the solution matching, which corresponds to one end of a chromosome, but the other end of this chromosome is not chosen yet. We call this chromosome an *opening chromosome* and this end an *open end*. Recursively, if the last operation is an edge fixing of (x, j), we proceed the enumeration by fixing in the solution, in turn, edge (k, l) where k is the other end of the edge (j, k) in $\mathcal{H}$ incident to j, for all l with no incident edge fixed so far or fix k as an end. Two cases exist if the last operation is an end fixing of x:

1. x is the other end of the current opening chromosome. We call x a *closed end* and the chromosome is closed. If there exist nodes that are not fixed in the solution, we enumerate by fixing in the solution, in turn, all available nodes as an open end in the solution.
2. x is an open end. We enumerate the cases as follows: each edge (y, l) is fixed in the solution one at a time, over all edges such that y is the node incident to x in $\mathcal{H}$ and l has no incident edge fixed so far; or we take y as a closed end.

4.3   Improved Branch-and-Bound Algorithm for PMP

The above scheme checks all concatenates of a genome thus will enumerate a genome more than once, which costs considerable computing time. We define the *canonical chromosome ordering* to overcome this problem. Let chromosome $X = \langle X_1, \ldots, X_M \rangle$. The *canonical flipping* of X is: $\langle X_1, \ldots, X_j, \ldots, X_M \rangle$, if $|X_1| < |X_M|$; otherwise the flipping is $\langle -X_M, \ldots, -X_j, \ldots, -X_1 \rangle$. For single-gene chromosomes, say $\langle g \rangle$, the canonical flipping is $\langle |g| \rangle$.

For a signed gene g, if $g > 0$, $l(g) = 2g - 1, r(g) = 2g$; if $g < 0$, $l(g) = 2|g|, r(g) = 2|g| - 1$. Let $\langle X_1, \ldots, X_M \rangle$ be the canonical flipping of chromosome X, the *smaller end* of X is $l(X_1)$. The *larger end* of X is $r(X_M)$. After the canonical flipping of each chromosome, we order the chromosomes by their smaller ends in increasing order to obtain the canonical chromosome ordering of a genome. Obviously, one genome has only one form of canonical ordering, and it can be uniquely represented by the canonical ordering along with the markers of start and end points. Fig. 1 shows an example.

We improve the basic branch-and-bound algorithm to enumerate each genome only once by using the canonical ordering. There are two states in the algorithm: (1) open chromosome and (2) build chromosome. The algorithm is in state "open chromosome" when it starts. It enumerates all the possible smaller ends of the first chromosome by fixing, in turn, either node 1, or 2,..., or $2n - 2$ as an end in the solution. When a smaller end is selected, the state is changed to "build chromosome". Recursively, in state "build chromosome", let the last edge fixed in the current partial solution be (i, j) and the edge in $\mathcal{H}$ incident to j be (j, k). There exist two branches if k can be a larger end of the current opening chromosome:

1. "closed chromosome": it fixes k as a larger end and closes the current opening chromosome. The state is changed to "open chromosome".
2. "build chromosome": it proceeds the enumeration by fixing in the solution, in turn, edge (k, l), for all l not be fixed in solution so far.

If k cannot be the larger end then only the "build chromosome" branch is permitted. If the state is "open chromosome", it will enumerate all the available smaller ends of the next chromosome by fixing, in turn, all available smaller ends.

We proceed in a depth-first order again. With this scheme we can perform the lower bound test after each edge fixing. When a node is fixed as an end, no operation is applied on the input genomes, thus the lower bound equals that of the previous step. At each end fixing, we record the number of newly generated SG_k-path, $k \in Q$, of length 0, where S denotes the partial solution. Thus when a complete solution is available, we will also know the total SG_k-paths in $\mathcal{M}(S) \cup \mathcal{M}_1, \ldots, \mathcal{M}(S) \cup \mathcal{M}_q$.

In the implementation, the initial lower bound LD is computed form the input genomes. The branch-and-bound starts searching for a PMP solution of target value $T = LD$ and tries another gene as soon as the lower bound for the current partial solution is greater than T. If a solution of value T is found, it is optimal and we stop, otherwise there is no solution of value LD. The algorithm then restarts with an increased target value $T = LD + 1$, and so on. The search stops as soon as we find a solution with the target value. All the partial solutions tested under target value T will be reconsidered under $T + 1$. The computation with $T + 1$ is typically much longer than the previous one, therefore the running time is dominated by the running time of the last target value.

4.4   Branch-and-Bound Algorithm for RMP

The above PMP algorithm can easily be modified to find the optimal RMP solutions. There are two modifications. First, the initial target value of the median score is computed as $\frac{\sum_{k=1}^{q-1}\sum_{l=k+1}^{q} d_{HP}(G_k,G_l)}{q-1}$. Second, when a complete solution whose lower bound is not greater than T is available, we compute the sum of HP distances between this solution and the input genomes. If the sum equals T, the algorithm stops, and the current solution is optimal. If there is no such solution, no genome exists whose sum of HP distances between it and the input genomes is better than $T+1$. We increase the target value by 1 and start the algorithm again. Though some non-optimal solutions can pass the lower bound test, they can not pass the HP distance test. But any optimal genome that passes the lower bound test will also pass the HP distance test according to Theorem 1. The first optimal genome (there may be several) encountered will be outputted as the optimal RMP solution.

5   Experimental Results

We have implemented the algorithm and conduct simulations to assess its performance. Our implementation is based on Caprara's unichromosomal median solver and uses MGR's code for multichromosomal reversal distance computation.

In our simulation study, each genome has 100 and 200 genes, with 2 and 4 chromosomes respectively. We create each dataset by first generating a tree topology with three leaves, assigning it with different edge lengths. We assign a genome G_0 to the root, then evolve the signed permutation down the tree, applying along each edge a number of operations equal to the assigned edge length. We test a large range of evolutionary rates: letting r denote the expected number of evolutionary events along an edge of the model tree, we used values of r in the range of 4 to 32 for datasets with 100 genes, and 4 to 40 for datasets with 200 genes. The actual number of events along each edge is sampled from a uniform distribution on the set $\{1, 2, \ldots, 2r\}$. We compare our new method with MGR and use two criteria to assess the accuracy: the median score which can be computed by summing the three edge lengths, and the multichromosomal reversal distance from the inferred median to the true ancestor which is known in our simulation.

Table 1 shows the median score for these two methods. When the genomes are closed (smaller r values), both methods return the same score. When r increases,

Table 1. Comparisons of the average median scores for 100 genes/2 chromosomes (top) and 200 genes/4 chromosomes (bottom)

	r=4	r=8	r=12	r=16	r=20	r=24	r=28	r=32
Our Method	11.1	22.6	39.0	55.2	63.4	72.6	76.1	84.2
MGR	11.1	22.6	39.0	55.2	63.5	73.3	77.4	86.5

	r=4	r=8	r=16	r=24	r=32	r=40
Our Method	11.8	22.0	45.2	78.0	98.8	111.6
MGR	11.8	22.0	45.2	78.0	98.8	112.2

Table 2. The average reversal distances from the inferred median to the true ancestor, for 100 genes/2 chromosomes (top) and 200 genes/4 chromosomes (bottom)

	r=4	r=8	r=12	r=16	r=20	r=24	r=28	r=32
Our Method	0	0.1	0	0.6	1.0	3.1	1.8	4.4
MGR	0	0.1	0	0.6	1.7	3.4	3.7	5.4

	r=4	r=8	r=16	r=24	r=32	r=40
Our Method	0	0.1	0	0	0	0.8
MGR	0	0.2	0	0	0	1.8

Table 3. The average time (in seconds) used for 100 genes/2 chromosomes (top) and 200 genes/4 chromosomes (bottom)

	r=4	r=8	r=12	r=16	r=20	r=24	r=28	r=32
Our Method	<1	<1	<1	12	6	149	203	411
MGR	<1	2	6	17	27	32	46	95

	r=4	r=8	r=16	r=24	r=32	r=40
Our Method	<1	<1	11.0	115.2	361.5	866.2
MGR	< 1	1.2	11.1	57.6	134.3	184.4

our method performs better by returning solutions with smaller scores. Although the difference of median scores seems small, in genome rearrangement analysis based on parsimony, such difference will have a big impact on the accuracy of phylogenies [11].

Table 2 shows the reversal distance of the inferred median to the true ancestor. Both MGR and our new method return solutions that are very close to the true ancestors, especially for datasets with 200 genes. Our new method is superior to MGR when the genomes become distant (for example, $r \geq 16$ for 100 genes).

Table 3 shows the average run time. Surprisingly, our method is faster when the genomes are not distant. However, it is much slower when the edge lengths increase and it cannot finish many datasets with r larger than the values in this experiment. For datasets with very large evolutionary rates, the edit distance will severely under-estimate the true distance, hence all median-based approaches will become unreliable.

6   Conclusions

In this paper we present a new branch-and-bound method for the multichromosomal reversal median problem. Our extensive experiments show that this method is more accurate than existing methods. However, this method is still primitive needs further improvements. In recent years, the double-cut-and-join (DCJ) distance has attracted much attention. We find that the lower bound used in this paper is indeed very similar to the DCJ distance [2], thus it may be relatively easy to extend our work and develop a new DCJ median solver.

Acknowledgments

WA and JT are supported by US National Institutes of Health (NIH grant number R01 GM078991-01). MZ is supported by NSF of China No.60473099.

References

1. Bergeron, A., Mixtacki, J., Stoye, J.: On sorting by translocations. In: Miyano, S., Mesirov, J., Kasif, S., Istrail, S., Pevzner, P.A., Waterman, M. (eds.) RECOMB 2005. LNCS (LNBI), vol. 3500, pp. 615–629. Springer, Heidelberg (2005)
2. Bergeron, A., Mixtacki, J., Stoye, J.: A unifying view of genome rearrangements. In: Bücher, P., Moret, B.M.E. (eds.) WABI 2006. LNCS (LNBI), vol. 4175, pp. 163–173. Springer, Heidelberg (2006)
3. Bergeron, A., Mixtacki, J., Stoye, J.: Hp distance via double cut and join distance. In: Ferragina, P., Landau, G.M. (eds.) CPM 2008. LNCS, vol. 5029, Springer, Heidelberg (2008)
4. Bourque, G., Pevzner, P.: Genome-scale evolution: reconstructing gene orders in the ancestral species. Genome Research 12, 26–36 (2002)
5. Caprara, A.: Formulations and hardness of multiple sorting by reversals. In: Proc. 3rd Ann. Int'l Conf. Comput. Mol. Biol. (RECOMB 1999), pp. 84–93. ACM Press, New York (1999)
6. Caprara, A.: On the practical solution of the reversal median problem. In: Gascuel, O., Moret, B.M.E. (eds.) WABI 2001. LNCS, vol. 2149, pp. 238–251. Springer, Heidelberg (2001)
7. Downie, S.R., Palmer, J.D.: Use of chloroplast DNA rearrangements in reconstructing plant phylogeny. In: Soltis, P., Soltis, D., Doyle, J.J. (eds.) Plant Molecular Systematics, pp. 14–35. Chapman and Hall, Boca Raton (1992)
8. Hannenhalli, S., Pevzner, P.A.: Transforming mice into men (polynomial algorithm for genomic distance problems). In: Proc. 36th Ann. IEEE Symp. Foundations of Comput. Sci. (FOCS 1995), pp. 581–592. IEEE Press, Piscataway (1995)
9. Pe'er, I., Shamir, R.: The median problems for breakpoints are NP-complete. Elec. Colloq. on Comput. Complexity 71 (1998)
10. Swenson, K.W., Arndt, W., Tang, J., Moret, B.M.E.: Phylogenetic reconstruction from complete gene orders of whole genomes. In: Proc. 6th Asia Pacific Bioinformatics Conf. (APBC 2008), pp. 241–250 (2008)
11. Tang, J., Wang, L.: Improving genome rearrangement phylogeny using sequence-style parsimony. In: Proc. 5th IEEE Symp. on Bioinformatics and Bioengineering BIBE 2005. IEEE Press, Los Alamitos (2005)
12. Tesler, G.: Efficient algorithms for multichromosomal genome rearrangements. J. Comput. Syst. Sci. 63(5), 587–609 (2002)
13. Yancopoulos, S., Attie, O., Friedberg, R.: Efficient sorting of genomic permutations by translocation, inversion and block interchange. Bioinformatics 21(16), 3340–3346 (2005)

Decompositions of Multiple Breakpoint Graphs and Rapid Exact Solutions

Andrew Wei Xu and David Sankoff

Department of Mathematics and Statistics, University of Ottawa, Canada K1N 6N5

Abstract. The median genome problem reduces to a search for the vertex matching in the multiple breakpoint graph (MBG) that maximizes the number of alternating colour cycles formed with the matchings representing the given genomes. We describe a class of "adequate" subgraphs of MBGs that allow a decomposition of an MBG into smaller, more easily solved graphs. We enumerate all of these graphs up to a certain size and incorporate the search for them into an exhaustive algorithm for the median problem. This enables a dramatic speedup in most randomly generated instances with hundreds or even thousands of vertices, as long as the ratio of genome rearrangements to genome size is not too large.

1   Introduction

The median problem underlies one approach to phylogenetics based on genomic distance. The idea, illustrated in Figure 1, is to optimize each ancestral node of an unrooted phylogeny in terms of its three or more immediate neighbours, modern or ancestral, and to iterate across the tree until convergence of the objective function (to a local optimum) at all nodes. This approach to the "small phylogeny" problem (i.e., the graph structure of the tree is given and does not need to be inferred, in contrast to the "big phylogeny problem") has a decade of history in the study of genome rearrangement [7,6,2,1], though its use in sequence-based phylogenetics dates to the 1970s [8].

In the study of genome rearrangement, genomes are treated as signed permutations on $1, \ldots, n$, either circular or linear, sometimes fragmented into chromosomes. The metric d on the set of genomes is an edit distance that counts the minimum number of operations required to transform one genome into another. The allowed operations may include the reversal of a contiguous chromosomal fragment, which also switches the sign on each term in the scope of the reversal; translocation, which involves the exchange of suffixes or prefixes of two chromosomes; transposition, or the excision of a contiguous chromosomal fragment and its re-insertion elsewhere on the chromosome; and a limited number of other operations. While distances involving reversals and translocations only can be calculated in time linear in n [4,10], the complexity of allowing transpositions in the distance calculation, either alone or in combination with reversals and translocations, is unknown. Recently, by generalizing the operation of transposition to that of block interchange [12], it became possible to include transpositions with

K.A. Crandall and J. Lagergren (Eds.): WABI 2008, LNBI 5251, pp. 25–37, 2008.

© Springer-Verlag Berlin Heidelberg 2008

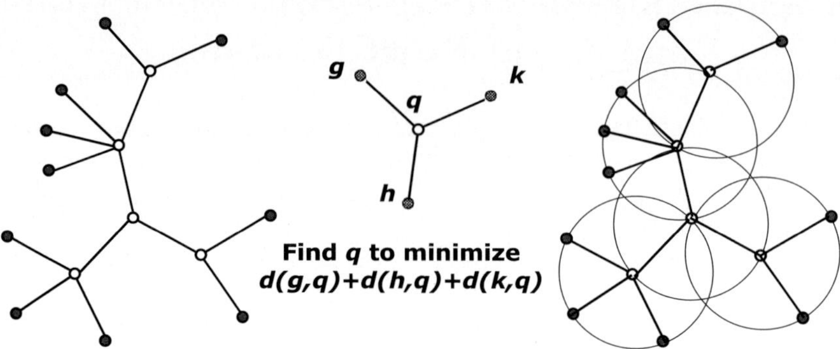

Fig. 1. Left: unrooted phylogeny with open dots representing ancestral genomes to be inferred. Middle: median problem with three given genomes g, h and k and median q to be inferred. Right: decomposition of phylogeny into overlapping median problems.

reversals and translocations in genomic distance calculations, within a framework known as "double cut and join" (DCJ). Moreover, the DCJ framework allows for substantial mathematical simplification of the distance calculation.

The median problem for genomic rearrangement distances in NP-hard [3,9]. Algorithms have been developed to find exact solutions for small instances [3,6] and there are rapid heuristics of varying degrees of efficiency and accuracy [2,1,5]. In the present paper, we explore the hypothesis that although there are no worst-case guarantees, it is worthwhile to develop methods to rapidly detect instances which are easily solved exactly.

Because of its simple structure, we choose to work with DCJ distance d as most likely to yield non-trivial mathematical results. We require genomes to consist of one or more circular chromosomes, but this is for simplicity of presentation, and our results could fairly easily be extended to genomes with multiple linear chromosomes. Then the median problem is to find a genome q with the smallest total distance $\sum_{g \in G} d(q, g)$, for a given set of genomes G.

The mathematical analysis of genomic distances generally invokes the *breakpoint graph*, which we will describe in Section 2. For DCJ, we have $d(g, h) = n - c$, where n is the number of genes in genomes g and h, and c is the number of cycles in the breakpoint graph. We define *adequate* subgraphs of the breakpoint graph, and key graph transformations in Section 2, and we demonstrate in Section 3 how to decompose large instances of median problems into smaller instances. This effectively reduces the search space of the median problem and makes it possible to design algorithms applicable to most instances of interest to biologists. In Sections 4 and 5, we sketch some of the considerations involved in these algorithms and describe the results of simulations on various data sets. The full development of the algorithm and its application to them are detailed in reference [11].

2   Graph and Subgraph Structures

2.1   Breakpoint Graph

We construct the breakpoint graph of two genomes as in Figure 2 by representing each gene by an ordered pair of vertices, adding coloured edges to represent the adjacencies between two genes, red edges for one genome and blue for the other.

In a genome, every gene has two adjacencies, one incident to each of its two endpoints, since it appears exactly once in that genome. Then in the breakpoint graph, every vertex is incident to one red edge and one blue one. Thus the breakpoint graph is a 2-regular graph which automatically decomposes into a set of alternating-colour cycles.

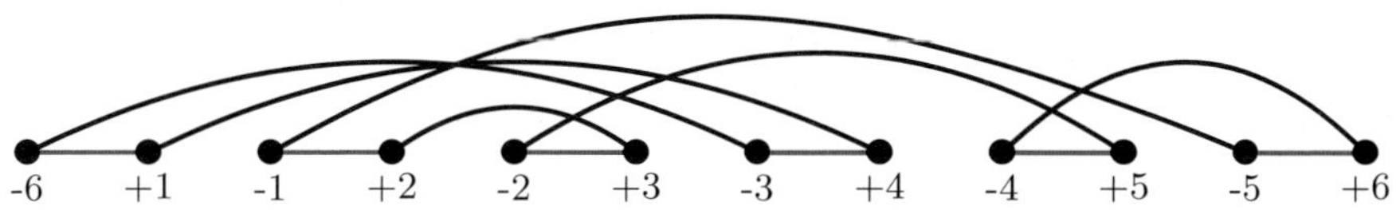

Fig. 2. Breakpoint graph for blue genome 1 -5 -2 3 -6 -4 (in gray) and red genome 1 2 3 4 5 6 (in black)

The edges of one colour form a perfect matching of the breakpoint graph, which we will simply refer to as a *matching*, unless otherwise specified. By the red matching, we mean the matching consisting of all the red edges.

The size for breakpoint graphs, multiple breakpoint graphs and median graphs is defined as half the number of vertices in it, which also equals to the number of genes in each genome and the size of each perfect matching.

2.2   Multiple Breakpoint Graph and Median Graph

The breakpoint graph extends naturally to a multiple breakpoint graph (MBG), representing a set $\mathcal{G}$ of three or more genomes. The number of genomes $N_{\mathcal{G}} \geq 3$ in $\mathcal{G}$ is also the edge chromatic number of the MBG. The colours assigned to the genomes are labeled by the integers from 1 to $N_{\mathcal{G}}$. We will use $B(\mathcal{G})$ or B throughout to refer to the MBG of the genomes $\mathcal{G}$.

For a given distance d, the median problem for $\mathcal{G} = \{g_1, \ldots, g_{N_{\mathcal{G}}}\}$ is to find a genome q which minimizes $\sum_{i=1}^{N_{\mathcal{G}}} d(g_i, q)$. For a candidate median genome, we use a different colour for its matching E, namely colour 0. Adding E to the MBG $B(\mathcal{G})$ results in the *median graph* $M_E(\mathcal{G}) = B(\mathcal{G}) \cup E$.

The set of all possible candidate matchings is denoted by $\mathcal{E}$. The set of all possible median graphs is $\mathcal{M}(\mathcal{G}) = \{M = B(\mathcal{G}) \cup E : E \in \mathcal{E}\}$.

The 0-i cycles in a median graph with matching E, numbering $c(0, i)$ in all, are the cycles where 0-edges and i edges alternate. Let $c_E(B) = \sum_{i=1}^{N_{\mathcal{G}}} c(0, i)$. Then $c_{\max}(B) = \max\{c_E(B) : E \in \mathcal{E}\}$ is the maximum number of cycles that can be constructed from B.

Minimizing the total distance in the median problem is equivalent to finding an optimal matching E, i.e., with $c_E(B) = c_{\max}(B)$. Let $\mathcal{E}^\star(B)$ be the set of all optimal matchings.

2.3   MBG Subgraphs and Connecting Edges

Let $\mathbf{V}(G)$ and $\mathbf{E}(G)$ be the sets of vertices and edges of a regular graph G. A *proper subgraph* H of G is one where $\mathbf{V}(H) = \mathbf{V}(G)$ and $\mathbf{E}(H) = \mathbf{E}(G)$ do not both hold at the same time. An *induced subgraph* H of G is the subgraph which satisfies the property that if $x, y \in \mathbf{V}(H)$ and $(x, y) \in \mathbf{E}(G)$, then $(x, y) \in \mathbf{E}(H)$.

In this paper, we will focus on the induced proper subgraphs, with an even number of vertices, of an MBG. Half of the number of these vertices is defined as the size of the subgraph H, denoted by m. $\mathcal{E}(H)$ is the set of all perfect 0-matchings $E(H)$, the cycle number determined by H and $E(H)$ is $c_{E(H)}(H)$, and $c_{\max}(H)$ is the maximum number of cycles that can be constructed from H by adding some $E(H)$. A 0-matching $E^\star(H)$ with $c_{E^\star(H)}(H) = c_{\max}(H)$ is called an optimal local matching, and $\mathcal{E}^\star(H)$ is the set of such matchings.

The *connecting edges* of a subgraph H in an MBG $B(\mathcal{G})$ are the edges of $B(\mathcal{G})$ incident to H exactly once, and are denoted by $K(H)$. The complementary induced subgraph of H in $B(\mathcal{G})$, denoted as $\overline{H}$, is the subgraph of $B(\mathcal{G})$ induced by $\mathbf{V}(B) - \mathbf{V}(H)$. Note that $B(\mathcal{G}) = H + K(H) + \overline{H}$, as illustrated in Figure 3.

2.4   Crossing Edges and Decomposers

For an MBG B and a subgraph H, a potential 0-edge would be *H-crossing* if it connected a vertex in $\mathbf{V}(H)$ to a vertex in $\mathbf{V}(\overline{H})$. A candidate matching containing one or more H-crossing 0-edges is an H-crossing candidate. A MBG subgraph H is called a *decomposer* if for any MBG containing it, there is an optimal matching that is not H-crossing. It is a *strong decomposer* if for any MBG containing it, all the optimal matchings are not H-crossing.

For an MBG B, the search space for an optimal matching is $\mathcal{E}$, which is of size $(2n - 1)!! = \frac{(2n)!}{2^n n!}$. If B contains a (strong) decomposer H of size m, then the search can be limited to the smaller space $\mathcal{E}(H) \times \mathcal{E}(\overline{H}) = \{E = E_H \cup E_{\overline{H}} : E_H \in \mathcal{E}(H),\ E_{\overline{H}} \in \mathcal{E}(\overline{H})\}$, which is of size $(2m - 1)!! \cdot (2n - 2m - 1)!!$.

2.5   Adequate and Strongly Adequate Subgraphs

In an MBG for a set of genomes $\mathcal{G}$, a connected subgraph H of size m is an *adequate subgraph* if $c_{\max}(H) \geq \frac{1}{2} m N_{\mathcal{G}}$; it is *strongly adequate* if $c_{\max}(H) > \frac{1}{2} m N_{\mathcal{G}}$.

A (strongly) adequate subgraph H is *simple* if it does not contain another (strongly) adequate subgraph as an induced subgraph; deleting any vertex from H will destroy its adequacy. In addition, a simple (strong) adequate subgraph H is *minimal* if we cannot even delete any edges without destroying its adequacy, i.e., for any edge $e \in \mathbf{E}(H)$, $c_{\max}(H - e) < \frac{1}{2} m N_{\mathcal{G}}$ ($c_{\max}(H - e) \leq \frac{1}{2} m N_{\mathcal{G}}$).

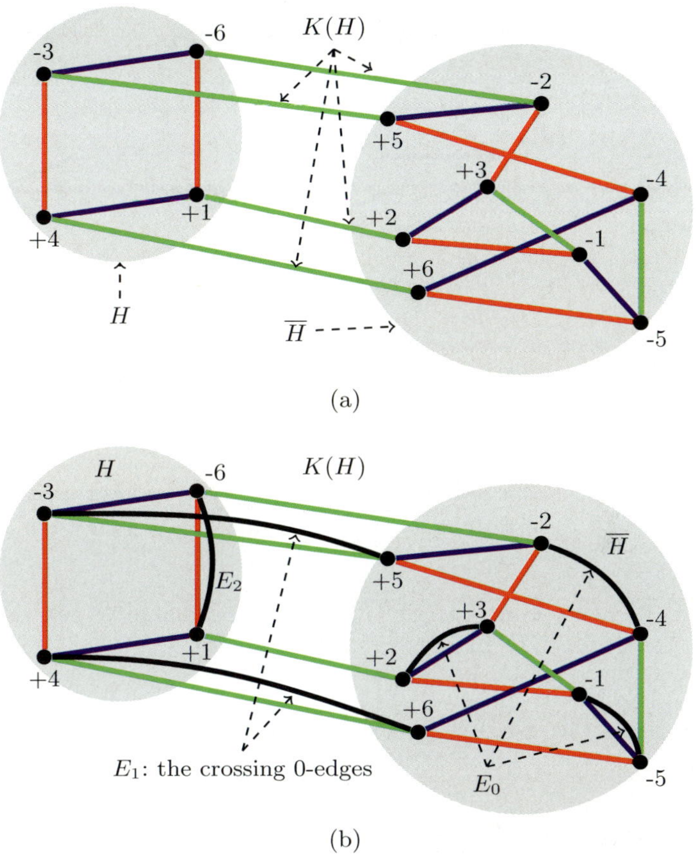

Fig. 3. MBG and median graph. Thick, gray, double and thin edges denote the edges with colours 1, 2, 3 and 0 correspondingly. (a) An MBG based on three genomes, (1 2 3 4 5 6), (1 -5 -2 3 -6 -4) and (1 3 5 -4 6 -2). A subgraph H, the connecting edge set $K(H)$ and the complementary subgraph $\overline{H}$ are illustrated. (b) A median graph. The candidate matching is divided into three 0-edge sets: E_0, E_1 and E_2.

2.6   Edge Shrinking, Expansion and Contraction

To shrink an edge e in a graph B, delete its two end vertices and any edges (including e) parallel to e, then for the edges incident to the deleted vertices, replace each pair of edges of same colour by a single edge of that colour, producing a new graph $B \circ e$, as illustrated by Fig 4(a)–(c). To shrink a set of edges A, shrink the edges in A one by one in any order, producing $B \circ A$.

To expand a 0-edge (a, b) in a graph B, remove that edge, add two new vertices i and j to the graph, connect i and j by $N_{\mathcal{G}}$ edges with colours ranging from 1 to $N_{\mathcal{G}}$, and add 0-edges (a, i) and (b, j), as illustrated by Fig 4(c) following the upward arrow.

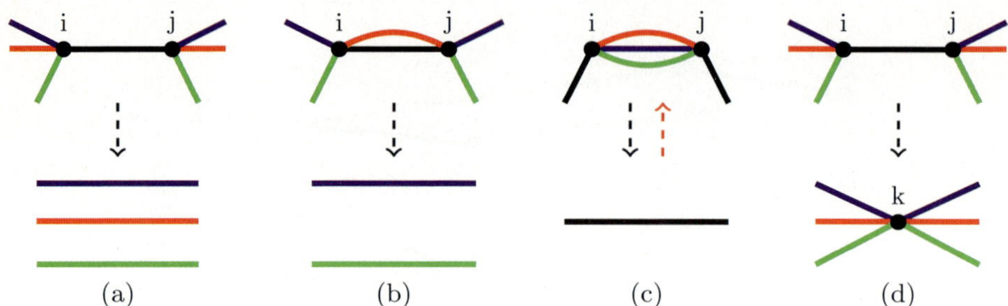

Fig. 4. Edge shrinking, expansion and contraction in a median graph based on 3 genomes: the downward arrows in (a), (b) and (c) illustrate edge shrinking in various situations; (c) the upward arrow illustrates an expansion of a thin edge; (d) illustrates a contraction of a thin edge

Proposition 1. *If median graph M' is obtained from another median graph M by expanding some 0-edge, then they contain the same number of cycles, i.e. $c(M') = c(M)$.*

To contract a 0-edge e from a graph G, delete e and merge its two end vertices, resulting in the graph G/e, as illustrated by Fig 4(d).

3   An Adequate Subgraph Is a Decomposer

In this section, we prove our main result: every (strongly) adequate subgraph is a (strong) decomposer. The general idea of the proof is that if H is a (strongly) adequate subgraph of MBG $B(\mathcal{G})$, for any H-crossing candidate matching E, we can always find another candidate matching E' that is not crossing, with $c_{E'}(B) \geq c_E(B)$ (or $c_{E'}(B) > c_E(B)$).

We partition the 0-edges in E among three sets: E_0, the set of 0-edges not incident to H; E_1, those incident to H exactly once; and E_2, those incident to H twice. In the median graph $M = B \cup E$, we shrink the 0-edge set E_0 and expand each 0-edge in E_2. The resultant median graph illustrated by Fig 5(a) is called the *twin median graph*, denoted by $\overset{\circ\bullet}{M} = \overset{\circ\bullet}{B} \cup \overset{\circ\bullet}{E}$.

If the 0-edges of a cycle in M are all in E_0, then after shrinking all 0-edges in E_0, this cycle does not appear in $\overset{\circ}{M}$. If a cycle in M contains 0-edges in E_1 or E_2, then with only part of the cycle being shrunk, this cycle does appear in $\overset{\circ\bullet}{M}$. Denote $c_{E_0}(B)$ as the number of cycles formed by B and 0-edges in E_0 only. Then

Proposition 2

$$c_E(B) = c_{E_0}(B) + c_{\overset{\circ\bullet}{E}}(\overset{\circ\bullet}{B}) \tag{1}$$

Since E_0 is not incident to the subgraph H, shrinking E_0 does not affect H. So H remains in $\overset{\circ\bullet}{M}$. Denote the subgraph in $\overset{\circ\bullet}{M}$ induced by $\mathbf{V}(H)$ as F. If a pair of connecting edges with colour i in M, is connected by a 0-i alternating colour

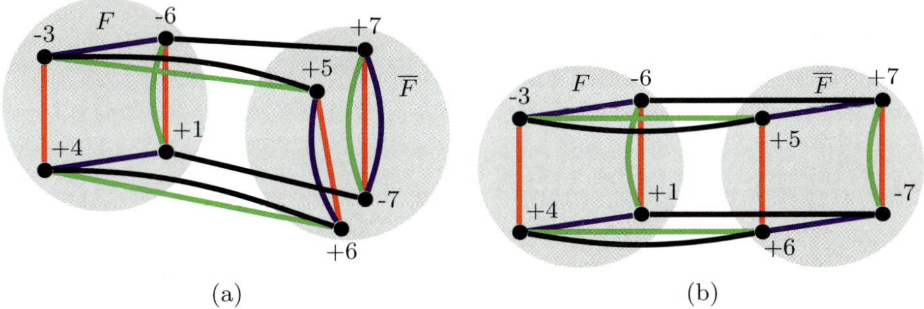

(a)   (b)

Fig. 5. Twin median graph and symmetrical median graph. (a) The twin median graph is obtained from the median graph in Figure 3b by shrinking the 0-edge set E_0 and expanding the 0-edge set E_2. (b) is the corresponding symmetric graph, with the left part mirror-symmetric to the right part.

path, with all 0-edges in E_0, then after shrinking E_0, this pair of i-edges are merged into a new i-edge e, with both ends incident to $\mathbf{V}(H)$. Edges like e are contained in F but not in H. Thus

Proposition 3. *Suppose $\overset{\circ\bullet}{B}$ is a twin MBG constructed from B based on a subgraph H of size m, and F is the subgraph in $\overset{\circ\bullet}{B}$ induced by $\mathbf{V}(H)$. Then F is of size m and $F \supseteq H$. If H is a (strongly) adequate subgraph, then so is F.*

Suppose the number of connecting edges in $K(F)$ of the twin MBG $\overset{\circ\bullet}{B}$ is $2k$. The 0-edges in $\overset{\circ\bullet}{M}$ denoted by $\overset{\circ\bullet}{E}$ are either from E_1 or the new added ones when expanding E_2. All of them are incident to F exactly once, so each 0-edge in $\overset{\circ\bullet}{E}$ is F-crossing. Then F and $\overline{F}$ must be of the same size.

The 0-edges in $\overset{\circ\bullet}{E}$ can be viewed as a mapping from the vertex set $\mathbf{V}(F)$ to $\mathbf{V}(\overline{F})$. If under this mapping, F is isomorphic to $\overline{F}$, as illustrated by Fig 5(b) then we call the twin median graph a *symmetrical median graph*, and we denote it by $\overset{\circ\circ}{M}$.

In any twin median graph, the size of an alternating colour cycle is at least 1, which is only possible when a 0-edge is parallel to a connecting edge. All other cycles have minimum size 2. We have

Proposition 4. *If in a twin median graph $\overset{\circ\bullet}{M}$, any cycle containing a connecting edge is of size 1 and any other cycle is of size 2, then $\overset{\circ\bullet}{M}$ contains the largest possible number of cycles among all twin median graphs formed from $\overset{\circ\bullet}{B}$. The maximum cycle number is $mN_G + k$. This can be achieved only when $\overset{\circ\circ}{M}$ is a symmetrical median graph $\overset{\circ\circ}{M}$.*

Proof. Since there are $2k$ connecting edges, the number of cycles of size 1 must be $2k$. Then the number of remaining non-0 edges is $2mN_G - 2k$. Hence there are $mN_G - k$ cycles of size 2. The maximum total number of cycles is $mN_G + k$. Because of the symmetry of $\overset{\circ\circ}{M}$, the other cycles can only be of size 2. Hence $\overset{\circ\circ}{M}$ is the only twin median graph containing the maximum number of cycles.   □

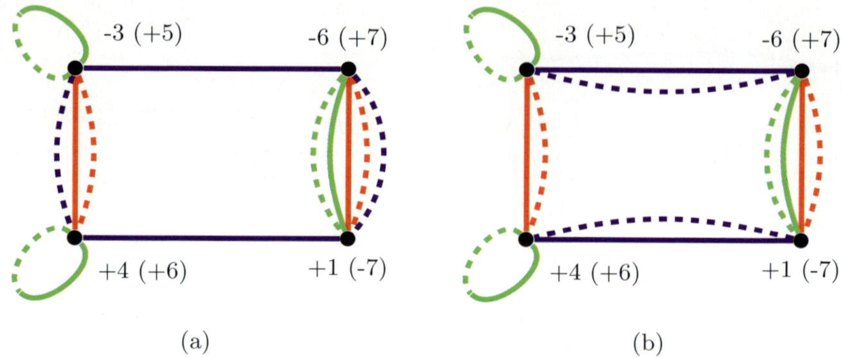

(a)   (b)

Fig. 6. The contracted twin graph (a) and contracted symmetric graph (b). The contracted graphs are generated from a twin median graph by contracting 0-edges. Dashed edges are from the complementary subgraphs and the half-solid-half-dashed ones are the connecting edges.

Next we investigate the difference between a twin median graph $\overset{\circ\bullet}{M}$ and a symmetric median graph $\overset{\circ\circ}{M}$, in terms of the number of DCJ operations needed to transform one into another.

Lemma 1. *If $\overset{\circ\bullet}{M}$ is a twin median graph and $\overset{\circ\circ}{M}$ is the symmetric median graph, then we can transform one into the other by exactly $mN_G + k - c(\overset{\circ\bullet}{M})$ DCJ operations on non 0-edges.*

Proof. We construct the *contracted graph*, illustrated in Figure 6, by contracting 0-edges of a median graph $\overset{\circ\bullet}{M}$, where edges in $\overline{F}$ are represented by dashed lines and the connecting edges are represented by half-dashed, half-solid lines with the solid end incident to F and the dashed end incident to $\overline{F}$. For conciseness, when we say *solid edges* (*dashed edges*), we mean the solid (dashed) edges contained by F ($\overline{F}$) or the solid (dashed) ends of connecting edges. The contracted graph for $\overset{\circ\bullet}{M}$ is denoted by $\overset{\circ\bullet}{M}$ and the contracted graph for $\overset{\circ\circ}{M}$ is denoted by $\overset{\circ\circ}{M}$.

Comparing the median graph $\overset{\circ\bullet}{M}$ and the contracted graph $\overset{\circ\bullet}{M}$, it easy to see that each vertex in $\overset{\circ\bullet}{M}$ has degree $2N_G$, incident to N_G solid edges and N_G dashed edges. The 0-i alternating colour cycle in $\overset{\circ\bullet}{M}$ becomes the alternating pattern (solid/dashed) cycle with colour i. The number of alternating pattern cycles is equal to the number of alternating colour cycles. Thus there are $c(\overset{\circ\bullet}{M})$ pattern alternating cycles in $\overset{\circ\bullet}{M}$ and $mN_G + k$ cycles in $\overset{\circ\circ}{M}$.

To transform $\overset{\circ\circ}{M}$ to $\overset{\circ\bullet}{M}$, we can show that there always exists a DCJ operation on two dashed edges with the same colour that increases the cycle number by one. When a connecting edge does not form a loop, apply a DCJ operation to loop it. Then arbitrarily select a solid edge from a cycle with size more than 2, apply a DCJ operation to make a dashed edge parallel to it. Thus with a number $mN_G + k - c(\overset{\circ\bullet}{M})$ DCJ operations, we can transform $\overset{\circ\bullet}{M}$ to $\overset{\circ\circ}{M}$ or *vice versa*. $\square$

Proposition 5. *An arbitrary DCJ operation on non-0 edges in a median graph changes the cycle number by 1, 0, or -1.*

Proof. If the two edges belong to one cycle, it will either split into two cycles or remain as a single cycle. If the two edges belong to two cycles, then they will be joined into one cycle. $\square$

Theorem 1. *If H is a (strongly) adequate subgraph of MBG B and E is a H-crossing candidate matching, then there is a candidate matching E' which is not H-crossing, with $c_{E'}(B) \geq c_E(B)$ (or $c_{E'}(B) > c_E(B)$).*

Proof.   1. From the median graph $M = B \cup E$, construct the twin median graph $\overset{\circ}{\overset{\bullet}{M}}$ and twin MBG $\overset{\circ}{\overset{\bullet}{B}}$ by shrinking 0-edges not incident to H (E_0) and expanding 0-edges incident to H twice (E_2). Denote the subgraph of $\overset{\circ}{\overset{\bullet}{M}}$ induced by $\mathbf{V}(H)$ as F. Then $c_E(B) = c_{E_0}(B) + c_{\overset{\circ}{\overset{\bullet}{E}}}(\overset{\circ}{\overset{\bullet}{B}})$.

 2. Construct the symmetrical median graph $\overset{\circ}{\overset{\circ}{M}}$ with $F = \overline{F}$ and F also a (strongly) adequate subgraph.

 3. Since F is a (strongly) adequate subgraph, there exists a 0-matching D of F satisfying $c_D(F) \geq \frac{1}{2}mN_\mathcal{G}$ (or $c_D(F) > \frac{1}{2}mN_\mathcal{G}$).

 4. Replace the 0-matching in $\overset{\circ}{\overset{\circ}{M}}$ by two copies of D, one on F and one on $\overline{F}$. Denote the 0-matching as $2D$ and denote the resultant median graph as $\overset{\circ}{\overset{\circ}{B}} \cup 2D$, with $c_{2D}(\overset{\circ}{\overset{\circ}{B}}) \geq mN_\mathcal{G}$ (or $> mN_\mathcal{G}$).

 5. Transform $\overset{\circ}{\overset{\circ}{B}}$ to $\overset{\circ}{\overset{\bullet}{B}}$ by $mN_\mathcal{G} + k - c(\overset{\circ}{\overset{\circ}{M}})$ DCJ operations on $\overline{F}$ in $\overset{\circ}{\overset{\circ}{B}}$. So $c_{2D}(\overset{\circ}{\overset{\bullet}{B}}) \geq c_{\overset{\circ}{\overset{\bullet}{E}}}(\overset{\circ}{\overset{\bullet}{B}})$ (or $c_{2D}(\overset{\circ}{\overset{\bullet}{B}}) > c_{\overset{\circ}{\overset{\bullet}{E}}}(\overset{\circ}{\overset{\bullet}{B}})$).

 6. Shrink the newly added sets of $N_\mathcal{G}$ parallel edges in $\overset{\circ}{\overset{\bullet}{B}}$ and reverse the shrinking operations on E_0 in step 1, to recover the MBG B. Then the 0-matching $2D$ becomes the candidate matching E' and the new median graph becomes $M' = B \cup E'$. Then $c_{E'}(B) = c_{2D}\overset{\circ}{\overset{\bullet}{B}} + c_{E_0}(B)$. Thus $c_{E'}(B) \geq c_E(B)$ (or $c_{E'}(B) > c_E(B)$). $\square$

Theorem 2. *Any adequate subgraph is a decomposer. A strongly adequate subgraph is a strong decomposer.*

Proof. For an adequate subgraph there must be a optimal matching that is not crossing. Otherwise by Theorem 1, from the optimal crossing matching, we can construct a candidate matching that is not crossing and has at least as many cycles. Thus the adequate subgraph is a decomposer.

   For a strongly adequate subgraph, the non-crossing candidate matchings are always better than the corresponding crossing candidate matchings. Then the optimal matchings cannot be crossing matchings. The strongly adequate subgraph is thus a strong decomposer. $\square$

4   Median Calculation Incorporating MBG Decomposition

As adequate subgraphs are the key to decompose the median problems, we need to inventory them before making use of them. It turns out that it is most useful to limit this project to simple adequate graphs. Non-simple adequate graphs are both harder to enumerate and harder to use, and are likely to have simple ones

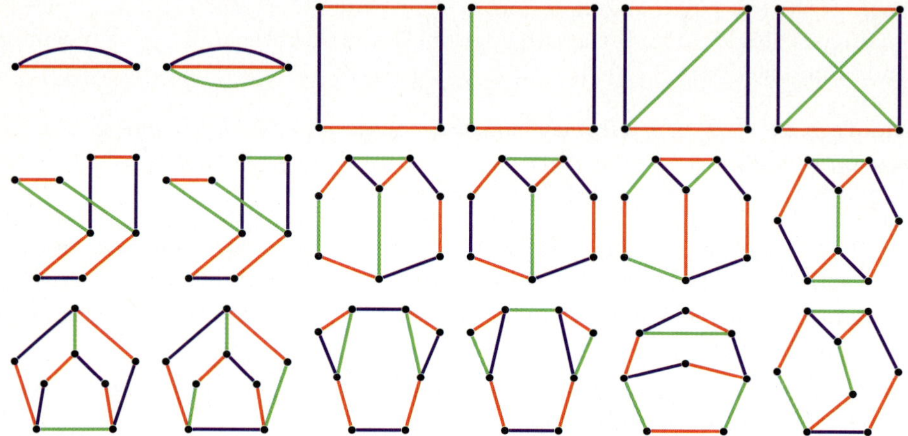

Fig. 7. Simple adequate subgraphs of size 1, 2 and 4 for MBGs on three genomes. See reference [11] for how they were identified.

embedded in them, which serve the same general purpose [11]. By exhaustive search, we have found all simple adequate graphs of size < 6; these are depicted in Figure 7. Though we have some of size 6, it would be a massive undertaking to compile the complete set with current methods.

Our basic algorithm for solving the median problem is a branch and bound, where edges of colour 0 are added at each step; we omit the details of procedures we use to increase the effectiveness of the bounds. To make use of the adequate subgraph theory we have developed, at each step we search for such an inventoried subgraph before adding edges, and if one is found, we carry out a decomposition and then solve the resulting smaller problem(s) [11].

Table 1. The number of runs, out of ten, where the median was found in less than 10 minutes on a MacBook, 2.16GHz, on one CPU

ρ/n n	10	20	30	40	50	60	80	100	200	300	500	1000	2000	5000
0.1	10	10	10	10	10	10	10	10	10	10	10	10	10	10
0.2	10	10	10	10	10	10	10	10	10	10	10	10	10	10
0.3	10	10	10	10	10	10	10	10	10	10	10	10	1	
0.4	10	10	10	10	10	10	10	10	0	0				
0.5	10	10	10	10	10	10	4	0						
0.6	10	10	10	10	9	6								
0.7	10	10	10	10	6									
0.8	10	10	10	10	6									
0.9	10	10	10	10	4									
1.0	10	10	10	8	2									

Table 2. Speedup due to discovery of larger adequate subgraphs (AS2, AS4). Three genomes are generated from the identity genome with $n = 100$ by 40 random reversals. Time is measured in seconds. Runs were halted after 10 hours. AS1, AS2, AS4, AS0 are the numbers of edges in the solution median constructed consequent to the detection of adequate subgraphs of sizes 1, 2, 4 and at steps where no adequate subgraphs were found, respectively.

run	speedup factor	run time with AS1,2,4	with AS1	AS1	AS2	AS4	AS0
1	41,407	4.5×10^{-2}	1.9×10^{3}	53	39	8	0
2	85,702	3.0×10^{-2}	2.9×10^{3}	53	34	12	1
3	2,542	5.4×10^{0}	1.4×10^{4}	56	26	16	2
4	16,588	3.9×10^{-2}	6.5×10^{2}	58	42	0	0
5	$> 10^{6}$	5.9×10^{2}	stopped	52	41	4	3
6	199,076	6.0×10^{-3}	1.2×10^{3}	56	44	0	0
7	6,991	2.9×10^{-1}	2.1×10^{3}	54	33	12	1
8	$> 10^{6}$	4.2×10^{1}	stopped	57	38	0	5
9	1,734	8.7×10^{0}	1.5×10^{4}	65	22	8	5
10	855	2.1×10^{0}	1.8×10^{3}	52	38	8	2

5   Experimental Results

To see how useful our method is on a range of genomes, we undertook experiments on sets of three random genomes. Our JAVA program included a search for adequate subgraphs followed by decomposition at each step of a branch and bound algorithm to find the maximum number of cycles. We varied the parameters n and $\pi = \rho/n$, where ρ was the number of random reversals applied to the ancestor $I = 1, \ldots, n$ independently to derive three different genomes.

5.1   The Effects of n and $\pi = \rho/n$ on the Proportion of Rapidly Solvable Instances

Table 1 shows that relatively large instances can be solved if ρ/n remains at 0.3 or less. It also shows that for small n, the median is easy to find even if ρ/n is large enough to effectively scramble the genomes.

5.2   The Effect of Adequate Subgraph Discovery on Speed-Up

Table 2 shows how the occurrence of larger adequate subgraphs (AS2 and AS4) can dramatically speed up the solution to the median problem, generally from more than a half an hour to a fraction of a second.

5.3   Time to Solution

Our results in Section 5.1 suggest a rather abrupt cut-off in performance as n or ρ/n become large. We explore this in more detail by focusing on the particular

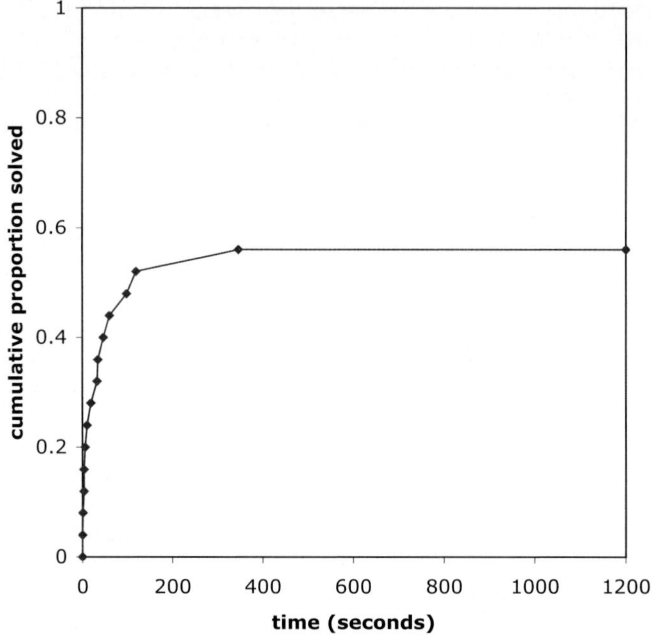

Fig. 8. Cumulative proportion of instances solved, by run time. $n = 1000$, $\rho/n = .31$. More than half are solved in less than 2 minutes; almost half take more than 20 minutes.

parameter values $n = 1000$ and $\rho/n = .31$. Figure 8 shows how the instances are divided into a rapidly solvable fraction and a relatively intractable fraction, with very few cases in between.

6   Conclusion

In this paper we have demonstrated the potential of adequate subgraphs for greatly speeding up the solution of realistic instances of the median problem. Many improvements seem possible, but questions remain. If we could inventory non-simple adequate graphs, or all simple adequate graphs of size 6 or more, could we achieve significant improvement in running time? It may well be that the computational costs of identifying larger adequate graphs within MBGs would nullify any gains due to the additional decompositions they provided.

Acknowledgments

We thank the reviewers for their helpful comments and suggestions. Research supported in part by a grant to DS from the Natural Sciences and Engineering Research Council of Canada (NSERC). DS holds the Canada Research Chair in Mathematical Genomics.

References

1. Adam, Z., Sankoff, D.: The ABCs of MGR with DCJ. Evol. Bioinform. 4, 69–74 (2008)
2. Bourque, G., Pevzner, P.: Genome-scale evolution: Reconstructing gene orders in the ancestral species. Genome Res. 12, 26–36 (2002)
3. Caprara, A.: The reversal median problem. INFORMS J. Comput. 15, 93–113 (2003)
4. Hannenhalli, S., Pevzner, P.: Transforming cabbage into turnip: Polynomial algorithm for sorting signed permutations by reversals. JACM 46, 1–27 (1999)
5. Lenne, R., Solnon, C., Stützle, T., Tannier, E., Birattari, M.: Reactive stochastic local search algorithms for the genomic median problem. In: van Hemert, J., Cotta, C. (eds.) EvoCOP 2008. LNCS, vol. 4972, pp. 266–276. Springer, Heidelberg (2008)
6. Moret, B.M.E., Siepel, A.C., Tang, J., Liu, T.: Inversion medians outperform breakpoint medians in phylogeny reconstruction from gene-order data. In: Guigó, R., Gusfield, D. (eds.) WABI 2002. LNCS, vol. 2452. Springer, Heidelberg (2002)
7. Sankoff, D., Blanchette, M.: Multiple genome rearrangement and breakpoint phylogeny. J. Comput. Biol. 5, 555–570 (1998)
8. Sankoff, D., Morel, C., Cedergren, R.: Evolution of 5S RNA and the non-randomness of base replacement. Nature New Biol. 245, 232–234 (1973)
9. Tannier, E., Zheng, C., Sankoff, D.: Multichromosomal median and halving problems. In: WABI 2008. LNBI, vol. 5251. Springer, Heidelberg (2008)
10. Tesler, G.: Efficient algorithms for multichromosomal genome rearrangements. JCSS 65, 587–609 (2002)
11. Xu, A.W.: A fast and exact algorithm for the median of three problem—a graph decomposition approach (submitted, 2008)
12. Yancopoulos, S., Attie, O., Friedberg, R.: Efficient sorting of genomic permutations by translocation, inversion and block interchange. Bioinform. 21, 3340–3346 (2005)

Read Mapping Algorithms for Single Molecule Sequencing Data

Vladimir Yanovsky[1], Stephen M. Rumble[1], and Michael Brudno[1,2]

[1]Department of Computer Science and
[2]Donnelly Centre for Cellular and Biomolecular Research
University of Toronto
{volodyan,rumble,brudno}@cs.toronto.edu

Abstract. Single Molecule Sequencing technologies such as the Helicope simplify the preparation of DNA for sequencing, while sampling millions of reads in a day. Simultaneously, the technology suffers from a significantly higher error rate, ameliorated by the ability to sample multiple reads from the same location. In this paper we develop novel rapid alignment algorithms for two-pass Single Molecule Sequencing methods. We combine the Weighted Sequence Graph (WSG) representation of all optimal and near optimal alignments between the two reads sampled from a piece of DNA with k-mer filtering methods and spaced seeds to quickly generate candidate locations for the reads on the reference genome. We also propose a fast implementation of the Smith-Waterman algorithm using vectorized instructions that significantly speeds up the matching process. Our method combines these approaches in order to build an algorithm that is both fast and accurate, since it is able to take complete advantage of both of the reads sampled during two pass sequencing.

1 Introduction

Next generation sequencing (NGS) technologies are revolutionizing the study of variation among individuals in a population. While classical, Sanger-style sequencing machines were able to sequence 500 thousand basepairs per run at a cost of over \$1000 per megabase, new sequencing technologies, such as Solexa/Illumina and AB SOLiD can sequence 4 billion nucleotides in the same amount of time, at the cost of only \$6 per megabase. The decreased cost and higher throughput of NGS technologies, however, are offset by both a shorter read length and a higher overall sequencing error rate.

Most NGS technologies reduce the cost of sequencing by running many sequencing experiments in parallel. After the input DNA is sheared into smaller pieces, DNA libraries are prepared by a Polymerase Chain Reaction (PCR) method, with many identical copies of the molecules created, and attached to a particular location on a slide. All of the molecules are sequenced in parallel using sequencing by hybridization or ligation, with each nucleotide producing a specific color as it is sequenced. The colors for all of the positions on the slide are recorded via imaging and are then converted into base calls.

K.A. Crandall and J. Lagergren (Eds.): WABI 2008, LNBI 5251, pp. 38–49, 2008.
© Springer-Verlag Berlin Heidelberg 2008

In this paper we address a specific type of NGS technology – Single Molecule Sequencing (SMS). While the origins of SMS date back to 1989 [3], it is only now becoming practical. The Heliscope sequencer, sold by Helicos, is the first commercial product that allows for the sequencing of DNA with SMS. In a recent publication, the complete genome of the M13 virus was resequenced with the Heliscope [4]. The key advantage of the SMS methods over other NGS technologies is the direct sequencing of DNA, without the need for the PCR step described above. PCR has different success rates for different DNA molecules, and introduces substitutions into the DNA sequence as it is copied. Additionally, the direct sequencing of DNA via SMS significantly simplifies the preparation of DNA libraries. SMS methods should also lead to more accurate estimates of the quantity of DNA which is required for detection of copy number variations and quantification of gene expression levels via sequencing. SMS technologies have very different error distribution than standard Sanger-style sequencing. Because only one physical piece of DNA is sequenced at a time, the sequencing signal is much weaker, leading to a large number of "dark bases": nucleotides that are skipped during the sequencing process leading to deletion errors. Additionally, a nucleotide could be mis-read (substitution errors), or inserted, however these errors are more typically the result of imaging, rather than chemistry problems, and hence are rarer. A full description of SMS errors is in the supplementary material of [4].

One advantage of the Heliscope sequencer, as well as several other proposed SMS technologies (e.g. Pacific Biosciences' expected method) is the ability to read a particular piece of DNA multiple times (called multi-pass sequencing). The availability of multiple reads from a single piece of DNA can help confirm the base calls, and reduce the overall error rate. In practice these methods usually use two passes, as this offers a good tradeoff between error rate and cost. In this paper we introduce a novel rapid alignment algorithm for multi-pass single molecule sequencing methods. While we will use the most common case of two-pass sequencing to describe our algorithms, they can be easily extended to a larger number of passes. We combine the Weighted Sequence Graph (WSG) [11] representation of all optimal and near optimal alignments between the two reads sampled from a piece of DNA with k-mer filtering methods [9] and spaced seeds [1] to quickly generate candidate locations for the reads on the reference genome. We also propose a novel, fast implementation of the Smith-Waterman algorithm [12] using vectorized instructions that significantly speeds up the matching process. Our algorithms are implemented as part of the SHort Read Mapping Package (SHRiMP). SHRiMP is a set of tools for mapping reads to a genome, and includes specialized algorithms for mapping reads from popular short read sequencing platforms, such as Illumina/Solexa and AB SOLiD. SHRiMP is freely available at **http://compbio.cs.toronto.edu/shrimp**.

2   Algorithms

In this section we will describe the alignment algorithms we use for mapping two reads generated from a single DNA sequence with SMS to a reference genome.

We will start (Sections 2.1 and 2.2) by reviewing how the Weighted Sequence Graph data structure [11] can be used to speed up the inherently cubic naive alignment algorithm to near-quadratic running time. We will then demonstrate how to combine the WSG alignment approach with standard heuristics for rapid alignment, such as seeded alignment and k-mer filters in Section 2.3. Finally, we will describe our implementation of the Smith-Waterman alignment algorithm with SSE vector instructions to further speed up our algorithm, making it practical for aligning millions of reads against a long reference genome.

2.1 Alignment with Weighted Sequence Graphs

Given unlimited computational resources, the best algorithm for mapping two reads sampled from the same location to the reference genome is full three-way alignment. This algorithm would require running time proportional to the product of the sequence lengths. Because the reads are typically short ($\sim$30bp), the overall running time to map a single read pair to a genome of length n may be practical ($30 * 30 * n = 900n$), however the naive algorithm will not scale to aligning millions of reads that an SMS platform produces every day.

An alternative approach, suggested in [4], is to align the two reads to each other, thus forming a profile, which could then be aligned to the reference

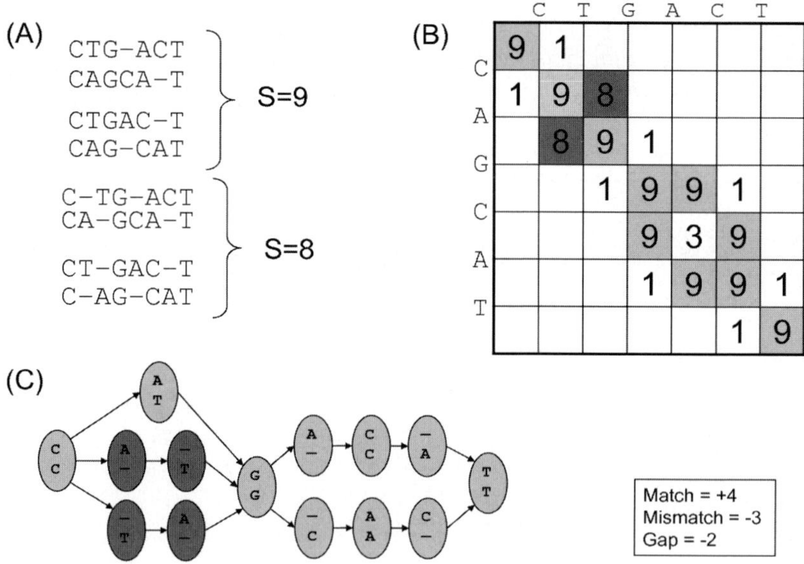

Fig. 1. An example of a WSG graph representing 1-suboptimal alignments of CTGACT with CAGCAT. (A). Optimal alignments of score 9 and two 1-suboptimal of score 8; two more 1-suboptimal alignments are possible. (B). Cells of the dynamic programming matrix such that the best path through them is 9 are shaded in light grey, if it is 8 – in dark grey. (C). WSG corresponding to at worst 1-suboptimal alignments.

genome. This approach, akin to the standard "progressive" alignment approaches used, e.g., in CLUSTALW [2], has the advantage of significantly decreased runtime, as the running time becomes proportional to the product of the lengths of the genome and the longer read ($\sim$30 times the length of the genome), however, because the profile only incorporates a single optimal alignment, it loses important information about possible co-optimal and suboptimal alignments between the sequences. For example, Figure 1 demonstrates two short sequences with two optimal alignments, as well as 4 more near-optimal ones. While the total number of optimal alignments between two sequences could be exponential in their lengths, Naor and Brutlag [7] and Hein [6] have suggested that all of these alignments can be represented compactly using a directed acyclic graph (DAG). Furthermore, Schwikowski and Vingron [11] have shown how to generalize the standard sequence alignment paradigms to alignment of DAGs, which they call weighted sequence graphs. Their original algorithm was stated in the context of the tree alignment problem, but is easily generalizable to any progressive alignment scoring scheme. Here we reintroduce the idea of representing alignments as graphs, and extend it for the SMS mapping problem.

Definitions. The following definitions will prove useful:

- Given the sequence S of length $|S|$ over an alphabet $\Sigma = \{A, C, T, G\}$, we refer to its ith symbol as $S[i]$
- We say that a sequence S' over alphabet $\overline{\Sigma} = \Sigma \bigcup `-`$, where character $`-`$ is called a gap, *spells* S if removing the gaps from S' results in sequence S. When this does not cause ambiguity, we will write S instead of S'.
- A k-mer is any sequence of length k.
- When addressing matrix indices, we use a notations "$i_1 : i_2$", "$: i_2$", "$i_1 :$" and "$:$" to represent the sets of indices j such that $i_1 \leq j \leq i_2$, $j \leq i_2$, $i_1 \leq j$ and all possible indices respectively. For example, given matrix M of size $K \times N$ we shall denote the i'th row of the matrix by $M[i; :]$ and its j'th column by $M[:; j]$.
- A *global alignment* of 2 sequences S_1 and S_2 over alphabet Σ is defined as a matrix $M = [2 \times N]$ such that $M[i; :]$ spells S_i for $i = 1, 2$.
  Given a real-valued score function $sc : \overline{\Sigma} \times \overline{\Sigma} \mapsto \Re$ we define the score SC of the alignment $M = [2 \times N]$ as

$$SC(M) = \Sigma_{j=1}^{N} sc(M[:; j])$$

- The *global alignment problem* is to find an alignment M such that $SC(M)$ is maximum. It is known [8] that for any cost function sc satisfying condition $sc(-, -) < 0$ the problem is well-defined though its solution is not necessarily unique. We denote the maximum score with *Opt*.
- We call an alignment M ε-suboptimal if $SC(M) \geq Opt - \varepsilon$.

While we defined the alignment for two sequences, the problem for more than two sequences is defined similarly. In particular, the problem we consider in this work – an alignment of a pair of sequences S_1 and S_2 to a reference sequence R – is to find a high scoring alignment matrix $M = [3 \times N]$ such that $M[i; :] = S_i$ for i=1 and 2 respectively and $M[3; :]$ is a substring of R.

Representing (Sub)-optimal Alignments in a Graph. A sequence S can be represented by a labeled directed graph $G'(S) = (V, E)$ with $|S| + 1$ nodes $\{s = V[0] \cdots V[|S|] = t]\}$ and $|S|$ edges: for every $0 \leq i < |S|$ there is an edge $V[i] \to V[i+1]$ labeled with $S[i]$ – the i'th symbol of sequence S; vertices s and t will be called the source and the sink. We obtain the graph $G(S)$ from graph G' by adding a self loop edge $V[i] \to V[i]$ labeled with a gap symbol '–' for every i. There is a one-to-one correspondence between the sequences over alphabet $\overline{\Sigma}$ spelling sequence S and the sequences produced by reading edge labels along the paths in $G(S)$ from s to t.

Given two strings, S_1 and S_2 and two graphs $A = G(S_1)$ and $B = G(S_2)$, their cross product $A \times B$ is defined as a graph with the vertex set $V = \{(v_A, v_B)\}$, for any v_A vertex of A and v_B vertex of B, and edge set $E = \{v_1 \to v_2\}$, for any $v_1 = (v_A^1, v_B^1)$ and $v_2 = (v_A^2, v_B^2)$, such that there is an edge $v_A^1 \to v_A^2$ in A and an edge $v_B^1 \to v_B^2$ in B. The edges of all graphs considered in this work will be labeled with strings over alphabet $\overline{\Sigma}$; the edge $v^1 \to v^2$ of the cross product graph will receive a label which is the concatenation of the labels of an edge $v_A^1 \to v_A^2$ in A and an edge $v_B^1 \to v_B^2$ in B.

This cross product graph, sometimes referred to as an edit graph, corresponds to the Smith-Waterman dynamic programming matrix of the two strings. It is easy to prove that there is a one-to-one correspondence between the paths from *source* $s = s_1 \times s_2$ to *sink* $t = t_1 \times t_2$ and alignments: if the sequence of edges along the path is $e_0 \cdots e_{N-1}$, then the corresponding alignment M is defined as $M(:, j) = label(e_j)$ for all values of j – recall that $label(e)$ is a pair of symbols. Note that the only cycles in the cross-product graph we just defined are the self loops labeled with '– –' at every node of the graph.

Now a *WSG* is defined as a subgraph of the cross-product graph such that its vertex set V contains vertices s, t and for every vertex $v \in V$ it is on some path from s to t. Intuitively, one may think of the WSG as a succinct representation of a set of alignments between the two strings. We use WSGs in Section 2.2 for the purpose of representing all high-scoring alignments.

2.2   Alignment of Read Pairs with WSG

During two-pass DNA sequencing, two reads are produced from every fragment of DNA. Both of the reads may have sequencing errors, the most common of which are skipped letters. These errors are nearly ten-fold more common than mis-calls or insertions [4]. Our algorithm for aligning two-pass SMS reads is based on the intuition that in a high-scoring three-way alignment of a pair of reads, S_1 and S_2, to a reference sequence, the alignment of sequences S_1 and S_2 to each other should also be high-scoring. The algorithm proceeds as follows:

- Build a cross-product graph $G' = (V, E)$ representing the set of all possible alignments of S_1 and S_2, s is the source of G', t is the sink.
- For every edge $e = u \to v$ in G' we compute the score of the highest scoring path from source s to sink t through the edge e; we denote this score as $w(e)$, the weight of the edge. To do this we use dynamic programming to

compute the scores of the highest scoring paths from s to every vertex of G' and of the highest scoring paths, using the edges of G' reversed, from t to every vertex of G. Time complexity of this step is $O(E)$.

- For a given suboptimality parameter ε, we build WSG G_2 from G' by discarding all edges such that

$$w(e) < Opt - \varepsilon$$

where Opt denotes the score of the highest scoring path from s to t in G'. Observe that while all ε-suboptimal alignments of S_1 and S_2 correspond to paths from s to t, in G_2, the converse is not true: not all paths from s to t are ε-suboptimal.

- Align WSG G_2 to the reference genome: compute the cross product of G_2 and the linear chain that represents the reference sequence, obtaining a WSG spelling all possible three way alignments M, such that $M[1:2;:]$ is one of the alignments, represented by G_2. Finally, use dynamic programming (similar to Smith and Waterman [12]) to search this graph for the locally optimal path. The score of this path is the score of mapping the pair of reads at the given location.

While the use of the WSG leads to a significant speedup compared with the full 3-dimensional dynamic programming, this is still insufficient for the alignment of many reads to a long reference genome, as the resulting algorithm is at best quadratic. In the subsequent section we combine the WSG alignment approach with standard heuristics for rapid sequence alignment in order to further improve the running time.

2.3   Seeded Alignment Approach for SMS Reads

Almost all heuristic alignment algorithms start with seed generation – the location of short, exact or nearly exact matches between two sequences. Whenever one or more seeds are found, the similarity is typically verified using a slow, but more sensitive alignment algorithm. The seeds offer a tradeoff between running time and sensitivity: short seeds are more sensitive than long ones, but lead to higher running times. Conversely, longer seeds result in faster searches, but are less sensitive. Our approach for SMS seeded alignment differs in two ways from standard methods. First, we generate potential seeds not from the observed reads, but rather from the ϵ-suboptimal WSGs representing their alignments. Second, we combine spaced seeds [1] with k-mer-filtering techniques [9] to further speed up our algorithm.

Seed Generation. Intuitively, given a read pair and an integer k, the seeds we want are k-long substrings (k-mers) likely to appear in the DNA segment represented by the pair and segment's matching location on the reference sequence. While k-mers present in each individual read may not match the genome directly due to the deletion errors introduced during the sequencing process, it is much less likely that both reads have an error at the same location. Consequently, we generate the potential seeds not from the the reads directly, but from the high-scoring alignments between them. Because every high-scoring alignment between the reads corresponds to a path through the WSG, we take all k-long subpaths in the WSG and output the sequence of edge labels along the path. Recall that each edge of the WSG is labeled by a pair of letters (or gaps), one from each read. If one of the reads is gapped at the position, we output the letter from the other read. If neither read is gapped we output a letter from an arbitrary read. While the number of k-mers we generate per pair of reads can be large if the two reads are very different, in practice the reads are similar and the WSG is small and "narrow". In our simulations (described in the Results section) we saw an average of 27 k-mers per read-pair for reads of approximately 30 nucleotides.

Genome Scan. Given the sets of k-mers generated from the WSGs of read-pairs, we build a lookup table, mapping from the k-mers to the reads. We used spaced seeds [1], where certain pre-determined positions are "squeezed" from the k-mers to increase sensitivity. We then scan the genome, searching for matches between k-mers present in the genome and the reads. If a particular read has as many or more than a specified number of k-mer matches within a given window of the genome, we execute a vectorized Smith-Waterman step, described in the next section.

Unlike some local alignment programs that build an index of the genome and then scan it with each query (read), we build an index of the reads and query this index with the genome. This approach has several advantages: first, it allows us to control memory usage, as our algorithm never needs memory proportional to the size of the genome, while the large set of short reads can be easily divided between many machines in a compute cluster. Secondly, our algorithm is able to rapidly isolate which reads have several k-mer matches within a small window by using a circular buffer to store all of the recent positions in the genome that matched the read. The genome scan algorithm is illustrated in Figure 2A.

2.4   Vectorized Smith-Waterman Implementation

While the genome scan described in the previous section significantly reduces the number of candidate regions, many false positives are encountered. To further reduce the number of potential mapping positions given to the WSG-based aligner, we first align both reads to the candidate region using a vectorized implementation of the Smith-Waterman algorithm. Most contemporary mobile, desktop and server-class processors have special vector execution units, which perform multiple simultaneous data operations in a single instruction. For example, it is possible to add the eight individual 16-bit elements of two 128-bit vectors in one machine instruction. Over the past decade, several methods have been devised to significantly enhance the execution speed of Smith-Waterman-type algorithms by parallelizing the computation of several cells of the dynamic programming matrix. The simplest such implementation by Wozniak [13] computes the dynamic programming matrix using diagonals (See Figure 2B). Since each cell of the matrix can be computed once the cell immediately above, immediately to the

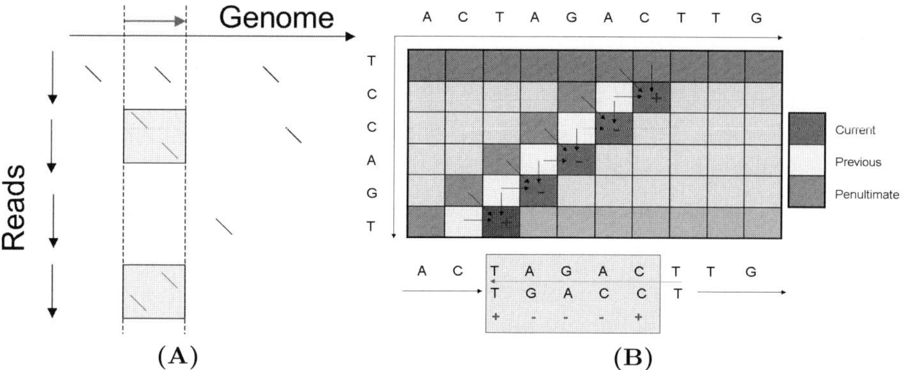

Fig. 2. A. Overview of the k-mer filtering stage within SHRiMP: A window is moved along the genome. If a particular read has a preset number of k-mers within the window the vectorized Smith-Waterman stage is run to align the read to the genome. **B.** Schematic of the vectorized-implementation of the Smith-Waterman algorithm. The red cells are the vector being computed, on the basis of the vectors computed in the last step (yellow) and the next-to-last (green). The match/mismatch vector for the diagonal is determined by comparing one sequence with the other one reversed (indicated by the red arrow below). To get the set of match/mismatch positions for the next diagonal the lower sequence needs to be shifted to the right.

left, and at the upper-left corner have been computed, one can compute each successive diagonal once the two prior diagonals have been completed. The problem can then be parallelized across the length of supported diagonals, often leading to a vectorization factor of 4 to 16. The only portion of such an approach that cannot be parallelized is the identification of match/mismatch scores for every cell of the matrix. These operations are done sequentially, necessitating 24 independent, expensive data loads for 8-cell vectors. The approach becomes increasingly problematic as vector sizes increase because memory loads cannot be vectorized; when the parallelism grows, so does the number of lookups.

Rognes and Seeberg [10] developed an alternative algorithm with the following innovations: first, they avoided the aforementioned loads by pre-calculating a 'query profile' before computing the matrix. By pre-allocating vectors with the appropriate scores in memory, they needed only load a single score vector on each computation of a vector of cells. The up-front cost in pre-calculation greatly reduced the expense of computing scores in the crucial, tight inner-loop of the algorithm. The query profile, however, requires that scores be computed per-column, rather than for each diagonal. This creates data dependencies between several cells within the vector. However, the authors took advantage of the fact that often cells contain the value 0 when computing the Smith-Waterman recurrence, and any gap continuation would therefore be provably suboptimal. This enables the conditional skipping of a significant number of steps. Although highly successful for protein sequences, where only 1/20 elements on average have a positive score, it is significantly less beneficial for DNA sequences where usually at least 1/4 elements in the matrix match. Farrar [5] most recently introduced a superior method: by striping the query profile layout in memory, he significantly reduced the performance impact of correcting data dependencies.

We propose an alternate method, where the running time of the fully vectorized algorithm is independent of the number of matches and mismatches in the matrix, though it only supports fixed match/mismatch scores (rather than full scoring matrices). Since SHRiMP only applies a vectorized Smith-Waterman scan to short regions of confirmed k-mer hits, alternative approaches that benefit by skipping computation in areas of dissimilarity are unable to take significant advantage. Figure 2B demonstrates the essence of our algorithm: by storing one of the sequences backwards, we can align them in such a way that a small number of logical instructions obtain the positions of matches and mismatches for a given diagonal. We then construct a vector of match and mismatch scores for every cell of the diagonal without having to use expensive and un-vectorizable load instructions or pre-compute a "query profile". In our tests (see Table 1), our approach surpasses the performance of Wozniak's original algorithm [13] and performs on par with Farrar's method [5]. The advantages of our method over Farrar's approach are simplicity, independence of the running time and the scores used for matches/mismatches/gaps, and linear scalability for larger vector sizes. The disadvantages of our method are that it cannot support full scoring matrices (i.e. it is restricted to match/mismatch scores) and is slower for queries against large reference sequences where significant areas of dissimilarity are expected. However, the former is less important for DNA alignment and the latter does not apply to SHRiMP.

The vectorized Smith-Waterman approach described above is used to rapidly determine if both of the reads have a strong match to a given region of the genome. The locations of the top n hits for each read on the genome are stored in a heap data structure, which is updated after every invocation of the vectorized Smith-Waterman algorithm if the heap is not full, or if the attained score is greater than or equal to the lowest scoring top hit. Once the whole genome is processed, the highest scoring n matches are re-aligned using the full WSG alignment algorithm described in Section 2.2.

3   Results

In order to test the efficacy of our read mapping algorithm we designed a simulated dataset with properties similar to those expected from the first generation SMS sequencing technologies, such as Helicos [4]. We sampled 10,000 pairs of reads from human chromosome one (the total length of the chromosome is approximately 1/10th of the human genome). The first read was sampled to have a mean length of 30 bases, with standard deviation of 8.9, the second one had a mean length of 26 bases with standard deviation 8.6. The lengths of both reads ranged from 12 to 58 bases. We introduced errors into each of the the reads, with 7% deletion errors and 0.5% insertion and substitution errors uniformly at random. We also introduced Single Nucleotide Polymorphisms (SNPs) at a rate of 1%.

Table 1. Performance (in millions of cells per second) of the various Smith-Waterman implementations, including a regular implementation, Wozniak's diagonal implementation with memory lookups, Farrar's striped algorithm and our diagonal approach without score lookups (SHRiMP). We used each method within SHRiMP to align 50 thousand reads to a reference genome with default parameters. The improvement of the Core 2 architecture for vectorized instructions lead to a significant speedup for our and Farrar's approaches, while the Wozniak algorithm's slight improvement is due to the slow memory lookups.

Processor type	Unvectorized	Wozniak	Farrar	SHRiMP
P4 Xeon	97	261	335	338
Core 2	105	285	533	537

Table 2. Comparison of the WSG-based alignment with the two alternative approaches. The first two rows describe the percentage of reads not mapped anywhere on the genome or mapped in multiple places (with equal alignment scores). The third row is the percentage of the unique hits that are correct. The sub-columns under each method are the number of matching positions in the spaced seed (we allowed for a single non-matching character in the middle of the seed). The last row shows the running time in minutes and seconds on a 2.66 GHz Core2 processor. The best result in each category is in boldface.

Type	SEPARATE			PROFILE			WSG		
seed weight	8	9	10	8	9	10	8	9	10
no hits %	**0.000**	**0.131**	**2.945**	1.741	4.905	10.52	1.609	4.314	10.15
multiple %	30.12	26.45	21.13	**10.20**	9.342	8.353	10.44	**9.127**	**8.258**
unique cor %	63.90	63.00	61.09	78.96	74.90	69.66	**79.17**	**75.84**	**70.85**
runtime	28m16	**8m48**	**4m22**	**27m17**	11m32	6m58	30m59	12m13	7m13

We used three approaches to map the read pairs back to the reference. Our main approach was the WSG-based alignment algorithm described in Section 2.2. For comparison we also implemented two alternatives: mapping the reads individually, using the version of SHRiMP for Illumina/Solexa, but with the same parameters as were used for WSG-based mapping in order to properly capture the high expected deletion rates, and a profile-based approach, which follows the overall framework of the WSG approach, but only considers a single top alignment (one path in the WSG) for further mapping to the reference genome. The methods are labeled *WSG*, *SEPARATE*, and *PROFILE*, respectively, in Table 2.

We evaluated the performance of these algorithms based on three criteria: the fraction of the reads that were not mapped at all (lower is better), the fraction of the reads that were not mapped uniquely (lower is better: while some reads have several equally good alignments due to repetitive DNA, others map with equal scores to non-repetitive DNA segments, and this should be minimized), and the percentage of the reads mapped uniquely and correctly (higher is better). When evaluating the two reads separately, we considered a hit unique if either of the two reads had a unique top hit, and we considered a unique top hit correct if either of the top hits was correct. We ran all algorithms with three seed weights: 8, 9, and 10, with each spaced seed having a single wildcard character (zero) in the middle of the seed (e.g. 1111101111 was the spaced seed of weight 9).

As one can see in Table 2, the approach of mapping the reads separately, while leading to very good sensitivity (almost all reads aligned), has poor specificity (only 70-78% of the reads have a unique top hit, with 61-64% both unique and correct). The WSG and $PROFILE$ approaches perform similarly, with the WSG approach slightly more sensitive and having a higher fraction of unique correct mappings for all seed sizes.

4   Discussion

In this paper we propose a novel read alignment algorithm for next-generation Single Molecule Sequencing (SMS) platforms. Our algorithm takes advantage of spaced k-mer seeds and effective filtering techniques to identify potential areas of similarity, a novel, vectorized implementation of the Smith-Waterman dynamic programming algorithm to confirm the similarity and a weighted sequence graph-based three way final alignment algorithm. Our approach is implemented as part of SHRiMP, the SHort Read Mapping Package, and is freely available at **http://compbio.cs.toronto.edu/shrimp**.

While the overall emphasis of the current paper was read mapping, we believe that several of our methods may have broader applications. For example, our implementation of the Smith-Waterman algorithm can be used in other sequence alignment methods, and unlike previous vectorized implementations the running time of our algorithm is independent of the scoring parameters. The weighted sequence graph model could also be useful for ab initio genome assembly of SMS data, where a variant of our method could be used for overlap graph construction.

References

1. Califano, A., Rigoutsos, I.: Flash: A fast look-up algorithm for string homology. In: Proceedings of the 1st International Conference on Intelligent Systems for Molecular Biology, pp. 56–64. AAAI Press, Menlo Park (1993)
2. Chenna, R., Sugawara, H., Koike, T., Lopez, R., Gibson, T.J., Higgins, D.G., Thompson, J.D.: Multiple sequence alignment with the clustal series of programs. Nucleic Acids Res. 31(13), 3497–3500 (2003)
3. Jettand, J.H., et al.: High-speed DNA sequencing: an approach based upon fluorescence detection of single molecules. Journal of biomolecular structure and dynamics 7(2), 301–309 (1989)
4. Harris, T.D., et al.: Single-molecule DNA sequencing of a viral genome. Science 320(5872), 106–109 (2008)
5. Farrar, M.: Striped Smith-Waterman speeds database searches six times over other SIMD implementations. Bioinformatics 23(2), 156–161 (2007)

6. Hein, J.: A new method that simultaneously aligns and reconstructs ancestral sequences for any number of homologous sequences, when the phylogeny is given. Molecular Biology and Evolution 6(6), 649–668 (1989)
7. Naor, D., Brutlag, D.L.: On near-optimal alignments of biological sequences. Journal of Computational Biology 1(4), 349–366 (1994)
8. Needleman, S.B., Wunsch, C.D.: A general method applicable to the search for similarities in the amino acid sequences of two proteins. JMB 48, 443–453 (1970)
9. Rasmussen, K.R., Stoye, J., Myers, E.W.: Efficient q-gram filters for finding all epsilon-matches over a given length. Journal of Computational Biology 13(2), 296–308 (2006)
10. Rognes, T., Seeberg, E.: Six-fold speed-up of smith-waterman sequence database searches using parallel processing on common microprocessors. Bioinformatics 16(8), 699–706 (2000)
11. Schwikowski, B., Vingron, M.: Weighted sequence graphs: boosting iterated dynamic programming using locally suboptimal solutions. Discrete Appl. Math. 127(1), 95–117 (2003)
12. Smith, T.F., Waterman, M.S.: Identification of common molecular subsequences. Journal of Molecular Biology 147, 195–197 (1981)
13. Wozniak, A.: Using video-oriented instructions to speed up sequence comparison. Computer Applications in the Biosciences 13(2), 145–150 (1997)

Exact Transcriptome Reconstruction from Short Sequence Reads

Vincent Lacroix[1], Michael Sammeth[1], Roderic Guigo[1], and Anne Bergeron[2]

[1] Genome Bioinformatics Research Group - CRG, Barcelona, Spain
vincent.lacroix@crg.es, roderic.guigo@crg.es, micha@sammeth.net
[2] Comparative Genomics Laboratory, Université du Québec à Montréal, Canada
bergeron.anne@uqam.ca

Abstract. In this paper we address the problem of characterizing the RNA complement of a given cell type, that is, the set of RNA species and their relative copy number, from a large set of short sequence reads which have been randomly sampled from the cell's RNA sequences through a sequencing experiment. We refer to this problem as the transcriptome reconstruction problem, and we specifically investigate, both theoretically and practically, the conditions under which the problem can be solved. We demonstrate that, even under the assumption of exact information, neither single read nor paired-end read sequences guarantee theoretically that the reconstruction problem has a unique solution. However, by investigating the behavior of the best annotated human gene set, we also show that, in practice, paired-end reads – but not single reads – may be sufficient to solve the vast majority of the transcript variants species and abundances. We finally show that, when we assume that the RNA species existing in the cell are known, single read sequences can effectively be used to infer transcript variant abundances.

1 Introduction

The genome sequence is said to be an organism's blueprint, the set of instructions specifying the biological features of the organism. The first step in the unfolding of these instructions is the transcription of the cell's DNA into RNA, and the subsequent processing of the primary RNA to functional RNA molecules. A typical eukaryotic cell contains tens of thousands of RNA species, each of them present in one or multiple copies. Usually, only a fraction of the DNA is transcribed to RNA in a given cell type. On the other hand, multiple species can be generated from the same primary transcripts through *alternative splicing*, a process in which fragments of the primary RNA molecule, the *exons*, can be spliced out in different combinations, often in a cell-type specific manner, to generate the mature RNA molecule. Thus, while the DNA complement of a cell is essentially constant across time, and across cells in multicellular organisms, the RNA complement is, in contrast, highly variable and is determinant of the specific phenotype of each cell type and condition. The determination of the RNA complement is therefore essential for the characterization and understanding of the biology of the cell.

K.A. Crandall and J. Lagergren (Eds.): WABI 2008, LNBI 5251, pp. 50–63, 2008.
© Springer-Verlag Berlin Heidelberg 2008

Systematic sequencing of cDNA libraries has probably been the main approach to the identification and quantification of the different RNA species in the cell. One popular strategy is based on the identification, by a "single pass" sequencing, of random cDNA clones, that results in the production of short partial sequences identifying a specific transcript. This technology is generally known as Expressed Sequence Tag, or EST [1,2,9,10]. Since different RNA species in the cell, originating from alternative splicing of the same primary transcript, may share a substantial fraction of their sequence, the partial nature of the ESTs makes it difficult to unequivocally establish the RNA molecule specifically responsible for a given EST. In other words, identical ESTs can originate from different alternative spliced RNA molecules. To address this issue, a number of methods, notably based on *splicing graphs* [8,12,13,17], a data structure used to describe the connectivity between adjacent exons in the alternative spliced RNAs originating from a given primary transcript, have been employed in an attempt to reconstruct splice variants from partial EST sequences obtained from non-normalized libraries [19]. A second problem with this approach is due to the large dynamic range of RNA abundances in the cell that makes sampling of the low abundant species very inefficient. In practice, these species can only be effectively recovered once their abundance has been selectively increased through normalization procedures. These procedures, however, destroy the relative abundance of the species in the cell, and are therefore inadequate for complete transcriptome reconstruction. Recently, next generation sequencing technologies have dramatically augmented the capacity for sequencing DNA and RNA molecules. In particular, they provide for an unprecedented capacity for deep sampling of the transcriptome of the eukaryotic cell. However, the most cost-effective such technologies, such as Solexa/Illumina [3], AB SOLiD (www.appliedbiosystems.com) and HELICOS (www.helicosbio.com), among others, typically produce very short sequence reads – from 25 to 35 bp – exacerbating enormously the problem of reconstructing complete RNA molecules from partial sequences.

In this paper we address the problem of characterizing the RNA complement of a given cell type, that is the set of RNA species, or *transcript variants*, and their relative copy number, called their *abundance*, from a large set of short sequence reads, which have been randomly sampled from the cell's RNA sequences through a sequencing experiment. We refer to this problem as the *transcriptome reconstruction problem*, and we specifically investigate, both theoretically and practically, the conditions under which the problem can be solved. Through all the paper we will assume that the number of sequence reads originating from each mRNA species is a positive integer proportional to the number of copies of such species in the cell, we will refer to this assumption as to the *exact information* hypothesis. We will also assume that the RNA molecules are uniformly sampled along their entire length, and that there is no 3' or 5' sampling bias. While these assumptions may not be realistic at present time, we argue that they do not affect the overall conclusions of our work. We specifically investigate whether solutions to the transcriptome reconstruction problem exist when the sequencing experiment has produced single sequence reads, and when it has produced paired-end reads – that is, pairs of non-adjacent sequence reads coming from the same RNA molecule. We also investigate the particular case in which the RNA species are assumed to be known. The reconstruction problem then simplifies to the estimation of the abundances.

We first show that only sequence reads spanning block junctions are relevant for transcriptome reconstruction, and that the remaining sequence reads are superfluous. Then, we formulate the problem of transcriptome reconstruction as a system of linear equations in which the abundance of such junction reads, or pairs of junctions reads, which are known from the sequencing experiment, is simply the sum of the abundances of the transcript variants in which they occur (the unknowns). In this framework, when the linear system has a unique solution, the transcriptome can be completely reconstructed. We demonstrate that, even under the exact information hypothesis, neither single read nor paired-end read sequences guarantee theoretically that the linear system has a unique solution. However, by investigating the behavior of the best annotated human gene set, we also show that, in practice, paired-end reads – but not single reads – may be sufficient to solve the vast majority of the transcript variants species and abundances. We finally show that, when we assume that the RNA species existing in the cell are known, single read sequences can effectively be used to infer transcript variant abundances.

In our approach, the system of linear equations is derived from the underlying splicing graph of the transcriptome to be reconstructed, which is assumed to be known *a priori* or inferred from the sequencing experiment. Inferring the splicing graph from the sequencing experiment requires that all junctions are covered, which is guaranteed in our case because we assume exact information. It also requires that the reads covering a junction are identified as such. This is related to the problem of spliced alignment of short reads on the genome, which we do not address in this paper. This will be the focus of the next step of this project. Available methods addressing this issue include [5].

2   The Transcriptome Reconstruction Problem

A *gene* can be regarded as a non-redundant set of *exons*, *i.e.*, continuous segments of the genomic sequence that are expressed in at least one of its variants. Consequently, transcript variants are typically described by the respective sequence of exons they incorporate from 5' to 3'. Clearly, exons from different variants of the same gene may overlap when projecting them on the genome, tokenizing them into different *exon fragments*. Since, in this work, we are more generally interested in common parts of transcripts rather than simply in common exons, we introduce the concept of a *block*.

Definition 1. *A* block *is a maximal sequence of adjacent exons or exon fragments that always appear together in a set of variants.*

Therefore, variants can be represented by sequences of blocks. Figure 1 illustrates how blocks are inferred from exons. A *block junction* is a pair of consecutive

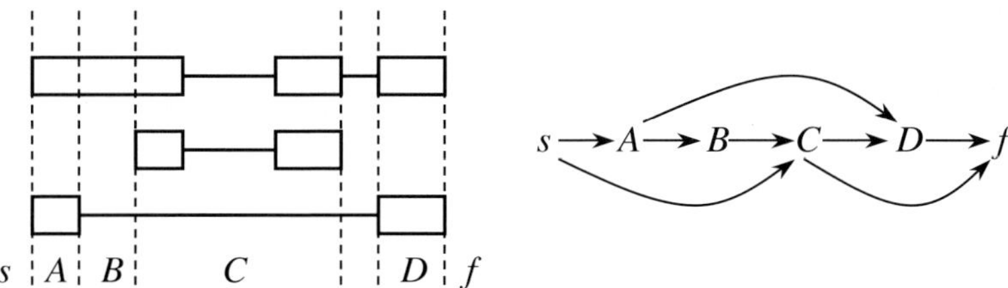

Fig. 1. The left part corresponds to a diagram of a gene with three variants, where rectangles represent exons or exon fragments that belong to a particular variant, and horizontal lines are the intervening parts that are not contained in a particular variant (introns). Vertical dashed lines delimit four blocks, that are labelled A, B, C, and D. A block either completely belongs to a variant, or is skipped. The depicted variants can thus be described by the sequences $ABCD$, C, and AD. The corresponding splicing graph is drawn in the right part where vertices are blocks and edges are block junctions pointing in the direction of transcription, from 5' to 3'. The graph is completed by two additional vertices s and f, for start and finish. Vertex s is connected to all first blocks of the variants, and vertex f is connected to all last blocks.

blocks in a variant, including pairs such as sA, or Df, to denote the 'start' and 'finish' junctions. As shown in the example, a block may even be an exon-intron structure (such as block C) and a block junction does not necessarily correspond to an exon junction (such as block junction AB).

Given a set S_o of variants, the *splicing graph* is a directed acyclic graph whose vertices are the blocks, and whose directed edges join two blocks if the corresponding junction exists in at least one of the variants in the set S_o [8]. Additionally, two vertices labelled s and f are included that respectively connect to the first and from the last block of each transcript as in [15]. These splicing graphs play a crucial role in analyzing alternative splicing [16,17].

A variant, in this splicing graph, corresponds to a directed path that goes from vertex s to vertex f reproducing the sequence of blocks it contains. However, paths from vertex s to vertex f do not necessarily correspond to expressed variants. For instance, in the graph of Fig. 1, the set of annotated transcripts is $S_o = \{ABCD, C, AD\}$ but the set of paths also includes ABC and CD. In general, if there are N blocks, there is a theoretical possibility of $2^N - 1$ different paths in a splicing graph.

We now formulate the *transcriptome reconstruction* problem as follows: Given a set $S = \{x_1, \ldots, x_k\}$ of candidate variants, and a set of constraints $\{C_1, \ldots, C_m\}$, each indicating the total abundance of a subset of variants of S, is it possible to assign a unique abundance to each variant such that all the constraints are satisfied?

In practice, a constraint C_j, will reflect, for example, the number of sequence reads mapping to a particular block junction j, called the *abundance* of the block junction, and the corresponding subset of variants will be all variants of S that contain junction j.

In Section 4, we will investigate a particular case of this problem in which the set S of candidate variants is the set of all possible paths of a given splicing graph, and the constraints are the abundances of block junctions. Section 5 discusses the case when only a known subset of all possible paths are candidate variants, and, in Section 6, we will consider the general problem when the constraints are the abundances of pairs of block junctions.

3   Preliminaries

In the following, we will assume that the information given by the constraints is *exact*, in the sense that the measure of total abundance of a subset of variants is an integer that faithfully reflects the corresponding number of molecules.

We are aware that this assumption is not realistic at present time. However, we argue that the results obtained under this assumption give a good indication of the results we should expect when we deal with partial information. In particular, when we show that a variant cannot be solved with exact information, it will obviously also be the case with partial information. Furthermore, given the pace at which new sequencing technologies are developing, it seems reasonable to predict that a typical transcriptome of size 6.10^8 nucleotides could be affordably sequenced to great depth. As an indication, a single Solexa experiment performed today (one flow cell) would produce a $2X$ coverage.

Given the technologies currently at hand, we consider constraints of two types: measures of the total abundance of block junctions, such as those obtained from single sequence reads, and measures of the total abundance of pairs of junctions, such as that obtained from paired end sequence reads. We refer to the first type as 'local information', and to the second as 'paired information'. In both cases it is always possible to recover the abundance of blocks from the abundance of block junctions, under the exact information hypothesis. To see this, denote by C_Y the total abundance of variants containing block Y, and by C_{XY} the total abundance of variants containing junction XY, then we have, for each block Y:

$$C_Y = \sum_{X \in [s..Y[} C_{XY} = \sum_{Z \in]Y..f]} C_{YZ}.$$

Similarly, it is possible to recover the abundance of a block junction using the data on the abundance of pairs of junctions. Let $C_{TU,XY}$ be the total abundance of variants containing both block junctions TU and XY, then, for each junction XY:

$$C_{XY} = \sum_{T \in [s..X[} C_{TX,XY} = \sum_{U \in]Y..f]} C_{XY,YU}.$$

To illustrate the main techniques and pitfalls of the reconstruction problem, consider the following example. We will consider the splicing graph of Fig. 1, but without the knowledge of which variants are expressed because we are now trying to predict them. Let $S = \{ABCD, ABC, AD, C, CD\}$ be the set of paths of the splicing graph, and let C_{XY} be the abundance of a block junction XY for each edge of the splicing graph.

The transcriptome reconstruction problem can be formulated using the following system of linear equations, where the abundance of a variant is denoted by the label of the variant:

$$ABCD + ABC + AD = C_{sA} \quad (1)$$
$$C + CD = C_{sC} \quad (2)$$
$$ABCD + ABC = C_{AB} \quad (3)$$
$$AD = C_{AD} \quad (4)$$
$$ABCD + ABC = C_{BC} \quad (5)$$
$$ABCD + CD = C_{CD} \quad (6)$$
$$ABC + C = C_{Cf} \quad (7)$$
$$ABCD + AD + CD = C_{Df} \quad (8)$$

The problem can be solved if the system has a unique solution. Unfortunately, many equations are linearly dependent, namely (3) = (5), (1) = (3) + (4), (2) + (5) = (6) + (7), and (8) = (4) + (6). These identities correspond to integrity constraints that impose that the sum of the abundances of edges that goes in a vertex V is equal to the sum of abundances on edges that go out of V. This reduces the number of equations to the following four:

$$C + CD = C_{sC}$$
$$AD = C_{AD}$$
$$ABCD + ABC = C_{AB}$$
$$ABC + C = C_{Cf}$$

If each variant of the original set $S_o = \{ABCD, C, AD\}$ used to construct the splicing graph in Fig. 1 is expressed, then the values of C_{sC}, C_{AD}, C_{AB}, and C_{Cf} are all positive. The abundance of only one of the variants, namely AD, can be correctly predicted, and the values of the four others are linked by only three equations and are thus impossible to determine. In this example, with the posterior knowledge of the original variants, we would conclude that we can correctly predict the abundance of $1/3$ of the variants, which is quite a poor performance!

The unsolvable instances of the preceding example result from the fact that in the two sets of variants $\{ABCD, C\}$ and $\{ABC, CD\}$, each block junction, and therefore each block, has the same number of occurrences: in this case sA, AB, BC, CD, sC, Cf and Df appear all exactly once in both sets of variants. Any method that measures only abundance of blocks, or block junctions, will be blind to the substitution of members of one set by members of the other. This leads to the following definition:

Definition 2. *Two disjoint sets of variants S_1 and S_2 are* interchangeable under block junctions *if each block junction has the same number of occurrences in each set. Similarly two disjoint sets of variants S_1 and S_2 are* interchangeable under pairs of block junctions *if each pair of block junctions has the same number of occurrences in each set.*

The smallest example of two sets interchangeable under block junctions is given by the sets $\{ABC, B\}$ and $\{AB, BC\}$, and the smallest example of interchangeability under pairs of block junctions is:

$$\{ABCDE, ABD, BCD, BDE\} \text{ and } \{ABCD, ABDE, BCDE, BD\}.$$

In order to find this set, we explicitly solved the systems of linear equations with the $2^N - 1$ possible variants for $N = 2, 3, 4$ and 5 blocks. In the first three cases, there was a unique solution for all variants. In the last case, we could predict the abundance of 23 of the 31 possible variants, and found these two interchangeable sets.

If enough variants belonging to interchangeable sets are expressed in a cell, then it is impossible to determine the abundance of any of the variants of the sets. Formally, we have:

Proposition 1. *If the set of candidate variants contains interchangeable sets S_1 and S_2, and all the variants of at least one of the two sets S_1 or S_2 have a strictly positive abundance, then it is impossible to predict the abundance of any variant in $S_1 \cup S_2$.*

Proof. Suppose that S_1 and S_2 are interchangeable under block junction – the proof is similar for interchangeability under pairs of block junctions. Suppose that all variants in S_1 have a positive abundance of at least a, and let C_{XY} be the total abundance of variants containing junction XY. If junction XY is present in k variants of S_1, it is thus present in k variants of S_2, then C_{XY} can be written as

$$C_{XY} = A + x_1 + \ldots + x_k + y_1 + \ldots y_k$$
$$= A + (x_1 - a) + \ldots + (x_k - a) + (y_1 + a) + \ldots (y_k + a)$$

where A is the total abundance of variants that have the junction XY but do not belong to $S_1 \cup S_2$, and $x_1, \ldots, x_k$ and $y_1, \ldots, y_k$ represents the abundances of variants in S_1 and S_2 that contain junction XY. Since the above equation is true for any constraint, subtracting the value a to all abundances of variants in S_1, and adding a to all abundances of variants in S_2 yields a solution that satisfies all the constraints.

In the work of Xing et al. [20], variants belonging to interchangeable sets are treated as equiprobable. This means that the experimental evidence is equally distributed among all variants of the interchangeable sets, yielding ungrounded predictions. Although, in the case of ESTs, the consequences may be less dramatic than with short reads, since interchangeable sets can be identified beforehand, it would be preferable to remove the corresponding variants from the list of candidate variants.

4   Transcriptome Reconstruction with Local Information

Local information is the type of data that one can obtain with short read sequencing but also with splicing microarrays. In the case of short reads, it is theoretically possible to reconstruct the actual splicing graph from experimental data, but, in the case of microarray detection, where only a certain number of block junctions are tested, the splicing graph can be incomplete. In this section, we will assume that the splicing graph is given, together with the abundance of each block junction.

The *reconstruction problem with local information* is the following: Given a splicing graph G and the abundance of each block junction, if S is the set of all possible paths in G, when is it possible to assign a unique abundance to a variant in S?

Proposition 2. *Given the abundances of block junctions, if S is the set of all possible paths in a splicing graph, the abundance of a variant $x \in S$ can be predicted if and only if x has a unique block junction among all elements of S.*

Proof. If x has a unique block junction, then its abundance is immediately given. On the other hand, if a path x shares all its junctions with other paths, then there exists a path y that shares a longest prefix with x, meaning that there exists a, x', y' sequences of blocks, and $B \neq C$ such that $x = aABx'$, and $y = aACy'$. The path x shares its junction AB with a word z that does not begin with aA, since y was a longest such prefix. Thus z is of the form $bABz'$, with $a \neq b$, see Figure 2. This implies that the four words $aABx'$, $aACy'$, $bABx'$ and $bACy'$ are all paths in the splicing graph, but $\{aABx', bACy'\}$ and $\{aACy', bABx'\}$ are interchangeable under block junction, thus the abundance of x cannot be predicted.

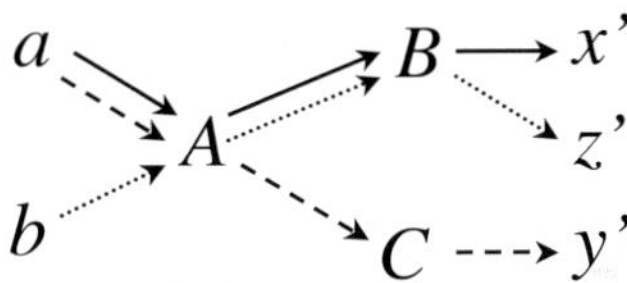

Fig. 2. The existence of three paths $x = aABx'$, $y = aACy'$ and $z = bABz'$, with $a \neq b$ and $B \neq C$, implies the existence of a fourth path $bACy'$. Note that the sequences x', y' and z' can be empty.

In order to evaluate the impact of this result in practice, we have quantified, in a set of exhaustively annotated human genes, the proportion of candidate variants among all paths in the splicing graph that have unique block junctions, and that therefore can be recovered using local information alone. We focused on the GENCODE dataset [7], a high quality data annotation of the protein coding genes in the regions covering the 1% of the human genome selected within the ENCODE project [6]. The GENCODE annotation was obtained through manual curation of cDNA and EST alignments onto the human genome, augmented with computational predictions that were subsequently experimentally tested by RT-PCR and RACE. The current version of GENCODE contains 681 genes and 2981 variants. We focused on the 335 genes that are annotated with at least 2 variants, corresponding to a total of 2635 variants. The first three lines of Table 1 present the distribution of the number of variants per gene.

Table 1. Lines A, B and C give statistics on the the 335 GENCODE genes annotated with at least two variants, grouped by their number of variants. Line D gives the number of these variants whose abundance could have been predicted using local information (Section 4). Line E gives the number of these variants whose abundance could have been predicted using local information, and prior knowledge of the variants (Section 5). Line F gives the number of these variants whose abundance could have been predicted using paired information (Section 6).

A	Number of genes	55	47	35	25	24	19	18	19	11	14	14	10	5	39
B	Number of variants per gene	2	3	4	5	6	7	8	9	10	11	12	13	14	≥ 15
C	Total number of variants	110	141	141	124	144	133	144	171	110	154	168	130	70	895
D	Solved with local information	8	7	4	1	0	0	2	2	0	0	1	1	0	4
E	Solved with fixed list	110	141	141	124	144	133	144	171	110	154	168	130	70	891
F	Solved with paired information	110	141	141	124	144	133	144	171	110	154	168	130	70	895

For each GENCODE gene, we constructed the splicing graph, and computed how many candidate variants had unique junctions among the set of all possible paths in the splicing graph. Note that in this section we are not assuming any knowledge of the existing variants. The splice graph could in principle be inferred from the RNA sequencing experiment. The computation in this experiment is quite straightforward. However, even if most splicing graphs are of modest dimension, the worst case in the dataset was a gene with 72 transcripts yielding 191 blocks, 352 block junctions, and 36.6 million paths. Results appear in line D of Table 1.

Only 30 out of 2,635 GENCODE variants (1.14%) have unique block junctions and these are then the only ones that can actually be solved. That is, even in the case of exact information, RNA sequencing by short single reads will only help to unequivocally quantify the abundances of a handful of variants. This was expected, since interchangeable sets with respect to block junctions exists even with only 3 blocks, and their number grows rapidly with the number of blocks. The only way to improve the performance is to reduce the number of candidate variants to consider. This can be done either by explicitly assuming that all potential variants are already known, and the problem reduces to the estimation of their abundances, which we will explore in the next section, or by using additional information on the variants that may actually exist, which we will discuss in Section 6.

5   Transcriptome Reconstruction with Local Information and a Fixed Set of Candidates

In this section, we investigate the problem of measuring the abundance of transcript variants with local information when the candidates are not necessarily all the paths of a splicing graph, but are given as a fixed list. This is an interesting alternative problem since we may consider that, in the coming years, all existing variants should be known, at least for extensively studied organisms.

In this case, it is sufficient to solve the corresponding set of equations. The solution is not unique when there are interchangeable sets in the given list of candidates, and they can then be identified from the set of variants that admit multiple solutions. Algorithms to do this are adapted from well-known algorithms in linear algebra. Consider the example of Section 3, and suppose that the set of candidates is $S = \{ABCD, C, AD, ABC\}$, then the corresponding – reduced – set of equations is:

$$C = C_{sC} \qquad\qquad ABCD + ABC = C_{AB}$$
$$AD = C_{AD} \qquad\qquad ABC + C = C_{Cf}$$

This system, with four unknowns and four independent equations has a unique solution.

Working again with the GENCODE dataset, we computed how many variants would have their abundance correctly predicted. In this experiment the list of candidates is the list of annotated GENCODE variants. Line E of Table 1 gives the number of correctly predicted variants. Overall, the total number of correctly predicted variants is 2631 out of 2635, which represents 99.84 % of the GENCODE variants. Clearly, the prior knowledge of possible variants makes a huge difference. These results show that technologies such as short read sequencing or microarrays, that use only local information, should eventually be able to reconstruct transcriptomes, even if they are not able to predict new variants. In this experiment, only one gene, ITSN1, was problematic. Out of the 21 known variants of this gene, 17 can be resolved. The four remaining variants can be split into two interchangeable sets, and predicting their relative abundance must rely on other sources of information.

6 Transcriptome Reconstruction with Paired Information

In Section 3 we saw that, even with exact information on pairs of block junctions, interchangeable sets could be a problem. However, the combinatorial complexity of the smallest interchangeable sets is a good indication that most of real examples should be solvable.

With the current technology, paired information can only be obtained through sequencing. In the process, molecules are randomly sheared and the two extremities of molecule fragments are sequenced. For *de novo* sequencing, the distance between the two sequenced fragments, called the *insert length*, must be controlled. This is achieved by migrating the fragments on a gel and selecting for specific sizes. In the case of transcriptome reconstruction, knowing the insert length is not necessary and this selection step can be skipped. This allows, in theory, a uniform distribution of insert lengths within an RNA molecule. Such paired reads may link two blocks of a variant, a block and a block junction, or two block junctions. With the exact information hypothesis, we only need to consider paired reads that link two block junctions, which are said to *co-occur* in a variant.

Given a splicing graph and observed pairs of block junctions, we first identify all paths in the graph that contain only observed pairs. This is done by Algorithm 1. The output of this algorithm is a list S of candidates. We then solve the corresponding system of equations, and report the number of variants whose abundance is uniquely determined. As shown in line F of Table 1, with paired information, we could have predicted the abundance of every single GENCODE variant.

Algorithm 1. Enumeration of candidates with observed pairs of block junctions. For each block b, the set of junctions whose source is b is denoted Incident(b); for each junction j, the set of junctions that may co-occur with j is denoted by Co_occur(j), and End(j) denotes the second block of a block junction. The procedure is initially called with the parameter 'candidate' equal to the empty string, the parameter 'block' as s, and the parameter 'allowedJunctions' as the set of all the edges of the splice graph.

```
SplicingGraphTraversal(candidate, block, allowedJunctions)
if block = f then
    Output candidate {Successful traversal}
else
    nextJunctions ← allowedJunctions ∩ Incident(block)
    if nextJunctions ≠ ∅ then
        for each junction ∈ nextJunctions do
            allowed ← allowedJunctions ∩ Co_occur(junction)
            SplicingGraphTraversal(Concat(candidate, block), End(junction), allowed)
        end for
    end if
end if
```

The fact that all GENCODE variants can be solved with paired information is very promising, since it means that with this kind of information, and assuming that the GENCODE annotations are representative, variant discovery will be possible using paired-end short read sequencing.

To assess to what extent the results on the GENCODE set, which cover 1% of the human genome, can be extrapolated to the entire genome, we have extended our analysis to the REFSEQ genome-wide gene and transcript reference set [14]. Results are shown in Table 2. Unlike the GENCODE dataset, genes in RefSeq are

Table 2. Number of Refseq variants whose abundance could have been predicted using paired information

Number of genes	2914	856	349	139	75	44	22	13	5	2	3	3	1	11
Number of variants per gene	2	3	4	5	6	7	8	9	10	11	12	13	14	≥ 15
Total number of variants	5828	2568	1396	695	450	308	176	117	50	22	36	39	14	239
Solved with paired information	5828	2568	1396	695	450	308	169	117	50	22	36	32	6	224

not extensively annotated for transcript variants. Most genes are annotated for 2 or 3 variants. Of the 11938 variants present in this dataset, we could solve 11901, (99.7%). Overall, we found 5 genes out of 4437 containing interchangeable sets of putative variants. This means that, even with paired information, these variants cannot be solved unless some of them are ruled out by additional experimental evidence. In 4 cases, the interchangeable sets contained four elements, and, in one case, six elements. These genes are: DMD, ENSA, CTNND1, CD46, NR1I3.

As an example, gene ENSA has eight annotated variants in Refseq: EFH, $ACDFH, ACDFG, BCDFH, ACFG, ACFH, BCFG, BCFH$. Of these 8 variants, one can be solved, EFH, but the last four form a subset interchangeable with $\{ACDFG, ACDFH, BCDFG, BCDFH\}$.

7   Discussion and Conclusions

In this study, we worked under the exact information hypothesis. This means that we assume that a sequencing experiment can give us all the abundances of block junctions.

We are aware that this assumption is not realistic at present time since, in practice, coverage will be limited, and we would therefore have to deal with partial information on the abundances of blocks and block junctions. In particular, variants which are present in low abundances may not be covered at all. However, this study still enables to explore the limits of various approaches. Indeed, if prediction is impossible with exact information, it cannot become possible with partial information.

The techniques developed in this paper can therefore be used to identify sets of existing variants whose relative abundance *cannot* be measured with local information, or with paired information. This is the case of gene ITSN1, for local information with fixed set of candidates, and of genes DMD, ENSA, CTNND1, CD46, NR1I3, for paired information. This information could for example be used to modulate the output of algorithms such as those in [20].

Another aspect of the transcriptome reconstruction problem which is highlighted in this study is the fact that information is generally highly redundant, *i.e.* many constraints concern the same set of variants. This can be seen either as an advantage, since missing information is almost unavoidable, or as a complication, in the case of inconsistent measures. However, these issues can be addressed with probabilistic or linear programming techniques, once the interchangeable sets are removed from the set of candidates.

In this paper, we assumed that, during a sequencing experiment, the RNA molecules were uniformly sampled along their entire length, and that there was no 3' or 5' sampling bias. On the other hand, it has been reported in the supplementary material of [18] that this is not the case, at least for 454 platforms and preliminary experiments indicate that this bias is also observed for other platforms. We think that this additional knowledge, if taken into account in the reconstruction problem, may be useful and allow to sort out cases which we have considered unsolvable.

In the case of paired information, we deliberately ignored all data – or absence of data – on pairs of blocks or pairs of block and block junctions. In a real experiment, a pair of block junctions might well be missed, and its absence from the data is not a very strong indication of its absence in reality. On the other hand, the absence from the data of a pair of blocks is a stronger signal and can be used to rule out many paths in a splicing graph. In this sense, Algorithm 1 is a canvas that can be extended to incorporate all available information.

References

1. Adams, M.D., et al.: Complementary DNA sequencing: expressed sequence tags and human genome project. Science 252(5013), 1651–1656 (1991)
2. Bellin, D., Werber, M., Theis, T., Schulz, B., Weisshaar, B., Schneider, K.: EST Sequencing, Annotation and Macroarray Transcriptome Analysis Identify Preferentially Root-Expressed Genes in Sugar Beet. Plant Biology 4(6), 700–710 (2002)
3. Bennett, S.T., Barnes, C., Cox, A., Davies, L., Brown, C.: Toward the 1,000 dollars human genome. Pharmacogenomics 6, 373–382 (2005)
4. Chen, J., Skiena, S.: Assembly For Double-Ended Short-Read Sequencing Technologies. In: Mardis, E., Kim, S., Tang, H. (eds.) Advances in Genome Sequencing Technology and Algorithms. Artech House Publishers (2007)
5. De Bona, F., Ossowski, S., Schneeberger, K., Rätsh, G.: Optimal Spliced Alignments of Short Sequence Reads. Bioinformatics (in press, 2008)
6. ENCODE Project Consortium. Identification and analysis of functional elements in 1 % of the human genome by the ENCODE pilot project. Nature 447, 799–816 (2007)
7. Harrow, J., et al.: GENCODE: producing a reference annotation for ENCODE. Genome Biology 7, S4 (2006)
8. Heber, S., Alekseyev, M., Sze, S.H., Tang, H., Pevzner, P.A.: Splicing graphs and EST assembly problem. Bioinformatics 18 (suppl. 1), 181–188 (2002)
9. Hoffmann, K.F., Dunne, D.W.: Characterization of the Schistosoma transcriptome opens up the world of helminth genomics. Genome Biology 5, 203 (2003)
10. Houde, M., et al.: Wheat EST resources for functional genomics of abiotic stress. BMC Genomics 7, 149 (2006)
11. Lander, E.S., et al.: Initial sequencing and analysis of the human genome. Nature 409(6822), 860–921 (2001)
12. Mironov, A.A., Fickett, J.W., Gelfand, M.S.: Frequent alternative splicing of human genes. Genome Research 9, 1288–1293 (1999)
13. Modrek, B., Resch, A., Grasso, C., Lee, C.: Genome-wide detection of alternative splicing in expressed sequences of human genes. Nucleic Acids Research 29, 2850–2859 (2001)
14. Pruitt, K.D., Tatusova, T., Maglott, D.R.: NCBI Reference Sequence (RefSeq): a curated non-redundant sequence database of genomes, transcripts and proteins. Nucleic Acids Research. 35, D61-D65 (2007)
15. Sammeth, M., Foissac, S., Guigo, R.: A General Definition and Nomenclature for Alternative Splicing Events. PLoS Computational Biology (in press, 2008)
16. Sammeth, M., Valiente, G., Guigo, R.: Bubbles: Alternative Splicing Events of Arbitrary Dimension in Splicing Graphs. In: Vingron, M., Wong, L. (eds.) RECOMB 2008. LNCS (LNBI), vol. 4955, pp. 372–395. Springer, Heidelberg (2008)

17. Sugnet, C.W., Kent, W.J., Ares, M., Haussler, D.: Transcriptome and genome conservation of alternative splicing events in humans and mice. In: Proceedings of the Pacific Symposium on Biocomputing, vol. 9, pp. 66–77 (2004)
18. Weber, A., Weber, K., Carr, K., Wilkerson, C., Ohlrogge, J.: Sampling the Arabidopsis Transcriptome with Massively Parallel Pyrosequencing. Plant Physiology 144, 32–42 (2007)
19. Xing, Y., Resch, A., Lee, C.: The multiassembly problem: reconstructing multiple transcript isoforms from EST fragment mixtures. Genome Research 14(3), 426–441 (2004)
20. Xing, Y., Yu, T., Wu, Y.N., Roy, M., Kim, J., Lee, C.: An expectation-maximization algorithm for probabilistic reconstructions of full-length isoforms from splice graphs. Nucleic Acids Research 34, 3150–3160 (2006)
21. Zerbino, D., Birney, E.: Velvet: Algorithms for de novo short read assembly using de Bruijn Graphs. Genome Research 18, 821–829 (2008)

Post-Hybridization Quality Measures for Oligos in Genome-Wide Microarray Experiments

Florian Battke[1], Carsten Müller-Tidow[2], Hubert Serve[3], and Kay Nieselt[1]

[1] Center for Bioinformatics Tübingen, Department of Information and Cognitive Sciences, University of Tübingen, Sand 14, 72076 Tübingen, Germany
[2] IZKF - Inderdisciplinary Center for Clinical Research at the University of Münster, Domagkstr. 3, 48129 Münster, Germany
[3] Department of Internal Medicine II, University Hospital, Johann Wolfgang Goethe-University, Theodor Stern-Kai 7, 60590 Frankfurt am Main, Germany

Abstract. High-throughput microarray experiments produce vast amounts of data. Quality control methods for every step of such experiments are essential to ensure a high biological significance of the conclusions drawn from the data. This issue has been addressed for most steps of the typical microarray pipeline, but the quality of the oligonucleotide probes designed for microarrays has only been evaluated based on their *a priori* properties, such as sequence length or melting temperature predictions. We introduce new oligo quality measures that can be calculated using expression values collected in direct as well as indirect design experiments. Based on these measures, we propose combined oligo quality scores as a tool for assessing probe quality, optimizing array designs and data normalization strategies. We use simulated as well as biological data sets to evaluate these new quality scores. We show that the presented quality scores reliably identify high-quality probes. The set of best-quality probes converges with increasing number of arrays used for the calculation and the measures are robust with respect to the chosen normalization method.

1   Introduction

The adoption of high-throughput microarray experiments as a standard method for gene expression analyses, and, more recently, for studies of chromatin modifications in ChIP-Chip experiments highlights the importance of quality control methods for these kinds of experiments. Methods for quality control of samples before and after hybridization as well as of hybridised arrays as a whole have been proposed [1,2,3]. Also, pre-production quality of oligonucleotide probes is extensively optimized [4,5].

Factors considered for optimization of oligonucleotide probes are sequence length, complexity, melting temperature, GC content, and low probability for crosshybridization and formation of secondary structures. These factors are evaluated *in silico* by the algorithms used for probe design. Where an optimal solution cannot be found, heuristics are used to find a sufficiently "good" set of

K.A. Crandall and J. Lagergren (Eds.): WABI 2008, LNBI 5251, pp. 64–75, 2008.
© Springer-Verlag Berlin Heidelberg 2008

probes [6]. On the other hand, these factors can be used to evaluate the quality of any given probe set.

Yet there is one problem inherent in any pre-production quality measure: It is derived entirely from data available *before* any experiment with those probes has been performed and thus can not incorporate knowledge gained during experimentation. A method to assess the quality of oligonucleotides *a posteriori*, that is after hybridization, has been missing so far. We present several oligo quality measures that can be calculated from measured expression values collected from indirect design experiments with a common reference sample in one channel or from absolute design experiments with a uniform sample.

A high-quality oligo probe produces reliable expression data. It fulfills three necessary requirements for this type of high-throughput experiment: Sensitivity, specificity and comparability. A sensitive oligo shows a signal if the target molecule is in the sample. A specific oligo only represents its target and no other target, i.e. will not show any cross hybridization. It will also produce signals that are directly related to the amount of target molecules in the sample. The aim of many microarray experiments is to compare the signal intensities produced by different spots. Good oligos should therefore report signals of the same strength if the hybridised sample contains the same amounts of their respective target molecules.

These three requirements are also the basis for the quality measures we present in this work. We define several measures that individually can be used to assess the sensitivity, specificity and comparability of an oligo, or in a combined score to subsequently rank all oligos of an array. The performance of the new oligo quality measures is evaluated with a simulated data set. Furthermore we show that application of these measures to normalized data from a large ChIP-Chip study using oligo microarrays interrogating 10,000 human promoter regions produces robust results for differing normalization strategies or when only a subset of all arrays is considered.

2   Methods

Microarray data is usually represented as a two-dimensional matrix of *expression values* where each row represents one spot on the array (usually one genomic target) and each column represents one experiment, i.e. one hybridized array. For background and details on microarray analyses see e.g. [7]. For a fixed microarray design let n_O be the number of oligos (matrix rows) and n_A be the number of experiments (matrix columns). We use s_{ij} to represent the value of the ith row (target) in the jth column (experiment), where $1 \leq i \leq n_O$ and $1 \leq j \leq n_A$. Furthermore, s_{i*} denotes the whole i^{th} row of the expression matrix, i.e. a vector of signal values for spot i on all arrays, while s_{**} designates the whole matrix s. Finally, we use r_{**} for a matrix obtained from the expression matrix by randomly permuting cells, i.e. $r_{ij} = s_{k_i \ell_j}$ where k and ℓ are random permutations of $1 \ldots n_O$ and $1 \ldots n_A$ respectively. Finally, o_i is defined as the oligo placed on the ith spot, i.e. the probe used to obtain the values of the ith matrix row.

2.1 The Flag Measure

During spot quantification, the imaging software usually assigns flags when spots are missing or when spots have a deviating morphology. If the hybridised sample contained the target for the flagged spot, such a flag indicates a failed hybridization. By counting the number of times a spot is flagged in the set of all arrays, a simple measure can be obtained of how often hybridization fails using this oligo. However, sometimes a large number of spots are flagged on an array, indicating that the microarray experiment itself failed for example due to problematic buffer conditions, problems during sample preparation or even scanning.

Taking these considerations into account, the oligo flag measure, ofm, is defined as

$$\mathrm{ofm}(o_i) = \sum_{k=1}^{n_A} f_{ik} w_k$$

with

$$w_j = \frac{\hat{w}_j}{\sum_{k=1}^{n_A} w_k}, \qquad \hat{w}_j = 1 - \left(\frac{1}{n_O} \sum_{\ell=1}^{n_O} f_{\ell j} \right)$$

and

$$f_{ij} = \begin{cases} 1 & \text{if there is a flag for spot } i \text{ on array } j \\ 0 & \text{else} \end{cases}$$

The weights w_j are chosen such that the badness of a flag is a linear function of the number of flags on the array and normalized such that their sum is one. The flag measure takes values between 0 (no flags) and 1 (maximal number of flags).

2.2 Oligo Stability Measures

Assessing the reliability of the signal produced by an oligo requires knowledge about the abundance of target DNA. ChIP-Chip experiment data can be used here, since ChIP-Chip experiments usually employ a so-called indirect design. This means that different conditions or treatments are not compared *directly* to one another but rather *indirectly* by way of an intermediate, i.e. the genomic DNA in the second channel that is the same in all array hybridizations of the experiment.

Since all arrays contain the same genomic DNA in the reference channel, the signals in that channel can be expected to be identical between two arrays after the data has been normalized, assuming that the normalization method was chosen correctly. The following measures were designed on the assumption of identical reference channel data in all arrays of the experiment.

Oligo Stability. We use the coefficient of variation (CV), defined as the standard deviation divided by the mean of a vector, to measure the deviation of a probe from its mean across all hybridizations independent of its absolute value.

The oligo stability measure, osm, for an oligo probe o_i is defined as follows:

$$\mathrm{osm}(o_i) = -\log_2 \frac{CV(s_{i*})}{\mathrm{median}\left(CV(r_{1*}), \ldots, CV(r_{n_O*})\right)}$$

Oligo Rank Stability. A variant of the oligo stability measure uses the *rank* function. The vector of signal values is converted into a vector of ranks, with the smallest value being assigned to rank one.

The oligo rank stability measure uses the same basic idea as the oligo stability measure, by defining an oligo to be stable if its value has the same rank for each array. To quantify the (in)stability, the standard deviation of the vector of rank values is used. While the oligo stability measure uses the coefficient of variation to be independent of absolute signal strength, the oligo rank stability measure does not require such normalization. We use the geometric mean, $\mathrm{gmean}(\boldsymbol{v})$, of a vector $\boldsymbol{v}$, defined as $\mathrm{gmean}(\boldsymbol{v}) = \sqrt[|v|]{\prod_{j=1}^{|v|} v_j}$, the standard deviation is computed on the geometric mean. Then the oligo rank stability, ors, is defined as

$$\mathrm{ors}(o_i) = -\log_2 \frac{\mathrm{gsd}(\tilde{s}_{i*})}{\mathrm{median}\left(\mathrm{gsd}(\tilde{r}_{1*}), \ldots, \mathrm{gsd}(\tilde{r}_{n_O *})\right)}$$

with

$$\mathrm{gsd}(\boldsymbol{v}) = \frac{1}{|v|} \sum_{i=1}^{|v|} (v_i - \mathrm{gmean}(\boldsymbol{v}))$$

where $\tilde{s}_{ij}$ is the rank of s_{ij} and $\tilde{r}$ is the randomized version of $\tilde{s}$.

Group Stability Measures. Both measures introduced in the previous section assume that, ideally, reference channel data should be identical on all arrays, reflecting the fact that the sample creating the reference signal is identical. If a spot shows different signal values for one or several arrays, the oligo stability measures as defined above assign lower values to the corresponding oligo.

There is, however, the possibility that the difference in signal values reflects a real difference in the abundance of the respective target DNA in the reference sample. One possible reason would be the use of independent reference sample preparations for individual arrays.

Furthermore, the abundance of DNA representing a certain genomic location could be different between two genomic reference samples. In that case one would expect that oligos from adjacent genomic regions show the same changes. Thus, if the signal produced for one genomic location changes between two arrays but at the same time, genomically adjacent probes produce similar signals, it is more likely that the changed signals reflect a real change in target abundance.

The group stability measure scores every oligo according to how similar its signal values are to those of oligos in the same group, where group is defined by genomic adjacency, for example representing a promoter region. Let $G(o_i) \subset [1, n_O]$ be the set of oligos belonging to the same group as oligo o_i and let $R(o_i)$ be a random set of the same size containing oligos unrelated to o_i. Then the group stability measure, ogs, is defined as

$$\mathrm{ogs}(o_i) = \mathrm{mean}\left(\hat{f}(s_{i1}), \ldots, \hat{f}(s_{in_A})\right)$$

with

$$\hat{f}(s_{ij}) = -\log_2 \frac{f(s_{ij}, G)}{\text{median}\left(f(s_{1j}, R), \ldots, f(s_{n_O j}, R)\right)}$$

and

$$f(s_{ij}, G) = \min_{i \neq k \in G(i)} |s_{ij} - s_{kj}|$$

The definition of the function f is the crucial part. It specifies how the similarity between the oligo and its group neighbors is computed. Although the minimum distance that we use is computationally more expensive than using the distance to the group mean or median, it is superior to both. Consider a group of three oligos where two have identical values and the third one differs strongly. The intended behaviour of the oligo group stability measure would be to give the best score to the two identical oligos and to give a bad score to the outlier.

The mean is very susceptible to outliers, thus the scores of the two good oligos would depend on how bad the outlier is. The distance to the group median, on the other hand, would always result in a score of zero for one of the three oligos. This means that one oligo would get an extremely good score just because its value lies between those of its group neighbors. If the oligos' values are very similar, experimental noise would decide which one gets the zero score.

Normalization Measure. After normalization of all arrays, it is expected that the signal of a probe in the reference channel is the same for all arrays in the experiment, i.e. $s_{ij} = s_{ik} \forall k \in \{1, \ldots, n_A\}$.

Based on this assumption, the oligo normalization measure, onm, computes the mean of all pairwise comparisons between the spots' signal strength on all arrays:

$$\text{onm}(o_i) = \text{mean}_{k < \ell} \left| \log_2 \frac{s_{ik}}{s_{i\ell}} \right|$$

2.3 Majority Measure

Ideally, all probes should produce signals of identical strength if the abundance of their respective targets is the same. Every experiment that aims at comparing the expression of two different genes relies on this relationship. A simple measure could be derived from the distance of the probe's signal strength to the median or mean of all probe signals on an array.

The oligo majority measure of an oligo is defined as the percentage of oligos whose signal is the same (allowing for small differences because of experimental noise and for computational considerations). Using the percentage of similar oligos gives better results if the underlying distribution of signal strengths is bi- or multimodal, for instance due to copy number variation [8].

The ideal definition of the majority measure would be based on the percentage of spots s_{kj} with $|s_{kj} - s_{ij}| \leq \delta$. This is very expensive to compute. A less time-consuming method is to first sort the spot signal values into bins of size 2δ and then to take the percentage of oligos falling into the same bin. This reduces the measure's accuracy by a small but acceptable amount. Using binning, the oligo majority measure is defined as:

$$\mathrm{omm}(o_i) = \mathrm{median}\left(f(s_{i1}),\ldots,f(s_{in_A})\right)$$

with $\qquad f(s_{ij}) = |\{k : b(s_{kj}) = b(s_{ij})\}|$ and $b(s_{ij}) = \mathrm{round}\left(\frac{\hat{s}_{ij}}{\delta}\right).$

2.4 Combined Score

Six oligo quality measures were presented so far. Applying all of them to a data set creates six quality values for each oligo, which could be difficult to interpret. To draw useful conclusions from quality measures, their values have to be combined into a single, descriptive quality value for each oligo. The selection of measures and of the method to combine them has to be made based on a model of what constitutes a "good" oligo.

The oligo stability, rank stability and group stability measures are log-odds values, the normalization measure is an absolute measure derived from signal intensities and the majority measure is a percentage. We propose using ranks to combine these different kinds of values into one quality score. The combined quality score, oq, is computed as follows: Calculate the flag measure $\mathrm{ofm}(o_i)$ and select a threshold value (e.g. 0.5). For all o_i with $\mathrm{ofm}(o_i) \leq$ threshold, compute the oligo quality, $\mathrm{oq}(o_i)$ on m measures $f_1,\ldots,f_m$ as

$$\mathrm{oq}(o_i) = \frac{1}{\sum_{k=1}^{m} w_k} \sum_{\ell=1}^{m} w_\ell \cdot \mathrm{rank}\left(f_\ell(o_i)\right)$$

where w_i is the weight associated with measure f_i. The result is an average rank value in the interval $[1, n_O]$.

Finally, we propose one further quality score, coined the best$k$ measure. It takes values in the range $[0, 100]$ which can be interpreted more easily. The idea behind this measure is that while the very best oligos are those that have high values in all quality measures, oligos that score badly in only one measure can still be among the best. The value returned by the best$k$ measure is the percentile of good values that the oligo falls into according to its k (of $m = k+1$) best quality measure values. To clarify, consider an oligo that is among the best $4, 5, 6, 9$ and 22% according to five quality measures, respectively. The best4 value would be the *fourth best* percentile, i.e. 9.

Using the quality measures $f_1,\ldots,f_m$, and a function $p_j(o_i)$ that computes the percentile of $f(o_i)$ with respect to all values in $\{f(o_j) : j = 1\ldots n_O\}$, best$k$ is calculated as

$$\mathrm{best}k(o_i) = \max\left\{p_j(o_i) : p_j(o_i) \neq \arg\max(p_1(o_i),\ldots,p_m(o_i))\right\}.$$

3   Results

3.1   Simulated Data Sets

We created simulated microarray signal data according to the following model: Each simulated gene is represented by three spots on the array, forming a *group*. Our simulated experiment contains 10,000 genes with 30,000 oligos in total. 100 hybridizations were simulated. We define a base signal level for each gene/group by drawing the log expression value according to a gamma distribution $\Gamma(k = 30, \theta = 3)$. The set of genes was split into two subsets ("good" vs. "bad" groups) of equal size. Using the base signal levels, an expression profile was created for each gene. The values for the 100 hybridizations were taken from a normal distribution with μ set to the base signal level and σ set to 0.01μ or 0.20μ for good and bad groups, respectively.

Now the set of all genes was split into four subsets of equal size. In the first group, all oligos report the true expression value of the gene (plus some noise). In the second group, one oligo may be further from the true value, in the third group two oligos are allowed to be wrong and in the fourth group, no oligo is accurate. From the gene expression values for each array, the values of the individual oligos are created by adding values drawn from a normal distribution (with $\mu = 0$) to the respective gene expression profile. The standard deviations σ are determined in the same way as described above, i.e. using a factor of 0.01 for the accurate oligos and 0.20 for the inaccurate ones.

Our simulated data set thus contains values for 30000 oligos on 100 arrays. Half of the oligos are accurate, half are inaccurate. Half of them come from a "good" gene, half do not. A quarter of them each come from a group with zero, one, two or three accurate oligos. We assigned each oligo to one of three classes: Accurate oligos from good groups form the "good" class, accurate oligos from bad groups and inaccurate oligos from good groups belong to the "intermediate" class and the inaccurate oligos from bad groups are the "bad" class.

We computed all oligo quality measures on this simulated data set except for the flag measure, as flags were not simulated. From these five measures, the combined oligo scores, oq and best$k$, were computed for each oligo as well as the rank according to oq. We expected the ranks of "good" oligos to be in the first quartile ($x \leq 7500$), those of "bad" oligos to be in the fourth quartile ($22500 < x$) and those of "intermediate" oligos to occupy the interquartile range ($7500 < x \leq 22500$).

Figure 1 shows how nicely the combined quality score oq reflects the simulated quality of the oligos. More than 98% of the "good" oligos' ranks are in the first quartile. Furthermore, the first quartile contains no "bad" oligos and only 3.0% "intermediate" oligos. The same tendency was observed for ranks derived from the best$k$ score (data not shown).

3.2   Biological Data Set

We tested our oligo quality measures on a novel data set generated on spotted arrays of a custom design. The arrays contain 31200 spots, of which 29367 interrogate the promoter regions of human genes known or suspected to be linked to cancer. Most (9345) promoter regions are covered by three probes. The data set was derived from 126 specimens (patients and controls) with different forms of acute myeloid leukemia and healthy hematopoietic progenitor cells. Chromatin-IP was performed using an antibody against acetylated Histone H3. Subsequent

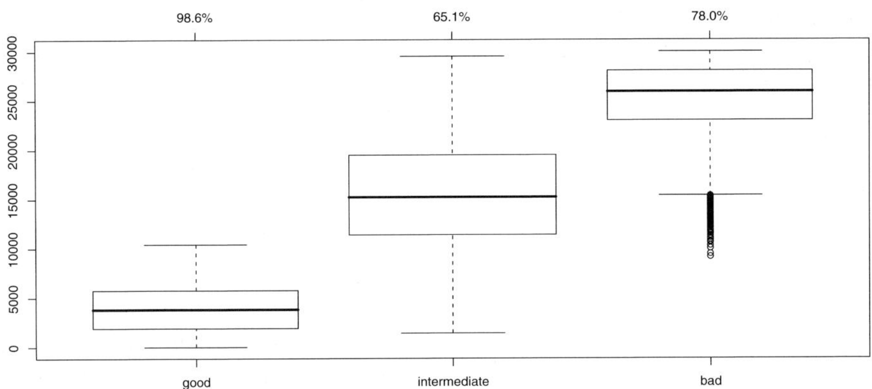

Fig. 1. Separation of simulated oligos − The boxplot shows the distribution of the oligos' ranks according to the combined quality measure (oq) for the three oligo classes. The percentage values represent the number of ranks falling into the expected range.

to the chromatin-IP, DNA was amplified using whole genome amplification before being labeled with Cy3. In addition, input DNA derived from a pool of DNA donors was simultaneously amplified and labeled with Cy5 before being hybridized as common reference.

Arrays were scanned using an AXON 4100B (Molecular Devices) scanner. Spots were identified and measured using the Spotreader (Niles Scientific) software. The raw data was normalized using functions provided by the `limma` [9] package for R [10]: The "normexp" method was used for background correction. Within-array and between-array normalization were performed using the "printtiploess" and "rquantile" methods, respectively. An additional normalization step was introduced between these two to remove a dependency of the probe signal on its column within the print tip: The probe signals were adjusted so that the median signal of each column was the same.

The flag measure was computed and a threshold of 0.5 applied for filtering, removing 4035 spots (12.9%). The reference channel data of the remaining spots was used to compute the other quality measures which were then combined into the quality scores, oq and best4, giving the majority measure twice the weight of the other measures.

3.3 Robustness of the Quality Score

We evaluated the robustness of our quality scores with respect to the number of available arrays. Furthermore, we evaluated the effect of using different normalization methods prior to the quality score calculation.

To assess the score's robustness, we use the disagreement between the sets of top-100 spots, defined as follows. Consider two vectors of quality scores, `data`

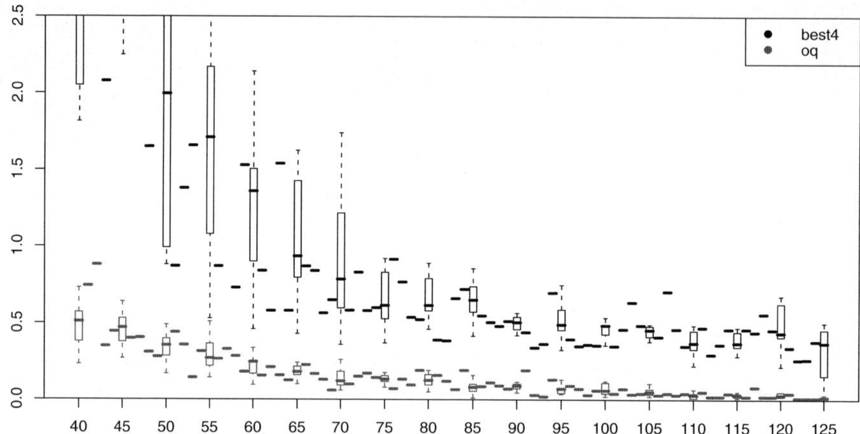

Fig. 2. Number of arrays vs. disagreement − The quality measures were computed on randomly chosen subsets with increasing size of all arrays. The disagreement between the quality score computed on the random subset and that computed on the full reference set of 126 arrays is plotted on the y-axis.

and `ref`, each assigning a quality score to every spot on our array. Let `data_top` be the set of the 100 best probes according to `data`. Their average rank is 50.5. We compute the difference of this *expected* rank to the observed average rank of `data_top` according to the quality scores in `ref`. The disagreement value is normalized by the number n of oligos used (e.g. 100).

$$\text{disagreement}_n(\text{data}, \text{ref}) = \frac{1}{n}\left(\text{median}(\text{rank}(\text{ref}[\text{topX}_n(\text{data})])) - (n+1)/2\right)$$

with
$$\text{topX}_n(\text{data}) = \{i : \text{rank}(\text{data}[i]) < n\}.$$

Thus, two perfectly matching quality score vectors would have a disagreement score of zero, a score of one would indicate that the top 100 spots according to one measure have an average rank of 101 in the second measure.

Figure 2 shows the disagreement when using a smaller subset of all arrays. Obviously, the quality score converges with an increasing number of arrays. The quality score oq performs very well even for relatively small sets of arrays, the best$k$ measure converges more slowly.

The quality measures are computed on normalized reference channel signals. To ensure that the choice of normalization method has no large influence on the resulting quality scores, we applied different background correction and normalization methods with subsequent computation of the two combined oligo scores. We used the methods as provided by the `limma` package [9].

The influence of the background correction method is very small, except for not using any background correction at all. The same holds for most of

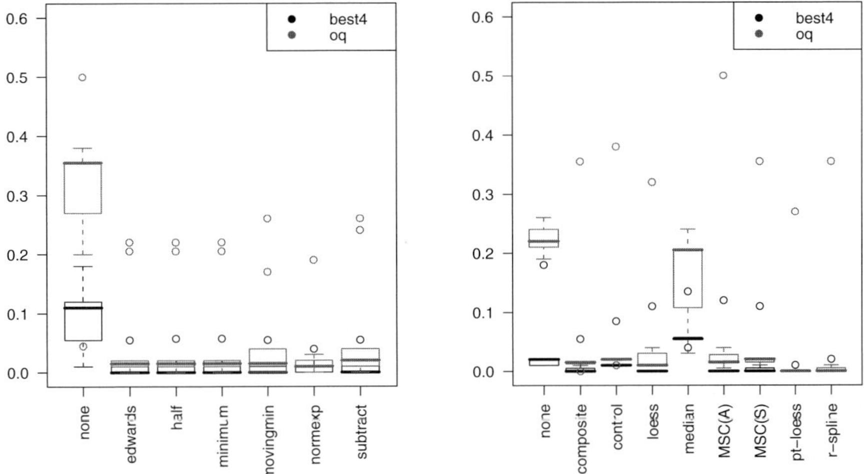

Fig. 3. Influence of different normalization methods on the quality measures – 63 combinations of seven background correction and nine within-array normalization methods were applied and the disagreement of the resulting oligo quality scores with the scores resulting from the normexp-printtiploess pipeline computed. The disagreement values are grouped by background correction in the left plot, and by within-array normalization in the right plot.

the normalization methods. Only the median scaling method produces a larger deviation.

4   Discussion

Ensuring high quality for microarray data is of great importance. Like other high-throughput methods, microarray experiments generate large amounts of data in parallel. These data are then first processed by the usual pipeline of background correction, intra- and inter-array normalization.

A list of significant probes is computed and presented to the researcher. Due to the nature of the statistical tests used to compute this list, a certain rate of false positives is always present. On the other hand, relevant probes remain undetected (false negatives). In addition, the results heavily depend on the quality of the input data. Quality control methods are one way to improve the biological significance of experiment results.

In silico methods for probe set optimization approach the problem by ensuring that, in theory, all probes on a microarray have very similar properties with respect to their binding specificities. These methods incorporate thermodynamic and basic chemical knowledge in combination with cross-hybridization predictions based on sequence databases. In addition to this pre-production optimization, post-hybridization quality measures for hybridised arrays as a whole have been proposed. *In vivo* quality control methods for microarray data have so far been limited to pre-hybridization tests of the labelled samples. Thus, while the quality of hybridised samples can be checked before and after hybridization, the quality of the probes on a microarray is only tested in a theoretical way, and data generated by those probes is not taken into consideration.

This gap is closed by our oligo quality measures, providing a very detailed, wide-angle view of microarray data. Using simulated data we showed that they perform very good separation of high-quality oligos from those of lower quality. These scores can be used in several ways.

Firstly, they allow researchers to optimize microarray designs, e.g. by replacing questionable probes. Secondly, they can be used to annotate experimental results with confidence values. Finally, our oligo scores can be incorporated into data normalization pipelines to select a set of high-quality oligos as a basis for fitting regression functions, similar to the established practice of using a set of oligos from spike-in experiments or dilution series [11].

Combining individual quality measures merges the multi-faceted view of the data into one oligo quality score, allowing different weights for the individual quality measures. While this poses the problem of deciding on an appropriate weighting scheme, it makes the quality score a very flexible tool. Furthermore, new combined measures can be derived from the presented oligo measures so that researchers can make use of the existing quality measures if they need to adhere to another definition of "good" oligos.

Regardless of refined or additional measures, any definition of good oligos should always contain two demands: Samples should reliably hybridize to the oligo and the produced signal should be consistent and proportional to the amount of sample. The former is covered by our flag measure, the latter by the stability measures. Most applications will also require that any pair of good oligos show equally strong signals if hybridised with equal amounts of sample, i.e. their proportionality factors linking hybridization and sample abundance should be very similar, which in this work is covered by the majority measure.

One important aspect to consider when calculating quality scores is the number of arrays used. More arrays are always better yet the *minimal* number of arrays needed to produce reliable quality scores is hard to specify. If the array data is of very high quality, a small number can be sufficient though in general we would suggest to use at least 20 arrays. To compensate for a small number of arrays, a more lenient threshold should be chosen when filtering oligos based on quality scores. Selecting appropriate thresholds is not trivial. Methods for deriving thresholds based on the characteristics of the actual data require further research. The stability, rank stability and group stability methods already are log-odds values and modifying the combined measures to behave likewise would be desirable. On the other hand, permutation simulations can provide background data for statistical evaluation of quality scores.

Our methods address the need for measuring data quality in order to draw biologically relevant conclusions. The high importance of this issue merits further

research into individual measures as well as overall quality scores and the choice
of quality thresholds.

Acknowledgements

We are grateful to Michael McClelland, Yipeng Wang and Anke Becker for design
and spotting of the oligonucleotide array and to the lab members of the Münster
lab for the array hybridizations.

References

1. Chen, D.T.: A graphical approach for quality control of oligonucleotide array data.
   J. Biopharm. Stat. 14(3), 591–606 (2004)
2. Degenkolbe, T., Hannah, M.A., Freund, S., Hincha, D.K., Heyer, A.G., Köhl,
   K.I.: A quality-controlled microarray method for gene expression profiling. Anal.
   Biochem. 346(2), 217–224 (2005)
3. Dumur, C.I., Nasim, S., Best, A.M., Archer, K.J., Ladd, A.C., Mas, V.R., Wilkin-
   son, D.S., Garrett, C.T., Ferreira-Gonzalez, A.: Evaluation of quality-control criteria for microarray gene expression analysis. Clin. Chem. 50(11), 1994–2002 (2004)
4. Gräf, S., Nielsen, F.G., Kurtz, S., Huynen, M.A., Birney, E., Stunnenberg, H.,
   Flicek, P.: Optimized design and assessment of whole genome tiling arrays. Bioinformatics 23(13), 195–204 (2007)
5. Kreil, D.P., Russell, R.R., Russell, S.: Microarray oligonucleotide probes. Methods
   Enzymol 410, 73–98 (2006)
6. Li, F., Stormo, G.D.: Selection of optimal dna oligos for gene expression arrays.
   Bioinformatics 17(11), 1067–1076 (2001)
7. Drăghici, S.: Data Analysis Tools for DNA Microarrays. Chapman & Hall/CRC,
   Boca Raton (2003)
8. Redon, R., Ishikawa, S., Fitch, K.R., Feuk, L., Perry, G.H., Andrews, T.D., Fiegler,
   H., Shapero, M.H., Carson, A.R., Chen, W., Cho, E.K., Dallaire, S., Freeman,
   J.L., González, J.R., Gratacós, M., Huang, J., Kalaitzopoulos, D., Komura, D.,
   MacDonald, J.R., Marshall, C.R., Mei, R., Montgomery, L., Nishimura, K., Oka-
   mura, K., Shen, F., Somerville, M.J., Tchinda, J., Valsesia, A., Woodwark, C.,
   Yang, F., Zhang, J., Zerjal, T., Zhang, J., Armengol, L., Conrad, D.F., Estivill, X.,
   Tyler-Smith, C., Carter, N.P., Aburatani, H., Lee, C., Jones, K.W., Scherer, S.W.,
   Hurles, M.E.: Global variation in copy number in the human genome. Nature 444,
   444–454 (2006)
9. Smyth, G.K.: Limma: linear models for microarray data, pp. 397–420. Springer,
   New York (2005)
10. R Development Core Team: R: A Language and Environment for Statistical Com-
    puting. R Foundation for Statistical Computing, Vienna, Austria (2006)
11. Hill, A., Brown, E., Whitley, M., Tucker-Kellogg, G., Hunter, C., Slonim, D.: Eval-
    uation of normalization procedures for oligonucleotide array data based on spiked
    crna controls. Genome Biology 2(12) (November 2001)

NAPX: A Polynomial Time Approximation Scheme for the Noah's Ark Problem

Glenn Hickey[1], Paz Carmi[1], Anil Maheshwari[1], and Norbert Zeh[2]

[1] School of Computer Science, Carleton University, Ottawa, Ontario, Canada
[2] Faculty of Computer Science, Dalhousie University, Halifax, Nova Scotia, Canada
ghickey@scs.carleton.ca, carmip@gmail.com, anil@scs.carleton.ca,
nzeh@cs.dal.ca

Abstract. The Noah's Ark Problem (NAP) is an NP-Hard optimization problem with relevance to ecological conservation management. It asks to maximize the phylogenetic diversity (PD) of a set of taxa given a fixed budget, where each taxon is associated with a cost of conservation and a probability of extinction. NAP has received renewed interest with the rise in availability of genetic sequence data, allowing PD to be used as a practical measure of biodiversity. However, only simplified instances of the problem, where one or more parameters are fixed as constants, have as of yet been addressed in the literature. We present NAPX, the first algorithm for the general version of NAP that returns a $1-\epsilon$ approximation of the optimal solution. It runs in $O\left(\frac{nB^2h^2\log^2 n}{\log^2(1-\epsilon)}\right)$ time where n is the number of species, and B is the total budget and h is the height of the input tree. We also provide improved bounds for its expected running time.

Keywords: Noah's Ark Problem, phylogenetic diversity, approximation algorithm.

1 Introduction

1.1 Motivation

Measures of biodiversity are commonly used as indicators of environmental health. Biodiversity is presently being lost at an alarming rate, due largely to human activity. It is speculated that this loss can lead to disastrous consequences if left unchecked [11]. Consequently, the discipline of conservation biology has arisen and a considerable amount of resources are being allocated to research and implement conservation projects around the world.

A conservation strategy will necessarily depend on the measure of biodiversity used. Traditionally, indices based on species richness and abundance have been used to quantify the biodiversity of an ecosystem [9]. These indices are based on counting and do not account for genetic variance. Phylogenetic diversity (PD) [4] addresses this issue by taking into account evolutionary relationships derived from DNA or protein samples. The use of PD in biological conservation has become increasingly widespread as more phylogenetic information becomes

K.A. Crandall and J. Lagergren (Eds.): WABI 2008, LNBI 5251, pp. 76–86, 2008.
© Springer-Verlag Berlin Heidelberg 2008

available [8]. It is also used to determine diverse sequence samples in comparative genomics [12].

The Noah's Ark Problem (NAP) [15] is an abstraction of the fundamental problem of many conservation projects: how best to allocate a limited amount of resources to maximally conserve phylogenetic diversity. This is in turn a generalization of the Knapsack Problem [5] and is therefore NP-Hard. Several algorithms have been proposed to solve special cases of the problem but, as yet, no non-heuristic solutions have been proposed to solve general instances of NAP. Given that NAP itself is an abstraction of realistic scenarios, it is important to have a general solution in order to be able to extend this framework to useful applications. For this reason, we present an algorithm that can be used to compute an approximate solution for NAP in polynomial time, so long as the approximation factor is held constant, and total budget is polynomial in the input size.

1.2   Definitions

Throughout this paper, we use the following definition of a phylogenetic tree $\mathcal{T}$, with notation consistent with that of [7]. $\mathcal{T}$ has a root of degree 2, interior vertices of degree 3, and n leaves, each associated with a species from set X. If an edge e of $\mathcal{T}$ is incident to a leaf, it is called a pendant edge. Otherwise e has exactly two adjacent edges, l and r, below it (not on the path from e to the root) and these are referred to as e's children. λ is a function that assigns a non-negative branch length to each edge in $\mathcal{T}$. The phylogenetic diversity of $\mathcal{T}$, $PD(\mathcal{T})$ is defined as

$$PD(\mathcal{T}) = \sum_{e} \lambda(e), \tag{1}$$

where the summation is over each edge e of the tree. Intuitively, this measure corresponds to the amount of evolutionary history represented by $\mathcal{T}$.

The Noah's Ark Problem has the objective of maximizing the expected PD, $\mathbb{E}(PD)$, under the following constraints. Each taxon $i \in X$ is associated with an initial survival probability a_i, which can be increased to b_i at some integer cost c_i; and the total expenditure cannot exceed the budget B. Since B is a factor in the running time, we assume that the budget and each cost have been divided by the greatest common divisor of all the costs. In the original formulation of NAP, each species was also associated with a utility value. However, in [7] it was shown that these values are redundant as they can be incorporated into the branch lengths without altering the problem. To avoid accounting for degenerately small probability values, we make the assumption that the conserved survival probabilities are not exponentially small in n. In other words, there exists a constant k such that $b_i \geq n^{-k}$ for each $i \in X$. We feel this assumption is reasonable as it is unrealistic that money would be allocated to obtain such a negligible probability of survival.

If a species survives, the information represented by its path to the root is conserved. Consequently, the probability that an edge survives is equivalent to the probability that at least one leaf below it in $\mathcal{T}$ survives. Let C_e be the set of leaves below e in the tree and S be the set of species selected for protection. $\mathbb{E}(PD|S)$, can be derived from (1) as follows:

$$\mathbb{E}(PD|S) = \sum_e \lambda(e) \left(1 - \prod_{i \in C_e \cap S} (1 - b_i) \prod_{j \in C_e - S} (1 - a_j) \right), \qquad (2)$$

where the summation is over all edges. NAP asks to maximize $\mathbb{E}(PD|S)$ subject to

$$\sum_{s \in S} c_s \leq B.$$

Our algorithm is based on decomposing $\mathcal{T}$ into clades which are associated with the edges of the tree. A clade corresponding to edge e, denoted $\mathcal{K}_e$, is the minimal subtree of $\mathcal{T}$ containing e and C_e, the set of leaves below it. The $\mathbb{E}(PD)$ of $\mathcal{K}_e$ can be computed as in (2) but summing only over edges in the clade. The entire tree can be considered a clade by attaching an edge of length 0 to its root. If e has two descendant edges l and r, then we say $\mathcal{K}_e$ has two child clades $\mathcal{K}_l$ and $\mathcal{K}_r$.

1.3 Related Work

Let $a_i \xrightarrow{c_i} b_i$ NAP refer to the problem as described above, where the survival probabilities and cost of each taxon are input variables. Fixing one or more of these variables as constants produces a hierarchy of increasingly simpler sub-problems [13]. The simplest, $0 \xrightarrow{1} 1$ NAP, is equivalent to finding the set of B leaves whose induced subtree (including the root) has maximum PD and can be solved by a greedy algorithm [14] [12]. $0 \xrightarrow{c_i} 1$ NAP on ultrametric (all leaves equidistant from the root) trees and $(1 - x_i) \xrightarrow{1} (1 - \kappa x_i)$ for general trees where x_i is a variable probability and κ is a constant factor such that $0 \leq \kappa \leq 1$ can likewise be solved in polynomial time by greedy algorithms [7]. Given that $0 \xrightarrow{c_i} 1$ NAP is itself a generalization of the Knapsack problem which is NP-Hard, it is extremely unlikely that an exact, polynomial-time solution for this kind of NAP or any generalizations will ever be found. Pardi and Goldman [13] did find a pseudopolynomial-time dynamic programming algorithm for the $0 \xrightarrow{c_i} 1$ NAP on general (non-ultrametric) trees that makes the realistic assumption that B is polynomial in n. They also show that any instance of $a_i \xrightarrow{c_i} 1$ NAP can be transformed to an instance of $0 \xrightarrow{c_i} 1$ NAP, allowing their algorithm to solve such instances as well.

This algorithm relies upon the observation that the solution to $0 \xrightarrow{c_i} 1$ NAP for any clade can be obtained from the solutions to its two child clades [13]. Which solutions to use depends on how the budget is allocated to the two sub-problems. If the budget at $\mathcal{K}_e$ is b, then there are $b + 1$ ways to split it across $\mathcal{K}_l$ and $\mathcal{K}_r$. By solving these $b + 1$ pairs of subproblems, the optimal solution can be found in the pair with maximum total $\mathbb{E}(PD)$ (plus the expected contribution of e). Recursively proceeding in this fashion from the root down would not yield an efficient algorithm as the number of possible budget divisions increases expo-nentially with each level of the tree. Instead, the clades are processed bottom-up

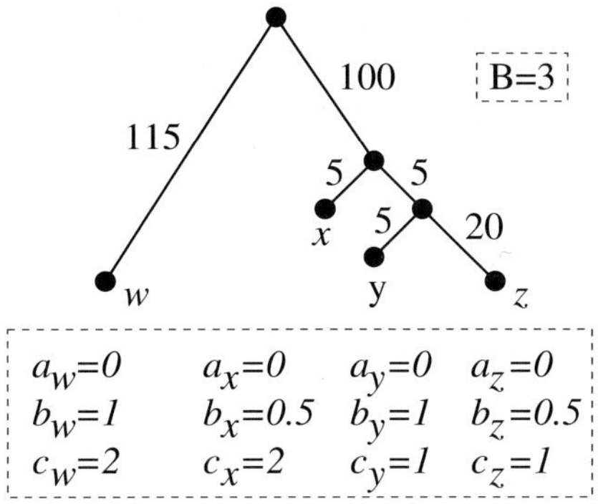

Fig. 1. An example why the dynamic programming algorithm of [13] does not work for general instances of NAP. The optimal allocation for the clade containing y and z is not part of a globally optimal solution.

from the leaves. All $b+1$ scores are computed and stored in a dynamic programming table for each clade. Each score can be determined by taking the maximum of $b+1$ possible scores of its child clades, which are already computed or computed directly from (2) if the clade contains a single leaf. Each table entry can therefore be computed in $O(B)$ time. There are $O(B)$ entries per clade and $O(n)$ clades in the tree giving a total running time of $O(nB^2)$.

This procedure does not work for $a_i \xrightarrow{c_i} b_i$ NAP because this version of the problem does not display the same optimal substructure [13]. In $0 \xrightarrow{c_i} 1$, the dynamic programming algorithm implicitly maximizes the survival probability of the clade in addition to its $\mathbb{E}(PD)$ value. The total score of the tree is a function of both of these values which is why the algorithm works for this case. In $a_i \xrightarrow{c_i} b_i$ NAP, a budget assignment that maximizes survival probability of the clade does not guarantee that it will have maximal $\mathbb{E}(PD)$ and vice versa. The correct allocation cannot be made without knowledge of the entire tree; hence, the optimal substructure exploited by [13] for $0 \xrightarrow{c_i} 1$ NAP is not present. As an example, consider the instance of NAP in Figure 1 with $B = 3$. The optimal solution is to conserve w and y for $\mathbb{E}(PD) = 225$. However, locally computing the best allocation of budget 1 for the clade containing y and z will select z for conservation, and any chance of obtaining the optimal solution will be lost. In this case, it is more important to maximize the survival probability of the clade rather than $\mathbb{E}(PD)$, but there is no way for an algorithm to be aware of this without globally solving the entire tree.

2   NAPX Algorithm

2.1   Description

In this section we present NAPX, an $O\left(\frac{nB^2 h^2 \log^2 n}{\log^2(1-\epsilon)}\right)$ dynamic programming algorithm for $a_i \xrightarrow{c_i} b_i$ NAP that produces a $(1-\epsilon)$-approximation of the optimal solution, where h denotes the height of $\mathcal{T}$. As that in [13], our algorithm is only polynomial if B is polynomial in n. This assumption is justifiable if, for example, B is expressed in millions of dollars and its value will be a reasonably small integer. Without loss of generality, we also assume that no single cost exceeds the budget.

NAPX essentially generalizes the dynamic programming table of [13] by computing for each clade, each desired survival probability of the clade, and any budget between 1 and B, the maximum $\mathbb{E}(PD)$ score achievable while guaranteeing this survival probability. This way, we need not make the choice between maximizing $\mathbb{E}(PD)$ or probability as the tables are constructed. From the definition of $\mathbb{E}(PD)$ in (2), the probability of survival of an edge can be written as a function of its two children. Let P_e denote the survival probability of edge e, and l and r be e's children. Then

$$P_e = P_l + P_r - P_l P_r. \tag{3}$$

In the optimal solution for NAP on $\mathcal{T}$, assume b dollars are assigned to clade $\mathcal{K}_e$ and e survives with probability p. It follows that i and $b - i$ dollars are assigned to $\mathcal{K}_l$ and $\mathcal{K}_r$ respectively where $0 \leq i \leq b$. These subclades must survive with probabilities j and $\frac{p-j}{1-j}$ (or 0 when $p = j = 1$), for some $0 \leq j \leq p$, in order to satisfy (3). Because the probability is continuous, we discretize it into intervals by rounding it down to the nearest multiple of a chosen constant α. Probabilities less than a chosen cutoff value p_{min} are rounded to zero.

$$p \in \left\{ 0, \alpha^{\lceil \log_\alpha p_{min} \rceil}, ..., \alpha^2, \alpha, 1 \right\}$$

If two non-zero probabilities lie in the same interval, their ratio is at most α. If they are in consecutive intervals, their ratio is likewise bounded by α^2. For notational convenience, we define a mapping $\pi(\cdot)$ that rounds a probability to the lower bound of its corresponding interval.

$$\pi(p) = \begin{cases} 0 & \text{if } p < p_{min} \\ \alpha^{\lfloor \log_\alpha p \rfloor} & \text{otherwise.} \end{cases}$$

We now formally describe our algorithm. For each edge e, we construct a two-dimensional table T_e where $T_e(b, p)$ stores the optimal expected diversity of $\mathcal{K}_e$ given that b dollars are assigned to it and it survives with a probability that lies no less than p. The table is constructed in the following manner if e is a pendant edge incident to the leaf for species s.

$$T_e(b, p) = \begin{cases} a_s \lambda(e) & \text{if } b < c_s \text{ and } p = \pi(a_s), \\ b_s \lambda(e) & \text{if } b \geq c_s \text{ and } p = \pi(b_s), \text{ or} \\ -\infty & \text{otherwise.} \end{cases} \tag{4}$$

Otherwise, T_e is computed from the tables of its two children, T_l and T_r.

$$T_e(b, p) = p\lambda(e) + \max_{i,j,k}\{T_l(i, j) + T_r(b - i, k)\} \tag{5}$$

subject to

$$i \in \{0, 1, 2, ..., b\},$$
$$j, k \in \{0, \alpha^{\lceil \log_\alpha \ p_{min} \rceil}, ..., \alpha^2, \alpha, 1\},$$
$$\pi(j + k - jk) = p$$

The $\mathbb{E}(PD)$ score for the entire tree can be obtained by attaching an edge e_r of length 0 to the root and finding $\max_j\{T_{e_r}(B, j)\}$. The tables are computed from the bottom up, and each time an entry is filled, pointers are kept to the two entries in the child tables from which it was computed. This way the optimal budget allocation can be obtained by following the pointers down from the entry for the optimal score for e_r.

2.2 Approximation Ratio

In this section, we express the worst-case approximation ratio as a function of the constants p_{min} and α introduced above, beginning with p_{min}. Note that since any species s with $c_s > B$ can be transformed into a new species s' with $c_{s'} = 0, b_{s'} = a_s$ and $a_{s'} = a_s$ without affecting the outcome, we can safely assume that $c_s \leq B$ for all $s \in X$.

Lemma 1. *Let I be an instance of NAP for which there exists a constant k such that all b_i values are greater than n^{-k}. Consider a transformed instance I' where all a_i values in the range $(0, p_{min})$ are rounded to 0. Let $OPT(I)$ and $OPT(I')$ be the expected PD scores of the optimal solutions to I and I' respectively. Then the ratio of these scores is bounded as follows:*

$$OPT(I') \geq (1 - n^{k+1}p_{min})OPT(I)$$

Proof. Let $\text{path}(s)$ be the set of edges comprising the path from leaf s to the root. We define $w(s)$ as the expected diversity of the path from s to the root if s is conserved:

$$w(s) = b_s \sum_{e \in \text{path}(s)} \lambda(c).$$

Let $w_{max} = \max_{s \in X}\{w(s)\}$. This value allows us to place a trivial lower bound on the optimal solution (recalling that we can assume that $c_s \leq B$).

$$w_{max} \leq OPT(I). \tag{6}$$

We also observe that if any species s survives with a non-zero probability smaller than p_{min} in the optimal solution, its contribution to $OPT(I)$ will be bounded by $\dfrac{p_{min}w(s)}{b_s}$. It follows that

$$OPT(I') \geq OPT(I) - \sum_{s \in X} \frac{p_{min}w(s)}{b_s}.$$

Since $b_s \geq n^{-k}$ and $w(s) \leq w_{max}$, we can express the bound as

$$OPT(I') \geq OPT(I) - n\frac{p_{min}w_{max}}{n^{-k}}.$$

Dividing by $OPT(I)$ yields

$$\frac{OPT(I')}{OPT(I)} \geq 1 - \frac{n^{k+1}p_{min}w_{max}}{OPT(I)}.$$

From (6) we obtain

$$\frac{OPT(I')}{OPT(I)} \geq 1 - n^{k+1}p_{min},$$

which completes the proof. $\qquad\square$

The size of the probability intervals in the tables, determined by α, also affects the approximation ratio. This relationship is detailed in the following lemma.

Lemma 2. *Let $OPT_e(b, p)$ denote the optimal expected PD score for clade $\mathcal{K}_e$ if e survives with probability exactly p and b dollars are allocated to it. Now consider an instance of NAP such that all a_s and b_s are either 0 or at least p_{min}. For any $OPT_e(b, p)$ where e is at height h in the tree, there exists a table entry $T_e(b, p')$ constructed by NAPX such that the following conditions hold:*

$$i)\ T_e(b, p') \geq \alpha^h OPT_e(b, p)$$
$$ii)\quad p' \quad \geq \alpha^h p$$

Proof. If $p = 0$, then $OPT_e(b, p) = 0$ and the lemma holds. For the remainder of the proof, we assume $p \geq p_{min}$. The proof will proceed by induction on h, the height of e in the tree, beginning with the base case where $h = 1$ and e is a pendant connected to leaf s. We need only consider the cases where the optimal solution is defined. So without loss of generality, assume we have $OPT_e(b, a_s) = \lambda(e)a_s$. From (4), we know there is an entry $T_e(b, \pi(a_s)) = a_s\lambda(e)$ and therefore both $i)$ and $ii)$ hold.

We now assume that the lemma holds for $h \leq x$ and consider some edge e at height $x + 1$. By definition, $OPT_e(b, p)$ can be expressed in terms of its children l and r.

$$OPT_e(b, p) = p\lambda(e) + OPT_l(i, j) + OPT_r(b - i, k)$$

where $j + k - jk = p$. From the induction hypothesis, there exist

$$T_l(i, j') \geq \alpha^x OPT_l(i, j)\ \text{and}$$
$$T_r(b - i, k') \geq \alpha^x OPT_r(b - i, k)$$

where $j' \geq \alpha^x j$ and $k' \geq \alpha^x k$. Let $q = j' + k' - j'k'$. It follows that

$$q \geq \alpha^x j + \alpha^x k - \alpha^{2x}jk \geq \alpha^x p. \qquad (7)$$

The left inequality in (7) holds because $j' + k' - j'k'$ increases as j' or k' increase, so long as their values do not exceed 1. This can be checked by observing that the partial derivatives with respect to j' and k' are $1 - k'$ and $1 - j'$, respectively.

$T_l(i, j')$ and $T_r(b - i, k')$ will be considered when computing the entry $T_e(b, p')$ where $p' \geq \pi(q)$. Since $q \geq p_{min}$, we have $\pi(q) \geq \alpha q$ because it simply rounds q to the nearest multiple of α. Therefore, $p' \geq \alpha^{x+1} p$ and $T_e(b, p')$ can be expressed as follows.

$$\begin{aligned}
T_e(b, p') &\geq p'\lambda(e) + T_l(i, j') + T_r(b - i, k') \\
&\geq \alpha^{x+1} p\lambda(e) + \alpha^x OPT_l(i, j) + \alpha^x OPT(b - i, k) \\
&\geq \alpha^{x+1}(p\lambda(e) + OPT_l(i, j) + OPT(b - i, k)) \\
&\geq \alpha^{x+1} OPT_e(b, p)
\end{aligned}$$

$\square$

Combining Lemmas 1 and 2 allows us to state that NAPX returns a solution that is at least a factor of $(1 - n^{k+1} p_{min})\alpha^h$ of the optimal solution. In this section we show that these results also imply that a $(1 - \epsilon)$ approximation can be obtained in polynomial time for an arbitrary constant ϵ.

Lemma 3. $O\left(\dfrac{h \log n}{|\log(1 - \epsilon)|}\right)$ *probability intervals are required in the table in order to obtain a $1 - \epsilon$ approximation.*

Proof. The number of probability intervals, t, required for the table is bounded by the number of times 1 must be multiplied by α to reach p_{min}. Hence $\alpha^t \leq p_{min}$ and

$$t = \left\lceil \frac{\log p_{min}}{\log \alpha} \right\rceil. \tag{8}$$

From Lemmas 1 and 2 we can obtain the desired approximation ratio by selecting $\alpha = \sqrt{(1 - \epsilon)^{\frac{1}{h}}}$ and $p_{min} = \frac{1 - \sqrt{1 - \epsilon}}{n^{k+1}}$. Plugging these values into (8) gives

$$t = \left\lceil \frac{\log\left(\frac{1 - \sqrt{1 - \epsilon}}{n^{k+1}}\right)}{\log\left(\sqrt{(1 - \epsilon)^{\frac{1}{h}}}\right)} \right\rceil = \left\lceil \frac{2h(\log(1 - \sqrt{1 - \epsilon}) - (k + 1)\log n)}{\log(1 - \epsilon)} \right\rceil$$

In the numerator, $\log(1 - \sqrt{1 - \epsilon})$ is dominated by $-\log n$ so we can express t asymptotically as

$$t \in O\left(\frac{-h \log n}{\log(1 - \epsilon)}\right)$$

$\square$

Theorem 1. *NAPX is a $(1 - \epsilon)$-approximation with time complexity* $O\left(\dfrac{nB^2 h^2 \log^2 n}{\log^2(1 - \epsilon)}\right)$.

Proof. For each table entry $T(b, p)$, we must find the maximum of all possible combinations of entries in the left and right child tables that satisfy b and p. These combinations correspond to the possible $\{i, j, k\}$ triples from (5). There are $O(Bt^2)$ such combinations as i corresponds to the budget and j and k correspond to probability intervals. Furthermore, for fixed values of p and j, there are potentially $O(t)$ different values of k that could satisfy $\pi(j + k - jk)$ due to rounding. It follows that a naive algorithm would have to compare all $O(Bt^2)$ combinations when computing the maximum in (5) for each table entry.

Fortunately, because $\pi(j+k-jk)$ is monotonically nondecreasing with respect to either j or k, we can directly compute for any fixed p and j the interval of k entries that satisfy $\pi(j + k - jk) = p$:

$$\left[\left\lceil \log_\alpha \left(\frac{\alpha p - j}{1 - j}\right)\right\rceil, \left\lceil \log_\alpha \left(\frac{p - j}{1 - j}\right)\right\rceil\right].$$

Finding the value of k in the interval such that $T(b - i, k)$ is maximized is effectively a range maxima query (RMQ) on an array. Regardless of the size of the interval, such a query can be performed in constant time if instead of an array, the values are stored in a RMQ structure as described in [2]. Such structures are linear both in space and the time they take to create, meaning that we can use them to store each column in the table (corresponding to budget value i) without adversely affecting the complexity. Now, when given a pair $\{i, j\}$, the optimal value of k can be computed in constant time, bringing the complexity of filling a single table entry to $O(Bt)$, the number of combinations of the pair $\{i, j\}$.

There are $O(Bt)$ entries in each table, and a table for each of the $O(n)$ edges in the tree. The space complexity is therefore $O(nBt)$ and the time complexity is $O(nB^2t^2)$. Substituting t for the value that yields a $(1 - \epsilon)$ approximation ratio shown in Lemma 3 gives $O\left(\dfrac{nB^2h^2\log^2 n}{\log^2(1 - \epsilon)}\right)$.

2.3   Expected Running Time

Since in general the height of a phylogenetic tree with n leaves is $O(n)$, the running time derived above is technically cubic in n. Fortunately, for most inputs we can expect the height to be much smaller. In this section, we will provide improved running times for trees generated by the two principal random models. Additionally we will show that caterpillar trees, which should be the pathological worst-case topology according to the above analysis, actually have a much lower complexity.

The Yule-Harding model [16][6], also known as the equal-rates-Markov model, assumes that trees are formed by a succession of random speciation events. The expected height of trees formed in this way, regardless of the speciation rate, is $O(\log n)$ [3] giving a time complexity of $O\left(\frac{nB^2\log^4 n}{\log^2(1-\epsilon)}\right)$. A large-scale study of published phylogenies has shown that in practice, trees are less balanced than the Yule-Harding model predicts [10] and may better fit a Uniform random model [1]. Under the uniform model, all possible topologies are equally likely and the expected tree height is $O(\sqrt{n})$ [3]. Such trees can therefore be processed by NAPX in $O\left(\frac{n^2 B^2 \log^2 n}{\log^2(1-\epsilon)}\right)$ time.

A caterpillar tree is a tree where all internal nodes are on a path beginning at the root, and therefore has height n. This implies that every internal edge has at least one child edge that is incident to a leaf. Suppose edge e has child l that is incident to the leaf for species s. This table only contains two meaningful values: $T_l(0, a_s)$ and $T_l(c_s, b_s)$. Therefore to compute entry $T_e(b, p)$, only $O(1)$ combinations of child table entries need to be compared and the time complexity is improved to $O\left(\frac{n^2 B \log n}{\log(1-\epsilon)}\right)$.

3   Conclusion

NAPX is, to our best knowledge, the first polynomial-time algorithm for $a_i \xrightarrow{c_i} b_i$ NAP that places guarantees on the approximation ratio. While there are still some limitations, especially for large budgets or tree heights, our algorithm still significantly increases the number of instances of NAP that can be solved. Moreover, our expected running time analysis shows that the algorithm will usually be much more efficient than its worst-case complexity suggests. This work towards a more general solution is important if the Noah's Ark Problem framework is to be used for real conservation projects. Some interesting questions do remain, however. Does NAP remain NP-Hard when the budget is constrained to be polynomial in n? We conjecture that it is, but the usual reduction from Knapsack is clearly no longer valid. We would also like to find an efficient algorithm whose complexity is independent of h and/or B.

References

1. Aldous, D.J.: Stochastic Models and Descriptive Statistics for Phylogenetic Trees, from Yule to Today. Statistical Science 16(1), 23–34 (2001)
2. Bender, M.A., Colton, M.I.F.: The LCA Problem Revisited. In: Gonnet, G.H., Viola, A. (eds.) LATIN 2000. LNCS, vol. 1776. Springer, Heidelberg (2000)
3. Erdos, P.L., Steel, M.A., Szekely, L.A., Warnow, T.J.: A few logs suffice to build (almost) all trees: Part II. Theoretical Computer Science 221(1), 77–118 (1999)
4. Faith, D.P.: Conservation evaluation and phylogenetic diversity. Biological Conservation 61, 1–10 (1992)
5. Garey, M.R., Johnson, D.S.: Computers and Intractability: A Guide to the Theory of NP-Completeness. W.H. Freeman and Company, New York (1979)
6. Harding, E.F.: The Probabilities of Rooted Tree-Shapes Generated by Random Bifurcation. Advances in Applied Probability 3(1), 44–77 (1971)
7. Hartmann, K., Steel, M.: Maximizing Phylogenetic Diversity in Biodiversity Conservation: Greedy Solutions to the Noah's Ark Problem. Systematic Biology 55(4), 644–651 (2006)

8. Heard, S.B., Mooers, A.O.: Phylogenetically patterned speciation rates and extinction risks change the loss of evolutionary history during extinctions. Proc. R. Soc. Lond. B 267, 613–620 (2000)
9. Magurran, A.E.: Measuring Biological Diversity. Blackwell Publishing, Malden (2004)
10. Mooers, A.O., Heard, S.B.: Inferring Evolutionary Process from Phylogenetic Tree Shape. The Quarterly Review of Biology 72(1), 31–54 (1997)
11. Nee, S., May, R.M.: Extinction and the Loss of Evolutionary History. Science 278(5338), 692 (1997)
12. Pardi, F., Goldman, N.: Species choice for comparative genomics: Being greedy works. PLoS Genetics 1(6), 71 (2005)
13. Pardi, F., Goldman, N.: Resource-aware taxon selection for maximizing phylogenetic diversity. Syst. Biol. 56(3), 431–444 (2007)
14. Steel, M.: Phylogenetic diversity and the greedy algorithm. Systematic Biology 54(4), 527–529 (2005)
15. Weitzman, M.L.: The Noah's Ark problem. Econometrica 66, 1279–1298 (1998)
16. Yule, G.U.: A Mathematical Theory of Evolution, Based on the Conclusions of Dr. JC Willis, FRS. Philosophical Transactions of the Royal Society of London. Series B, Containing Papers of a Biological Character 213, 21–87 (1925)

Minimum Common String Partition Parameterized[*]

Peter Damaschke

Department of Computer Science and Engineering
Chalmers University, 41296 Göteborg, Sweden
ptr@cs.chalmers.se

Abstract. Minimum Common String Partition (MCSP) and related problems are of interest in, e.g., comparative genomics, DNA fingerprint assembly, and ortholog assignment. Given two strings with equal symbol content, the problem is to partition one string into k blocks, k as small as possible, and to permute them so as to obtain the other string. MCSP is NP-hard, and only approximation algorithms are known. Here we show that MCSP is fixed-parameter tractable in suitable parameters, so that practical instances can be efficiently solved to optimality.

1 Parameterization of MCSP

String z is called a *substring* of string x, if there exist strings u, v (possibly empty) such that $x = uzv$. The same substring z may appear at several positions in x. By a *segment* of x we mean an occurence of a substring at a specific position in x. A substring or segment may be empty, where an empty segment is defined by its location between two consecutive symbols. Length $|x|$ is the number of symbol occurences in x.

In the Minimum Common String Partition (MCSP) problem, two strings x, y of length n, that contain the same symbols equally many times, shall be partitioned each into k segments called *blocks*, so that the blocks in x and y constitute the same multiset of substrings. Cardinality k shall be minimized. An equivalent formulation is: Split x into k segments so that y equals the concatenation of some permutation of them. A position between two consecutive blocks is called a *break*, hence we have $k - 1$ breaks in each of x and y. MCSP is of interest in comparative genomics [4], to answer questions like: Is a DNA string possibly obtained by rearrangements of another DNA string? The problem is NP-hard even in very restricted cases [7], and several approximation algorithms and non-approximability results are known [7,5,8]. A related problem is Minimum Common Integer Partition (MCIP): Given two multisets of at most k integers, split these integers into a minimum number of integer summands so that the two resulting multisets are equal. Biological motivations include ortholog assignment and DNA fingerprint assembly. MCIP also appears as a

---

[*] Supported by the Swedish Research Council (Vetenskapsrådet), grant no. 2007-6437, "Combinatorial inference algorithms – parameterization and clustering".

K.A. Crandall and J. Lagergren (Eds.): WABI 2008, LNBI 5251, pp. 87–98, 2008.

© Springer-Verlag Berlin Heidelberg 2008

subproblem in certain instances of MCSP. The complexity status is similar to that of MCSP [3,12]. Due to lack of space we refer to [3,4] for details about the biological background.

Since k and other input parameters are expected to be much smaller than n in biological applications, it is sensible to ask whether MCSP is fixed-parameter tractable (FPT). To our best knowledge, this issue has not been addressed before in the literature. A problem with input length n and one or more input parameters is in FPT, if it can be exactly solved in time polynomial in n, for any fixed parameter values, where the exponent in the polynomial bound must be constant, i.e., independent of the parameters. For an introduction to FPT see, e.g., [6]. Many problems in computational biology, among them sequence comparison problems, are NP-hard but in FPT, here we only refer to [2,9].

Besides the number k of blocks we consider two more parameters for MCSP: We define the *distance ratio* $d = n/m$, where m is the minimum distance between breaks. A limited d means that we ignore (or do not expect) solutions where some blocks are very short fragments. We define the *repetition number* r of a string x as the maximum i so that $x = uv^i w$ holds for some strings u, v, w, where v is nonempty. Although repeats in genomes are not uncommon, the repetition number in biological sequences will often be small. (See the remark below.)

Contributions and organization of the paper

The obvious question is whether MCSP is fixed-parameter tractable in k. While this remains open, we show fixed-parameter tractability in combined parameters k, r. For ease of presentation we step two paces back and derive first a weaker result in Section 2: an FPT algorithm for MCSP with combined parameters k, d, r. Actually, parameters d, r suffice, since obviously $k \leq d$ must be assumed. The algorithm has a simple structure: First we roughly guess the breaks and the matching of blocks "in parameterized time". Then we are left with a rather special linear system of equations that enables us to check whether a solution of this shape exists. Since we test all possible branches, the same approach can also be used to enumerate all solutions that respect the given parameters. (If alternative solutions exist, it is not clear which of them is the correct interpretation of data, hence one should consider them all. See [1] for another example.) The time is $O(n)$ for any fixed parameters. We remark that linear-time *approximation* algorithms for MCSP and the related reversal distance problem are known if other parameters are bounded, namely if every symbol may appear only constantly many times [10,11]. This case is interesting, e.g., if the symbols represent genes. Since r cannot be larger than the number of appearances of every symbol, our result implies also that MCSP, with bounded number of appearances of every symbol, is in FPT.

We also state a simple explicit upper bound for the parametric part of the complexity bound, however it seems overly pessimistic. The restriction that aligned substrings be equal is so strong that the actual number of branches to consider is typically much smaller, although there might be malicious cases. In Section 3 we illustrate the issue and add some $O(n)$ time heuristics that discard hopeless branches quickly. Simplicity of the scheme, and the observation that almost no branching takes place, should make the approach very practical.

In Section 4 we get rid of parameter d and replace it with $k \leq d$. That is, we also identify arbitrarily short blocks. The difficulty is to separate the close breaks, which requires different techniques and arguments. Our $O(n \log n)$ time algorithm (for any fixed k, r) applies the one from Section 2 in an iterative fashion on a decreasing scale. In Section 5 we point out that MCIP is polynomial for any fixed k (which is simple). Since MCIP is related to MCSP instances with high r, this gives rise to the conjecture that MCSP might be fixed-parameter tractable with k as the only parameter. Section 6 gives some outlook. Proofs are basically complete, but due to space limitations, some passages are only sketched and not much formalized, and no pictures to illustrate the algorithms could be included.

2   An FPT Algorithm for MCSP

A few definitions and basic facts around periodic strings will be needed. A non-empty string p is called a *period* of string z if $z = sp^i t$, where $i > 0$ is an integer, s a suffix and t a prefix of p (both may be empty). Every string has a shortest period, uniquely determined up to cyclic shifts. If string z overlaps itself on $(1 - \alpha)|z|$ symbols (i.e., prefix and suffix of z of that length are equal strings), where $0 < \alpha \leq 1$, then z has a period of length $\alpha|z|$. We will call certain segments of x or y *fragile* (*conserved*), when we decided to put (not to put) breaks there. Two segments without a symbol in between are *adjacent*.

Now we start describing our algorithm for parameters k, d, r. For some $c > d$ specified below, we split both x and y into c segments called *pieces*, of length roughly $n/c < n/d$. Since we consider only blocks of length at least n/d, a piece can contain at most one break. (Breaks exactly between two adjacent pieces are arbitrarily assigned to, e.g., the piece to the left.) We guess the $k - 1$ pieces that contain breaks, these are our fragile segments. We branch on these choices. We also guess the *matching* between the resulting k blocks of x and y that shall be aligned, and branch on the possible matchings, i.e., permutations of blocks. Clearly, the number of branches is bounded by some function of k and c. Note that only the exact positions of breaks in the designated fragile segments are yet unknown in every branch.

If we choose $c \geq 2d$, at least one entire piece in every block is conserved. Now consider any pair of matched blocks s from x, and t from y. Let s' be the concatenation of all conserved pieces in s. Although the exact borders of s are yet undetermined, we know that s' is a subsegment of s. Let t^+ be the segment of y consisting of all pieces that are certainly in t or may contribute some symbols to t. (The latter ones are the delimiting fragile pieces. Informally, t^+ is the "maximal possible t".) Clearly, s' must be aligned to some segment of t^+. If s' occurs several times in t^+, we guess the alignment partner of s' in t^+, and branch on these choices. Assume that the start positions of two occurrences of s' in t^+ have a distance $\alpha|s'|$, $0 < \alpha \leq 1$. Then s' has a period of length $\alpha|s'|$, which is repeated $1/\alpha$ times. It follows $\alpha \geq 1/r$, hence the number of branches is bounded by some function of c and r.

As $c \geq 2d$ was the only condition on c, we may choose $c = 2d$. Now the *total* number of branches (deciding on: pieces with breaks, matching of blocks, and alignment of some conserved segments in every pair of matched blocks) is bounded by some function of d and r. Still it remains to determine the exact breaks. We call a pair of conserved segments of x and y an *aligned pair* if we have already decided to align them (as we did above in every pair of matched blocks). Now we are, in every branch, in a situation addressed by:

Lemma 1. *Consider an instance x, y of MCSP. Suppose that we have already fixed the following items:*
(i) exactly $k - 1$ fragile segments in both x and y, which are pairwise disjoint, non-adjacent, do not include the first or last symbol of x or y, and have length at most n/k each,
(ii) a matching between the k conserved segments of x and y which are defined by cutting out these fragile segments,
(iii) an aligned pair in every matched pair of blocks given by (ii).
Then we can, in linear time, construct a solution to this constrained instance of MCSP, or prove that no such solution exists.

Proof. In the following, the terms *fragile segment* and *conserved segment* refer only to the $2(k - 1) + 2k$ distinguished segments specified in (i) and (ii). Since only $k - 1$ breaks are allowed, every fragile segment contains exactly one break. The left (right) *semi-block* of a block is the segment between the member of the aligned pair and the left (right) break delimiting this block.

We define an auxiliary graph as follows. For every fragile segment f, we create two vertices that represent the two segments of f to the left and to the right of the break that will be placed in f. We call them the *left and right vertex* of f. We create edges that join
(1) the left and right vertex of every fragile segment,
(2) the left vertices of any two fragile segments in x and y whose conserved segments adjacent to the left are matched,
(2') the right vertices of any two fragile segments in x and y whose conserved segments adjacent to the right are matched.

The *weight* of a vertex is the length of the segment it represents. These weights are unknown in the beginning (as the breaks are yet unknown), but we get a linear equation for the weights of any two vertices joined by any edge, due to the following observations:
(1) The total length of any fragile segment is known.
(2) We have an aligned pair in the conserved segments to the left, and the number of symbols in their right semi-blocks must be the same in x and y.
(2'): similar to (2).

For a solution to the whole constrained problem instance, it is sufficient that:
(a) every symbol from x and y gets into exactly one block, and
(b) the matched semi-blocks to the left or right of aligned pairs have equal lengths, and are also equal as strings.

Equations from (1) ensure (a), and equations from (2) and (2') ensure the length condition in (b). Equality of strings in (b) must be tested separately, and not every solution to the linear system yields already a valid solution to the constrained MCSP instance. However, thanks to the structure of our linear system we need not search the entire solution space. Instead we do the following in every connected component of the graph.

We guess the weight of one vertex v. Then, the weights of all other vertices are uniquely determined by a trivial elimination procedure. Any solution specifies the lengths of certain semi-blocks. If no value of the weight of v yields a valid solution, i.e., if some weights become negative or some symbols that get aligned are not equal, then no valid solution can exist at all. On the other hand, since the semi-blocks to the left or right of an aligned pair are always associated to the same connected component, condition (b) is fulfilled if all aligned semi-blocks pass the equality test *in their respective connected components*. It follows that *any* combination of valid solutions from the connected components is a valid overall solution. Thus we can solve the subproblems and establish (a) and (b) in the connected components independently.

Analysis: We show the linear time bound. In an $O(n)$ time preprocessing in x and y we index the symbols from left to right, so that the length of any one segment can be computed by one subtraction in $O(1)$ time. Then the "graphical" linear system of equations can be set up in $O(k)$ time. Connected components are determined in $O(k)$ time as well. Once the weight of one vertex in a connected component of size k' is fixed, we can solve the linear system trivially in $O(k')$ time. The test whether the corresponding partitioning of x and y is valid can also be executed in $O(k')$ time, but this step needs some more preparation:

Since we have already fixed an aligned pair in every matched pair of blocks, we can easily determine the maximum number of symbols in any pair of aligned semi-blocks: Note that aligned symbols have to be equal. This gives a threshold for the number of symbols in every fragile segment that can be attached to the block to the left and right, respectively. For any solution to the linear system it suffices now to check whether each variable is nonnegative and below its threshold. All thresholds can be computed once in the beginning, in another $O(n)$ time preprocessing for the instance, just by symbol comparisons and counting.

In every connected component we need to try at most n/k possible weights of the start vertex, since this is the maximum length of a fragile segment. As we saw above, for every initial vertex weight we can solve the system and validate the solution in $O(k')$ time. Thus we need $O(k'n/k)$ time in the component. For all components together this sums up to $O(kn/k) = O(n)$. $\qquad\square$

Theorem 1. *MCSP is fixed-parameter tractable in the combined parameters d, r (distance ratio and repetition number). More specifically, MCSP can be solved in $O(n)$ time for any fixed values of d, r.*

Proof. In the branching phase we fix the items specified in Lemma 1. The number of branches is bounded by a function of d, r. The assumptions of Lemma 1 are in fact satisfied: The selected fragile segments are pairwise disjoint by construction, and $c \geq 2d$ ensures that none of them are adjacent or include the beginning or the end of x or y. Fragile segments have length $n/c \leq n/2d < n/k$. Now we solve every constrained instance in $O(n)$ time. Any solution found in a branch is a solution to the unconstrained instance, and if all branches are unsolvable, then so is the unconstrained instance.   □

The scheme can be modified to efficiently compute a concise representation of *all* solutions to an instance of MSCP that comply with the given parameter values. We consider all branches anyhow (i.e., no potential solution is lost), and the solution space of the linear system of equations and inequalities (for nonnegativity, thresholds) can be concisely described.

The graphs in Lemma 1 have actually a very special structure, but we did not make use of this fact, because linear time is already optimal for trivial reasons. However, in practice one can save physical running time also by accelerating some parts of a linear-time algorithm, therefore we go into some more details now. Note that the graphs corresponding to our linear systems merely consist of even-length cycles. We call an edge, corresponding to an equation $u + v = a$ and $u - v = a$ (where u, v are variables and a some constant), an *additive and subtractive edge*, respectively. Now observe that, in every cycle, additive and subtractive edges alternate. Thus we can easily divide the vertices in two sides so that edges between vertices on different sides (on the same side) are additive (subtractive). If a linear system with two variables per equation, where each variable is integer and ranges from 0 to an individual threshold, has such a "quasi-bipartite" graph, we can find a solution in every connected component already by a logarithmic number of guesses of one variable: All variables on one side can only increase simultaneously, while all variables on the other side decrease. Hence we can find two extremal values of our probe variable (such that all other variables are still in their ranges) by binary search. As we said, this does not help the worst-case time bound, but it speeds up the search if the graph has only a few connected components.

3   Branching Heuristics

A closer look at the above FPT algorithm gives an explicit bound on the hidden factor in $O(n)$: The $k - 1$ (out of $2d$) pieces in x with breaks can be chosen in $(2d - k)^{k-1}/(k - 1)!$ ways, since no adjacent pieces have breaks. Blocks can be permuted in $k!$ ways. Since the block lengths are now fixed, up to a tolerance of 2 piece lengths, we can choose the pieces in y with breaks, successively from left to right, in 3^{k-1} ways. In every pair of matched blocks, at most r possible alignment partners exist. In total we get at most $kr(3(2d - k)r)^{k-1}$ branches. However, this is a crude bound. For typical, rather irregular input strings we expect that the vast majority of branches can be quickly discarded: Note that already one symbol mismatch in the segments we try to align renders a branch impossible. Therefore it is worthwhile to add to the basic algorithm some simple and fast heuristics that recognize many hopeless branches *before* we run the procedure of Lemma 1. Some ideas have been already brought up in approximation algorithms [11], but not in the context of parameterized exact algorithms. To apply the linear-time procedure we need enough "well-separated" fragile segments (see Lemma 1 for details). A simple observation is that a segment of x not occuring as a substring in y must be fragile in x. We can search for all occurrences in y of a particular substring in $O(n)$ time, using suffix trees or linear-time string matching.

We suggest a top-down approach to find shorter and shorter fragile segments: Halve the string x recursively, and in the jth iteration check which minimal concatenations of the obtained 2^j pieces have no occurences in y. Moreover, from the obtained list F of fragile segments we can erase those containing shorter fragile segments detected later. Clearly, the time for this procedure remains bounded by n times a polynomial of distance ratio d. We need not even fix a parameter value d in advance, but we may also run the process until it succeeds with enough fragile segments, or some time limit is reached. There is a good chance that many long segments have only a few occurences in y: Intuitively, many occurences of many different segments must heavily overlap, hence they can exist only if y contains periodic segments with many repetitions. Therefore we should, in this way, quickly obtain $k - 1$ disjoint fragile segments. (Here it is also good to notice that the maximum number of pairwise disjoint segments in a given set F of segments is easily determined by a linear-time greedy algorithm.) Next, for aligning pairs of conserved segments it is barely necessary to try all $k!$ matchings. For the same reason as above, our conserved segments should typically have few occurences in y, so that relatively few matchings remain.

We do not only keep the number of matchings small, but also the combinations of breaks, based on another simple fact: Once we know some set F of fragile segments, the set of breaks must be a hitting set of size at most $k - 1$ for F. In particular, breaks can only lie in the union U of all hitting sets of size $k-1$. Often U will be considerably smaller than the segments covered by F, and consist of several disjoint segments (ideally $k - 1$). On the other hand, we can compute U in linear time, so that the $O(n)$ time bound is not sacrificed. We give the details now. In the following, a "point" is a position between two symbols.

Theorem 2. *Given a set F of segments in a string of length n (none of them contained in another segment of F) and an integer k, we can compute the union U of all hitting sets of size $k - 1$ for F in $O(n)$ time.*

Proof. First note that $|F| \leq n$. All segments in F not containing a specific point p are divided in two sets $LF(p)$ and $RF(p)$: the segments to the left and to the right of p. A minimum hitting set of $LF(p)$ is obtained by an obvious greedy algorithm that scans the points from left to right and always adds the latest possible point to the hitting set. This yields the size $lf(p)$ of a minimum hitting set for $LF(p)$, for *all* p together, in $O(n)$ time. Similarly we proceed with all $LR(p)$ and get all $lr(p)$. Finally observe that $p \in U$ if and only if $lf(p) + rf(p) \leq k - 2$. $\qquad\square$

-	01	02	03	04	05	06	07	08	09	10	11	12	13	14	15	16	17	18	19	20	21	22
x	c	a	b	c	c	d	a	a	a	a	b	a	b	c	c	a	b	a	b	d	c	d
y	b	d	c	d	c	a	b	c	c	a	b	a	b	d	a	a	a	c	c	a	b	a

We illustrate the heuristics on this small example taken from [5]. Refining the "grid" of segment endpoints on string x by successive halving, we find the following segments that do not appear in y and must be fragile: $[1, 22]$, $[1, 11]$, $[12, 22]$, $[1, 6]$, $[7, 11]$, $[17, 22]$, $[4, 6]$, $[9, 11]$, $[5, 7]$, $[7, 10]$, $[11, 14]$, $[18, 21]$, etc. Taking only the minimal fragile segments found until now, in left-to-right order, we retain the following set F: $[4, 6]$, $[5, 7]$, $[7, 10]$, $[9, 11]$, $[11, 14]$, $[18, 21]$. This F has 4 disjoint intervals, hence $k \geq 5$. We try whether $k = 5$ is enough. The simple algorithm from Theorem 2 restricts the possible breaks to segments $[5, 6]$, $[9, 10]$, $[11, 14]$, $[18, 21]$. That is, under the assumption $k \geq 5$ we obtained 4 disjoint fragile segments, where the 2 leftmost breaks are even uniquely determined. Next we align the conserved segments between the breaks with segments of y. Segments $[1, 5] = cabcc$, $[6, 9] = daaa$, and $[21, 22] = cd$ appear only once in y. Segment $[14, 18] = ccaba$ appears twice, but one such segment in y overlaps $cabcc$, hence this alignment is also unique. Now $[10, 11] = ab$ must be placed between $cabcc$ and $daaa$ in y. Here we have a case with repetition number 2, but we never had to branch on the permutations of blocks in this small example. Finally, the algorithm from Theorem 1 applied to these constellations finds the optimal solution $x = cabcc|daaa|abab|ccaba|bdcd$ and $y = bdcd|cabcc|abab|daaa|ccaba$.

4 Separating Close Breaks

Consider a constrained instance of MCSP with aligned pairs and fragile segments as in Lemma 1. The linear algorithm in Lemma 1 works only under the assumption that each of the designated fragile segments has exactly one break. Now we relax this assumption and suppose that, in each of the fragile segments, all breaks are within a segment of length w called a *window*. Here, w is the maximum window size in all fragile segments. The case solved by Lemma 1 is $w = 0$. In the following, *long blocks* of a partitioning of x, y are those containing the conserved segments that we identified so far, in the constrained instance. The others are *short blocks*. In order to apply a procedure similar to Lemma 1, we will need windows that have a uniform length t and are also aligned, therefore we have to enlargen the windows first.

Lemma 2. *If there is a solution to the MCSP instance constrained as above, then there exist windows in the fragile segments with the following properties:*
(1) All breaks are contained in the windows.
(2) All windows have exactly the same length $t < 2kw$.
(3) The long blocks in the partitioning are aligned in such a way that also the ends of windows are aligned. (But still the long blocks may end inside the windows.) Equivalently, symbols in windows are only aligned to symbols in windows in the other string.

Proof. We will only extend the given windows, to obtain properties (2) and (3). Thus, (1) is preserved automatically. First we extend every window to the left and to the right (if necessary) to establish (3). Since all long blocks end inside the given windows, we never have to move the end of a window by more than

w symbols outwards. Thus we have amplified the windows to length at most $3w$. In the next phase we extend them further to equalize their lengths, without destroying (3). The difficulty is that any two windows from x and y whose left (right) ends are aligned must be extended simultaneously to the left (right).

Since (3) is already true, the sum of lengths of all windows in x and y are equal. We define a graph, with windows as vertices, where edges connect windows whose left or right ends are aligned. Every vertex is labeled with the length of the window it represents. Clearly, this graph is just a disjoint union of even cycles, and vertices come alternatingly from x and y. (The graph is related to the one in Lemma 1, but not identical.) Consider any cycle. We are allowed to add 1 to the labels of any pair of adjacent vertices. As long as two adjacent vertices have non-maximum labels, we simply increment them. If no vertices with non-maximal labels are adjacent, we choose some vertex with minimum label. Suppose this vertex represents a segment of x (the other case is symmetric). Because of the equal sums of labels in x and y, there must exist also a vertex from y with non-maximum label. The two chosen vertices have an odd distance in the cycle and split it in two paths, in the natural sense. Now it is easy to increase the labels of our two special vertices by 2, and the all other labels by only 1. Iterating the procedure we arrive at property (2). To bound the number of steps, consider the sum of differences between the maximum label and all vertex labels. The above procedure decreases this quantity by 2, and initially it was no larger than $2(k-1)w$. Hence the final window length is less than $2kw$. $\qquad\square$

Theorem 3. *MCSP is fixed-parameter tractable in the combined parameters k, r (number of blocks and repetition number). More specifically, MCSP can be solved in $O(n \log n)$ time for any fixed values of k, r.*

Proof. We modify the linear system of Lemma 1 as follows: For every fragile segment, the sum of the two assigned variables *plus t* must equal the length of this fragile segment. Similarly as in Lemma 1 we solve the linear system and test whether some solution is *valid,* in the sense that the parts of long blocks *outside the windows* are aligned. At this stage we do not yet attempt to determine the breaks inside the windows, let alone the matching of short blocks.

First we set $t = 0$ and test whether a solution with exactly one break per fragile segment exists. If this fails, we need more breaks. In this case we determine the smallest $t > 0$ that gives a valid solution. A crucial observation for $t > 0$ is that the window positions in *all* valid solutions differ by some $O(t) = O(kw)$. Namely, if two valid solutions with remote window positions exist, then any two windows whose (e.g.) left ends are aligned must contain equal strings in one of the solutions, because these two windows are aligned segments within long blocks in the other solution. But then we can append these equal windows to the aligned long blocks to the left, and this can be done for all pairs of aligned long blocks. But this means that one break per fragile segment would be enough, which contradicts the failure for $t = 0$. In conclusion, if a solution with window size w exists, there is *one* window of length $t := O(kw)$ in every fragile segment, so that all breaks are inside these windows, where t is the same everywhere. (We need not branch on $O(n)$ possible window positions!) The next difficulty is that we might have found a valid solution to the linear system for some $t > 0$, but a solution to the constrained MCSP instance exists only for some larger t, that is, some breaks are actually outside our windows. We have to analyze this situation.

Consider a partitioning where the total length of long blocks is maximal. In this partitioning, consider any pair of matched long blocks b, c which are not the rightmost blocks in x, y. Let u and v be the blocks next to the right of b and c, respectively. (The reasoning for the left-hand side is similar.) Assume $|u| \le |v|$, the other case is symmetric. Consider the windows to the right of b and c. Note that the segments to the left of the window in x and y, until the long blocks, are equal substrings, and the left ends of u and v are aligned. Hence, if u is entirely to the left of the window, then u is a prefix of v. Now we can transform the partitioning as follows. We append u and the prefix of length $|u|$ of v to the long blocks. If $u = v$, and u, v are matched, then the number of blocks decreases. Otherwise, we split the matching partner of v in two blocks: the prefix of length $|u|$ and the (possibly empty) rest. The former is matched with the old alignment partner of u, and the latter (if not empty) with the rest of v. This yields a new partitioning, where the number of blocks is not increased, but the long blocks have larger total length, contradicting the maximality. Hence there exists a partitioning where, in every fragile segment, at most one break is to the left of the window we have determined. By symmetry, the same statement holds "to the right".

Now consider a fragile segment where the outermost breaks have the maximum distance (in all fragile segments), denoted w. Recall that our $t = O(kw)$ may be too small, but as shown above, we can assume that at most one break is to the left and to the right, respectively, of our window of length t in this fragile segment. If an outermost break is away from the window by more than, say, t symbols, there must still exist a segment of length $\Omega(t/k)$ without breaks, in the segment of length t next to the window, on the same side. The other case is that all breaks are still in an extended window of length $O(t)$. Since $w = \Omega(t/k)$, there must be a segment of length $\Omega(t/k^2)$ without breaks in the window.

Finally we are able to guess a window and therein a new conserved segment that separates two breaks. Due to the minimum distance guarantee for two of the breaks, the number of branches is bounded by a function of parameter k. We perform this branching step in both x and y, to guess at least one new aligned pair of conserved segments. Details are similar as in Section 2 (guessing pieces with breaks, all possible matchings, etc.). This step increases the number of blocks, hence, after less than k such branching steps we have finally separated all breaks. For any fixed value of the parameters, the time is $O(n \log n)$: We always find the smallest suitable t by binary search, solving $O(\log n)$ times a system of equations and checking validity in $O(n)$ time. $\square$

5   MCIP with a Fixed Number of Integers

MCIP appears as a subcase of MCSP, in a sense: It is not hard to transform any instance of MCIP to an equivalent instance of MCSP where the integers are represented as periodic strings of those lengths. (However, this is not a polynomial transformation, since integers are represented unary.) The obtained MCSP instances have large repetition numbers, a case that is not captured by our FPT results. Therefore it is interesting to notice the following "pseudo FPT" result:

Theorem 4. *For any fixed cardinality of multisets, MCIP can be solved in polynomial time.*

Proof. As observed in [3], MCIP is equivalent to the following problem: Partition both multisets into the same number of sub-multisets, as *many* as possible, and match them one-to-one, so that the integers in any matched pair have the equal sums. From any such partition of the multisets, one can construct in polynomial time an optimal solution to the MCIP instance, by a simple elimination process. (Due to lack of space we omit the straightforward details of this equivalence.)

It remains to find a pair of partitionings that can be matched. The number of partitionings is obviously bounded by a function of cardinality of multisets (by the Stirling number, to be specific). We may naively generate all partitionings of both multisets, and check whether a pair of them have the same sums of sub-multisets. The last step can be executed in polynomial time (in the number of partitionings), with help of lexicographic sorting. $\square$

Due to the above correspondence to repetition numbers, this gives hope that MCSP might be fixed-parameter tractable in k as the only parameter: Theorem 4 might be applied, inside some sophisticated extension of the previous FPT algorithms, to any substring with many repeats in x, y. But we have to leave this as an open question.

6   Conclusions and Further Research

We proved that MCSP is fixed-parameter tractable in some natural parameters, indicating that MCSP is exactly solvable in practice. A simple algorithm needs distance ratio d (length ratio of blocks) and repetition number r of periodic segments as parameters, a second algorithm with weaker parameters k (number of blocks) and r is more intricate. We also proposed some simple heuristics that probably discard the vast majority of conceivable branches in most instances. Still it would be interesting to prove good worst-case bounds on the number of branches, using deeper combinatorics of strings. An intriguing open question is whether MCSP is fixed-parameter tractable in k. Since repeats are quite possible, it would be nice to get rid of parameter r. Implementing the FPT algorithms and testing their time performance of on real biological strings would complement the theoretical findings. A natural extension of MCSP is to tolerate a limited number of mismatches in the solutions. Apparently our FPT scheme can nicely accommodate this extra parameter: In the validity tests of solutions we just have to count the mismatches.

References

1. Braga, M.D.V., Sagot, M.F., Scornavacca, C., Tannier, E.: The solution space of sorting by reversals. In: Măndoiu, I.I., Zelikovsky, A. (eds.) ISBRA 2007. LNCS (LNBI), vol. 4463, pp. 293–304. Springer, Heidelberg (2007); journal version to appear in IEEE/ACM Transactions on Computational Biology and Bioinformatics
2. Cai, L., Huang, X., Liu, C., Rosamond, F., Song, Y.: Parameterized complexity and biopolymer sequence comparison. The Computer Journal, Special Issue in Parameterized Complexity 51, 270–291 (2008)
3. Chen, X., Liu, L., Liu, Z., Jiang, T.: On the minimum common integer partition problem. In: Calamoneri, T., Finocchi, I., Italiano, G.F. (eds.) CIAC 2006. LNCS, vol. 3998, pp. 236–247. Springer, Heidelberg (2006)
4. Chen, X., Zheng, J., Fu, Z., Nan, P., Zhong, Y., Lonardi, S., Jiang, T.: Assignment of orthologous genes via genome rearrangement. IEEE/ACM Transactions on Computational Biology and Bioinformatics 2, 302–315 (2005)
5. Chrobak, M., Kolman, P., Sgall, J.: The greedy algorithm for the minimum common string partition problem. In: Jansen, K., Khanna, S., Rolim, J., Ron, D. (eds.) APPROX 2004. LNCS, vol. 3122, pp. 84–95. Springer, Heidelberg (2004)
6. Downey, R.G., Fellows, M.R.: Parameterized Complexity. Springer, Heidelberg (1999)
7. Goldstein, A., Kolman, P., Zheng, J.: Minimum common string partitioning problem: Hardness and approximations. In: Fleischer, R., Trippen, G. (eds.) ISAAC 2004. LNCS, vol. 3341, pp. 484–495. Springer, Heidelberg (2004); also in The Electronic Journal of Combinatorics 12, paper R50 (2005)
8. He, D.: A novel greedy algorithm for the minimum common string partition problem. In: Măndoiu, I.I., Zelikovsky, A. (eds.) ISBRA 2007. LNCS (LNBI), vol. 4463, pp. 441–452. Springer, Heidelberg (2007)
9. Hüffner, F., Niedermeier, R., Wernicke, S.: Developing fixed-parameter algorithms to solve combinatorially explosive biological problems. In: Bioinformatics, vol. II: Structure, Function and Applications. Methods in Molecular Biology, vol. 453. Humana Press (2008)
10. Kolman, P.: Approximating reversal distance for strings with bounded number of duplicates in linear time. In: Jedrzejowicz, J., Szepietowski, A. (eds.) MFCS 2005. LNCS, vol. 3618, pp. 580–590. Springer, Heidelberg (2005)
11. Kolman, P., Walén, T.: Reversal distance for strings with duplicates: Linear time approximation using hitting set. The Electronic Journal of Combinatorics 14 (2007); paper R50, also in: 4th Workshop on Approximation and Online Algorithms WAOA. LNCS, vol. 4368, pp. 281–291. Springer, Heidelberg (2007)
12. Woodruff, D.P.: Better approximations for the minimum common integer partition problem. In: Díaz, J., Jansen, K., Rolim, J., Zwick, U. (eds.) APPROX 2006. LNCS, vol. 4110, pp. 248–259. Springer, Heidelberg (2006)

Hardness and Approximability of the Inverse Scope Problem

Zoran Nikoloski[1,*], Sergio Grimbs[2], Joachim Selbig[1,2], and Oliver Ebenhöh[2]

[1]Institute for Biochemistry and Biology, University of Potsdam, Potsdam, Brandenburg, Germany
[2]Max-Planck Institute for Molecular Plant Physiology, Potsdam, Brandenburg, Germany

Abstract. For a given metabolic network, we address the problem of determining the minimum cardinality set of substrate compounds necessary for synthesizing a set of target metabolites, called the *inverse scope problem*. We define three variants of the inverse scope problem whose solutions may indicate minimal nutritional requirements that must be met to ensure sustenance of an organism, with or without some side products. Here, we show that the inverse scope problems are NP-hard on general graphs and directed acyclic graphs (DAGs). Moreover, we show that the general inverse scope problem cannot be approximated within $n^{1/2-\epsilon}$ for any constant $\epsilon > 0$ unless P $=$ NP. Our results have direct implications for identifying the biosynthetic capabilities of a given organism and for designing biochemical experiments.

Keywords: scope, metabolic networks, completeness.

1   Introduction

Availability of fully sequenced genomes for several organisms has rendered it possible to reconstruct their metabolic networks and further characterize their biosynthetic capabilities. Identifying the biosynthetic capabilities of a given organism is crucial for the development of cost-efficient energy sources, as they are directly related to plant biomass [20], [16], [3]. On the other hand, knowing the compounds necessary for obtaining a desired product can be employed in designing optimal environmental conditions, in the sense of minimizing the nutrients for biosynthesis, and for effective altering of bioprocesses to assist the industrial manufacture of chemicals [4].

Several mathematical methods have been developed to study the biosynthetic capabilities of metabolic networks, including: metabolic control analysis [21], flux balance analysis [2], metabolic pathway analysis [19], cybernetic modeling [15], biochemical systems theory [18], to name just a few. Many of these methods require detailed kinetic information to carry out the analysis—a condition which is often impossible to satisfy.

---

* Z.N., J.S., and O.E. are supported by the GoFORSYS project funded by the Federal Ministry of Education and Research, Grant Nr. 0313924.

K.A. Crandall and J. Lagergren (Eds.): WABI 2008, LNBI 5251, pp. 99–112, 2008.

© Springer-Verlag Berlin Heidelberg 2008

A method which relies only on an available metabolic network and limited knowledge about the stoichiometry of the included biochemical reactions has been recently developed and applied to study the biosynthetic capabilities of various organisms [6], [11]. This method is based on the concept of a *scope*: The basic principle is that a reaction can only operate if and only if all of its substrates are available as nutrients or can be provided by other reactions in the network. Starting from the nutrients, called *seed compounds*, operable reactions and their products are added to an expanding subnetwork of a given metabolic network. This iterative process ends when no further reaction fulfills the aforementioned condition. The set of metabolites in the expanded subnetwork is called the scope of the seed compounds and represents all metabolites that can be in principle synthesized from the seed by the analyzed metabolic network [11].

The scope concept has been applied to a variety of problems, such as: hierarchical structuring of metabolic networks [12], comparison of metabolic capabilities of organism specific networks [7], metabolic evolution [8], and changes of metabolic capacities in response to environmental perturbations [9].

In [10], the inverse problem was addressed as that of determining *minimal* sets of seed compounds from which metabolites that are essential for cellular maintenance and growth can be produced by a given metabolic network. There, a greedy algorithm was applied and heuristics inspired by biological knowledge were introduced to determine biologically relevant minimal nutrient requirements. Whereas by this approach a large number of minimal solutions may be obtained, the minimum cardinality set of seed compounds remains unknown and moreover, it is unclear how well this minimum was approximated by the proposed heuristic.

For a given metabolic network, we investigate the general inverse scope problem of determining the *minimum* cardinality set of seed compounds necessary for the synthesis of a specific compound or a set of compounds. In particular, the latter set may comprise metabolic precursors that an organism requires for maintenance or growth. Therefore, solving this inverse problem may indicate minimal nutritional requirements that must be met to ensure sustenance of the organism. The nutrients which can be provided in synthesis are often restricted to a specific set, in which case we address the inverse problem with a forbidden set. In addition, we address the problem of finding the minimum cardinality set of seed compounds that are necessary for synthesis of a given set of compounds and, at the same time, guarantee that a specific set of compounds are not created as side products. This is the inverse problem with two forbidden sets.

The problems addressed here have applications that span various fields: In a sensor network with directed communications, one is interested in finding the minimum number of nodes that can be used for fast delivery of information. In the field of computational geometry, one may formulate the problem of determining the minimum number of flood-lights that can illuminate a given polygon [1], while in automated reasoning one may seek automated deduction with minimum number of axioms [5]. We note that none of the related variants has the constraint that all precursors must be present for an action to take place. **Contributions.** Here, we show that the inverse scope problems are NP-hard on general graphs and directed acyclic graphs (DAGs). We also demonstrate that the inverse scope problem with two forbidden sets on general graphs cannot be approximated within $n^{1/2-\epsilon}$ for any constant $\epsilon > 0$ unless P = NP. In addition, we discuss the practical implications of the hardness of approximation results.

2   Problem Definition

A *metabolic network* is typically represented by a directed bipartite graph $G = (V, E)$. The vertex set of G can be partitioned into two subsets: V_r, containing *reaction nodes*, and V_m, comprised of *metabolite nodes*, such that $V_r \cup V_m = V(G)$. The edges in $E(G)$ are directed either from a node $u \in V_m$ to a node $v \in V_r$, in which case the metabolite u is called a *substrate* of the reaction v, or from a node $v \in V_r$ to a node $u \in V_m$, when u is called a *product* of the reaction v.

The scope concept is related to reachability in the metabolic network graph G: A reaction node $v \in V_r$ is reachable if all of its substrates are reachable. Given a subset S of metabolite nodes, a node $u \in V_m$ is reachable either if $u \in S$ or if u is a product of a reachable reaction. With these clarifications, we can present a precise mathematical formulation for the scope of a set of seed compounds:

Definition 1. *Given a metabolic network $G = (V, E)$ and a set $S \subseteq V_m$, the scope of S, denoted by $R(S)$, is the set of all metabolite nodes reachable from S.*

For a given metabolic network $G = (V, E)$ and a set $S \subseteq V_m$, the scope $R(S)$ can be determined in polynomial time of the order $O(|E| \cdot |V|)$, as can be established by analyzing the following algorithm:

We define the inverse scope problem as follows:

INVERSE SCOPE (IS)
INSTANCE: Given a metabolic network $G = (V, E)$ and a subset of metabolites $P \subseteq V_m$.
PROBLEM: Find a subset of metabolites $S \subseteq V_m$ such that $P \subseteq R(S)$.
MEASURE (min): Cardinality of S.

Often, there is a restriction to the subset of metabolites from which we would like to identify S, the seed compounds synthesizing P. In that case, we address the inverse scope problem with a forbidden set $V(G) - S'$, such that $S \subseteq S'$, defined below:

INVERSE SCOPE WITH A FORBIDDEN SET (ISFS)
INSTANCE: Given a metabolic network $G = (V, E)$ and two subsets of metabolites $V(G) - S', P \subseteq V_m$, where $V(G) - S'$ is the forbidden set and P is the set of products.
PROBLEM: Find a subset of metabolites $S \subseteq S'$ such that $P \subseteq R(S)$.
MEASURE (min): Cardinality of S.

Algorithm 1. Scope for a set of seed metabolites S in a metabolic network G

 Input: $G = (V_m \cup V_r, E)$, metabolic network
 S, set of seed metabolites, $S \subseteq V_m$
 Output: $R(S)$, scope of S
1 mark all nodes in $V(G)$ **unreachable**
2 mark all nodes in V_r **unvisited**
3 mark all nodes in S **reachable**
4 **repeat**
5   **foreach** *node* $v \in V_r$ **do**
6   **if** v *is reachable* **then**
7   mark v as **reachable**
8   **end**
9   **end**
10   **if** *there is a **reachable unvisited** node* $v \in V_r$ **then**
11   mark v **visited**
12   mark successors of v **reachable**
13   **end**
14 **until** *no **reachable unvisited** nodes in* V_r
15 $R(S) \leftarrow$ all **reachable** nodes in V_m

It is interesting to also consider the problem of determining the set of nutrients to be provided in the synthesis of a given set of products while not yielding a pre-specified set of side products. The study of this problem may, for instance, indicate how to design a biochemical experiment to minimize the effect of some undesirable compounds. To ensure that a specific set of metabolites is not synthesized, we modify ISFS as follows:

> INVERSE SCOPE WITH TWO FORBIDDEN SETS (IS2FS)
> INSTANCE: Given a metabolic network $G = (V, E)$ and three subsets of metabolites $V(G) - S', F, P \subseteq V_m$, where $V(G) - S'$ and F are the forbidden sets and P is the set of products.
> PROBLEM: Find a subset of metabolites $S \subseteq S'$ such that $P \subseteq R(S)$ and $F \cap R(S) = \emptyset$.
> MEASURE (min): Cardinality of S.

Remark 1. Note that by taking $S' = V(G)$ in ISFS, every instance of IS becomes an instance of ISFS, and ISFS can be restricted to IS. In addition, every instance of IS2FS with $F = \emptyset$ is an instance of ISFS. Therefore, IS2FS is the most general of the three problems.

For completeness, we show the decision versions of the three problems defined above:

> INVERSE SCOPE DECISION (ISD)
> INSTANCE: Given a metabolic network $G = (V, E)$, subset of metabolites $P \subseteq V_m$, and an integer K.

PROBLEM: Does there exist a subset of metabolites $S \subseteq V_m$ such that $P \subseteq R(S)$ and $|S| \leq K$.

INVERSE SCOPE WITH A FORBIDDEN SET DECISION (ISFSD)
INSTANCE: Given a metabolic network $G = (V, E)$, two subsets of metabolites $V(G) - S', P \subseteq V_m$, where $V(G) - S'$ is the forbidden set and P is the set of products, and an integer K.
PROBLEM: Does there exist a subset of metabolites $S \subseteq S'$ such that $P \subseteq R(S)$ and $|S| \leq K$.

INVERSE SCOPE WITH TWO FORBIDDEN SETS DECISION (IS2FSD)
INSTANCE: Given a metabolic network $G = (V, E)$, three subsets of metabolites $V(G) - S', F, P \subseteq V_m$, where $V(G) - S'$ and F are the forbidden sets and P is the set of products, and an integer K.
PROBLEM: Does there exist a subset of metabolites $S \subseteq S'$ such that $P \subseteq R(S)$, $F \cap R(S) = \emptyset$, and $|S| \leq K$.

In the next section we present the results regarding the NP-hardness of the three inverse scope problems.

3   Hardness Results

An optimization problem Π is shown to be NP-hard by establishing a polynomial time reduction from a problem known to be NP-complete to the decision version of Π. First, we show that IS2FS is NP-hard on a general graph by providing a reduction from MINIMUM DISTINGUISHED ONES (MIN-DONES). We also show that ISFS is NP-hard even on DAGs by providing a reduction from the SET COVER (SC) problem. In a similar way, we show that IS, too, is NP-hard when restricted to DAGs. These results will later be used for obtaining the approximation results for the three problems.

Theorem 1. INVERSE SCOPE WITH TWO FORBIDDEN SETS DECISION *problem is NP-complete.*

Proof. First, we need to show that IS2FSD is in NP. Given an instance of IS2FSD with three subsets of nodes $S, F, P \subseteq V_m(G)$, one can find $R(S)$, by employing Algorithm 1, and check whether $P \subseteq R(S)$, $F \cap R(S) = \emptyset$ and $|S| \leq K$ in polynomial time.

Next, we provide a reduction from the MIN-ONESD problem. An instance of the decision version of MIN-ONES is given by a set of n variables Z, collection C of disjunctive clauses of 3 literals, and an integer K' (a literal is a variable or a negated variable in Z). The problem is then to find a truth assignment for Z that satisfies every clause in C such that the number of variables in Z that are set to true in the assignment is at most K'.

Given an instance of MIN-ONESD, we can construct an instance of IS2FSD, a bipartite directed graph $G = (V, E)$ with $V(G) = V_m \cup V_r$, three subsets of nodes $S', F, P \subseteq V_m(G)$, and an integer K as follows: For each variable $x_i \in Z$, we use the gadget shown in Figure 1. The gadget is composed of six nodes $y_{i,1}^T$, $y_{i,2}^T$, x_i^T, x_i^F, p_i, and f_i connected through four reactions—r_i^1 with x_i^T and x_i^F as substrates and f_i as a product, r_i^2 with $y_{i,1}^T$ and $y_{i,2}^T$ as substrates and x_i^T as a product, r_i^3 with x_i^T as substrate and p_i as a product, and r_i^4 with x_i^F as substrate and p_i as a product. Moreover, for each clause in C we add a node c_j. A node x_i^T is connected to c_j via a reaction if variable x_i appears non-negated in c_j; similarly, a node x_i^F is connected to c_j via a reaction if variable x_i appears negated in c_j. Finally, we let S' be composed of all $y_{i,1}^T$, $y_{i,2}^T$, and x_i^F nodes, P be composed of all c_j and p_i nodes, while F be comprised of all f_i nodes.

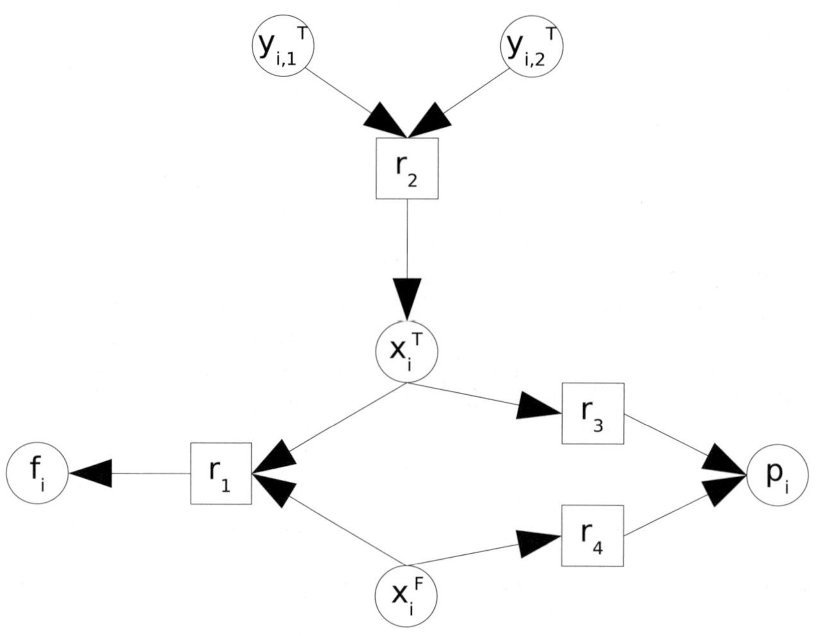

Fig. 1. Gadget for the construction of IS2FS$_D$ instance

Note that x_i^T is reached if and only if its two corresponding nodes $y_{i,1}^T$ and $y_{i,2}^T$ are included as substrates. Moreover, the inclusion of p_i nodes in P and f_i nodes in F ensures that exactly one of the x_i^T and x_i^F is chosen. Therefore, to complete the construction, we set $K = n + K'$.

A solution to MIN-ONES$_D$ can be transformed to a solution of IS2FS$_D$ of cardinality $K = n + K'$ by taking those nodes $y_{i,1}^T$ and $y_{i,2}^T$ in S' whose corresponding variable x_i is set to TRUE in the solution of MIN-ONES$_D$. Moreover, the solution also includes the x_i^F whose value can be set to FALSE. All c_j nodes can be reached, since a solution to MIN-ONES$_D$ guarantees that at least one literal in the clause c_j has value TRUE. Similarly, all p_i nodes are also reached.

Moreover, a valid truth assignment guarantees that no pair x_i^T and x_i^F is in the solution of MIN-ONESD; thus, the nodes in F cannot be accessed.

Given a solution S to IS2FSD on G, the solution to MIN-ONESD can be obtained by assigning value TRUE to that variable x_i in Z whose corresponding nodes $y_{i,1}^T$ and $y_{i,2}^T$ are in S; the remaining variables are assigned value FALSE. Since all c_j nodes are reachable, then each of them has at least one directed path from a node in S and thus the value of the corresponding clause is TRUE. Moreover, since no node in F is in the scope of S, the reconstructed truth assignment is valid.

Since IS2FSD can be solved if and only if there is a solution to MIN-ONESD, we have the NP-completeness of the problem in the theorem.

Corollary 1. *The* INVERSE SCOPE WITH TWO FORBIDDEN SETS *problem is NP-hard on general graphs.*

For the inverse scope with a forbidden set we have:

Theorem 2. INVERSE SCOPE WITH A FORBIDDEN SET DECISION *is NP-complete even on DAGs.*

Proof. The decision version of ISFS is in NP since for any given set of metabolites S, we can find $R(S)$, by using Algorithm 1, and check whether $P \subseteq R(S)$ in polynomial time.

We provide a polynomial time reduction from the SCD problem: An instance of the SCD problem is given by a collection C of subsets from a finite set U and an integer K'. The problem then is to determine whether there is a set cover for U of cardinality at most K', *i.e.* a subset $C' \subseteq C$ such that every element of U belongs to at least one member of C' and $|C'| \leq K'$.

Given an instance of the SCD problem we design an instance of the ISFSD problem as follows: First, we build the metabolic network G which must be bipartite. Let the number of subsets in the collection C be denoted by p. For every subset $C_i \in C$ we create a reaction node r_i, $1 \leq i \leq p$; thus, we have p reaction nodes. Let the number of elements in U be denoted by n. For every element $x_j \in U$ we create a metabolite node, denoted by x_j, $1 \leq j \leq n$. Furthermore, we create p additional metabolite nodes, denoted by y_i, $1 \leq i \leq p$. Therefore, $V_r = \{r_i \mid 1 \leq i \leq p\}$ and $V_m = \{x_j \mid 1 \leq j \leq n\} \cup \{y_i \mid 1 \leq i \leq p\}$. Finally, we set $P = \{x_j \mid 1 \leq j \leq n\}$.

A reaction r_i is connected via a directed edge to a metabolite x_j if and only if the subset C_i corresponding to the node r_i contains the element x represent by the node x_j. Additionally, we include a directed edge from y_i to r_i, $1 \leq i \leq n$.

Finally, we set $S' = \{y_i \mid 1 \leq i \leq p\}$ and let the integer K of the decision version of the ISFS problem be equal to K'. This construction can be completed in time polynomial in the size of the SC instance. The construction is illustrated in Figure 2.

If we have a solution to the SC problem and it is given by a subset C', $|C'| \leq K'$ then the solution to the ISFS problem, the subset $S \subseteq S'$, is comprised of the nodes in $\{y_i \mid 1 \leq i \leq p\}$ that are connected via a directed edge to reaction nodes r_i representing the subsets in C', since $R(S) = P$.

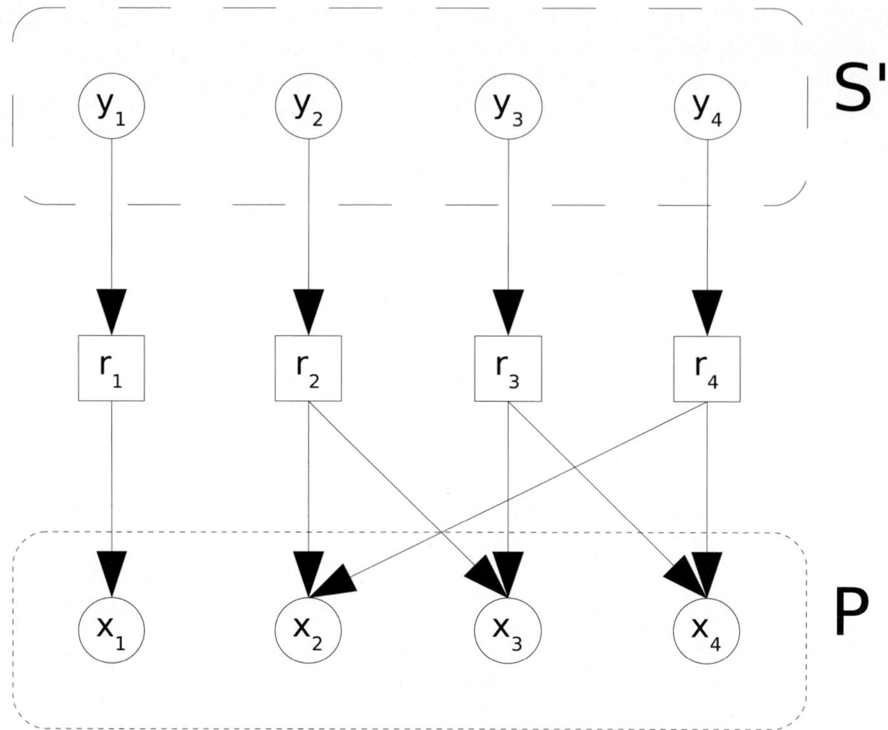

Fig. 2. Instance of ISFS obtained from the instance of SC with $U = \{1, 2, 3, 4\}$ and $C = \{\{1\}, \{2, 3\}, \{3, 4\}, \{2, 4\}\}$

Conversely, if we have a solution to the ISFS problem, *i.e.*, a subset $S \subset S'$, $|S| \leq K$ then R(S) = P. The solution to the SC problem, a subset $C' \subseteq C$, can be obtained by including those elements $C_i \in C$ corresponding to the reaction nodes r_i to which there exists an edge from $y_i \in S$ in G.

Since the ISFS problem can be solved if and only if there is a solution to the SC problem, we have the NP-completeness of the problem in the theorem. Furthermore, the polynomial time construction results in an acyclic directed graphs (DAGs), so ISFSD is NP-complete on DAGs.

We then have the following corollary:

Corollary 2. *The* INVERSE SCOPE WITH A FORBIDDEN SET *problem is NP-hard even on DAGs.*

We can use a similar approach as in the proof of Theorem 2 to obtain the following result:

Theorem 3. *The* INVERSE SCOPE *problem is NP-hard even when restricted to DAGs.*

Proof. The ISD problem is in NP even when restricted to DAGs: For any given set of metabolites S and an integer K, we can find $R(S)$, by using Algorithm 1, and check whether $P \subseteq R(S)$ and $|S| \leq K$ in polynomial time.

Given an instance of the SCD problem we design an instance of the ISD problem, a graph G and an integer K, on DAGs as follows: First, we build the metabolic network G which must be bipartite. Let the number of subsets in the collection C be denoted by p. For every subset $C_i \in C$ we create a metabolite node y_i; thus, we have p metabolite nodes from this step of the construction. Let the number of elements in U be denoted by n. For every element $x_j \in U$ we create a metabolite node, denoted by x_j^1, $1 \leq j \leq n$.

A metabolite y_i is connected via a directed path of length 2 with a middle reaction node r_i to a metabolite x_j^1 if and only if the subset C_i, corresponding to the node y_i, contains the element x represented by the node x_j. Additionally, we include $p - 1$ copies of each node x_j^1, denoted by x_j^i, $2 \leq i \leq p$, and connect them to the in-neighbors of x_j^1. Finally, we let P contain all x_j^i nodes and $K' = K$. We increase the number of elements per set to ensure that only nodes among y_i are chosen as a solution to ISD, so that it can be transformed to a solution of SCD.

If we have a solution to the SCD problem and it is given by a subset C', $|C'| \leq K'$ then the solution to the ISD problem, the subset $S \subseteq V_m$, is comprised of the nodes in $\{y_i \mid 1 \leq i \leq p\}$ representing the subsets in C', since $R(S) = P$.

Conversely, if we have a solution to the ISD problem, *i.e.*, a subset $S \subseteq V_m$, $|S| \leq K$, then $R(S) = P$. The set C' can be obtained in the following way: Let S contains a node u from $\{x_j^i \mid 1 \leq i \leq p, 1 \leq j \leq n\}$. There are two cases: If u can be reached by some y_i, then a solution to SCD excludes this element from S. If u cannot be reached by some y_i, then none of its remaining $p - 1$ copies can be reached ($p \geq 2$). Including one y_i representing a set that contains u can always decrease the cardinality of S by a at least two. The solution to the SCD problem, a subset $C' \subseteq C$, therefore includes those elements $C_i \in C$ corresponding to the elements in S as well as the nodes added by the algorithm to cover nodes from $\{x_j^i \mid 1 \leq i \leq p, 1 \leq j \leq n\}$ which are initially included in S but are removed by the previous algorithm. It follows that $|C'| \leq K'$.

Since the ISD problem can be solved if and only if there is a solution to the SCD problem, we have the NP-completeness of the problem in the theorem. Furthermore, the polynomial time construction results in an acyclic directed graphs (DAGs), so ISD is NP-complete on DAGs.

4   Approximation Results

Let us recall a few definitions about approximability. Given an instance x of an optimization problem A and a feasible solution y of x, we denote by $m(x, y)$ the value of the solution y, and by $opt_A(x)$ the value of an optimum solution of x. Here, we consider minimization problems. The performance ratio of the solution y for an instance x of a minimization problem A is

$$R(x, y) = \frac{m(x, y)}{opt_A(x)}.$$

For a constant $\rho > 1$, an algorithm is a ρ-approximation if for any instance x of the problem it returns a solution y such that $R(x, y) \leq \rho$. We say that an optimization problem is constant approximable if, for some $\rho > 1$, there exists a polynomial-time ρ-approximation for it. APX is the class of optimization problems that are constant approximable. An optimization problem has a polynomial-time approximation scheme (a PTAS, for short) if, for every constant $\epsilon > 0$, there exists a polynomial-time $(1 + \epsilon)$-approximation for it.

L-reduction was introduced as a transformation of optimization problems which keeps the approximability features [17]. L-reductions in studies of approximability of optimization problems play a similar role to that of polynomial reductions in the studies of computational complexity of decision problems. For completeness we include the following definition:

Definition 2. *Let A and B be two optimization problems. Then A is said to be L-reducible to B if there are two constants $\alpha, \beta > 0$ such that:*

1. *there exists a function, computable in polynomial time, which transforms each instance x of A into an instance x' of B such that $opt_B(x') \leq \alpha \cdot opt_A(x)$,*
2. *there exists a function, computable in polynomial time, which transforms each solution y' of x' into a solution y of x such that $|m(x, y) - opt_A(x)| \leq \beta \cdot |m(x', y') - opt_B(x')|$.*

Remark 2. This reduction preserves PTAS, *i.e.*, if A is L-reducible to B and B has a PTAS then A has a PTAS as well.

Remark 3. From the above, if δ is a lower bound of the worst-case approximation factor of A, then $\rho = \frac{\delta}{\alpha \cdot \beta}$ is a lower bound of the worst-case relative error of B.

We employ L-reduction to obtain results about the lower bound of the worst-case approximation factor for IS2FS.

Theorem 4. INVERSE SCOPE WITH TWO FORBIDDEN SETS *on a graph G with n nodes cannot be approximated to within a factor of $n^{1/2-\epsilon}$ in polynomial time for any constant $\epsilon > 0$, unless $P = NP$.*

Proof. To construct an L-reduction, we first choose A to be MIN-ONES and B, IS2FS. Jonsson [14] has shown that MIN-ONES is NPO PB-complete, and is not approximable within $|Z|^{1/2-\epsilon}$ for any $\epsilon > 0$. Given an instance x of MIN-ONES, we construct an instance x' of IS2FS the same as in Theorem 1. From the proof of Theorem 1, we have $opt_B(x') = n + opt_A(x)$, so $opt_B(x) \leq opt_A(x')$ and $\alpha = 1$. Moreover:

$$|m(x, y) - opt_A(x)| \leq |m(x', y') - opt_B(x')|,$$

so $\beta = 1$. Thus, we have $n^{1/2-\epsilon}$ for the lower bound of the worst-case approximation factor of IS2FS by Remark 3.

Hendorf *et al.* [10] developed a heuristic for finding minimal sets of seed compounds from which metabolites that are essential for cellular maintenance can be produced. As every minimum set of seed compounds is also minimal, the heuristic can approximate the IS problem. The heuristic takes as input an ordered list of all metabolites in a given metabolic network and a set of target metabolites. It then continually removes a metabolite from the beginning of the list, while recalculating the scope of the remainder of the list. If the resulting scope does not contain the full target set, the metabolite is inserted back in the list; otherwise, it remains permanently removed. Clearly, the set of metabolites contained in the list after the exhaustive search represents a minimal seed, as the removal of any metabolite would result in a scope that does not contain all target metabolites. Since different orderings of the list may result in a different minimal set of seed metabolites, it is not known how well this heuristic approximates the IS problem. It remains as an open problem to develop provably good approximation algorithms for all of the addressed inverse scope problems.

5   Instances of IS and ISFS in P

IS and ISFS are solvable in polynomial time on trees. In a tree metabolic network, each metabolite, other than the root, is a product of a reaction with only one substrate. In other words, a tree metabolic network is a tree rooted in a metabolite node. Given a tree metabolic network T and a node u, let S_u be the set of predecessors of u. Given two sets $S', P \subseteq V_m(T)$, the solution to ISFS on T is given by the following greedy algorithm:

Algorithm 2. Algorithm for ISFS on a tree

    Input: $T = (V_m \cup V_r, E)$, metabolic tree network
    $P \subseteq V_m$, set of metabolites
    $S' \subseteq V_m$, the set to choose seeds
    Output: S, $S \subseteq S'$, such that $P \subseteq R(S)$
1 **foreach** *node $u \in P$* **do**
2 | find the set S_u
3 **end**
4 $S \leftarrow \emptyset$
5 $L \leftarrow$ order list of nodes from P
6 **while** *there is a node $u \in L$* **do**
7 | **while** *there is node $v \neq u$, $v \in L$ with $S_v \cap S_u \cap S' \neq \emptyset$* **do**
8 | | $S_u \leftarrow S_u \cap S_v \cap S'$ remove v from L
9 | **end**
10 | $S \leftarrow S \cup$ last common ancestor in S_u
11 | remove u from L
12 **end**
13 output S

The algorithm works by finding the last common ancestor for as large a subset of P as possible (lines $6 - 12$). Since this subset is reachable from one node only, its cardinality cannot be decreased and the algorithm is optimal.

Given two directed graphs G and H, the Cartesian product $G \square H$ is a graph with node set $V(G) \times V(H)$ such that there is an edge $\{(u_1, v_1), (u_2, v_2)\} \in E(G \square H)$ if and only if: (1) $u_1 = u_2$ and $\{v_1, v_2\} \in E(H)$ or (2) $v_1 = v_2$ and $\{u_1, u_2\} \in E(G)$.

Given a tree metabolic network T in which each reaction has precisely one substrate and one product, let $\tilde{T}$ be the tree obtained by the following steps: (1) for each reaction node, connect the substrate with all the products, (2) remove reaction nodes. Note that $\tilde{T}$ includes only the metabolite nodes of T. These tree will be called *reduced*.

Given a directed graph G, let G_s be the graph in which each edge of G is subdivided (while keeping the direction of the edge). The nodes used in the subdivision can be treated as reaction nodes, and all nodes in $V(G)$ as metabolite nodes.

If there is a graph G which can be represented as subdivision of the directed product of $\tilde{T}_1$ and $\tilde{T}_2$ in the instance of a ISFS with a set P, such that ISFS has a non-empty solution on T_1 and T_2, then:

$$S((\tilde{T}_1 \square \tilde{T}_2)_s) = \min \{S(T_1), S(T_2)\},$$

where $S(T_1)$ is a solution to ISFS with P mapped onto $V_m(T_1)$, and $S(T_2)$ is a solution to ISFS with P mapped onto $V_m(T_2)$. Therefore, the problem is polynomially solvable if G can be obtained by subdividing the Cartesian product of two reduced tree metabolic networks.

We anticipate that similar constructions may lead to ways of decomposing metabolic networks into smaller parts on which the inverse scope problems may be polynomially solvable. However, we leave this as an open problem and a direction for future research.

6   Discussion

The inverse scope problem discussed here is of great importance for biological research since its solution allows to computationally predict minimal nutrient requirements for the cultivation of organisms or to identify cost efficient combinations of substrates for biotechnological applications.

The hardness of approximation results obtained in Section 4 bear some important implications to the application of the inverse scope problems. For a general graph, we show that IS2FS cannot be approximated within $n^{1/2-\epsilon}$ for any constant $\epsilon > 0$ unless P = NP. The hardness of approximation and parameterized complexity of IS and ISFS on general graphs remain as open problems. Our conjecture is that their complexity on general graphs strongly depends on the existence of directed cycles in the given metabolic network.

Our results imply that divising an efficient approximation algorithm will depend on finding biologically meaningful metabolic networks representations with no or a small number of cycles. Analyses of real metabolic networks have demonstrated the abundance of directed cycles, which result in high clustering and small average path length [13]. Furthermore, one may observe that the directed cycles are predominantly induced on the ubiquitous compounds, such as ATP and NADH. Under physiological conditions, a cell maintains such substances at rather constant levels guaranteeing their availability to the many processes in which they are required. It is therefore unrealistic to assume that these compounds have to be produced a priori. This allows for an alteration of the network structure reflecting that ubiquitous compounds are always available, while still describing the biochemical capabilities of the considered organism. Such a reduction will considerably reduce the number of cycles. Another type of cylces results from the representation of a metabolic network as bipartite graph, in which a reversible reaction is represented by two reaction nodes which are connected to an identical set of reactants with directions of all corresponding edges reversed. This results in cycles of length four for each reversible reaction. To remove such cycles without altering the biochemical capabilities of the network is more challenging. A possible approach is to study networks with flux balance analysis to identify those reactions which under physiological conditions always proceed in one direction. We are currently working on applying our findings to metabolic networks obtained from the KEGG database.

In addition, our analysis demonstrates that the concerted interrelation of biochemical processes responsible for efficient systematic adjustment of an organism to changing environmental conditions are indeed complex and not yet well-understood.

References

1. Bagga, J., Gewali, L., Glasser, D.: The complexity of illuminating polygons by alpha-flood-lights. In: Proceedings of the 8th Canadian Conference on Computational Geometry, pp. 337–342 (1996)
2. Bonarius, H.P.J., Schmid, G., Tramper, J.: Flux analysis of underdetermined metabolic networks: The quest for the missing constraints. Trends Biotechnology 15, 308–314 (1997)
3. Burchardt, G., Ingram, L.O.: Conversion of xylan to ethanol by ethanologenetic strains of escherichia coli and klebsiella oxytopa. Appl. Environ. Microbiol. 58, 1128–1133 (1992)
4. Burton, S.G., Cowan, D.A., Woodley, J.M.: The search for ideal biocatalyst. Nature Biotechnology 20, 37–45 (2002)
5. Duffy, A.D.: Principles of automated theorem proving. Wiley & Sons, Chichester (1991)
6. Ebenhöh, O., Handorf, T., Heinrich, R.: Structural analysis of expanding metabolic networks. Genome Informatics 15, 35–45 (2004)
7. Ebenhöh, O., Handorf, T., Heinrich, R.: A cross species comparison of metabolic network functions. Genome Informatics 16(1), 203–213 (2005)
8. Ebenhöh, O., Handorf, T., Kahn, D.: Evolutionary changes of metabolic networks and their biosynthetic capacities. IEE Proceedings in Systems Biology 153(5), 354–358 (2006)

9. Ebenhöh, O., Liebermeister, W.: Structural analysis of expressed metabolic subnetworks. Genome Informatics 17(1), 163–172 (2006)
10. Handorf, T., Christian, N., Ebenhöh, O., Kahn, D.: An environmental perspective on metabolism. Journal of Theoretical Biology (in press, 2007)
11. Handorf, T., Ebenhöh, O., Heinrich, R.: Expanding metabolic networks: scopes of compounds, robustness, and evolution. Journal of Molecular Evolution 61(4), 498–512 (2005)
12. Handorf, T., Ebenhöh, O., Kahn, D., Heinrich, R.: Hierarchy of metabolic compunds based on synthesizing capacity. IEE Proceedings in Systems Biology 153(5), 359–363 (2006)
13. Jeong, H., Tombor, B., Albert, R., Oltvai, Z.N., Barabási, A.L.: The large-scale organization of metabolic networks. Nature 407(6804), 651–654 (2000)
14. Jonsson, P.: Near-optimal nonapproximability results for some npo pb-complete problems. Information Processing Letters 68, 249–253 (1997)
15. Kompala, D.S., Ramkrishna, D., Jansen, N.B., Tsao, G.T.: Cybernetic modeling of microbial growth on multiple substrates. Biotechnology and Bioengineering 28, 1044–1056 (1986)
16. Nissen, T.L., Kielland-Brandt, M.C., Nielsen, J., Villadsen, J.: Optimization of ethanol production in saccharomyces cerevisiae by metabolic engineering of the ammonium assimilation. Metab. Eng. 2, 69–77 (2000)
17. Papadimitriou, C., Yannakakis, M.: Optimization, approximation and complexity classes. Journal of Computer and System Science 43, 425–440 (1991)
18. Savageau, M.A.: Biochemical systems analysis. i. some mathematical properties of the rate law for the component enzymatic reactions. Journal of Theoretical Biology 25, 365–369 (1969)
19. Schilling, C.H., Schuster, S., Palsson, B.O., Heinrich, R.: Metabolic pathway analysis: Basic concepts and scientific applications in the post-genomic era. Biotechnol. Prog. 15, 296–303 (1999)
20. Tsantilil, I.C., Karim, M.N., Klapal, M.I.: Quantifying the metabolic capabilities of engineered zymomonas mobilis using linear programming analysis. Microbial Cell Factories 6(8) (2007)
21. Wildermuth, M.C.: Metabolic control analysis: biological applications and insights. Genome Biology 1, 1031.1–1031.5 (2000)

Rapid Neighbour-Joining

Martin Simonsen, Thomas Mailund, and Christian N.S. Pedersen

Bioinformatics Research Center (BIRC), University of Aarhus,
C. F. Møllers Allé, Building 1110, DK-8000 Århus C, Denmark
{kejseren,mailund,cstorm}@birc.au.dk

Abstract. The neighbour-joining method reconstructs phylogenies by iteratively joining pairs of nodes until a single node remains. The criterion for which pair of nodes to merge is based on both the distance between the pair and the average distance to the rest of the nodes. In this paper, we present a new search strategy for the optimisation criteria used for selecting the next pair to merge and we show empirically that the new search strategy is superior to other state-of-the-art neighbour-joining implementations.

1   Introduction

The neighbour-joining method by Saitou and Nei [8] is a widely used method for phylogenetic reconstruction, made popular by a combination of computational efficiency combined with reasonable accuracy. With its cubic running time by Studier and Kepler [11], the method scales to hundreds of species, and while it is usually possible to infer phylogenies with thousands of species, tens or hundreds of thousands of species is infeasible. Various approaches have been taken to improve the running time for neighbour-joining. QuickTree [5] was an attempt to improve performance by making an efficient implementation. It outperformed the existing implementations but still performs as a $O\left(n^3\right)$ time algorithm. Quick-Join [6,7], instead, uses an advanced search heuristic when searching for the pair of nodes to join. Where the straight-forward search takes time $O\left(n^2\right)$ per join, QuickJoin on average reduces this to $O\left(n\right)$ and can reconstruct large trees in a small fraction of the time needed by QuickTree. The worst-case time complexity, however, remains $O\left(n^3\right)$, and due to a rather large overhead QuickJoin cannot compete with QuickTree on small data sets. Other approaches, such as "relaxed neighbour-joining" [3,10] and "fast neighbour-joining" [1] modify the optimisation criteria used when selecting pairs to join. The method "relaxed neighbour-joining" has a worst-case $O\left(n^3\right)$ running time while "fast neighbour-joining" has $O\left(n^2\right)$ running time.

In this paper we introduce a new algorithm, RapidNJ, to lower the computing time of canonical neighbour-joining. We improve the performance by speeding up the search for the pair of nodes to join, while still using the same optimisation criteria as the original neighbour-joining method. Worst-case running time remains $O\left(n^3\right)$, but we present experiments showing that our algorithm outperforms both QuickTree, QuickJoin and an implementation of relaxed neighbour-joining on all input sizes.

K.A. Crandall and J. Lagergren (Eds.): WABI 2008, LNBI 5251, pp. 113–122, 2008.
© Springer-Verlag Berlin Heidelberg 2008

2   Canonical Neighbour-Joining

Neighbour-joining [8,11] is a hierarchical clustering algorithm. It takes a distance matrix D as input, where $D(i,j)$ is the distance between cluster i and j. It then iteratively joins clusters by using a greedy algorithm, which minimises the total sum of branch lengths in the reconstructed tree. Basically the algorithm uses n iterations, where two clusters (i,j) are selected and joined into a new cluster. The two clusters are selected by minimising

$$Q(i,j) = D(i,j) - u(i) - u(j)\,, \tag{1}$$

where

$$u(l) = \sum_{k=0}^{r-1} D(l,k)/(r-2)\,, \tag{2}$$

and r is the number of remaining clusters. When the minimum Q-value $q_{\min} = \min_{0 \le i,j < r} Q(i,j)$ is found, D is updated, by removing the i'th and j'th row and column. A new row and column are inserted with the distances of the new cluster. Distance between the new cluster $a = i \cup j$ and an old cluster k, are calculated as

$$D(a,k) = \frac{D(i,k) + D(j,k) - D(i,j)}{2}\,. \tag{3}$$

The result of the algorithm is a unrooted bifurcating tree where each initial cluster corresponds to a leaf and each join creates an internal node. Finding the pair of clusters to join in each round takes time $O\left(n^2\right)$. The running time of canonical neighbour-joining thus becomes $O\left(n^3\right)$.

3   Rapid Neighbour-Joining

We seek to improve the performance of canonical neighbour-joining by speeding up the search for the pair of clusters to join, while still using the same optimisation criteria as the canonical neighbour-joining method. The overall aim is thus similar to that of QuickJoin, but the approach is different. The RapidNJ algorithm presented in this paper is based on the following observation.

– When searching for $q_{\min}$ in (1), $u(i)$ is constant in the context of row i.

   This observation can be used to create a simple upper bound on the values of each row in Q, thereby reducing the search space significantly. The upper bound is dynamically updated based on the Q-values searched. While the algorithm still has a worst-case running time of $O\left(n^3\right)$, our experiments, reported later, show that in practise the algorithm performs much better. To utilize the upper bound, two new data structures, S and I, are used. S is a sorted representation of D, and I maps S to D. Memory consumption is increased due to S and I. The actual increase depends on implementation choices, and can be minimised at the cost of speed. Worst case memory consumption remains $O\left(n^2\right)$.

3.1   Data Structures

The two new data structures, S and I, needed to utilize our upper bound, are constructed as follows. Matrix S contains the distances from D but with each row sorted in increasing order. Matrix I maps the ordering in S back to positions in D. Let $o_1, o_2, \ldots, o_n$ be a permutation of $1, 2, \ldots, n$ such that $D(i, o_1) \leq D(i, o_2) \leq \cdots \leq D(i, o_n)$, then

$$S(i, j) = D(i, o_j), \tag{4}$$

and

$$I(i, o_j) = j. \tag{5}$$

3.2   Search Heuristic

Matrix S is used for a bounded search of $q_{\min}$. First the maximum u-value $u_{\max}$ needs to be found. Recalculating all u-values, and finding the $u_{\max}$ can be done in time $O(n)$. The following algorithm can then be used to search for the pair (i, j) minimising $Q(i, j)$:

1. Set $q_{\min} = \infty$, $i = -1$, $j = -1$
2. for each row r in S and column c in r:
    (a) if $S(r, c) - u(r) - u_{\max} > q_{\min}$ then move to the next row.
    (b) if $Q(r, I(r, c)) < q_{\min}$ then set $q_{\min} = Q(r, I(r, c))$, $i = r$ and $j = I(r, c)$.

The algorithm searches each row in S and stops searching a row when the condition

$$S(r, c) - u(r) - u_{\max} > q_{\min} \tag{6}$$

becomes true or the end of a row is reached. If we reached an entry in a row where (6) is true, we are looking at a pair (i, j), where $D(i, j)$ is too large for (i, j) to be a candidate for $q_{\min}$, and the following entries in the row $S(i)$, can be disregarded in the search. This is easily seen by remembering that $u(i)$ is constant in context of a row, $u(j) = u_{\max}$ in (6) and $S(r, k) \geq S(r, l)$ when $k > l$. Whenever we see a new $Q(r, I(r, c))$ smaller than the previously smallest, we remember the entry, using I to obtain the right column index in D. Q is never fully calculated as this would require $O(n^2)$ time in each iteration. We only calculate entries as they are needed.

The number of entries searched in each row depends on the current value of the $q_{\min}$ variable. If $q_{\min}$ is close to the actual minimum value, more entries in a row can most likely be excluded from the search. To improve the overall performance, it is therefore important quickly to find a value for $q_{\min}$ as close as possible to the actual minimum. We propose two strategies for this:

- The first entry in each row can be searched for a good minimum. Intuitively $S(r, 0)$ is likely to have a small Q-value compared to entries in the rest of row i.

– Cache one or more rows containing good minimums in each iteration, and use these rows as a primer in the next iteration. The u-values do not change very much in each iteration, so a row with a small q value is likely to have a small q value in the succeeding iteration. Quite often, the minimum q value can be found in the row containing the second best minimum found in the previous iteration.

Since we can risk searching all entries in every row, the worst-case running time of the algorithm is $O\left(n^3\right)$.

3.3   Updating the Data Structures

The D matrix is updated as in the original neighbour-joining algorithm. Two rows and two columns, corresponding to the two clusters we merged, are removed. One row and column are inserted corresponding to the new cluster.

The S and I matrices need to be updated to reflect the changes made in the D matrix. Removing the two rows can easily be done, but because the rows of S are sorted, removing the columns is more complicated. Identifying and removing entries in S corresponding to the two columns removed in D, would require at least $O\left(n\right)$ time depending on implementation of S. By keeping a record of deleted columns, we can simply leave the obsolete entries in S. Identifying obsolete entries while searching S gives a small performance penalty. The penalty can be minimised by garbage collecting obsolete entries in S, when they take up too much space. Our experiments have shown that most rows in S are removed, before obsolete entries become a problem.

Let d be the number of remaining clusters after merging the cluster-pair (i,j), into a new cluster. A new row with length $d-1$, and sorted by increasing distance is inserted in S. The new row contains all new distances introduced by merging (i,j). A row which maps the new row in S to the corresponding row in D is created and inserted in I. Due to the time required for sorting, creating and inserting the two rows in S and I takes $O\left(d\log d\right)$ time.

4   Results and Discussion

To evaluate the performance of our search heuristic, we have implemented the RapidNJ algorithm using the second of the two strategies mentioned above to initialise $q_{\min}$ with a good value. This strategy can be implemented with a small overhead and it often initialises $q_{\min}$ with the actual minimum q value. RapidNJ only caches the row containing the q value closest to $q_{\min}$ when the row is not one of the two rows which are associated with $q_{\min}$. This row is searched before any other row thereby initialising $q_{\min}$. The current implementation of RapidNJ is available at http://www.birc.au.dk/Software/RapidNJ/.

QuickJoin [7], QuickTree [5] and Clearcut [10] have been used as reference in our experiments. We have not been able to locate an implementation of FastNJ [1] to include in our experiments. Clearcut is an implementation of relaxed neighbour-joining, while QuickJoin and QuickTree implements the canonical NJ method. QuickJoin and QuickTree are both fast implementations and use different heuristics to optimize the running time of the canonical neighbour-joining method. Clearcut uses a different algorithm, but it can reconstruct trees with almost the same accuracy as canonical neighbour-joining. Phylip formatted distance matrices were used as input, and Newick formatted phylogenetic trees as output.

4.1   Environment

All experiments were preformed on a machine with the following specifications:

- Intel Core 2 6600 2.4 GHz with 4 MB cache
- 2 GB RAM
- Fedora 6, kernel 2.6.22.9-61 OS

QuickTree and Clearcut were written in C while QuickJoin and RapidNJ were written in $C++$. All reference tools were compiled with the `make` script included in the source code. We used the standard `time` tool available in the Fedora 6 distribution for measurements of running time. The "real time" output of the `time` tool was used in all measurements. Three runs of each program on every input was made, and the best time of the three runs was used in order to avoid disturbance from other processes on the system.

4.2   Data

All four implementations were given distance matrices in phylip format as input. We used the following two sources of data to evaluate the performance of the four implementations.

The first data source was protein sequence alignments from Pfam [4]. The alignments were translated into phylip formatted distance matrices using Quick-Tree. Data sets from Pfam are real data, and we found that most data sets contained lots of redundant data, i.e. the distance matrix contains rows i and k where $\forall j : D(i, j) = D(k, j)$. We also found several data sets where the same taxa was represented twice. Apart from representing real data, Pfam data sets also test how resilient the different optimisations and algorithms used in the four implementations, are to irregular inputs. Figures 1, 2 and 3 show the results of the experiments on these data.

The second data source was based on simulated phylogenetic trees. The trees were generated using r8s [9], which is a tool for simulating molecular evolution on phylogenetic trees. As a feature r8s can also simulate random phylogenetic trees. We used the yule-c model to generate the data used in the experiments. Simulated data sets provide a good basis for testing the algorithmic complexity of the four implementations. They contain no redundant data, and the distance matrices are always additive. Since the trees generated by r8s using the yule-c model are clocklike, the distance matrices are actually ultrametric. This, how-ever, should not yield an advantage for any of the methods, so in this study, we do not make any effort to e.g. perturbe the branch lengths for obtaining

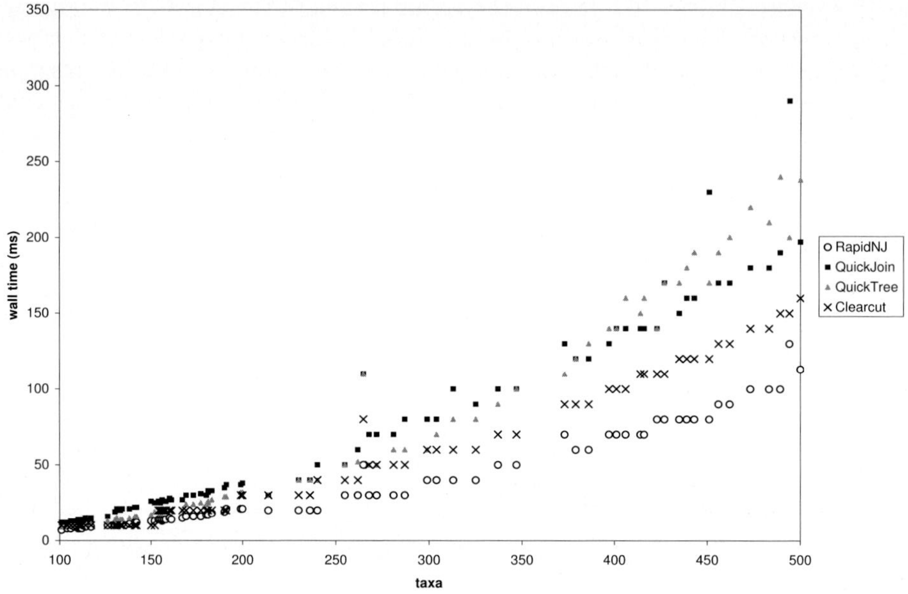

Fig. 1. Performance of RapidNJ compared to QuickJoin, QuickTree and Clearcut on small Pfam data (< 500 taxa)

non-ultrametric but additive distance matrices. Figure 5 shows the result of the experiments on simulated data sets.

Experiments on data sets larger than 11000 taxa were not performed due to the memory requirements of RapidNJ and QuickJoin. Both have $O\left(n^2\right)$ memory consumption, but with larger constants than Clearcut and QuickTree.

4.3   Results on Pfam Data

On small inputs, see Fig. 1, all four implementations have good performance, only minor differences separates the implementations. QuickJoin does suffer from a large overhead, and has the longest running time, while RapidNJ has the shortest. When we look at medium sized data sets (Fig. 2), QuickJoin starts to benefit from the heuristics used. Again we observe that RapidNJ has a consistently short running time compared to both QuickTree and QuickJoin. Clearcut comes quite close to RapidNJs running time, and only a small constant factor separates the two. QuickTree begin to suffer from the $O\left(n^3\right)$ time complexity on these input sizes, while the other three implementations have a much lower running time. Looking at large input sizes, see Fig. 3, QuickTree clearly suffers from its $O\left(n^3\right)$ time complexity, while Clearcut, QuickJoin and RapidNJ continue to perform much better. On most data sets RapidNJ has half the running time of QuickJoin, while the running times of Clearcut is nearly identical to those of RapidNJ on most data sets.

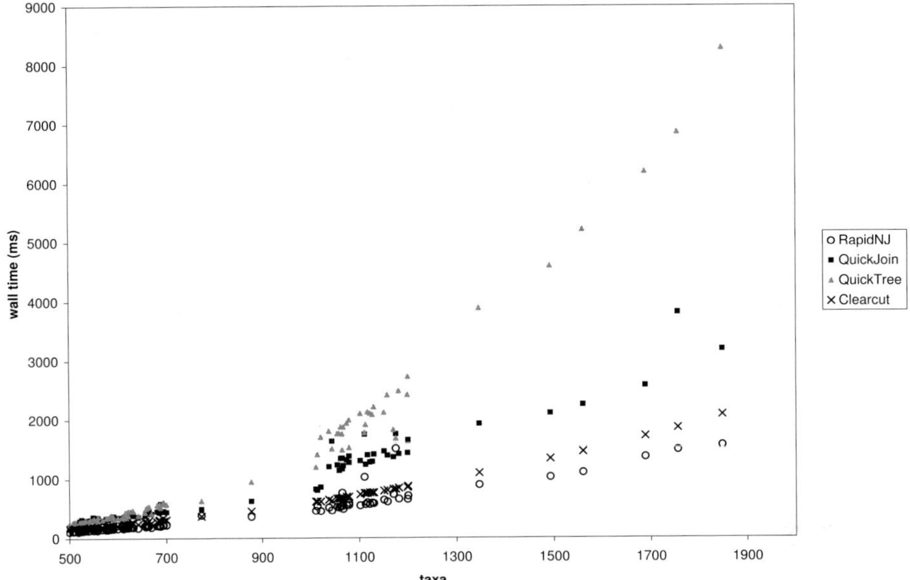

Fig. 2. Performance of RapidNJ compared to QuickJoin, QuickTree and Clearcut on medium Pfam data (500–2000 taxa)

Some outliers belonging to all four implementations can be observed in Fig. 1, 2 and 3. The most noticeable are observed on medium and large data sets, where the running time of RapidNJ on some data sets of approximately the same size differs quite a lot. We have investigated the problem, and found that it to some extent can be explained by redundant data. Redundant data are taxa with equal distances to all other taxa and a mutual distance of 0. They are quite common in Pfam data sets and affect the performance of RapidNJ negatively. In data sets where the mutual distance between taxa varies, the number of new entries in the Q matrix which fall below the upper bound is roughly the same as the entries which are joined. In data sets containing m redundant taxa the number of q values falling below the upper bound can suddenly explode. Every time an entry in the Q matrix which is related to a redundant taxa falls below the upper bound, at least $m - 1$ additional entries also fall below the upper bound. If m is large enough, the running time of RapidNJ increases significantly. Among the data sets which gave rise to outliers in the running time of RapidNJ, we found some where more than one quarter of all taxa where redundant. QuickJoin is also affected by redundant data but to a less extent than RapidNJ. Clearcut seems to be unaffected while QuickTree actually benefits from redundant data due to a heuristic which treats redundant taxa as a single taxon.

Not all deviating running times can be explained by redundant data. We found that a very few data sets from Pfam, contained large sets of taxa where the mutual distances are concentrated in small range. In rare cases these data sets increased the running time of RapidNJ by up to a factor two.

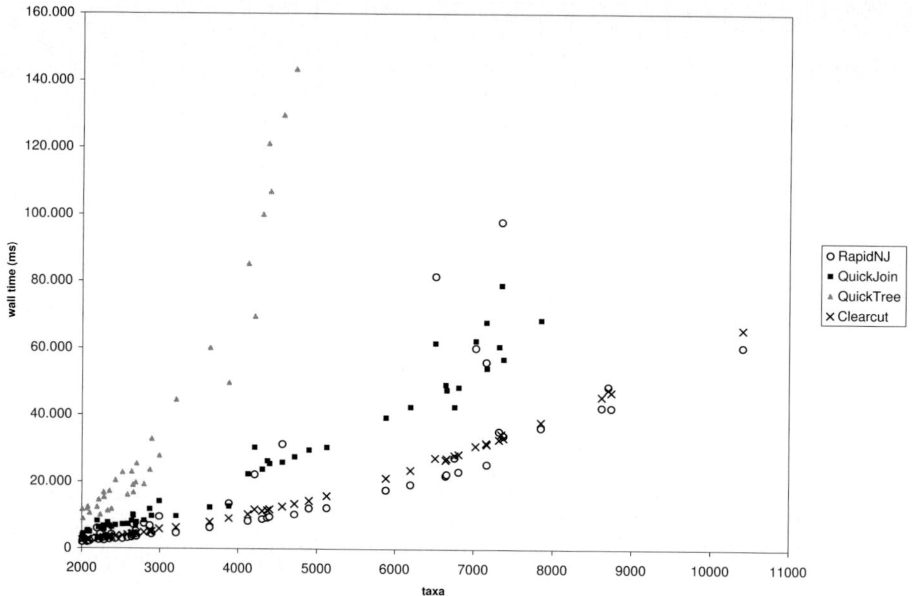

Fig. 3. Performance of RapidNJ compared to QuickJoin, QuickTree and Clearcut on large Pfam data (> 2000 taxa)

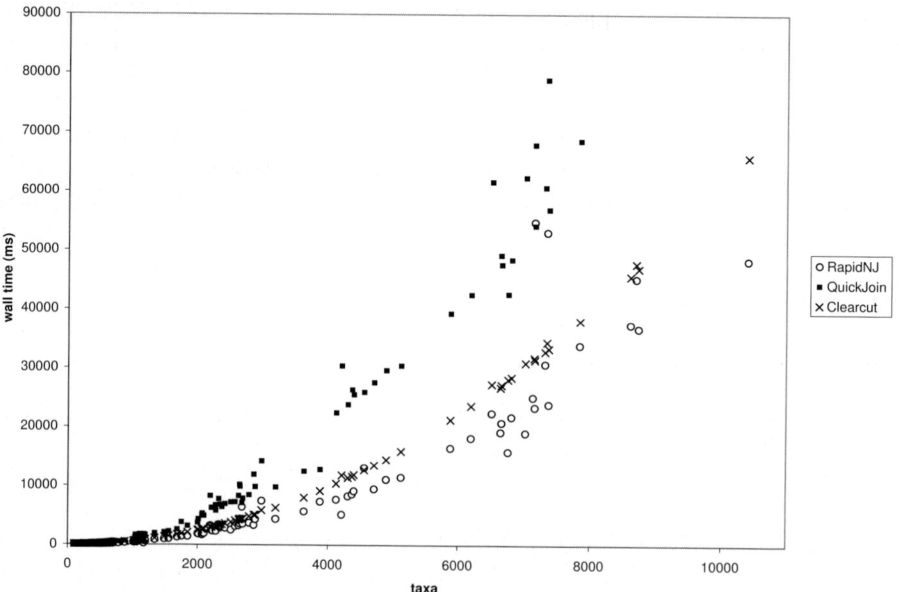

Fig. 4. Performance of RapidNJ with the preprocessing phase where redundant data is eliminated compared to QuickJoin and Clearcut on Pfam data sets

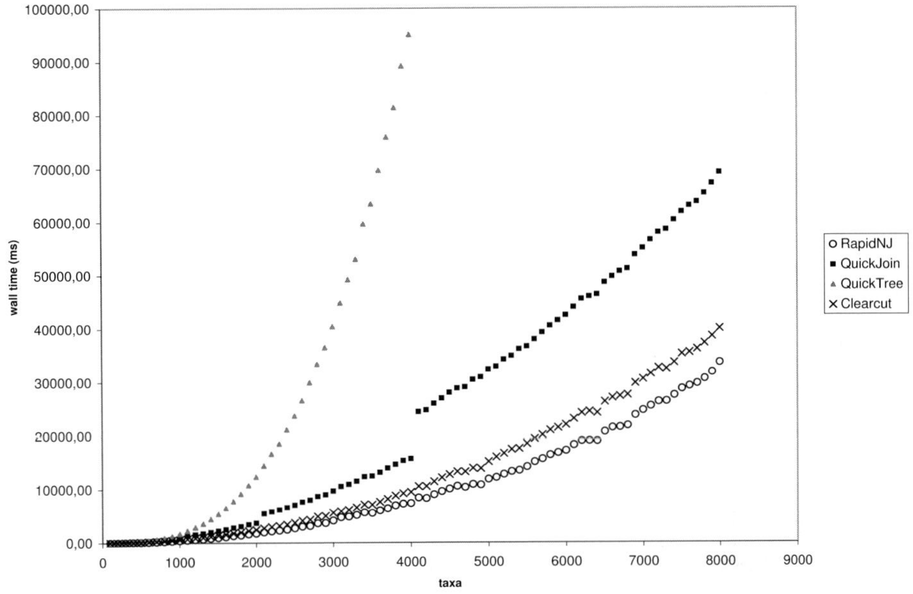

Fig. 5. Performance of RapidNJ compared to QuickJoin, QuickTree and Clearcut on simulated data sets

To test the impact of redundant data, we implement a preprocessing phase where identical rows are joined before neighbour-joining is applied. When using this preprocessing phase, the output is a tree with a slightly different topology than those produces by the canonical neighbour-joining algorithm. The difference lies mainly in the subtrees containing redundant data, and can be explained by the order which the redundant taxa were joined. It is a close approximation of a canonical neighbour-joining tree, but this implementation mainly serves to illustrate the problem with redundant data. The result of experiments with preprocessing is presented in Fig. 4, which shows that the number of deviating running times are reduced significantly.

4.4   Results on Simulated Data

Results on simulated data show the same tendencies as the results on Pfam data. These data sets contain no redundant rows, and we see no deviating running times. A clear trend line of all tree implementations can be observed on Fig. 5. It seems Clearcut, QuickJoin and RapidNJ have almost the same asymptotic time complexity, but with different constants. We can observe a number of jumps on QuickJoins curve. These can be explained by the way data structures are implemented in QuickJoin. Every time the input size reaches the next power of two, memory consumption increases by a factor four. This affects the running time thus creating the jumps seen in Fig. 5. The memory consumptions and running times of RapidNJ, Clearcut and QuickTree have no sudden jumps.

5   Conclusion

We have presented RapidNJ, a search heuristic used to speed up the search for pairs to be joined in the neighbour-joining method. The search heuristic searches for the same optimisation criteria as the original neighbour-joining method, but improves on the running time by eliminating parts of the search space which cannot contain the optimal node pair. While the worst-case running time of RapidNJ remains $O\left(n^3\right)$, it outperforms state of the art neighbour-joining and relaxed neighbour-joining implementations such as QuickTree, QuickJoin and Clearcut. Since the distance matrix used as input to neighbour-joining is typically derived from a multiple alignment, it would be interesting to investigate the overall performance of RapidNJ when combined with efficent methods such as [2] to obtain the distance matrix from a multiple alignment.

References

1. Elias, I., Lagergren, J.: Fast neighbour joining. In: Caires, L., Italiano, G.F., Monteiro, L., Palamidessi, C., Yung, M. (eds.) ICALP 2005. LNCS, vol. 3580, pp. 1263–1274. Springer, Heidelberg (2005)
2. Elias, I., Lagergren, J.: Fast computation of distance estimators. BMC Bioinformatics 8, 89 (2007)
3. Evans, J., Shcncman, L., Foster, J.A.: Relaxed neighbor joining: A fast distance-based phylogenetic tree construction method. Journal of Molecular Evolution 62(6), 785–792 (2006)
4. Finn, R.D., Mistry, J., Schuster-Böckler, B., Griffiths-Jones, S., Hollich, V., Lassmann, T., Moxon, S., Marshall, M., Khanna, A., Durbin, R., Eddy, S.R., Sonnhammer, E.L.L., Bateman, A.: Pfam: Clans, web tools and services. Nucleic Acids Research, Database Issue 34, D247–D251 (2006)
5. Howe, K., Bateman, A., Durbin, R.: QuickTree: Building huge neighbour-joining trees of protein sequences. Bioinformatics 18(11), 1546–1547 (2002)
6. Mailund, T., Brodal, G.S., Fagerberg, R., Pedersen, C.N.S., Philips, D.: Recrafting the neighbor-joining method. BMC Bioinformatics 7(29) (2006)
7. Mailund, T., Pedersen, C.N.S.: QuickJoin – fast neighbour-joining tree reconstruction. Bioinformatics 20, 3261–3262 (2004)
8. Saitou, N., Nei, M.: The neighbor-joining method: A new method for reconstructing phylogenetic trees. Molecular Biology and Evolution 4, 406–425 (1987)
9. Sanderson, M.L.: Inferring absolute rates of molecular evolution and divergence times in the absence of molecular clock. Bioinformatics 19, 301–302 (2003)
10. Sheneman, L., Evans, J., Foster, J.A.: Clearcut: A fast implementation of relaxed neighbor-joining. Bioinformatics 22(22), 2823–2824 (2006)
11. Studier, J.A., Kepler, K.J.: A note on the neighbour-joining method of Saitou and Nei. Molecular Biology and Evolution 5, 729–731 (1988)

Efficiently Computing Arbitrarily-Sized Robinson-Foulds Distance Matrices

Seung-Jin Sul, Grant Brammer, and Tiffani L. Williams

Department of Computer Science
Texas A&M University
College Station, TX 77843-3112 USA
{sulsj,grb9309,tlw}@cs.tamu.edu

Abstract. In this paper, we introduce the HashRF(p, q) algorithm for computing RF matrices of large binary, evolutionary tree collections. The novelty of our algorithm is that it can be used to compute arbitrarily-sized $(p \times q)$ RF matrices without running into physical memory limitations. In this paper, we explore the performance of our HashRF(p, q) approach on 20,000 and 33,306 biological trees of 150 taxa and 567 taxa trees, respectively, collected from a Bayesian analysis. When computing the all-to-all RF matrix, HashRF(p, q) is up to 200 times faster than PAUP* and around 40% faster than HashRF, one of the fastest all-to-all RF algorithms. We show an application of our approach by clustering large RF matrices to improve the resolution rate of consensus trees, a popular approach used by biologists to summarize the results of their phylogenetic analysis. Thus, our HashRF(p, q) algorithm provides scientists with a fast and efficient alternative for understanding the evolutionary relationships among a set of trees.

Keywords: phylogenetic trees, Robinson-Foulds distance, clustering, performance analysis.

1   Introduction

Bayesian analysis [1] is one of the most common approaches for reconstructing an evolutionary history (or phylogeny) of a set of organisms (or taxa). Such analysis can easily produce tens of thousands of potential, binary trees that later have to be summarized in some way. Currently, scientists use consensus trees to summarize the results from these multitudes of trees into a single tree. Yet, much information is lost by summarizing the evolutionary relationships between the trees into a single consensus tree [2], [3]. In this paper, we explore an alternative approach, which compliments consensus trees, to help scientists better understand the results of their phylogenetic analysis.

We developed the HashRF(p, q) algorithm as the basis for an alternative approach for understanding the relationships among a collection of t trees. The *novelty* of this algorithm is that it can compute arbitrarily-sized $(p \times q)$ RF matrices (see Figure 1). Moreover, it can be leveraged to minimize the memory requirements of computing extremely, large RF matrices and it can be used to take

K.A. Crandall and J. Lagergren (Eds.): WABI 2008, LNBI 5251, pp. 123–134, 2008.
© Springer-Verlag Berlin Heidelberg 2008

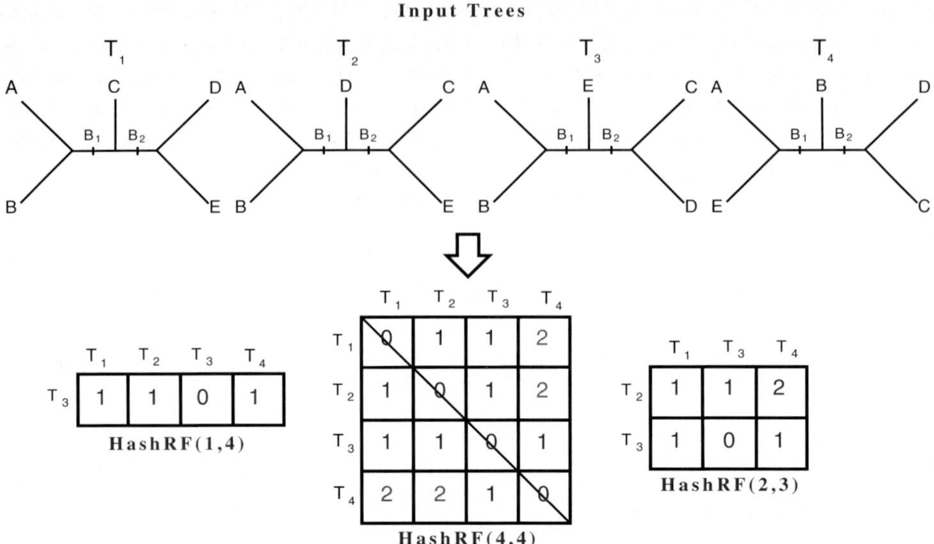

Fig. 1. Overview of computing the RF distance matrix using the HashRF(p, q) algorithm. The tree collection consists of four phylogenies: T_1, T_2, T_3, and T_4. Bipartitions (or internal edges) in a tree are labeled B_i, where i ranges from 1 to 2. Three different RF matrices are shown that can be produced by HashRF(p, q). HashRF(1,4) produces a *one-to-all* matrix, HashRF(4,4) computes an *all-to-all* matrix, and HashRF(2,3) computes a 2×3 matrix.

advantage of parallel platforms. Several software packages, such as PAUP* [4] and Phylip [5], support computing the topological distance between phylogenetic trees. However, these tools are only suitable for small distance matrices. Other RF algorithms, such as Day's algorithm [6], PGM-Hashed [7], and HashRF [8] cannot compute arbitrarily-sized matrices. They are limited to $t \times t$ matrices.

Our results are based on our large collection of biological trees (20,000 trees of 150 taxa and 33,306 trees of 567 taxa) obtained from a Bayesian analysis. Although the benefits of our HashRF(p, q) algorithm is that it can compute arbitrarily-sized matrices, we test its performance against other RF matrix algorithms (PAUP*, PGM-Hashed, and HashRF) that compute all-to-all distance matrices. Hence, in these experiments, the p and q parameters of the HashRF(p, q) algorithm are both set to t, the number of trees in the collection of interest. For large all-to-all matrices, the results clearly demonstrate that HashRF(p, q) is the best approach. It is over 200 times faster than PAUP*, and up to 40% faster than HashRF, one of the fastest algorithms for computing an all-to-all matrix.

Once a RF distance matrix is obtained for a tree collection, there are a number of data-mining techniques that can be used to help understand the relationship among the trees. In this paper, we cluster the RF distance matrices in order to improve the quality of the consensus trees, which are used to summarize large numbers of trees into a single phylogeny. As stated earlier, consensus trees are traditionally used in this way. However, the disadvantage of such an approach is potentially losing vital evolutionary relationships that are in the tree collection, but not depicted in the consensus tree. For our tree collections, we found mixed results. For our 150 taxa trees, clustering the resulting RF distance matrices made a significant impact on improving the quality of the consensus tree. However, the improvement for our 567 taxa trees was minimal since the trees are already quite similar.

Overall, HashRF(p, q) provides researchers with a technique to produce arbitrarily-sized RF distance matrices. With our new algorithm, life scientists have a way to apply further clustering and post-processing methods to understand their large collections of phylogenetic trees.

2   Basics

2.1   Phylogenetic Trees

In a phylogenetic tree, modern organisms (or taxa) are placed at the leaves and ancestral organisms occupy internal nodes, with the edges of the tree denoting evolutionary relationships. Oftentimes, it is useful to represent phylogenies in terms of their *bipartitions*. Removing an internal edge e from a tree separates the taxa (or leaves) on one side from the taxa on the other. The division of the taxa into two subsets is the non-trivial bipartition B associated with internal edge e. (Note: all trees have trivial bipartitions denoted by external edges.) In Figure 1, T_2 has two bipartitions (or internal edges): $AB|CDE$ represented by B_1 and $ABD|CE$ represented by B_2. An evolutionary tree is uniquely and completely defined by its set of $O(n)$ bipartitions.

2.2   Robinson-Foulds (RF) Distance

The Robinson-Foulds (RF) distance between two trees is the number of bipartitions that differ between them. Let $\Sigma(T)$ be the set of bipartitions defined by all edges in tree T. The RF distance between trees T_1 and T_2 is defined as:

$$d_{RF}(T_1, T_2) = \frac{|\Sigma(T_1) - \Sigma(T_2)| + |\Sigma(T_2) - \Sigma(T_1)|}{2} \tag{1}$$

In Figure 1, consider the RF distance between trees T_1 and T_2. The set of bipartitions defined for tree T_1 is $\Sigma(T_1) = \{AB|CDE, ABC|DE\}$. $\Sigma(T_2) = \{AB|CDE, ABD|CE\}$. The number of bipartitions appearing in T_1 and not T_2 (i.e., $|\Sigma(T_1) - \Sigma(T_2)|$) is 1, since $\{ABC|DE\}$ does not appear in T_2. Similarly, the number of bipartitions in T_2 but not in T_1 is 1. Hence, $d_{RF}(T_1, T_2) = 1$. The largest possible RF distance between two binary trees is $(n - 3)$, which results in a maximum RF value of 2 in the example shown in Figure 1.

In this paper, we are interested in computing the $p \times q$ matrix, where $1 \leq p, q \leq t$. Given two sets, S_1 and S_2, of input trees (where $|S_1| = p$ and $|S_2| = q$), the output is a $p \times q$ matrix of RF distances. If p and q are equal to t (the number of trees of interest), then the matrix represents the *all-to-all RF distance* (or $t \times t$ RF matrix) between every pair of t trees. When $p = 1$ and $q = t$, then the matrix represents the *one-to-all distance* between the trees. The all-to-all RF distance is quite useful if one is interested in feeding the resulting $t \times t$ matrix to a clustering algorithm. We show such an application for our collection of biological trees in Section 6.2. One-to-all RF distances are useful for comparing a single tree (such as the best tree found by a heuristic or a published tree) to a set of trees of interest.

2.3   Consensus Trees

Consensus trees are used to summarize the information from the set of trees usually resulting from phylogenetic search heuristics employed by popular software such as PAUP* [4] and MrBayes [9]. Among many different consensus methods, the strict consensus and majority consensus trees are widely used in phylogenetics. The strict consensus approach returns a tree such that the bipartitions of the tree are only those bipartitions that occur in every input tree. On the other hand, the majority consensus algorithm makes a consensus tree such that the bipartitions are only those that occur in over 50% of the input trees.

Oftentimes, the resulting consensus tree is not a binary tree. If this is the case, the consensus tree is considered to be multifurcating. We use *resolution rate* to measure the percentage of binary internal edges in a phylogenetic tree. The resolution rate is defined as the number of binary internal nodes (nodes of degree 3) divided by $(n - 3)$, which represents the maximum number of internal nodes in a binary tree of n taxa. Hence, a binary tree is 100% resolved. For the four trees in Figure 1, the majority consensus tree simply consists of a tree with the bipartition $AB|CDE$, which results in a resolution rate of 75%. On the other hand, the strict consensus trees is a star (i.e., it's completely unresolved) and results in a resolution rate of 0% resolved.

3   HashRF: Computing an All-to-All RF Matrix

Figure 2 provides an overview of the HashRF algorithm, which has a running time complexity of $O(nt^2)$, where n is the number of taxa and t is the number of trees. Each input tree, T_i, is traversed in post-order, and its bipartitions are fed through two hash functions, h_1 and h_2. Hash function h_1 is used to generate the location needed for storing a bipartition in the hash table. h_2 is responsible for creating bipartition identifiers (BIDs). For each bipartition, its associated hash table record contains its BID along with the tree index (TID) where the bipartition originated.

3.1   Organizing Bipartitions in the Hash Table

Our h_1 and h_2 universal hash functions are defined as follows.

$$h_1(B) = \sum b_i r_i \bmod m_1 \tag{2}$$

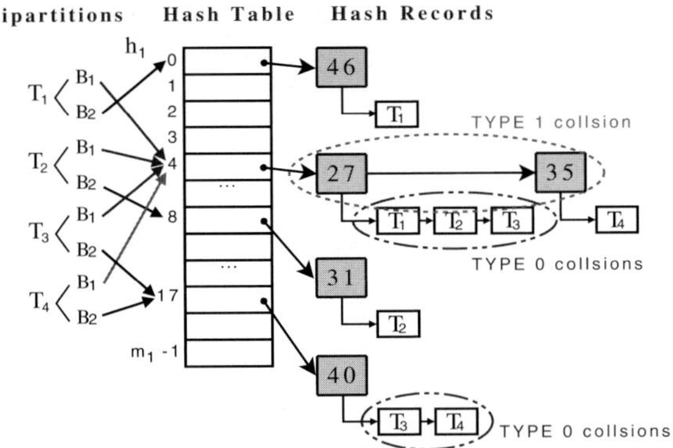

Fig. 2. Overview of the HashRF algorithm. Bipartitions are from Figure 1. (a) The implicit representation of each bipartition, B_i, is fed to the hash functions h_1 and h_2. The shaded value in each hash record contains the bipartition ID (or h_2 value). Each bipartition ID has a linked list of tree indexes that share that particular bipartition.

$$h_2(B) = \sum b_i s_i \bmod m_2 \tag{3}$$

m_1 represents the number of entries (or locations) in the hash table. m_2 represent the largest bipartition ID (BID) that we can be given to a bipartition. That is, instead of storing the n-bitstring, a shortened version of it (represented by the BID) will be stored in the hash table instead. $R = (r_1, ..., r_n)$ is a list of random integers in $(0, ..., m_1 - 1)$, $S = (s_1, ..., s_n)$ is a list of random integers in $(0, ..., m_2 - 1)$, and $B = (b_1, ..., b_n)$ is a bipartition represented by an n-bitstring. We can avoid sending the n-bitstring bipartition representations to our hash functions h_1 and h_2. Instead, we use an implicit bipartition representation to compute the hash functions quickly. An implicit bipartition is simply the integer value (instead of the n-bitstring) that provides the representation of the bipartition.

A consequence of using hash functions is that bipartitions may end up residing in the same location in the hash table. Such an event is considered a collision. There are three types of collisions that our hashing algorithms must resolve. *Type 0* collisions occur when the same bipartition (i.e. $B_i = B_j$), is shared across the input trees. Such collisions are not serious and are a function of the set of input trees. *Type 1* collisions result from two different bipartitions B_i and B_j (i.e., $B_i \neq B_j$) residing in the same location in the hash table. That is, $h_1(B_i) = h_1(B_j)$. (We note that this is the standard definition of collisions in hash table implementations.) *Type 2* collisions are serious and require a restart of the algorithm if such an event occurs. Otherwise, the resulting output will be incorrect. Suppose that $B_i \neq B_j$. A Type 2 collision occurs when B_i and B_j hash to the same location in the hash table and the bipartition IDs (BIDs) associated with them are also the same. In other words, $h_1(B_i) = h_1(B_j)$ and $h_2(B_i) = h_2(B_j)$. The probability of our hash-based approach having to restart because of a double collision among any pair of the bipartitions is $O\left(\frac{1}{c}\right)$. Since c

can be made arbitrarily large, the probability of a restart can be made infinitely small. In our experiments, $c = 1,000$.

3.2   Computing the RF Matrix

Once all the bipartitions are organized in the hash table, then the RF distance matrix can be calculated. For each non-empty hash table location i, we have a list of tree index (TID) nodes for each unique bipartition ID (BID) node. Consider the linked list of bipartitions in location 4 of the hash table in Figure 2 for BID 27 which contains TIDs $\{T_1\}$, $\{T_2\}$, $\{T_3\}$. The hash table shows that trees T_1, T_2, and T_3 share the same bipartition ($ABC|DE$ from Figure 1). HashRF uses a $t \times t$ dissimilarity matrix, D, to track the number of bipartitions that are different between all tree pairs. For each tree i, its row entries are initialized to b_i, the number of bipartitions present in tree i. Hence, $D_{i,j} = b_i$ for $0 \leq j < t$ and $i \neq j$. $D_{i,i} = 0$. In the case of binary trees, the $D_{i,j}$ entries are initialized to $n - 3$, the maximum number of internal edges in a binary tree.

For each BID node at location l, every pair of TID nodes in the linked list are compared to each other. Then, the counts of $D_{i,j}$ and $D_{j,i}$ are decremented by one. That is, we have found a common bipartition between T_i and T_j and decrement the difference counter by one. For example, trees with BID 27 at location 4 in the hash table shows that the pairs (T_1, T_2), (T_1, T_3), and (T_2, T_3) share a bipartition. Thus, entries $D_{1,2}$, $D_{1,3}$, and $D_{2,3}$ are decremented by one. Once we have computed D, we can compute the RF matrix quite easily. Thus, $RF_{i,j} = \frac{D_{i,j} + D_{j,i}}{2}$, for every tree pair i and j.

4   Our New Approach: The HashRF(p, q) Algorithm

4.1   Motivation

Our previous work with HashRF [8] focused on designing a fast algorithm to compute the all-to-all (or $t \times t$) RF matrix between every pair of t trees. Consider a collection of t binary trees. There are two distinct advantages to using our new HashRF(p, q) approach, which extends our HashRF algorithm.

1. We can now compute a $p \times q$ matrix, where $1 \leq p, q \leq t$.
2. There are no physical memory limitations regarding the size of a RF matrix we can compute.

The first advantage of our HashRF(p, q) algorithm results in being able to compute other types of RF matrices besides all-to-all matrices. Although all-to-all matrices are quite useful—we show their usefulness in Section 6.2—other matrix shapes are helpful. For example, one might be interested in a one-to-all RF matrix, where $p = 1$ and $q = t$, which shows how different a single tree (e.g., best-known or published tree) is from a collection of trees. The other important advantage of the HashRF(p, q) approach is that it can be used to compute extremely large matrices—especially those that are too large to fit entirely into physical memory. By dividing such large $t \times t$ matrices into smaller $p \times q$ chunks, it becomes feasible to compute any RF matrix size desirable. Furthermore, if a multi-core machine is available, then these independent blocks could be computed in parallel to achieve a significant increase in performance.

4.2 HashRF(p, q) vs. HashRF

HashRF(p, q) works similarly to HashRF, which was described in Section 3. The first major difference between the two algorithms is that HashRF(p, q) requires as input two sets of trees, S_p and S_q, where $|S_p| = p$ and $|S_q| = q$. HashRF requires only one set of trees, S, as input and $|S| = t$. HashRF(p, q) requires the sets S_p and S_q of trees since the trees in S_p will only be compared to trees in S_q. Trees within a set will not be compared to each other.

In the HashRF(p, q) approach, we assume that $p \leq q$. As a result, we place all of bipartitions of the trees in S_q into the hash table. Once this is done, then we process each tree T_i in the set S_p. For each bipartition B of tree T_i, we apply our h_1 and h_2 hashing functions as described in Section 3. Once we determine where bipartition B would be located (using the h_1 function) in the hash table, we compare it's bipartition ID (using the h_2 function) to those nodes that are at that location. Suppose this location or index is l in the hash table. At location l, for each tree T_j at location l with the same BID as tree T_i, we increment the counter in the matrix at locations (T_i, T_j) and (T_j, T_i) by one—assuming the full matrix is of interest and not the lower or upper triangle. We repeat the above steps for each remaining tree in the set S_p. Afterwards, since we are interested in the RF distance (and not similarity), we subtract $n - 3$ from the values since that is the maximum RF distance for a binary tree consisting of n taxa.

In terms of running time, the HashRF approach requires $O(nt^2)$ time. For HashRF(p, q) since we assume that $p \leq q \leq t$, then the running time is $O(nq^2)$. Hence, if a smaller sub-matrix is of interest, it will be significantly faster to compute than an all-to-all RF distance matrix.

5 Experimental Methodology

5.1 Biological Tree Collections

The biological trees used in this study were obtained from two recent Bayesian analysis, which we describe below.

- 20,000 trees obtained from a Bayesian analysis of an alignment of 150 taxa (23 desert taxa and 127 others from freshwater, marine, ands oil habitats) with 1,651 aligned sites [10].
- 33,306 trees obtained from an analysis of a three-gene, 567 taxa (560 angiosperms, seven outgroups) dataset with 4,621 aligned characters, which is one of the largest Bayesian analysis done to date [11].

In our experiments, for each number of taxa, n, we created different tree set sizes, t, to test the scalability of the algorithms. For $n = 150$ and 567, the entire

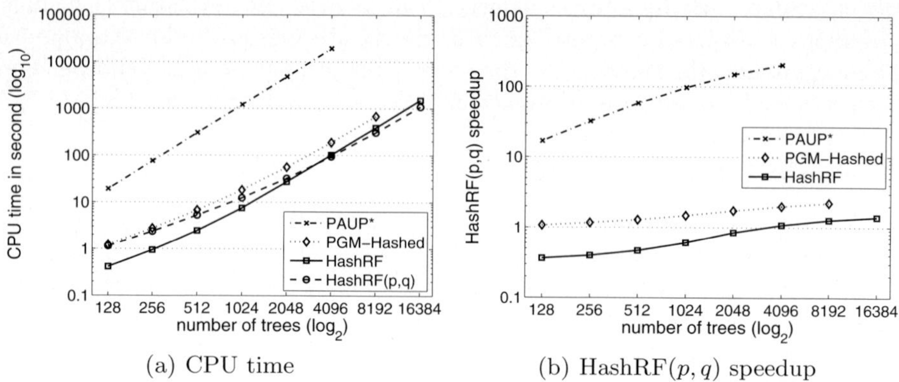

Fig. 3. The performance of the various RF matrix algorithms to compute a $t \times t$ matrix, where t is the number of trees, our 567 taxa dataset. For HashRF(p,q), p and q are both equal to t.

collection of trees is divided into smaller sets, where t is $128, 256, 512, \ldots, 16,384$ trees. Thus, for each (n,t) pair, t trees with n taxa were randomly sampled without replacement from the appropriate tree collection. We repeated the above randomly sampling procedure five times in order to plot the average performance of the RF algorithms.

5.2   RF Matrix Implementations

We obtained the source code for PGM-Hashed from the authors. We used version 4.0b10 of PAUP*. Experimental results were compared across the competing algorithms for experimental validation. All experiments were run on an Intel Pentium platform with 3.0GHz dual-core processors and a total of 2GB of memory. HashRF and PGM-Hashed were written in C++ and compiled with gcc 4.1.0 with the -O2 compiler option. Each plot shows the average performance over five runs.

6   Experimental Results

6.1   HashRF(p,q) Performance

First, we consider the running time of our HashRF(p,q) algorithm against its competitors for computing an all-to-all (or $t \times t$) RF matrix. We also show the performance of HashRF in comparison with HashRF(p,q). Figure 3 shows the running time and speedup of HashRF(p,q) over its competitors on our 567 taxa tree collection. The results for our 150 taxa tree collections are similar. To compute the speedup values, the running times of PAUP*, PGM-Hashed, and HashRF are divided by the running time of HashRF(p,q), the algorithm of most interest to us in our experiments. Both HashRF and HashRF(p,q) clearly outperform the other algorithms. PAUP* is the slowest RF matrix algorithm requiring

5.35 hours to compute a $4,096 \times 4,096$ RF distance matrix. HashRF(p,q) only requires 93.9 seconds for the same data set, which results in a speedup of over 200 in comparison to PAUP*. As the number of taxa is increased, the speedup of HashRF(p,q) is increases as well, where the closest competitor PGM-Hashed is up to two times slower than HashRF(p,q).

Figure 3 also depicts several other interesting points. Although HashRF(p,q) was designed to compute sub-matrices, it is the best choice for computing the all-to-all matrix when $t > 2,048$ trees. HashRF is already a fast approach, but HashRF(p,q) can improve performance by as much as 40% when $t = 16,384$. For computing smaller $t \times t$ matrices, HashRF is the preferred approach. Since PAUP* takes over 12 hours to compute $t \times t$ matrices, when $t \geq 8,192$, we choose to terminate the run. Finally, PGM-Hashed is unable to compute the $16,384 \times 16,384$ RF matrix.

Lastly, we consider computing the all-to-all matrix of our entire collection of trees, which is useful for the clustering of trees that is described in the next section. In other words, we are interested in computing the $20,000 \times 20,000$ matrix and the $33,306 \times 33,306$ matrix of our 150 and 567 taxa trees, respectively. Both the HashRF and HashRF(p,q) algorithms can compute the $20,000 \times 20,000$ matrix quite easily. However, both algorithms could not compute the all-to-all matrix for the 33,306 trees since the algorithms ran out of memory. In such a situation, we divided the larger matrix into 3 blocks, where each block was of size $11,102 \times 33,306$. We divided the input appropriately to feed to the HashRF(p,q) algorithm (i.e., one file contained 11,102 trees and the other had 33,306 trees). After each block was computed, we simply concatenated the results. Overall, 1 hour and 15 minutes was used to produce the $33,306 \times 33,306$ matrix.

6.2 HashRF(p,q) Application: Improving Consensus Tree Resolution Rates

Now, consider an application of the HashRF(p,q) algorithm. Given a collection of trees returned from a Bayesian analysis, phylogenetic researchers often reduce this information into a single consensus tree. However, one disadvantage of such an approach is losing vital evolutionary relationships, which can be quantified by the resolution rate of the consensus tree. The higher the resolution rate, the more in agreement the consensus tree is with the overall collection of trees. We explore the impact of clustering based on using RF distance matrices to improve the resolution rates of majority and strict consensus trees.

Clustering Methodology. Using HashRF(p,q) we generated the all-to-all RF distance matrices for our two biological collections of t trees, where t is 20,000 and 33,306 for 150 and 567 taxa trees, respectively. Next, we converted the distance matrices into similarity matrices by subtracting the value in each cell from $n-3$, the total possible number of bipartitions for n taxa. These similarity matrices are then used as the input to CLUTO [12], which is software designed for clustering high-dimensional datasets.

Unfortunately, we were unable to cluster the entire $t \times t$ matrices in CLUTO since its memory requirements were too large for our platform. Although we have an approach that can create very large RF matrices that isn't limited by physical memory space, CLUTO required more memory space than we had available. Hence, for each of our tree collections, we created smaller, random samples of trees to cluster. These samples were created by selected 5,000 trees at random without replacement from our collection of 20,000 and 33,306 trees. We then created a $5,000 \times 5000$ similarity matrix, which was fed to CLUTO. The above process was repeated five times.

For our clusterings, we mostly used the default settings in CLUTO. For example, we used repeated bipartition clustering, where the graph is partitioned multiple times to produce the requested number of clusters. We also chose to use the default optimization criterion which is CLUTO's $i2$ function. The $i2$ function is meant to optimize the similarity between the data points in the clusters. Moreover, we felt that optimizing the internal similarity of the data suited our goals of increasing consensus resolution rates through clustering.

Clustering 150 and 567 Taxa Trees. The only default setting we changed was specifying the number of clusters, K, to partition the 5,000 trees in the sample. The default setting for K is 10. Since the best value for K is unknown, we set $K = 2$ for the 150 taxa dataset since the trees were obtained by two distinct runs, where each run consisted of 10,000 trees. Moreover, we also wanted to minimize the number of different clusterings produced. If we take the resolution rate of the entire collection 20,000 trees, the resulting majority and strict consensus resolution rates are 85.7% and 34.0%, respectively. But, by clustering the data into two partitions, we were able to increase the average consensus resolution rate to 89.2% and 38.2% for majority and strict trees, respectively.

However, the 567 taxa dataset tells a different story concerning its 33,306 total trees. We initially set $K = 3$ for this data because it was derived from three distinct runs, each consisting of 11,102 trees. However, setting $K = 3$ resulted in two clusters of around 2,500 trees and a third cluster of size two. Hence, we repeated the experiment with $K = 2$. The resolution rate of the majority and strict trees for the entire collection of 33,306 trees is 92.2% and 51.7%, respectively. By clustering, we were able to improve the average resolution rate slightly to 92.7% for the majority trees and 53.4% for the strict trees. In this case the clustering algorithm was unable to make significant improvements to the consensus resolution rates since the level of similarity among the trees is already quite high.

7   Conclusions and Future Work

Phylogenetic analysis can produce a large number of binary trees, which present a tremendous data-mining challenge for understanding the evolutionary relationships between them. Currently, life scientists often use consensus trees to summarize their trees. However, a lot of information regarding the evolutionary relationships contained in the trees can be lost. One of the advantages of computing a RF distance matrix to summarize the information is that it can be fed to a clustering algorithm to explore the relationships among the trees, which we explore in our work.

We develop a new algorithm called $\mathrm{HashRF}(p,q)$ which is not limited to computing the all-to-all (or $t \times t$) matrices between a collection of t trees. $\mathrm{HashRF}(p,q)$ can compute arbitrarily-sized $p \times q$ matrices, where $1 \leq p, q \leq t$. Moreover, the $\mathrm{HashRF}(p,q)$ approach can be used to compute very large RF matrices, which are not bounded by the amount of physical memory available on a user's system. Our experimental study on large collections of Bayesian trees shows that $\mathrm{HashRF}(p,q)$ is the best performing algorithm for large $t \times t$ matrices, where the number of trees is greater than 2,048. Popular phylogenetic software, such as PAUP*, is up to 200 times slower than $\mathrm{HashRF}(p,q)$. Furthermore, $\mathrm{HashRF}(p,q)$ is around 40% faster than our HashRF approach for large all-to-all matrices. Such performance results suggests that we could combine the two hash-based approaches into one.

The main motivation for computing the RF distance matrix for such a large collection of trees is to provide researchers with input data for post-processing techniques. In this paper, we show clustering as a method for increasing the resolution rates of consensus trees. For the 150 taxa dataset, by creating two clusters of trees, we were able to improve the resolution rate of the majority and strict consensus trees by 3.5% and 4.2%, respectively. More specifically, clustering providing us with two majority (strict) trees that had an average resolution of 89.2% (38.2%). However, for the 567 taxa trees, clustering was not able to improve the quality of the consensus trees as the 33,306 trees are already quite similar.

Our work can be extended in many different directions. First, we plan to develop a parallel algorithm that uses $\mathrm{HashRF}(p,q)$ to compute an arbitrarily sized RF matrix faster. Now that we are able to obtain distance matrices of arbitrary size, we can use them for further exploration of clustering and other post-processing methods. Finally, we believe our approach is a good step toward handling larger sets of trees since the goal of a phylogenetics is to reconstruct the Tree of Life, which is estimated to contain up to 100 million taxa.

Acknowledgements

We would like to thank Matthew Gitzendanner, Bill Murphy, Paul Lewis, and David Soltis for providing us with the Bayesian tree collections used in this paper. Funding for this project was supported by the National Science Foundation under grants DEB-0629849 and IIS-0713618.

References

1. Huelsenbeck, J.P., Ronquist, F., Nielsen, R., Bollback, J.P.: Bayesian inference of phylogeny and its impact on evolutionary biology. Science 294, 2310–2314 (2001)
2. Hillis, D.M., Heath, T.A., John, K.S.: Analysis and visualization of tree space. Syst. Biol. 54(3), 471–482 (2005)

3. Stockham, C., Wang, L.S., Warnow, T.: Statistically based postprocessing of phylogenetic analysis by cluste ring. In: Proceedings of 10th Int'l Conf. on Intelligent Systems for Molecular Biology (ISMB 2002), pp. 285–293 (2002)
4. Swofford, D.L.: PAUP*: Phylogenetic analysis using parsimony (and other methods), Sinauer Associates, Underland, Massachusetts, Version 4.0 (2002)
5. Felsenstein, J.: Inferring Phylogenies. Sinauer Associates (2003)
6. Day, W.H.E.: Optimal algorithms for comparing trees with labeled leaves. Journal Of Classification 2, 7–28 (1985)
7. Pattengale, N., Gottlieb, E., Moret, B.: Efficiently computing the Robinson-Foulds metric. Journal of Computational Biology 14(6), 724–735 (2007)
8. Sul, S.J., Williams, T.L.: A randomized algorithm for comparing sets of phylogenetic trees. In: Proc. Fifth Asia Pacific Bioinformatics Conference (APBC 2007), pp. 121–130 (2007)
9. Huelsenbeck, J.P., Ronquist, F.: MRBAYES: Bayesian inference of phylogenetic trees. Bioinformatics 17(8), 754–755 (2001)
10. Lewis, L.A., Lewis, P.O.: Unearthing the molecular phylodiversity of desert soil green algae (chlorophyta). Syst. Bio. 54(6), 936–947 (2005)
11. Soltis, D.E., Gitzendanner, M.A., Soltis, P.S.: A 567-taxon data set for angiosperms: The challenges posed by bayesian analyses of large data sets. Int. J. Plant Sci. 168(2), 137–157 (2007)
12. Karypis, G.: CLUTO—software for clustering high-dimensional datasets. Internet Website (last accessed, June 2008),
http://glaros.dtc.umn.edu/gkhome/cluto/cluto/overview

Efficient Genome Wide Tagging by Reduction to SAT

Arthur Choi[1], Noah Zaitlen[2], Buhm Han[3], Knot Pipatsrisawat[1],
Adnan Darwiche[1], and Eleazar Eskin[1,4,*]

[1] Department of Computer Science, University of California Los Angeles,
Los Angeles, CA, 90095
[2] Bioinformatics Program, University of California,
San Diego, La Jolla, CA, 92093
[3] Department of Computer Science and Engineering, University of California,
San Diego, La Jolla, CA, 92093
[4] Department of Human Genetics, University of California Los Angeles,
Los Angeles, CA, 90095
Tel.: +1-310-594-5112, Fax: +1-310-825-2273
eeskin@cs.ucla.edu

Abstract. Whole genome association has recently demonstrated some remarkable successes in identifying loci involved in disease. Designing these studies involves selecting a subset of known single nucleotide polymorphisms (SNPs) or tag SNPs to be genotyped. The problem of choosing tag SNPs is an active area of research and is usually formulated such that the goal is to select the fewest number of tag SNPs which "cover" the remaining SNPs where "cover" is defined by some statistical criterion. Since the standard formulation of the tag SNP selection problem is NP-hard, most algorithms for selecting tag SNPs are either heuristics which do not guarantee selection of the minimal set of tag SNPs or are exhaustive algorithms which are computationally impractical. In this paper, we present a set of methods which guarantee discovering the minimal set of tag SNPs, yet in practice are much faster than traditional exhaustive algorithms. We demonstrate that our methods can be applied to discover minimal tag sets for the entire human genome. Our method converts the instance of the tag SNP selection problem to an instance of the satisfiability problem, encoding the instance into conjunctive normal form (CNF). We take advantage of the local structure inherent in human variation, as well as progress in knowledge compilation, and convert our CNF encoding into a representation known as DNNF, from which solutions to our original problem can be easily enumerated. We demonstrate our methods by constructing the optimal tag set for the whole genome and show that we significantly outperform previous exhaustive search-based methods. We also present optimal solutions for the problem of selecting multi-marker tags in which some SNPs are "covered" by a pair of tag SNPs. Multi-marker tags can significantly decrease the number of tags we need to select, however discovering the minimal number of

---

* Corresponding author.

K.A. Crandall and J. Lagergren (Eds.): WABI 2008, LNBI 5251, pp. 135–147, 2008.
© Springer-Verlag Berlin Heidelberg 2008

multi-marker tags is much more difficult. We evaluate our methods and perform benchmark comparisons to other methods by choosing tag sets using the HapMap data.

1   Introduction

Whole genome association is a powerful method for discovering the genetic basis of human diseases. Recently, it has been successfully employed to reveal novel loci correlated with risks for diseases including coronary artery disease, bipolar disorder, type 1 and type 2 diabetes, amongst many others[11]. A typical association study collects genotype information at a set of single nucleotide polymorphism (SNP) and compares the allele frequency at each SNP in a case and a control population using a statistical test in order to determine which loci are associated with the disease.

Even with the tremendous technological advances that have driven down the cost of collecting SNP genotypes, collecting all known SNPs is prohibitively expensive. Genetic association studies take advantage of the fact that genotypes at neighboring SNPs are often in linkage disequilibrium (LD) or are correlated with each other. This correlation allows for "indirect association" where a SNP which is associated with the disease is detected not by collecting genotypes at the SNP directly, but instead by collecting genotypes at a neighboring "tag" SNP or SNP that is correlated with the associated SNP. The availability of reference data sets such as those provided by the HapMap project[10] allow for us to measure the linkage disequilibrium patterns between SNPs and use this information when determining which SNPs to select as tags. *Tag SNP selection* is a central problem in designing association studies and has been extensively studied [27].

Research on the tag SNP selection problem can be roughly split into two categories: the statistical criteria used for selecting tag SNPs and the algorithms for choosing a tag set given this statistical criteria. Many statistical criteria have been proposed for tag SNP selection [27]. The most popular criterion considers the square correlation coefficient r^2 between SNPs. Under this formulation of the tag SNP selection problem the goal is to choose a subset of the SNPs as tags such that each SNP not selected in the tag set has an r^2 value with a tag SNP above a minimal threshold. This problem is commonly represented as a graph where each SNP is represented as a node and an edge connects two nodes if their corresponding SNPs have a correlation above the threshold. A set of tags which covers the remaining tag SNPs corresponds to a vertex cover in this graph. This problem is NP-complete [1] and it was widely believed that we cannot obtain the minimal solutions to this problem for the whole genome[27]. Relatively few algorithmic approaches have been proposed for this problem with the greedy algorithm being the most widely used[7]. While the greedy approach has been used to construct the current generation of commercial products used for whole genome association studies, the greedy algorithm provides no guarantees that the chosen tag set has the minimal number of tag SNPs necessary to cover the region. Relatively few approaches have been presented to obtaining minimal tag sets and these approaches reduce the problem in order to make it computationally feasible. Halldorson et al. [5] restrict LD patterns to a window and FESTA [24] solves the problem by partitioning the SNPs into precincts which do not have any linked SNPs in them and then exhaustively enumerating the solutions within the precinct.

In this paper we present a novel approach to solving the tag SNP selection problem which can discover all minimal solutions and can scale to the whole genome. Our method encodes the instance of the tag SNP selection problem as an instance of the satisfiability (SAT) problem. Here, our SAT instances are clauses in conjunctive normal form (CNF) where a variable assigned to **true** corresponds to the inclusion of a SNP into the tag set. We take advantage of the local structure inherent in human variation, as well as progress in knowledge compilation, and convert our CNF encoding into a representation known as DNNF, from which solutions to our original problem can be easily enumerated. A satisfying assignment of variables to truth values in the SAT instance yields a valid solution to the tagging problem (and vice versa). As we shall clarify, a "minimal" satisfying assignment yields a minimal set of tags.

We compare the results of our method on the single SNP r^2 tag SNP selection problem to FESTA [24] and Halldorson's method [5] over the ENCODE [10] data set. We also demonstrate that our methods scales to the whole genome HapMap data. Consistent with previous studies, minimal solutions for the tag SNP selection method use only slightly less tags than greedy solutions[24]. One advantage of our framework is that we can characterize the entire set of optimal solutions, but in a tractable form that allows for flexible designs. Another advantage is that our approach extends to more challenging variants of the tag SNP selection problem where we are selecting mutli-marker tags in which a SNP can be "covered" by a pair of tags. Multi-marker tags have been shown to significantly increase the power of association studies[23,30].

2 Methods

We present methods for choosing the minimal number of tags SNP for several variations of the *tagging* problem. First we show how to solve the single SNP r^2 tagging problem in which we search for a minimum set of tag SNPs which cover the remaining SNPs in a region of the genome with an r^2 above some minimum threshold. Second, we present a method for combining our optimal solutions in local regions to an optimal solution for the entire genome. Third, we extend our solution to multi-marker tags or tags which combine two or more SNPs. The use of multi-marker tags can significantly reduce the number of tags which need to be collected in order to cover a region, but the optimization procedure is much more difficult.

2.1 Single SNP r^2 Tagging

Let $S = \{s_i\}_{i=1}^n$ be a set of SNPs. We say SNP s_i "covers" SNP s_j if their correlation coefficient r^2, exceeds some threshold r_{min}^2. If $T' \subseteq S$, and $\forall s_j \in$

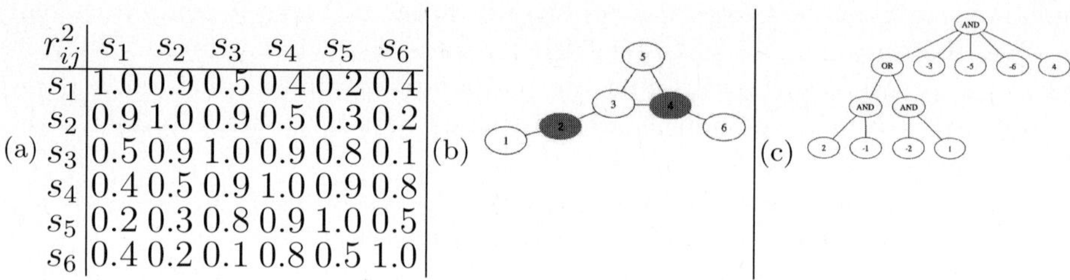

Fig. 1. (a) Single SNP r^2 table (b) Graph of cover problem (c) NNF equivalent to CNF

$S, \exists s_i \in T'$ such that $r_{ij}^2 \geq r_{min}^2$, we call T' a *valid* cover of S. Our goal is to select the smallest set $T' \subseteq S$ that is a *valid* cover of S.

Consider the example in Figure 1, where we have 6 SNPs $s_1, \ldots, s_6$, and the pairwise r^2 values described in the table in Figure 1(a). Suppose that we have the threshold $r_{min}^2 = 0.8$. We can represent the SNPs as the graph shown in Figure 1(b) where an edge denotes an r^2 above the minimum threshold. The standard greedy algorithm [7,18] picks tag SNPs by repeatedly selecting the SNP with the largest number of uncovered neighbors. We can easily see that there are two optimal solutions, $T = \{s_4, s_2\}$ and $T = \{s_4, s_1\}$. Note that one greedy solution will select SNP s_3 in the first step resulting in a non-optimal solution $T = \{s_3, s_1, s_6\}$. Our approach to the tag SNP selection problem will characterize all optimal solutions in a compact directed acyclic graph, which in this case happens to be a tree in the example of Figure 1(c).

We shall reduce the problem of identifying a valid selection of SNPs to the problem of identifying a satisfying assignment to a propositional sentence in conjunctive normal form (CNF). In particular, we want a sentence in CNF where satisfying assignments correspond to a valid selection of SNPs. We create a literal for every SNP and a clause for every SNP consisting of literals that can cover that SNP.

Given a threshold r_{min}^2, consider a sentence in CNF: $\Phi = \phi_1 \wedge \cdots \wedge \phi_n$ with as many clauses ϕ_i as there are SNPs s_i, where each clause is of the form:

$$\phi_i = \bigvee_{r_{ij}^2 \geq r_{min}^2} s_j$$

Each SNP $s_j \in S$ is a positive literal in the CNF sentence Φ, and appears in clause ϕ_i if and only if SNP s_j can cover SNP s_i. A valid selection T' of SNPs then corresponds precisely to a satisfying assignment of Φ.

In order to find a *minimally* valid selection T of SNPs, we seek a minimum cardinality model of our propositional sentence, where a minimum cardinality model is a satisfying assignment with a minimal number of positive literals.

Consider the example in Figure 1 with six SNPs $s_1, \ldots, s_6$. Given the threshold $r_{min}^2 = 0.8$ we have the following CNF formula $(s_1 \vee s_2) \wedge (s_1 \vee s_2 \vee s_3) \wedge (s_2 \vee s_3 \vee s_4 \vee s_5) \wedge (s_3 \vee s_4 \vee s_5 \vee s_6) \wedge (s_3 \vee s_4 \vee s_5) \wedge (s_4 \vee s_6)$ We have two minimum cardinality models, $(\neg s_1, s_2, \neg s_3, s_4, \neg s_5, \neg s_6)$ and $(s_1, \neg s_2, \neg s_3, s_4, \neg s_5, \neg s_6)$, corresponding to our two minimally valid selection of SNPs.

Not surprisingly, identifying a minimum cardinality model for a given sentence in CNF is also an NP–hard problem. Our approach is based on converting our sentence Φ in CNF into a logically equivalent sentence Δ in *decomposable negation normal form* (DNNF) [12,13,15,16]. DNNF is a logical representation that allows certain queries, which are in general intractable, to be computed in time polynomial in the size of the DNNF sentence. For example, if a conversion from CNF to DNNF does indeed result in a sentence of manageable size, we can efficiently test whether the original sentence is satisfiable, count and enumerate its models, and identify another sentence in DNNF that characterizes all its minimum cardinality models. By enumerating the models of the resulting sentence, we can enumerate all of the minimally valid selections of SNPs. In general, there are no guarantees that a CNF can be converted to a DNNF of reasonable size, but we demonstrate that for the tag SNP selection problem, due to the inherent local structure of the problem, our approach is tractable.

This conversion is performed here by the c2D compiler, which compiles CNF instances into DNNF [6].[1] The c2D compiler has already been successfully employed in a number of other applications, serving as a backbone reasoning system in support of higher level tasks. For example, c2D was used as the backbone for planning systems [4,20], for diagnostic systems [19,3,22,2], for probabilistic reasoning [28,9,25,8], and for query rewrites in databases [29]. In each one of these applications, high level reasoning problems were encoded into CNF, which was compiled into DNNF by c2D. The resulting compilation was then used to solve the original problem by posing polytime queries to it.

Decomposable negation normal form. A negation normal form (NNF) is a rooted and directed acyclic graph in which each leaf node is labeled either by a literal (say i for a positive literal, and $-i$ for a negative literal), or simply by true or false. Each internal node is labeled with a conjunction ($\wedge$ or AND) or a disjunction ($\vee$ or OR); Figure 1(c) depicts an example. A negation normal form is decomposable (DNNF) if it satisfies the *Decomposability* property: for each conjunction in the NNF, the conjuncts do not share variables. The NNF in Figure 1(c) is also in DNNF.

If we are able to efficiently compile a CNF instance into DNNF, many queries are straightforward to compute [12]. For example, we can test if a DNNF sentence is satisfiable by simply traversing the graph bottom-up, while visiting children before parents. If a leaf node is labeled by a literal (i or $-i$), or true, then it is satisfiable; otherwise, it is unsatisfiable (labeled with false). A subsentence rooted at an OR node is satisfiable iff any of its children are satisfiable. For an AND node, it is satisfiable iff all of its children are satisfiable (since AND nodes are decomposable). We can compute the minimum cardinality of a sentence and enumerate its models in a similar way [12,17].

Before we proceed to describe how to compile a sentence in CNF into DNNF, consider the *conditioning* of a sentence Δ on an instantiation α, denoted $\Delta \mid \alpha$. This operation yields a sentence that can be obtained by replacing every literal in

---

[1] c2D further enforces the *determinism* property, and more specifically, compiles CNF instances into d-DNNF [16].

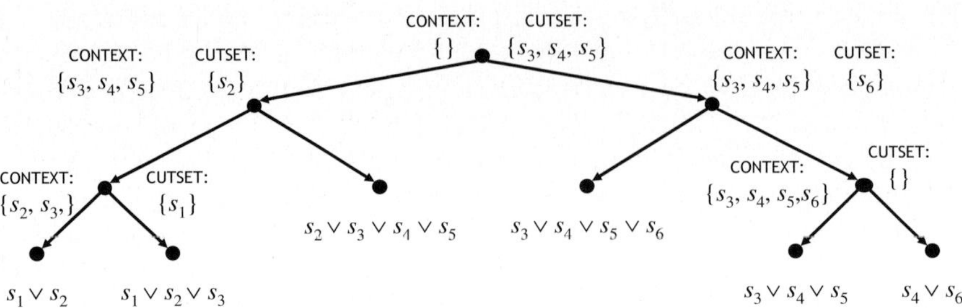

Fig. 2. A d-tree for the CNF used for the example in Figure 1. Internal nodes are labeled with their contexts and cutsets.

Δ with **true** (respectively, **false**) if it is consistent (inconsistent) with instantiation α. For example, conditioning the DNNF $(\neg a \wedge \neg b) \vee (b \wedge c)$ on instantiation $b \wedge d$ gives $(\neg a \wedge \mathsf{false}) \vee (\mathsf{true} \wedge c)$. Note that DNNF is closed under conditioning, i.e., conditioning a DNNF Δ on an instantiation α results in another DNNF. Moreover, the resulting sentence $\Delta \mid \alpha$ does not mention variables assigned by α.

Consider the following theorem, due to [12], which motivates the compilation procedure underlying C2D.

Theorem 1 (Case Analysis). *Let Δ_1 and Δ_2 be two sentences in DNNF, and let Δ be the sentence $\bigvee_\alpha (\Delta_1 \mid \alpha) \wedge (\Delta_2 \mid \alpha) \wedge \alpha$, where α are instantiations of variables mentioned in both Δ_1 and Δ_2. Then Δ is in DNNF, and is equivalent to $\Delta_1 \wedge \Delta_2$.*

This theorem suggests a recursive algorithm $\mathsf{DNNF1}(\Phi)$ that converts a sentence Φ in CNF into a sentence Δ in DNNF:

1. If Φ contains a single clause ϕ, return $\mathsf{DNNF1}(\Phi) \leftarrow \phi$. Note that a clause is vacuously decomposable.
2. Otherwise, return

$$\mathsf{DNNF1}(\Phi) \leftarrow \bigvee_\alpha \mathsf{DNNF1}(\Phi_1 \mid \alpha) \wedge \mathsf{DNNF1}(\Phi_2 \mid \alpha) \wedge \alpha,$$

where Φ_1 and Φ_2 is a partitioning of clauses in Φ, and α is an instantiation of the variables mentioned in both Φ_1 and Φ_2.

We can see that this procedure gives us the decomposability property, but at the expense of increasing the size of the original sentence. This increase is incurred primarily due to the case analysis performed, and the extent of this increase is sensitive to the way we decide to partition the clauses of the input sentence Φ. In particular, we would want to minimize the number of common variables between Φ_1 and Φ_2, as the complexity of case analysis is exponential in this number.

Partitioning can be guided by decomposition trees, or simply d-trees [12].

Definition 1. *A <u>d-tree</u> $\mathcal{T}$ for a CNF Φ is a binary tree whose leaves correspond to the clauses in Φ.*

An example d-tree for the CNF used for the example in Figure 1 is shown in Figure 2. Intuitively, the above compilation procedure traverses the d-tree, starting from the root, where case analysis is performed based on how the d-tree partitions the clauses of the CNF Φ. In particular, each interior node t is associated with the set of clauses that appear below it, and the partition is determined by the clauses of t's left and right children.

As we can see in Figure 2, each internal node is labeled with two variable sets: the *cutset* and the *context*. At a given node t, the cutset tells our compilation algorithm which variables to perform case analysis on. The context tells us those variables that appear in both of t's children, but have already been instantiated for case analysis by an ancestor. An instantiation α of the context variables can then be used as a key for a cache that stores the results of compiling the subset of clauses $\Phi \mid \alpha$. When a node is revisited with the same context, then the algorithm can simply return the DNNF sentence $\Delta \mid \alpha$ already computed. For example, at the root of the tree, the cutset contains $\{s_3, s_4, s_5\}$ since those variables appear in both children. If we follow the left branch twice, the context is now $\{s_2, s_3\}$, which was instantiated by the root and its left children. Note that this node will be visited multiple times for different instantiations of $\{s_4, s_5\}$, but only different instantiations of the context yield different subproblems. Thus, when this node is revisited with the same context instantiation, we simply fetch the result from the cache. It is this subformula re-use that allows compilation to moderate the exponential growth of the formula caused by case analysis.[2] The c2d compiler, while based on this approach, employs more advanced techniques to further improve on the efficiency of compilation [14].

Scaling to Whole Genome Tagging. The c2d compiler is capable of computing minimal tag sets, for several thousands of SNPs. Unfortunately, memory becomes an issue when we try to compile even larger regions of the genome. To encode the entire genome as a CNF, however, we must use 3.8 million literals and 3.8 million clauses. Clearly, we need new techniques to scale to this problem size. Roughly, our approach is to compile sufficiently small regions of the genome into DNNF, which are then "stitched" together to construct a DNNF for the entire genome. Due to space limitations, we omit the detail of this approach.

2.2   Multi-marker SNP Tagging

Recent work has shown that using statistical tests based on haplotypes over multiple SNPs improves the power of whole genome association studies[23,30]. In the context of tagging, this permits combinations of tag SNPs (multi-marker tags) to cover a SNP, allowing for a smaller set of tags to cover the SNPs.

In this situation, an even smaller set T' may be a valid cover of SNPs.

Again, we reduce the problem of identifying a valid set of SNPs to satisfiability. Given a threshold r^2_{min}, we now have two classes of clauses Φ and Ψ. Clauses Φ, as before, enforce constraints that require each SNP in S to be covered:

---

[2] In particular, the complexity of compilation can be bounded in terms of the size of the context and the cutset [12].

$\Phi = \phi_1 \wedge \cdots \wedge \phi_n$, where there are as many clauses ϕ_i as there are SNPs s_i, but where each clause is now of the form:

$$\phi_i = \left(\bigvee_{r^2_{j \to i} \geq r^2_{min}} s_j \right) \vee \left(\bigvee_{r^2_{j,k \to i} \geq r^2_{min}} p_{jk} \right).$$

In this case, either a positive literal s_j representing SNP s_j or a positive literal p_{jk} representing a SNP pair (s_j, s_k) can also satisfy clause ϕ_i and cover SNP s_i.

Clauses Ψ enforce the constraints that if a pair literal p_{jk} is true, then both s_j and s_k are true (i.e., in the tag set):

$$\Psi = \bigwedge_{r^2_{j,k \to i} \geq r^2_{min}} (p_{jk} \equiv s_j \wedge s_k)$$

where $p_{jk} \equiv s_j \wedge s_k$ are equivalence constraints that ensure that p_{jk} is selected iff the corresponding pair (s_j, s_k) is selected. In clausal form, this equivalence constraint is given by three clauses: $\neg p_{jk} \vee s_j$, $\neg p_{jk} \vee s_k$ and $p_{jk} \vee \neg s_j \vee \neg s_k$.

Consider the example from the previous section with six SNPs $s_1, \ldots, s_6$. Suppose that the pair (s_1, s_3) can cover s_6, and that $r^2_{1,3 \to 6} = 0.9$. Given the threshold $r^2_{min} = 0.8$, we gain a third optimal solution s_1, s_3, to go with the two solutions s_4, s_2 and s_4, s_1 from before. Encoding the problem, we have the following formula: $(s_1 \vee s_2) \wedge (s_1 \vee s_2 \vee s_3) \wedge (s_2 \vee s_3 \vee s_4 \vee s_5) \wedge (s_3 \vee s_4 \vee s_5 \vee s_6) \wedge (s_3 \vee s_4 \vee s_5) \wedge (s_4 \vee s_6 \vee p_{13}) \wedge (p_{13} \equiv s_1 \wedge s_3)$. We again want a minimum cardinality assignment, but minimizing only the number of positive s_i literals. We can introduce constraints $p_{jk} \equiv \neg q_{jk}$ to cancel out the contribution of the p_{jk}'s to the cardinality with the q_{jk}'s; we can then convert to DNNF and minimize, as before.[3]

3   Results

We downloaded the complete HapMap build 22 data including all ENCODE regions. These data are genotypes on 270 individuals in 4 populations and over 3.8 million SNPs. They represent the most complete survey of genotype data currently available and are used as our test data sets. The 10 ENCODE regions span 5 MB and are believed to have complete ascertainment for SNPs with frequency greater than 5%. They are commonly used to estimate the performance of association study design methods and tag SNP selection methods since there are still many unknown common SNPs in the rest of the genome.

3.1   ENCODE Single SNP r^2 Comparison

We compared the performance of our method to the two other optimal methods as well as the non-optimal greedy algorithm over each of the ENCODE regions

---

[3] We can also existentially quantify out the p_{jk} literals; this operation is supported by C2D [6].

Region	n_s	Greedy	Halldorson(W=15)	FESTA($L=10^7$)	Optimal
ENm010	567	159	243(12m)	157(0m)	157
ENm013	755	93	309(23m)	90(0m)	90
ENm014	914	164	393(33m)	157(0m)	157
ENr112	927	180	340(34m)	173(34m)	173
ENr113	1072	179	395(46m)	176(274m)	176
ENr123	937	174	463(35m)	172(0m)	172
ENr131	1041	228	414(44m)	221(0m)	221
ENr213	659	122	248(17m)	122(0m)	122
ENr232	533	142	181(11m)	140(36m)	140
ENr321	599	132	202(14m)	131(0m)	131

Fig. 3. Comparison of several *tagging* algorithms over the Encode regions in the CEU population. n_s is the number of SNPs in the region. The table shows the tag set size for each of the methods various methods (smaller is better). Running times are given in parentheses in minutes. All running times for Optimal are less than one minute (0m).

in each of the populations. Haldorsson *et al.*[5] restricts the maximum length of correlations and uses a dynamic programming procedure which guarantees to find an optimal solution. Given a window size W, Haldorsson *et al.*[5] examines all 2^W possible choices of tag SNPs in the window and then uses dynamic programming to extend this to a longer region. FESTA [24] extends the standard greedy algorithm in a natural way. Given an r^2 threshold, FESTA partitions the SNPs into precincts where SNPs are correlated only within the precinct. For a small precinct FESTA enumerates all possible tag sets in search of the minimal tag set. For a larger precinct, FESTA applies a hybrid exhaustive enumeration and greedy algorithm by first selecting some SNPs using exhaustive enumeration and then applying greedy algorithm. The user defines a threshold L so that the hybrid method is applied when $\binom{n}{k} > L$ where n is the number of SNPs in a precinct and k is the number of tags that need to be selected from the precinct. We use $L = 10^7$ in our experiments.

The results are presented in table 3. Surprisingly, the Halldorson *et al.*[5] method, at the maximum limit of what is computationally feasible ($W = 15$) performs worse than the simple greedy algorithm and is much slower than than both FESTA and our approach. FESTA and our approach both recover optimal solutions for the ENCODE regions, however, FESTA ends up spending a very large amount of computational time in very large precincts, taking several hours to complete some of the data sets while our approach requires less than a minute.

3.2   Whole Genome Single SNP r^2

We ran our approach and the greedy approach over the entire genome wide HapMap data for the CEU population in order to find the minimal tag set to cover all SNPs with MAF ≥ 0.05 and $r^2 \geq 0.8$. Greedy resulted in 472729 tag SNPs while our approach needed only 468967 over the entire 1692323 SNP data set. This modest decrease shows that in the single SNP r^2 tag SNP selection approach, greedy search performs almost as well as optimal search. Our program required less than one day on a single CPU to run over the whole genome. Once we compile into DNNF, which is done just once, we can perform a variety of queries and operations on the set of all optimal solutions [16], allowing for flexible design or optimization on a secondary criteria. For example, we can efficiently obtain the set of optimal solutions that contain some SNPs and do not contain other SNPs.

3.3   Mutli-marker Tag SNP Selection ENCODE Results

Although for the single SNP r^2 tag SNP selection problem, the greedy algorithm achieves a solution close to the optimal solution, this is not the case for multi-marker tag SNP selection. We compare our method to the popular Tagger [18] method over the Encode region ENm010 in the CEU population. This regions contains 567 SNPs with minor allele frequency (MAF) greater than 0.05. Tagger first computes a single SNP tag set using greedy search resulting in 159 tag SNPs and then uses a "roll back" procedure in which a SNP s_i is removed from the tag set if another pair of SNPs in the tag set cover s_i with $r^2 = 1.0$. That is, redundant SNPs are removed from the tag set. Tagger's multi-marker approach does reduce the number of SNPs required to cover an ENCODE region to 141 SNPs compared to 157 optimal single SNP tags, but the reduction is far from optimal. Our method requires only 72 SNPs to cover the ENCODE region a 54% and 40% reduction over single SNP tags and the Tagger's multi-marker tags.

4   Discussion

We have presented a novel method for solving a variety of tag SNP selection problems which guarantee a solution the the minimal number of tag SNPs. Since the problem is NP-hard, our methods provide no guarantees on the running time of the algorithms. However, we have demonstrated that in practice, our approach is very efficient and much faster than previously proposed exhaustive methods. Improvements over the classic single SNP r^2 tagging problem are modest compared to greedy search. The FESTA [24] algorithm also achieves these results over the ENCODE regions, but is not guaranteed to discover a minimal tag set in the general case. We outperform FESTA in terms of running time. In addition, our approach allows us to characterize *all* optimal solutions as opposed to just those containing perfectly linked SNPs. This permits flexible tag set choice that can be further optimized over secondary criteria. This method is also extensible to other measures of SNP coverage besides r^2.

Recent work has shown the multi-marker methods are more powerful than single SNP techniques in the context of association studies. While a variety of multi-marker statistical tests exist, the current optimal tagging methods such as FESTA are not able to tag for multiple markers efficiently. Our SAT based method is able to find optimal multi-marker tags for pairs of SNPs over the dense ENCODE regions. The gain for optimal tagging over greedy search in this context is significantly better than for the single SNP with improvement over the popular Tagger [18] method reaching 40%. Since the cost of custom genotype arrays remains high, this tool is valuable for follow up association studies. That is, once genome wide results have been found, further genotyping must be done to localize the region containing the causal variant. Intelligent choice of tag sets for follow up studies can greatly improve their power and until now, multi-marker tagging for follow up has been non-optimal.

In addition to r^2, other factors that influence power are the minor allele frequency of the causal variant and the strength of the effect of a variant on disease risk. We can also formulate a variant of the tag SNP selection problem in which the goal is to optimize the statistical power given that we are collecting a fixed number of tag SNPs. For this problem, formulating the problem as a SAT problem is difficult because the constraint of using a fixed number of tags creates long range dependencies which complicate the decomposition of the problem into a DNNF. A typical approach to solve problems with such constraints is to formulate the problem as a Max-SAT problem where each clause has a weight and we are trying to satisfy the largest number of clauses. In future work, we will explore formulating the maximum power tag SNP selection problem as a Max-SAT problem in order to choose tags which optimally maximize the statistical power.

Our method and whole genome optimal data sets are available for use via webserver at http://whap.cs.ucla.edu.

Acknowledgments

N.Z. is supported by the Microsoft Graduate Research Fellowship. N.Z., B.H. and E.E. are supported by the National Science Foundation Grant No. 0513612, and National Institutes of Health Grant No. 1K25HL080079.

References

1. Bafna, V., Halldorsson, B.V., Schwartz, R., Clark, A., Istrail, S.: Haplotypes and informative snp selection: Don't block out information. In: RECOMB, pp. 19–27 (2003)
2. Barrett, A.: From hybrid systems to universal plans via domain compilation. In: Proceedings of the 14th International Conference on Planning and Scheduling (ICAPS), pp. 44–51 (2004)
3. Barrett, A.: Model compilation for real-time planning and diagnosis with feedback. In: Proceedings of the Nineteenth International Joint Conference on Artificial Intelligence (IJCAI), pp. 1195–1200 (2005)
4. Bonet, B., Geffner, H.: Heuristics for planning with penalties and rewards using compiled knowledge. In: Proceedings of the International Conference on Principles of Knowledge Representation and Reasoning (KR), pp. 452–462 (2006)
5. Halldorsson, B.V., Bafna, V., Lippert, R., Schwartz, R., De La Vega, F.M., Clark, A.G., Istrail,: Optimal haplotype block-free selection of tagging snps for genome-wide assoaciation studies. Genome Research 14, 1633–1640 (2004)
6. The c2d compiler, http://reasoning.cs.ucla.edu/c2d/

7. Carlson, C.S., Eberle, M.A., Rieder, M.J., Yi, Q., Kruglyak, L., Nickerson, D.A.: Selecting a maximally informative set of single-nucleotide polymorphisms for association analyses using linkage disequilibrium. Am. J. Hum. Genet. 74(1), 106–120 (2004)

8. Chavira, M., Darwiche, A.: Compiling Bayesian networks with local structure. In: Proceedings of the 19th International Joint Conference on Artificial Intelligence (IJCAI), pp. 1306–1312 (2005)

9. Chavira, M., Darwiche, A., Jaeger, M.: Compiling relational Bayesian networks for exact inference. International Journal of Approximate Reasoning 42(1–2), 4–20 (2006)

10. The International HapMap Consortium. A haplotype map of the human genome 437(7063), 1299–1320 (2005)

11. The Wellcome Trust Case Control Consortium. Genome-wide association study of 14,000 cases of seven common diseases and 3,000 shared controls. Nature 447, 661–678 (2007)

12. Darwiche, A.: Decomposable negation normal form. Journal of the ACM 48(4), 608–647 (2001)

13. Darwiche, A.: On the tractability of counting theory models and its application to belief revision and truth maintenance. Journal of Applied Non-Classical Logics 11(1-2), 11–34 (2001)

14. Darwiche, A.: A compiler for deterministic, decomposable negation normal form. In: Proceedings of the Eighteenth National Conference on Artificial Intelligence (AAAI), pp. 627–634. AAAI Press, Menlo Park (2002)

15. Darwiche, A.: New advances in compiling CNF to decomposable negational normal form. In: Proceedings of European Conference on Artificial Intelligence, pp. 328–332 (2004)

16. Darwiche, A., Marquis, P.: A knowledge compilation map. Journal of Artificial Intelligence Research 17, 229–264 (2002)

17. Darwiche, A., Marquis, P.: Compiling propositional weighted bases. Artificial Intelligence 157(1-2), 81–113 (2004)

18. de Bakker, P.I.W., Yelensky, R., Pe'er, I., Gabriel, S.B., Daly, M.J., Altshuler, D.: Efficiency and power in genetic association studies. Nat. Genet. 37(11), 1217–1223 (2005)

19. Elliott, P., Williams, B.: Dnnf-based belief state estimation. In: Proceedings of the 21st National Conference on Artificial Intelligence (AAAI 2006) (2006)

20. Darwiche, A., Palacios, H., Bonet, B., Geffner, H.: Pruning conformant plans by counting models on compiled d-dnnf representations. In: Proceedings of the 15th International Conference on Planning and Scheduling (ICAPS), pp. 141–150. AAAI Press, Menlo Park (2005)

21. Huang, J.: Complan: A conformant probabilistic planner. In: Proceedings of the 16th International Conference on Planning and Scheduling (ICAPS) (2006)

22. Huang, J., Darwiche, A.: On compiling system models for faster and more scalable diagnosis. In: Proceedings of the 20th National Conference on Artificial Intelligence (AAAI), pp. 300–306 (2005)

23. Pe'er, I., de Bakker, P.I.W., Maller, J., Yelensky, R., Altshuler, D., Daly, M.: Evaluating and improving power in whole genome association studies using fixed marker sets. Nature Genetics 38, 663–667 (2006)

24. Qin, Z.S., Gopalakrishnan, S., Abecasis, G.R.: An efficient comprehensive search algorithm for tagSNP selection using linkage disequilibrium criteria. Bioinformatics 22(2), 220–225 (2006)

25. Sang, T., Beame, P., Kautz, H.: Solving Bayesian networks by weighted model counting. In: Proceedings of the Twentieth National Conference on Artificial Intelligence (AAAI 2005), vol. 1, pp. 475–482. AAAI Press, Menlo Park (2005)
26. Siddiqi, S., Huang, J.: Hierarchical diagnosis of multiple faults. In: Proceedings of the Twentieth International Joint Conference on Artificial Intelligence (IJCAI) (2007)
27. Halldorsson, B.V., Istraila, S., De La Vegab, F.M.: Optimal selection of snp markers for disease association studies. Human Heredity 58, 190–202 (2004)
28. Wachter, M., Haenni, R.: Logical compilation of bayesian networks. Technical Report iam-06-006, University of Bern, Switzerland (2006)
29. Yolifè Arvelo, M.-E.V., Bonet, B.: Compilation of query–rewriting problems into tractable fragments of propositional logic. In: Proceedings of AAAI National Conference (2006)
30. Zaitlen, N., Kang, H.M., Eskin, E., Halperin, E.: Leveraging the HapMap correlation structure in association studies. Am. J. Hum. Genet. 80(4), 683–691 (2007)

Computing the Minimal Tiling Path from a Physical Map by Integer Linear Programming

Serdar Bozdag[1], Timothy J Close[2], and Stefano Lonardi[1]

[1] Dept. of Computer Science and Eng., University of California, Riverside, CA
[2] Dept. of Botany and Plant Sciences, University of California, Riverside, CA
{sbozdag,stelo}@cs.ucr.edu, timothy.close@ucr.edu

Abstract. We study the problem of selecting the minimal tiling path (MTP) from a set of clones arranged in a physical map. We formulate the constraints of the MTP problem in a graph theoretical framework, and we derive an optimization problem that is solved via integer linear programming. Experimental results show that when we compare our algorithm to the commonly used software FPC, the MTP produced by our method covers a higher portion of the genome, even using a smaller number of MTP clones. These results suggest that if one would employ the MTP produced by our method instead of FPC's in a clone-by-clone sequencing project, one would reduce by about 12% the sequencing cost.

1 Introduction

A physical map is a linear ordering of a set of clones encompassing one or more chromosomes. Physical maps can be generated by first digesting clones with restriction enzymes and then detecting the clone overlaps by matching the lengths of the fragments produced by digestion. A minimum-cardinality set of overlapping clones that spans the region represented by the physical map is called *minimal tiling path* (MTP).

The problem of determining a good set of MTP clones is crucial step of several genome sequencing projects. For instance, in the sequencing protocol called *clone-by-clone*, first a physical map is constructed, then the MTP is computed and finally, the clones in the MTP are sequenced one by one [9]. The clone-by-clone sequencing method has been used to sequence several genomes including *A. thaliana* [15] and *H. sapiens* [12] among others. Also, in several recent whole-genome shotgun sequencing projects, the MTP obtained from a physical map has been employed to validate and improve the quality of sequence assembly [23]. This validation step has been used, for example in the assembly of *M. musculus* [10], *R. norvegicus* [13], and *G. gallus* [18].

With the introduction of next-generation sequencing machines (454, Solexa/Illumina, and ABI SOLiD) we expect the MTP computation to become an essential step in *de novo* sequencing projects of eukaryotic genomes. Next-gen sequencing technology produces massive amount of very short reads (about 250bps for 454, 35bps for Illumina and SOLiD) [4] and therefore the *de novo* assembly of the whole eukaryotic genomes is extremely challenging [17]. Arguably, the only feasible method at this time is a clone-by-clone approach, where each clone in the MTP is sequenced using next-gen technology, and the assembly is resolved separately on each clone (see [17,21,25] and references therein).

K.A. Crandall and J. Lagergren (Eds.): WABI 2008, LNBI 5251, pp. 148–161, 2008.
© Springer-Verlag Berlin Heidelberg 2008

If the exact locations of all clones in the physical map were known, computing its MTP would be straightforward; simply select the set of clones in the shortest path from the leftmost clone to the rightmost clone in the interval graph representing all the clones. This, however, is not a realistic solution. The noise in the fingerprinting data makes it impossible to build a perfect map. As a consequence, determining the minimal tiling path becomes a challenging computational problem. On one hand, a method that tends to select more clones as MTP might include many *redundant clones* (i.e., clones that do not provide additional coverage) and therefore it would waste time and money later in sequencing. On the other hand, an approach that tries to reduce the number of MTP clones may introduce gaps between the clones in some of the contigs, and thus reduce the coverage.

Although the problem of computing MTP has been studied extensively in the literature (see e.g. [22,15]), in practice there is only commonly used software tool, namely FingerPrinted Contigs (FPC) [7]. FPC provides three methods to compute an MTP, but only one uses solely restriction fingerprint data (hereafter called FPC-MTP). FPC-MTP computes the approximate overlap between clones in the contig and validates each overlap by using three extra clones, a *spanner* that verifies the shared fragments of the pair and two flanking clones that extend to the left and right of the pair and confirm fragments in the pair that are not confirmed by the spanner clone. Once the verification of the overlapping fragments between each clone pair is completed, a shortest path algorithm is used to find the minimal tiling path [2].

FPC-MTP's algorithm is quite good, but can be improved; our experimental results show that FPC-MTP is significantly distant from the optimal[1] MTP. In general, FPC-MTP selects fewer clones than necessary which in turns reduces the overall coverage. By changing parameters one can increase the coverage, but this comes at the cost of introducing many redundant clones. This limitation of FPC-MTP can be attributed to the fact that it checks the clone positions as an overall constraint when computing the MTP [2]. However, the positions of clones in contigs are known to be not very reliable [16].

Our contribution. We propose a new algorithm, called FMTP, that computes the MTP of a physical map based purely on restriction fingerprint data (and the contigs). In other words, our algorithm completely ignores the ordering of clones obtained by the physical map algorithm.

FMTP first computes a *preliminary* MTP by selecting the smallest set of clones that covers the genomic region that is covered by all clones in the contig. The problem of computing the preliminary MTP set is formulated in a combinatorial optimization framework as an Integer Linear Program (ILP). The preliminary MTP set may contain redundant clones. In the second phase, FMTP orders the clones in the preliminary MTP and computes the final MTP by using a shortest path algorithm.

We carried out an extensive set of experiments on the physical map of rice and barley. For the former dataset, the actual coordinates for the clones are known and therefore we could measure the accuracy of our algorithm. The experimental results show that the set of MTP clones computed by FMTP on the physical map for rice has higher coverage than the one produced by FPC (using approximately the same number of clones overall). This suggests that a larger portion of the genome could be obtained at the same cost when

---

[1] The optimal MTP is the one that we could compute if we knew the coordinates of all the clones.

FMTP is used instead of FPC. Our experimental results also show that if one fixes a target coverage, FMTP produces MTPs with about 12% fewer clones than MTPs produced by FPC. This suggests that our tool could reduce sequencing costs by about 12%.

2   Basic Concepts

We use the term *clone fragment* (or *fragment*) to indicate a portion of a clone obtained by digesting it with a restriction enzyme. Let $b(u)$ be the size for clone fragment u. We say that two clone fragments u and v *match* if their corresponding sizes are within the *tolerance* T, i.e., if $|b(u) - b(v)| \leq T$. The tolerance parameter depends on the finger-printing method, thus it should match the one used in the construction of the physical map [19]. Given a fragment u, let $N(u)$ be the name of the clone which u belongs to. This notion can be extended to a set of fragments as follows. Given a set of fragments U, $N(U)$ represents the set of all clones that contain at least one fragment in U.

We declare two clones c_i and c_j to be *overlapping* if their Sulston score S is lower than or equal to a user-defined *cutoff* threshold C, i.e., $S(c_i, c_j) \leq C$. The Sulston score measures the probability that two clones share a given number of restriction fragments by chance according to a binomial probability distribution [20]. FPC has an analogous cutoff parameter to build the physical map [19].

A *matching fragment graph* (MFG) is a weighted undirected graph $G = (V, E)$ in which V represents the set of all clone fragments and an edge $(u, v) \in E$ exists if fragment u matches fragment v. The weight on the edge $(u, v) \in E$ is defined as the negative logarithm of $S(N(u), N(v))$. In the ideal MFG, each connected component of the matching fragment graph G should correspond to a unique region in the target genome. Unfortunately, this in not realistic due to noise in the fingerprinting data and falsely matching fragments.

Given a connected component $U \subseteq V$ of G, we call $N(U)$ a *connected component cloneset* (or *cloneset* if it is clear from the context). We say that a clone c *covers* a connected component U of G if it belongs to the corresponding cloneset, i.e., $c \in N(U)$. Given an MFG G, the *minimal tiling path* of G is the smallest set M of clones that cover all connected components of G, i.e., $M \cap N(U) \neq \emptyset$ for all connected components U of G.

A toy example of an MFG is shown in Figure 1-LEFT. Each node is labeled as <clone name>-<fragment size>-<copy number of the fragment>. Clones K, L, M, N, and O are overlapping. Shared fragments are represented by the connected component of the MFG (namely, fragments of approximate size of 1870, 1805, 1255, and 1400 bases respectively). For example, given the connected component $U = \{$L-1803-1, M-1807-1, N-1802-1$\}$, the cloneset $N(U)$ is $\{$L,M,N$\}$. The minimum tiling path is $\{$L,O$\}$ because by selecting these two clones we cover all the connected components of the graph.

3   Methods

FMTP consists of two modules, namely MTP-ILP and MTP-MST. The module MTP-ILP computes a preliminary MTP by solving an integer linear program. The objective of MTP-ILP is to cover all connected components in the MFG using the smallest possible set of clones. The module MTP-MST computes the final MTP based on an overlap

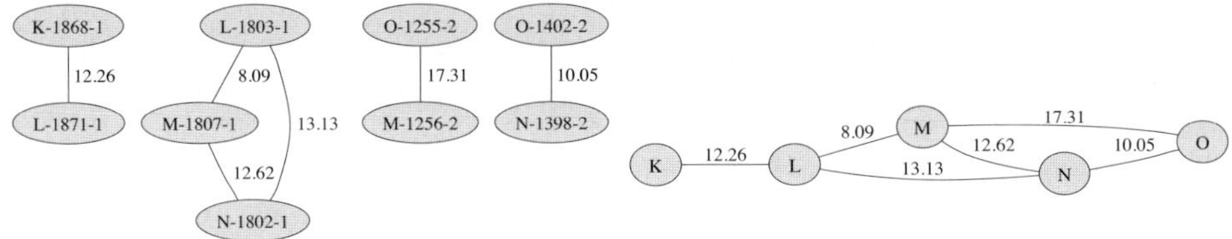

Fig. 1. (LEFT) An example of a matching fragment graph. Each node is a clone fragment and it is labeled as the <clone name>-<fragment size>-<copy number of the fragment>. Each edge is weighted by the negative logarithm of the Sulston score between the clones that contain the incident node fragments. (RIGHT) Overlap graph of clones in this MFG. There is an edge between two clones if their fragments are together in at least one connected component in the MFG.

graph. The goal of MTP-MST is first to order the clones in the preliminary MTP and then run a shortest path algorithm to compute the final MTP. In both modules, an MFG is constructed by performing the steps described below.

3.1   Constructing the MFG

For each contig, FMTP performs a pairwise alignment between all clones in that contig based on their restriction fingerprint data. The alignment produces the initial MFG. Then, some of the false positive edges (i.e. falsely matched fragments) are removed and the final MFG is constructed, as described below in more details.

Preprocessing: burying clones.   The purpose of this preprocessing step is to reduce the problem size by discarding clones that are almost completely contained in another clone. A clone is defined *buried* if $B\%$ or more of its fragments are matching fragments of another clone, where B is a user-defined parameter. According to our experiments, B should be at least 80 to avoid false positive buried clones. If two clones can be buried into each other, the smaller of the two (i.e., the one with fewer fragments) is buried into the other one. FPC also buries clones during the process of building the physical map, but it does not discard them during MTP computation.

Building the preliminary MFG.   First, we align the fingerprint data for each pair of overlapping clones. For each clone pair (c_i, c_j) for which $S(c_i, c_j) \le C$, we build a bipartite graph $G_{i,j} = (L_i \cup R_j, E_{i,j})$, where L_i and R_j consist of the fragments of c_i and c_j, respectively, and $E_{i,j} = \{(u, v)|u \in L_i, v \in R_j \text{ such that } |b(u) - b(v)| \le T\}$.

In order to align clones c_i and c_j, we search for the maximum bipartite matching in $G_{i,j}$. The matching of maximum cardinality is found by solving max flow on the corresponding flow network [6]. Let $M_{i,j}$ be set of matched edges.

For all clone pairs c_i and c_j for which $S(c_i, c_j) \le C$, the matching edges in $M_{i,j}$ are used to create the (preliminary) matching fragment graph G. Specifically, for each edge $(u, v) \in M_{i,j}$, nodes u, v and edge (u, v) are added to G (unless they have been already added). The weight of (u, v) is set to be the negative logarithm of the Sulston score between clone c_i and clone c_j.

The objective of the bipartite matching is to attempt to group together clone fragments that are located at the same location on the genome. Because of the noise in the fingerprint data, some of the matched fragments might not represent the same region in the target genome. In the following steps, we try to eliminate as many false matches as possible.

MFG pruning. In this step, some of the components of G that might represent more than one unique region in the target genome are split. Specifically, we examine all the connected components of G and mark the ones that satisfy at least one of the following conditions as *candidates*.

1. Extra fragment: The connected component contains multiple fragments of a clone.
2. Unmatched fragments: The difference between the length of the second shortest and the second longest fragment in the connected component is more than the tolerance value T. We ignore the shortest and the longest fragments to allow two outliers per component.
3. Weak overlap: The connected component contains at least one pair of clones that are very unlikely to overlap (i.e., have Sulston score of at least 1e-2.5 for MTP-ILP or 1e-1 for MTP-MST).

For each candidate component, a min-cut algorithm is used to partition it by removing the weakest set of edges (i.e., minimum total edge weight) [11]. After a component is partitioned into two subgraphs, we check both subgraphs against the three conditions. If at least one is satisfied, we partition that component again. This iterative process terminates as soon as there are no more components that need to be split.

Elimination of spurious components. In order to introduce the notion of spurious components, let us assume for the time being that the exact locations of the clones in a contig are known. Let U be a connected component of G, and let $N(U)$ be the corresponding connected component cloneset. We call U *spurious* if the overlapping region for the clones in $N(U)$ is spanned by another clone in the contig. An example is illustrated in the Appendix.

This step is crucial to reduce the number of redundant clones in the preliminary MTP produced by MTP-ILP. Recall that MTP-ILP selects the smallest set of clones that cover each connected component in the MFG. If the spurious components are not removed, at least one clone in each spurious component is added to the MTP without contributing to the overall coverage.

Since the exact ordering of the clones is not available, we employ a probabilistic method to detect spurious components. More specifically, we compute the probability that a connected component is observed purely by chance. If this probability is high then we mark the component as spurious. We expect the number of spurious components to be low, since missing fragments or extra fragments are unlikely to occur frequently for a specific cloneset.

In our model, we assume that fragment sizes are distributed uniformly, and that the probability that two clones share a fragment is $P = 2T/gellen$ [19], where T is the tolerance and *gellen* is the number of possible fragment size values. The probability that two clones share f fragments is P^f and the probability that c clones share f fragments is $P^{(c-1)f}$. This suggests that the probability that a connected component U is observed in the MFG is $P^{(|U|-1)f}$, where f is number of times that $N(U)$ forms a connected component. In our algorithm, if $(|U| - 1)f$ is lower than a threshold Q then we mark U as spurious. Spurious components are removed from the MFG permanently.

Removing clones that are buried into multiple clones. Recall that the goal of the first step in the construction of the MFG is to bury clones into other clones to reduce the problem size. In this step, we reduce the problem size further by removing clones that are buried into multiple clones. For example, in Figure 3 in the Appendix, clone C is buried into B $\cup$ D, thus C can be removed.

For each clone in a contig, we compute the ratio between the number of its fragments in the MFG and the total number of its fragments. If at least $B'\%$ of its fragments are already in the MFG, then we mark that clone buried. However, there is a complication. It is possible that two clones are *mutually buried* and therefore we cannot remove both. In order to solve this problem, we first identify all possible candidates that can be buried into multiple clones. Then, we examine each connected component U of G, and save $N(U)$ in a list L if all clones in $N(U)$ are candidates to be buried.

All candidates that do not exist in any $N(U) \in L$ can be buried. However, we need to make sure that at least one clone in each $N(U) \in L$ is not buried, otherwise the region in the target genome that is represented by U is not covered. So the task at hand is to remove as many candidate clones as possible with the constraint that at least one clone survives from each cloneset $N(U) \in L$. This problem is called *minimum hitting set* and it is NP-complete (polynomial reduction from the vertex cover problem [8]). Here, we solve it sub-optimally using a greedy approach [8]. At each iteration, we (1) select the clone c that occurs in the maximum number of clone sets in L, (2) remove all clone sets that contain c, (3) save c into minimum hitting set H, and (4) repeat. The iterative process terminates when the list L is empty.

Buried clones selected in this way are removed from the MFG permanently. If this removal introduces components with only one node, they are also removed from the MFG permanently. The algorithm is summarized as Algorithm 1 in the Appendix.

Adding mandatory clones to the MFG. Clones that have to be selected as MTP clones are called *mandatory* clones. More specifically, if at least 50% of the fragments of a clone are not present in the MFG, then that clone is marked as mandatory. When a clone is marked as mandatory, it is immediately stored in the MTP and ignored for further analysis by removing all of its fragments from the MFG.

3.2   Solving the MTP Via ILP

Recall that the MTP can be computed by selecting the smallest set of clones that cover all connected components of the MFG. This problem is a special case of the *minimum hitting set* problem. Although this problem is NP-complete on general graphs [8], it can be solved optimally in polynomial time on overlap (or interval) graphs [3]. Here, we solve this problem optimally by expressing it as an integer linear program (ILP), but first we remove any connected component from the MFG that does not affect the solution.

We reduce the problem size by removing some of the connected components of G according to the following Lemma (the proof is immediate). This simplification step drastically reduces the execution time required to solve the ILP (about 200-fold for rice).

Lemma 1. *Let U, V be two connected components of G, such that $N(V) \subset N(U)$ and G' is a graph obtained from G by removing U. If M is an MTP for G' then M is an MTP for G.*

The MTP problem can be expressed as follows

$$Minimize \quad \sum_{c \in N(V)} X[c]$$

$$Subject\ to \quad \sum_{c \in N(V)} X[c].CC[c, U] \geq 1 \quad \forall U \in G$$

$$X[c] \in \{0, 1\} \quad \forall c \in N(V)$$

The coefficient $CC[c, U]$ is 1 if clone $c \in N(V)$ has a fragment f that belongs to the connected component U of G, 0 otherwise. The integer variable $X[c]$ is 1 if $c \in N(V)$ is selected, 0 otherwise. When c is selected, all connected components U of G such that $c \in N(U)$ are covered.

We generate one ILP for each contig. The number of variables in each ILP is $|N(V)|$ (i.e., number of clones that contain fragments in the MFG). We do not need to relax the ILP to a linear program, because the problem size is small enough that can be solved optimally using GNU Linear Programming Package (GLPK). The clones in each solution set are then merged in the preliminary MTP. This preliminary MTP is very likely to contain several redundant clones and need to be processed further.

3.3   Solving the MTP Via MST

From now on, all the clones in the physical map that do not belong to the preliminary MTP are disregarded. A new MFG $G = (V, E)$ is constructed only on the clones in the preliminary MTP. Then, an overlap graph is built from G and the minimum spanning tree (MST) of the overlap graph is computed to order the clones. Finally, the shortest path from the first clone to the last clone in the ordering is computed. Clones on this path constitute the final MTP. Here are the details.

First, MTP-MST attempts to order the clones in the preliminary MTP. For this purpose, a weighted overlap graph $G_O = (V_O, E_O)$ is constructed for each contig, where V_O is the set of preliminary MTP clones in each contig and $E_O = \{(u, v)|$ There exists a connected component U of G such that $\{u, v\} \subseteq N(U)\}$. The weight of edge $(u, v) \in E_O$ is again the negative logarithm of the Sulston score between u and v.

Figure 1-RIGHT shows the overlap graph for the MFG in Figure 1-LEFT. For example, there is an edge between clones K and L because their fragments occur together in at least one connected component in the MFG (see Figure 1-LEFT).

In [24] we established that under some assumptions on the metrics (which are also met here), the MST of an edge-weighted overlap graph gives an ordering of nodes in the graph. One of the assumptions is that edge weights are proportional to the overlap size. Although the correlation between Sulston score and overlap size is not perfect, the MST still gives very accurate ordering because the clones come from the preliminary

MTP and not the original problem. Recall that in the preliminary MTP a clone is not expected to overlap to many clones with similar overlap size.

According to our experiments, the MST of G_O is usually a path. When the MST is not a path, the relative ordering of some of the clones may not be determined. However, this is not a serious problem if we can detect a pair of overlapping clones that cover the group of clones whose order is undetermined. Because of this overlap, clones with unknown relative ordering do not need to be in the MTP, hence they can be discarded in the analysis.

Once the MST of G_O is computed, we pick the longest path $\mathbb{P}$ in the tree. If there is more than one such path, we select the path with the smallest total weight. The rationale is to minimize the total overlap size between consecutive clones, and thus select the path with higher coverage. The clones in $\mathbb{P}$ cover almost the whole contig, but they may not be an MTP. In order to find the MTP, the path $\mathbb{P}$ must be augmented with high confidence overlap edges and a shortest path must be found.

To minimize the possibility of adding false negative and false positive edges to $\mathbb{P}$, we use several criteria based on the Sulston score and the MFG. The details of this algorithm are shown in Algorithm 2 in the Appendix. After augmenting $\mathbb{P}$, the shortest path from u_1 to $u_{|\mathbb{P}|}$ (in terms of number of hops) is computed. All nodes in this path are chosen as the MTP clones of this contig.

4   Experimental Results

We used the genomic data of two plants, namely barley and rice, to compare our software to FPC-MTP (hereafter called FPC to be clearer). We ran FPC and FMTP on the physical maps of rice and barley and obtained their MTPs. The physical maps were obtained using FPC and our compartmentalized method [5] on real restriction fingerprints of BACs. Restriction fingerprint data for rice and barley BACs were obtained from AGCoL [1] and our NSF-sponsored project [14], respectively.

The physical map that we used for rice contains 22,474 clones, 1,937 contigs and 1,290 singletons. The publicly available rice physical map [1] is a superset of our map because we selected only clones that could be uniquely mapped to the rice genome. Barley physical map contains 72,052 clones, 10,794 contigs, and 10,598 singletons.

Since the barley genome has not been fully sequenced yet, it is not possible to equivalently assess the quality of the barley MTP. However, we obtained genomic coordinates for all rice BACs in the physical map by locating their BAC end sequences (BESs) on the genome. Details of this procedure were given in [5]. Before we present the experimental results, we describe how to compute the best and the "average" MTP, which metrics are used to evaluate the quality of the MTP, and the parameters used.

4.1   Optimal and Random MTP

One of the first questions we asked ourselves was "Given a physical map, what is the best (and the average) coverage the MTP can achieve?". If the coordinates of the clones are known, then we can compute the optimal MTP as follows. Given a contig, we compute the optimal MTP by first building a directed interval graph where each node is a clone and there is an edge between two clones if their coordinates are overlapping. Buried clones are obviously disregarded. Then, we compute the shortest (unweighted) path from the leftmost clone to the rightmost clone. The clones in the path constitute the optimal MTP. We record the number of clones of the MTP and its genome coverage.

In order to have a fair comparison between the computed MTP and the optimal MTP, we also limit the number of MTP clones that can be selected as optimal MTP. The number of optimal MTP clones in a contig is bounded by the number of computed MTP clones in that contig.

If the optimal MTP is the best coverage we can hope for, the random MTP is the one we could achieve if we had no particular strategy. For the random MTP, we select a set of clones for each contig at random, according to a uniform distribution. The total number of MTP clones selected is matched to the total number of MTP clones computed by either FPC or FMTP. The number of MTP clones that are selected for each contig is determined by distributing the total of available MTP clones among the contigs proportional to their size.

4.2   Metrics

Several metrics have been devised to evaluate the quality of the MTP. Due to lack of space, we can only describe one.

Global and contig-wise coverage. The *global coverage* of an MTP is the ratio between the total length of the region spanned by all MTP clones and the genome size.

The *contig-wise* coverage of a contig is the ratio between the coverage of MTP clones and the coverage of all clones in the contig. The contig-wise coverage is computed for each contig and then an overall score is computed as the weighted average of contig-wise coverage, where the weight is the number of MTP clones in each contig. Although contig-wise coverage appears very similar to global coverage, it may produce different results when a region in the genome is covered by multiple contigs.

4.3   Parameters

Both FPC and FMTP have six parameters. Depending on the fingerprinting method (i.e., agarose or High Information Content Fingerprinting (HICF)), FMTP provides default values for its parameters. Using values for these parameters close to the defaults is crucial to obtain good performance. For example, the cutoff parameter C should be changed slightly. By default, MTP-ILP uses a low C value (1e-10 for agarose or 1e-40 for HICF). Since MTP-ILP processes the original contigs which usually contain many clones, a higher value of C would introduce many false positive overlaps. On the other hand, to detect shorter overlaps, MTP-MST uses a high C value (1e-2 for agarose or 1e-10 for HICF).

We have generated a large number of MTPs using both tools with several parameter sets, however we only recorded the best possible MTP for a given size (i.e., number of clones). If we obtained two MTPs M_i and M_j using different parameter values, if the size of M_i is greater than the size of M_j then the coverage of M_i must be greater than coverage of M_j, or otherwise M_i is disregarded. As a result, in the experimental results as the size of the MTP increases, the coverage increases monotonically.

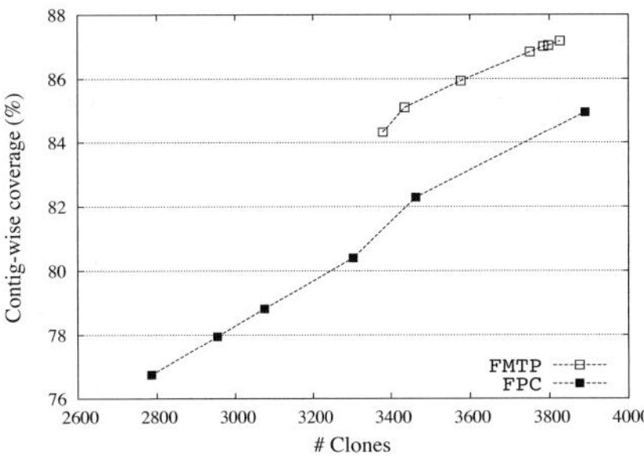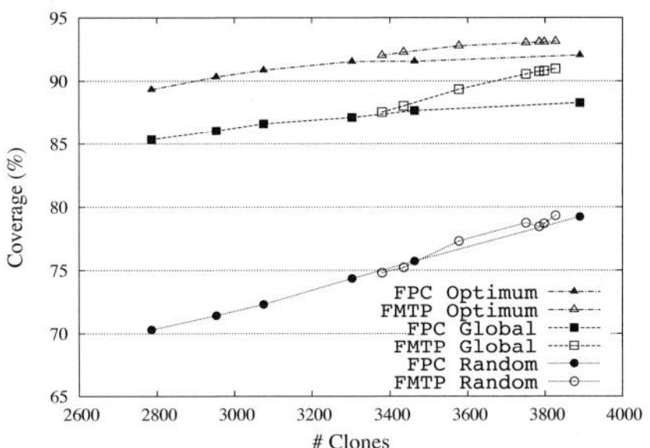

Fig. 2. Comparing the MTPs generated by FMTP and FPC with several parameter sets. Each point in the graph represents a unique MTP. (LEFT) Contig-wise coverage, (RIGHT) Optimal, global, and random coverage.

In order to have a fair comparison between FMTP and FPC, we used the same B and T values used by FPC when building the physical map ($B = 90$, $T = 7$ for rice, and $T = 3$ for barley). We set the C for the MTP-MST to values between 9e-2 and 5e-4. For all other parameters, we used the default values ($Q=3$, $B'=80$, $C'=$1e-2.5 for MTP-ILP, 1e-1 for MTP-MST, $C=$1e-10 for MTP-ILP).

4.4   Evaluation Results

The graphs summarizing the results are shown in Figure 2. Each point in the graphs represents an MTP. Figure 2-LEFT shows the contig-wise coverage as a function of the number of MTP clones; Figure 2-RIGHT illustrates optimal, global, and random coverage of the MTPs as a function of the MTP size.

First, observe that when all clones in the rice contigs are selected as MTP clones, the global coverage is only 94.45% of the genome. As shown in Figure 2-LEFT and RIGHT, FMTP produces MTPs with significantly better contig-wise and global coverage than FPC, sometimes even with fewer clones. For instance, the highest possible contig-wise coverage that we were able to obtain by using FPC is 84.96%, whereas FMTP's is 85.11% with about 460 (12%) less clones. This would imply 12% reduction in the sequencing costs. Also, global coverage of MTPs produced by FMTP converges to the optimum coverage much faster than FPC.

The number of redundant clones and gaps produced by FPC and FMTP are almost identical and very small. The average overlap size between consecutive clones is smaller in FMTP than FPC (data not shown). This explains the difference in coverages in Figure 2.

Another interesting observation can be made by comparing the optimal coverage in Figure 2-RIGHT. The optimal MTP for FMTP has higher coverage and has a smaller number of clones than the optimal MTP of FPC. Recall that the optimal coverage is computed by selecting a set of clones for each contig that covers the widest possible region. A higher coverage (even when the number of clones is smaller) suggests that FMTP selects relatively more MTP clones from the "big" contigs than FPC.

We ran FMTP and FPC on the barley physical map generated by our group at the University of California, Riverside and several other institutions [14]. FPC generated

MTPs that contain between 11,000 and 21,000 clones. When default values are used, FMTP generated MTPs that contain about 18,000 clones.

In terms of running time, FMTP and FPC are comparable. Both tools compute MTP in a couple of hours.

5 Conclusions

We presented a set of novel algorithms to compute the MTP of a physical map by using a two-step approach. In the first step, we used a stringent threshold to reduce the problem size by generating a preliminary MTP without compromising the coverage of the contigs. In the second step, we attempted to order the clones in the preliminary MTP by computing MST of an overlap graph. Then, we ran a shortest path algorithm to compute the MTP. Our experimental results show that our method generates MTPs with significantly higher coverage than the most commonly used software FPC, even using a smaller number of MTP clones. Our experimental results also show that FMTP could reduce substantially the cost of clone-by-clone sequencing projects.

Acknowledgements

The authors would like to thank Prof. Carol Soderlund, Dr. William Nelson, and the anonymous reviewers for insightful comments that helped improving the manuscript. This project was supported in part by NSF CAREER IIS-0447773 and NSF DBI-0321756.

References

[1] Rice physical map data, `ftp://ftp.genome.arizona.edu/pub/fpc/rice/`

[2] FPC-MTP tutorial webpage,
     `http://www.agcol.arizona.edu/software/fpc/userGuide/mtpdemo/`

[3] Asano, T.: Dynamic programming on intervals. In: Hsu, W.-L., Lee, R.C.T. (eds.) ISA 1991. LNCS, vol. 557, pp. 199–207. Springer, Heidelberg (1991)

[4] Bentley, D.R.: Whole-genome re-sequencing. Curr. Opin. Genet. Dev. 16(6), 545–552 (2006)

[5] Bozdag, S., Close, T.J., Lonardi, S.: A compartmentalized approach to the assembly of physical maps. In: Proceedings of BIBE, pp. 218–225 (2007)

[6] Edmonds, J., Karp, R.M.: Theoretical improvements in algorithmic efficiency for network flow problems. J. ACM 19(2), 248–264 (1972)

[7] Engler, F.W., Hatfield, J., et al.: Locating sequence on FPC maps and selecting a minimal tiling path. Genome Res. 13(9), 2152–2163 (2003)

[8] Garey, M.R., Johnson, D.S.: Computers and intractability; a guide to the theory of NP-completeness. W.H. Freeman, New York (1979)

[9] Green, E.: Strategies for the Systematic Sequencing of Complex Genomes. Nature Reviews Genetics 2, 573–583 (2001)

[10] Gregory, S., Sekhon, M., et al.: A physical map of the mouse genome. Nature 418, 743–750 (2002)

[11] Hao, J., Orlin, J.B.: A faster algorithm for finding the minimum cut in a directed graph. J. Algorithms 17(3), 424–446 (1994)

[12] International Human Genome Sequencing Consortium. A physical map of the human genome. Nature 409, 934–941 (2001)

[13] Krzywinski, M., Wallis, J., et al.: Integrated and Sequence-Ordered BAC- and YAC-Based Physical Maps for the Rat Genome. Genome Res. 14(4), 766–779 (2004)

[14] Madishetty, K., Condamine, P., et al.: Towards a physical map of the barley "gene space" (in preparation, 2008)

[15] Marra, M., Kucaba, T., et al.: A map for sequence analysis of the Arabidopsis thaliana genome. Nat. Genet. 22(3), 265–270 (1999)

[16] Nelson, W., Soderlund, C.: Software for restriction fragment physical maps. In: The Handbook of Genome Mapping: Genetic and Physical Mapping, pp. 285–306 (2005)

[17] Pop, M., Salzberg, S.: Bioinformatics challenges of new sequencing technology. Trends Genet. 24(3), 142–149 (2008)

[18] Ren, C., Lee, M., et al.: A BAC-Based Physical Map of the Chicken Genome. Genome Res. 13(12), 2754–2758 (2003)

[19] Soderlund, C., Humphray, S., et al.: Contigs Built with Fingerprints, Markers, and FPC V4.7. Genome Res. 10(11), 1772–1787 (2000)

[20] Sulston, J., Mallett, F., et al.: Software for genome mapping by fingerprinting techniques. Comput. Appl. Biosci. 4(1), 125–132 (1988)

[21] Sundquist, A., Ronaghi, M., et al.: Whole-genome sequencing and assembly with high-throughput, short-read technologies. PLoS ONE 2(5), 484 (2007)

[22] Venter, J.C., Smith, H.O., Hood, L.: A new strategy for genome sequencing. Nature 381(6581), 364–366 (1996)

[23] Warren, R.L., Varabei, D., et al.: Physical map-assisted whole-genome shotgun sequence assemblies. Genome Res. 16(6), 768–775 (2006)

[24] Wu, Y., Bhat, P., et al.: Efficient and accurate construction of genetic linkage maps from noisy and missing genotyping data. In: Giancarlo, R., Hannenhalli, S. (eds.) WABI 2007. LNCS (LNBI), vol. 4645, pp. 395–406. Springer, Heidelberg (2007)

[25] Zerbino, D., Birney, E.: Velvet: Algorithms for de novo short read assembly using de bruijn graphs. Genome Res. 18(5), 821–829 (2008)

Appendix

An Example of a Spurious Component

In Figure 3, an illustration of five clones in a contig are shown based on their real coordinates in the genome. Suppose now that the MFG of this contig contains a connected component that contains fragments only from clones B and D. That component is spurious (i.e., does not represent a region in the target genome) because the overlap between clones B and D is completely covered by clone C. Spurious components arise for two reasons, namely missing fragments (in clone C in the example) or extra fragments (in clone B or D in the example).

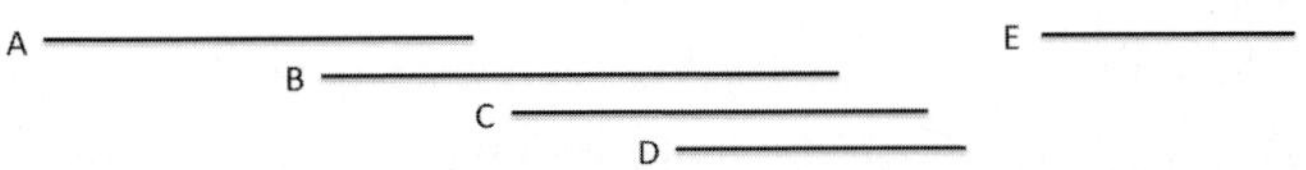

Fig. 3. A layout of five clones in a contig based on their actual genome coordinates

---

Algorithm 1

1: **Input:** Set of clones $\mathbb{C}$ in the contig
2: **Input:** $G = (V, E)$ {MFG of $\mathbb{C}$}
3: **Output:** G {Graph obtained by removing clones that are buried into multiple clones}
4: $A=\{\}$ {List of candidate clones}
5: **for all** clones $c \in \mathbb{C}$ **do**
6:    $m_c = 0$ {number of times c has a fragment in G}
7:    **for all** connected components $U \subseteq V$ **do**
8:       **if** $c \in N(U)$ **then**
9:          $m_c = m_c + 1$
10:    $r_c = m_c/|c|$ {$|c|$ is the number of fragments in c}
11:    **if** $r_c \geq B'$ **then**
12:       Add c to A
13: $L = \{\}$ {Set of clone sets}
14: **for all** connected components $U \subseteq V$ **do**
15:    **if** $N(U) \subseteq A$ **then**
16:       $L.insert(N(U))$
17: $H =$Solve Minimum Hitting Set(L)
18: **for all** clones $c \in A - H$ **do**
19:    **for all** nodes $v \in V$ **do**
20:       **if** $N(v) = c$ **then**
21:          Remove v from G
22: Remove connected components of size 1 from G

Algorithm 2

1: **Input:** The longest path $\mathbb{P}$ from the MST of G_O
2: **Input:** Clones $u_i \in \mathbb{P}$, $1 \leq i \leq |\mathbb{P}|$, where i is the order of u
3: **Input:** $G = (V, E)$ {MFG of the preliminary MTP clones in the contig}
4: **Output:** Augmented path $\mathbb{P}$
5: **for** $d = 2$ to $|\mathbb{P}| - 1$ **do**
6:    **for** $i = 1$ to $|\mathbb{P}| - d$ **do**
7:       Check if $S(u_i, u_{i+d-1}) \leq C$
8:       Check if $S(u_{i+1}, u_{i+d}) \leq C$
9:       Check if there exists a connected component U of G such that $\bigcup_{j=i}^{i+d} u_j \subseteq U$
10:       Check if $S(u_i, u_{i+d}) \leq C$
11:       **if** All conditions are true **then**
12:          Add (u_i, u_{i+d}) to $\mathbb{P}$.

Pseudocode for Removal of Clones that are Buried into Multiple Clones

A sketch of the algorithm that removes clones that are buried into multiple clones is presented as Algorithm 1.

Pseudocode for Clone Overlap Detection

A sketch of the algorithm that detects clone overlap is presented as Algorithm 2.

At each iteration four conditions are checked to determine if u_i and u_{i+d} are overlapping where u_i, $1 \leq i \leq |\mathbb{P}|$, is the *ith* clone in $\mathbb{P}$. All conditions have to be true to add the edge (u_i, u_{i+d}) to $\mathbb{P}$.

In line 7 and 8 of Algorithm 2 we check whether the clone pairs u_i, u_{i+d-1} and u_{i+1}, u_{i+d} are overlapping. Obviously, if at least one of these pairs do not overlap then u_i and u_{i+d} cannot be overlapping (assuming that no clone is completely contained in another clone). In line 9, we check if clones $u_i, u_{i+1}, \ldots, u_{i+d}$ have fragments together in at least one connected component of G. If u_i and u_{i+d} are overlapping then $u_i, u_{i+1}, \ldots, u_{i+d}$ have to be overlapping, and therefore they should share at least one fragment in G. At the end, we check if $S(u_i, u_{i+d}) \leq C$.

An Efficient Lagrangian Relaxation for the Contact Map Overlap Problem

Rumen Andonov[1,*], Nicola Yanev[2], and Noël Malod-Dognin[1]

[1] INRIA Rennes - Bretagne Atlantique and University of Rennes 1
[2] University of Sofia, Bulgaria
randonov@irisa.fr, choby@math.bas.bg, nmaloddg@irisa.fr

Abstract. Among the measures for quantifying the similarity between protein 3-D structures, contact map overlap (CMO) maximization deserved sustained attention during past decade. Despite this large involvement, the known algorithms possess a modest performance and are not applicable for large scale comparison.

This paper offers a clear advance in this respect. We present a new integer programming model for CMO and propose an exact B&B algorithm with bounds obtained by a novel Lagrangian relaxation. The efficiency of the approach is demonstrated on a popular small benchmark (Skolnick set, 40 domains). On this set our algorithm significantly outperforms the best existing exact algorithms. Many hard CMO instances have been solved for the first time. To assess furthermore our approach, we constructed a large scale set of 300 protein domains. Computing the similarity measure for any of the 44850 couples, we obtained a classification in excellent agreement with SCOP.

Keywords: Protein structure alignment, contact map overlap, combinatorial optimization, integer programming, branch and bound, Lagrangian relaxation.

1 Introduction

A fruitful assumption in molecular biology is that proteins sharing close three-dimensional (3D) structures are likely to share a common function and in most cases derive from a same ancestor. Computing the similarity between two protein structures is therefore a crucial task and has been extensively investigated [1,2,3,4]. Interested reader can also refer to [6,7,8,9,10]. We study here the *contact-map-overlap* (CMO) maximization, a scoring scheme first proposed in [5]. This measure is robust, takes partial matching into account, is translation-invariant and it captures the intuitive notion of similarity very well. Formally, a contact map is a $0 - 1$ symmetric matrix C where $c_{ij} = 1$ if the Euclidean distance between the alpha carbons (C_α) of the i-th and the j-th amino acid of a protein is smaller than a given threshold in the protein native fold. In the

---

[*] Corresponding author.

K.A. Crandall and J. Lagergren (Eds.): WABI 2008, LNBI 5251, pp. 162–173, 2008.
© Springer-Verlag Berlin Heidelberg 2008

CMO approach one tries to evaluate the similarity between two proteins by determining the maximum overlap (also called alignment) of their contact maps.

The counterpart of the CMO problem in the graph theory is the maximum common subgraph problem [11], which is APX-hard [12]. CMO is also known to be NP-hard [5]. Designing efficient algorithms that guarantee the CMO quality is an important problem that has eluded researchers so far. The most promising approach seems to be integer programming coupled with either Lagrangian relaxation [2] or B&B reduction technique [13,14]. This paper confirms once more the superiority of Lagrangian relaxation approach to solve CMO problems. Moreover, the Lagrangian relaxation has been successfully applied to RNA structures alignments [15].

The contributions of this paper are as follows. We propose a new mixed integer programming (MIP) formulation for the CMO problem. We present an efficient Lagrangian relaxation to solve our model and incorporate it into a B&B search. We also developed a second version of our algorithm which performs in agreement with the type of the secondary structure elements (SSE). To the best of our knowledge, such incorporation of biological informations in the CMO optimization search is done for the first time. We compare our approach to the best exact algorithms that exist [2,14] on a widely used benchmark (the Skolnick set), and we notice that it outperforms them significantly, both in time and in quality of the provided bounds. New hard Skolnick set instances have been solved. Finally, our method was used as a classifier on both the Skolnick set and the Proteus_300 set (a large benchmark of 300 domains that we extracted from SCOP [19]). Again, we are not aware of any previous attempt to apply a CMO approach on such large database. The obtained results are in perfect agreement with SCOP classification and clearly demonstrate that our algorithm can be used as a tool for large scale classification.

2   The Mathematical Model

Our interest in CMO was provoked by its resemblance with the protein threading problem (PTP) for which we have presented an approach based on the non-crossing matching in bipartite graphs [16]. It yielded a highly efficient algorithm by using the Lagrangian duality [17,18]. We aim to extend this approach in the case of CMO by presenting it as a matching problem in a bipartite graph, which in turn will be posed as a maximum weight augmented path in a structured graph.

Let us first introduce a few notations as follows. The contact maps of two proteins P1 and P2 are given by graphs $G_m = (V_m, E_m)$ for $m = 1, 2$, with $V_m = \{1, 2, \ldots, n_m\}$. The vertices V_m are better seen as ordered points on a line and correspond to the residues of the proteins. The sets of edges E_m correspond to the contacts. The right and left neighbours of node i are elements of the sets $\delta_m^+(i) = \{j | j > i, (i, j) \in E_m\}$ and $\delta_m^-(i) = \{j | j < i, (j, i) \in E_m\}$. Let $i \in V_1$ be matched with $k \in V_2$ and $j \in V_1$ be matched with $l \in V_2$. We will call a matching *non-crossing*, if $i < j$ implies $k < l$. Feasible alignments of two proteins P_1 and

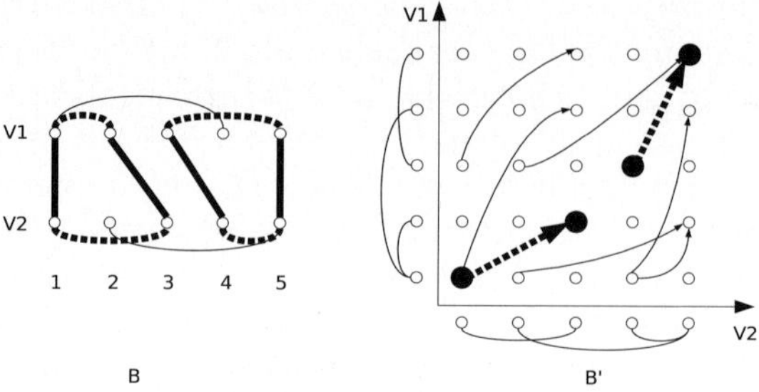

Fig. 1. Left: Vertex 1 from V1 is matched with vertex 1 from V2 and 2 is matched with 3: matching couple $(1,1)(2,3)$. Other matching couples are $(3,4)(5,5)$. This defines a feasible matching $M = \{(1,1)(2,3),(3,4)(5,5)\}$ with weight $w(M) = 2$. Right: The same matching is visualized in the graph B'.

P_2 are given by non-crossing matchings in the complete bipartite graph B with a vertex set $V_1 \cup V_2$.

Let the weight w_{ikjl} of the matching couple $(i,k)(j,l)$ be set as follows

$$w_{ikjl} = \begin{cases} 1 & \text{if } (i,j) \in E_1 \text{ and } (k,l) \in E_2 \\ 0 & \text{otherwise.} \end{cases} \tag{1}$$

For a given non-crossing matching M in B we define its weight $w(M)$ as the sum of weights over all couples of edges in M. CMO consists then in maximizing $w(M)$, where M belongs to the set of all non-crossing matchings in B.

In [16,17,18] we have already dealt with similar non-crossing matchings (in fact in the PTP they are many-to-one) and we have proposed for them a network flow presentation. This approach is adapted to CMO as follows (see for illustration Fig. 1). The edges of the bipartite graph B are mapped to the points of a $n_1 \times n_2$ rectangular grid $B' = (V', E')$ according to the rule: a point $(i,k) \in V'$ corresponds to the edge (i,k) in B and vice versa.

Definition 1. *A **feasible path** is an arbitrary sequence of points in B' (i_1, k_1), $(i_2, k_2), ..., (i_t, k_t)$ such that $i_j < i_{j+1}$ and $k_j < k_{j+1}$ for $j \in [1, t-1]$.*

The correspondence between a feasible path and a non-crossing matching is then obvious. Searching for feasible alignments of two proteins is in this way converted into searching for strictly increasing node sets in B'. We also add arcs $(i,k) \rightarrow (j,l) \in E'$ iff $w_{ikjl} = 1$. In B', solving CMO, corresponds to finding the densest (in terms of arcs) subgraph of B' whose node set is a feasible path.

To each node $(i,k) \in V'$ we associate now a 0/1 variable x_{ik}, and to each arc $(i,k) \rightarrow (j,l) \in E'$, a 0/1 variable y_{ikjl}. Denote by X the set of feasible paths. The problem can now be stated as follows :

$$v(CMO) = \max \sum_{(ik)(jl)\in E'} y_{ikjl} \tag{2}$$

subject to

$$x_{ik} \geq \sum_{l \in \delta_2^+(k)} y_{ikjl}, \qquad j \in \delta_1^+(i), \quad i \in [1, n_1 - 1], \quad k \in [1, n_2 - 1]. \tag{3}$$

$$x_{ik} \geq \sum_{l \in \delta_2^-(k)} y_{jlik}, \qquad j \in \delta_1^-(i), \quad i \in [2, n_1], \quad k \in [2, n_2]. \tag{4}$$

$$x_{ik} \geq \sum_{j \in \delta_1^+(i)} y_{ikjl}, \qquad l \in \delta_2^+(k), \quad i \in [1, n_1 - 1], \quad k \in [1, n_2 - 1]. \tag{5}$$

$$x_{ik} \geq \sum_{j \in \delta_1^-(i)} y_{jlik}, \qquad l \in \delta_2^-(k), \quad i \in [2, n_1], \quad k \in [2, n_2]. \tag{6}$$

$$x \in X \tag{7}$$

Actually, we know how to represent X with linear constraints. It is easily seen that definition 1 of a feasible path yields the following inequalities

$$\sum_{l=1}^{k} x_{il} + \sum_{j=1}^{i-1} x_{jk} \leq 1, \qquad i \in [1, n_1], \quad k \in [1, n_2]. \tag{8}$$

The same definition also implies that the j-th residue from $P1$ could be matched with at most one residue from $P2$ and vice-versa. This explains the sums on the right hand sides of (3) and (5)   (for arcs having their tails at vertex (i, k)); and (4) and (6)   (for arcs heading to (i, k)). Any $(i, k)(j, l)$ arc can be activated (i.e. $y_{ikjl} = 1$) iff $x_{ik} = 1$ and $x_{jl} = 1$ and in this case the respective constraints are active because of the objective function.

A tighter description of the polytope defined by (3)–(6) and $0 \leq x_{ik} \leq 1$, $0 \leq y_{ikjl}$ could be obtained by lifting the constraints (4) and (6) as it is shown in Fig. 2. The points shown are just the predecessors of (i, k) in the graph B' and they form a grid of $\delta_1^-(i)$ rows and $\delta_2^-(k)$ columns. Let $i_1, i_2, \ldots, i_s$ be all the vertices in $\delta_1^-(i)$ ordered according to the numbering of the vertices in V_1 and likewise $k_1, k_2, \ldots, k_t$ in $\delta_2^-(k)$. Then the vertices in the l-th column $(i_1, k_l), (i_2, k_l), \ldots (i_s, k_l)$ correspond to pairwise crossing matchings and at most one of them could be chosen in any feasible solution $x \in X$ (see (6)). This "all crossing" property holds even if we add to this set the following two sets: $(i_s, k_1), (i_s, k_2), \ldots, (i_s, k_{l-1})$ and $(i_1, k_{l+1}), (i_1, k_{l+2}), \ldots, (i_1, k_t)$. Denote by $col_{ik}(l)$ the union of these three sets, and analogously, by $row_{ik}(j)$ the corresponding union for the j-th row of the grid. When the grid is one column/row, only the set $row_{ik}(j)/col_{ik}(l)$ is empty.

Now a tighter LP relaxation of (3)–(6) is obtained by substituting (4) with (9), and (6) with (10).

$$x_{ik} \geq \sum_{(r,s) \in row_{ik}(j)} y_{rsik}, \qquad j \in \delta_1^-(i), \quad i \in [2, n_1], \quad k \in [2, n_2]. \tag{9}$$

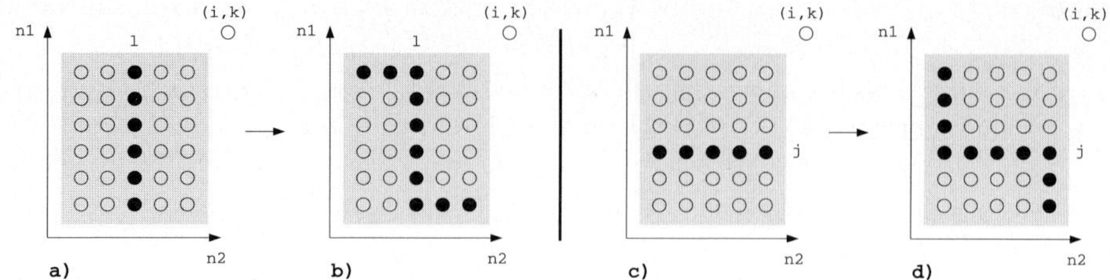

Fig. 2. The shadowed area represents the set of vertices in V' which are tails for the arcs heading to a given point (i, k). In a) : $\bullet$ correspond to the indices of y_{jlik} in (6) for l fixed. b) presents the same for the tightened constraint (10). In c) : $\bullet$ correspond to the y_{jlik} in (4) for j fixed. d) presents the same for the tightened constraint (9).

$$x_{ik} \geq \sum_{(r,s) \in col_{ik}(l)} y_{rsik}, \qquad l \in \delta_2^-(k), \quad i \in [2, n_1], \quad k \in [2, n_2]. \tag{10}$$

Remark: Since we are going to apply the Lagrangian technique there is no need neither for an explicit description of the set X neither for lifting the constraints (3) and (5).

3   Method

3.1   Lagrangian Relaxation

Here, we show how the Lagrangian relaxation of constraints (9) and (10) leads to an efficiently solvable problem, yielding upper and lower bounds that are generally better than those found by the best known exact algorithm [2].

Let $\lambda_{ikj}^h \geq 0$ (respectively $\lambda_{ikj}^v \geq 0$) be a Lagrangian multiplier assigned to each constraint (9) (respectively (10)). By adding the slacks of these constraints to the objective function with weights λ, we obtain the Lagrangian relaxation of the CMO problem

$$LR(\lambda) = \max \sum_{(ik)(jl) \in E_{B'}} y_{ikjl} + \sum_{i,k,j \in \delta_1^-(i)} \lambda_{ikj}^h \left(x_{ik} - \sum_{(r,s) \in row_{ik}(j)} y_{rsik} \right)$$
$$+ \sum_{i,k,l \in \delta_2^-(k)} \lambda_{ikl}^v \left(x_{ik} - \sum_{(r,s) \in col_{ik}(l)} y_{rsik} \right) \tag{11}$$

subject to $x \in X$, (3), (5) and $y \geq 0$.

Proposition 1. *$LR(\lambda)$ can be solved in $O(|V'| + |E'|)$ time.*

Proof. For each $(i, k) \in V'$, if $x_{ik} = 1$ then the optimal choice y_{ikjl} amounts to solving the following : The heads of all arcs in E' outgoing from (i, k) form a $|\delta^+(i)| \times |\delta^+(k)|$ table. To each point (j, l) in this table, we assign the profit

$\max\{0, c_{ikjl}(\lambda)\}$, where $c_{ikjl}(\lambda)$ is the coefficient of y_{ikjl} in (11). Each vertex in this table is a head of an arc outgoing from (i, k). Then the subproblem we need to solve consists in finding a subset of these arcs having a maximal sum $c_{ik}(\lambda)$ of profits(the arcs of negative weight are excluded as candidates for the optimal solution) and such that their heads lay on a feasible path. This could be done by a dynamic programming approach in $O(|\delta^+(i)||\delta^+(k)|)$ time. Once profits $c_{ik}(\lambda)$ have been computed for all (i, k) we can find the optimal solution to $LR(\lambda)$ by using the same DP algorithm but this time on the table of $n1 \times n2$ points with profits for (i, k)-th one given by: $c_{ik}(\lambda) + \sum_{j \in \delta_1^-(i)} \lambda^h_{ikj} + \sum_{l \in \delta_2^-(k)} \lambda^v_{ikl}$, where the last two terms are the coefficients of x_{ik} in (11). The inclusion $x \in X$ is explicitly incorporated in the DP algorithm.

3.2   Subgradient Descend

In order to find the tightest upper bound on $v(CMO)$ (or eventually to solve the problem), we need to solve in the dual space of the Lagrangian multipliers $LD = \min_{\lambda \geq 0} LR(\lambda)$, whereas $LR(\lambda)$ is a problem in x, y. A number of methods have been proposed to solve Lagrangian duals: dual ascent, constraint generation, column generation, etc... Here, we choose the subgradient descent method, because of our large number of lagrangian multipliers. It is an iterative method in which at iteration t, given the current multiplier vector λ^t, a step is taken along a subgradient of $LR(\lambda)$; then, if necessary, the resulting point is projected onto the nonnegative orthant. It is well known that practical convergence of the subgradient method is unpredictable. For some "good" problems, convergence is quick and fairly reliable, while other problems tend to produce erratic behavior. The computational runs on a rich set of real-life instances confirm that our approach belong to the "good" cases.

In our realization, the update scheme for λ_{ikj} (and analogously for λ_{ikl}) is $\lambda^{t+1}_{ikj} = \max\{0, \lambda^t_{ikj} - \Theta^t g^t_{ikj}\}$, where $g^t_{ikj} = \bar{x}_{ik} - \sum \bar{y}_{jlik}$ (see (9) and (10) for the sum definition) is the sub-gradient component $(0, 1, \text{or} -1)$, calculated on the optimal solution $\bar{x}, \bar{y}$ of $LR(\lambda^t)$. The step size Θ^t is $\Theta^t = \frac{\alpha(LR(\lambda^t) - Z_{lb})}{\sum(g^t_{ikj})^2 + \sum(g^t_{ikl})^2}$ where Z_{lb} is a known lower bound for the CMO problem and α is an input parameter. Into this approach the x-components of $LR(\lambda^t)$ solution provides a feasible solution to CMO and thus a lower bound also. If $LD \leq v(CMO)$ then the problem is solved. If $LD > v(CMO)$ holds, in order to obtain the optimal solution, one could pass to a B&B algorithm suitably tailored for such an upper and lower bounds generator.

3.3   Branch and Bound

From among various possible nodes splitting rules, the one shown in Fig. 3 gives good results (see section 4). Formally, a node of B&B is given by n_2 couples (b_k, t_k) for $k \in [1, n_2]$ representing the zone to be explored (the white area on Fig. 3). A vertex (j, l) of the graph B' belongs to this area if $b_l \leq j \leq t_l$. Let (r_b, c_b) be the $\text{argmax}_{(i,k) \in V'}[\min(D(i, k), U(i, k))]$, where $D(i, k) = \sum_{l \geq k} \max(i - b_l, 0)$ and $U(i, k) = \sum_{l \leq k} \max(t_l - i, 0)$. Now, the two descendants

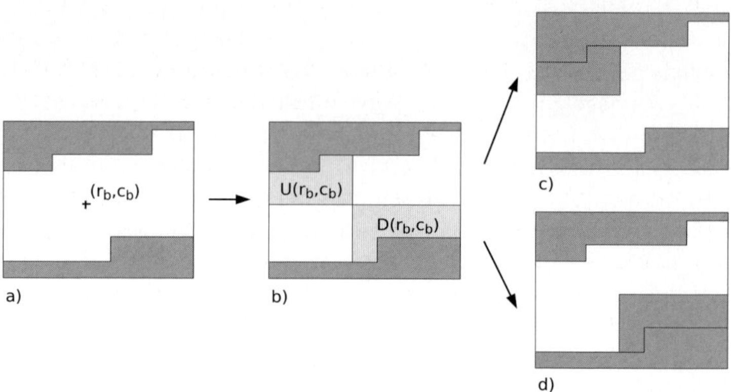

Fig. 3. Sketch of the B&B splitting strategy. a) The white area represents the current node feasible set; b) Fixing the point (r_b, c_b) creates the regions, $D(r_b, c_b)$ and $U(r_b, c_b)$; c) and d) are the descendants of the node a).

of the current node are obtained by discarding from its feasible set the vertices in $U(r_b, c_b)$ and $D(r_b, c_b)$ respectively. The goal of this strategy is twofold: to create descendants that are balanced in sense of feasible set size and to reduce maximally the parent node's feasible set.

Finally, the main steps of the B&B algorithm are as follows:

<u>Initialization:</u> Set $L=\{\text{root}\}$ (root is the original CMO problem, i.e. with no restrictions on the feasible paths).
<u>Problem selection and relaxation:</u> Select and delete the problem P from L having the biggest upper bound. Solve the Lagrangian dual of P.
<u>Fathoming and Pruning:</u> Follow classical rules.
<u>Partitioning :</u> Create and add to L the two descendants of P
<u>Termination :</u> if $L = \emptyset$, the solution (x^*, y^*) is optimal.

3.4   Adding Biological Informations

In it's native state, a protein contains secondary structure elements (SSE), which are highly regular substructures. There are two SSE types: α helix (H) and β strand (b). Residues which are not part of a SSE are said to be in a coil (c).

As defined in (2)-(7), CMO does not consider the SSE type. Potentially, this can conduct to biologically not acceptable matching, such as aligning α with β SSE. To avoid that, we enrich the feasible path definition with the SSE information. A node (i, k) will be not acceptable, if residue i is of type H, while residue k is of type b, or vice versa. The SSE knowledge is already used in non-CMO methods like VAST[1], but has been never used before in a CMO approach. The impact of such a filter on the CMO behavior is twofold: from one hand it directly leads to a biologically correct alignment, from the other hand, it makes the search space sparser, and accelerates about 2.5 times the solution process.

4   Numerical Results

The results presented here were obtained on a computer with AMD Opteron (TM) CPU at 2.4 GHz, 4 Gb Ram. The algorithm was implemented in C++. To generate contact maps we consider two residues to be in contact if their C_α are within 7.5 Å, without taking into account contacts between consecutive residues. When needed, the SSE information was computed using the publicly available software Kaksi[1]. The first version of our algorithm will be denoted by `A_purva`[2], while the version with the SSE filter will be indicated as `A_purva_sse`.

To evaluate these algorithms we performed two kinds of experiments. In the first one (section 4.1), we compared our approach - in terms of performance and quality of the bounds - with the best exact algorithms from the literature [2,14]. The former one is based on Lagrangian relaxation, and will be denoted here by `LAGR`. Our approach differs from it in two main points: i) in the proposed MIP formulation and ii) in the set of dualized constraints. This can explain the significant differences in the computational behavior of `A_purva` versus `LAGR`. The second algorithm, denoted here by `CMOS`, has been recently described in [14]. The comparison was done on a set of protein domains suggested by J. Skolnick. It contains 40 medium size domains from 33 proteins. The number of residues varies from 97 (2b3iA) to 256 (1aw2A), and the number of contacts varies from 320 (1rn1A) to 936 (1btmA). According to SCOP classification [19], the Skolnick set contains five families (see Table 1).

Table 1. The five families in the Skolnick set

SCOP Fold	SCOP Family	Proteins
Flavodoxin-like	CheY-related	1b00, 1dbw, 1nat, 1ntr, 3chy 1qmp(A,B,C,D), 4tmy(A,B)
Cupredoxin-like	Plastocyanin /azurin-like	1baw, 1byo(A,B), 1kdi, 1nin 1pla, 2b3i, 2pcy, 2plt
TIM beta/alpha-barrel	Triosephosphate isomerase (TIM)	1amk, 1aw2, 1b9b, 1btm, 1hti 1tmh, 1tre, 1tri, 1ydv, 3ypi, 8tim
Ferritin-like	Ferritin	1b71, 1bcf, 1dps, 1fha, 1ier, 1rcd
Microbial ribonuclease	Fungal ribonucleases	1rn1(A,B,C)

Afterwards (section 4.2), we experimentally evaluated the capability of our algorithm to perform as a classifier on two sets: Skolnick and Proteus_300. The latter benchmark was proposed by us. It contains more, and significantly larger proteins: 300 domains, with number of residues varying from 64 (d15bbA_) to 455 (d1po5A_). The maximum number of contacts is 1761 (d1i24A_). These domains are classified by SCOP in 30 families. All our data and results[3] are available on the URL: *http://www.irisa.fr/symbiose/softwares/resources/proteus300*.

---

[1] http://migale.jouy.inra.fr/outils/mig/kaksi/
[2] Apurva (Sanskrit) = not having existed before, unknown, wonderful, ...
[3] Contact maps, solved instances, upper and lower bounds, run time, classifications...

4.1   Performance and Quality of Bounds

The Skolnick set requires aligning 780 pairs of domains. A_purva and LAGR (whose code was kindly provided to us by G. Lancia) were executed on the same computer and with the same 7.5Å contact maps. For both algorithms, the computation time was bounded to 1800 sec/instance. Table 2 shows the number of instances solved by each algorithm. A_purva succeeded to solve 502 couples, while LAGR solved only 161 couples. All these 161 instances are "easy", i.e. they align domains from the same SCOP family. In table 2, we also give the number of solved instances by LAGR and CMOS taken from [14]. These results where obtained on a similar workstation, with the same time limit (this information was kindly provided to us by N. Sahinidis), but with different contact maps (the threshold used was 7Å). CMOS solves 161 easy instances. Note that A_purva is the only one to solve 338 "hard" instances, i.e. couples with domains from different families. On the other hand, A_purva was outperformed by its SSE version, which demonstrated the usefulness of integrating this filter.

Table 2. Number of instances solved by the different CMO methods, with a time limit of 1800 sec/instance. A_purva is the only one able to solve all easy instances, as well as many of the hard instances. The advantage of adding a SSE filter is noticeable.

	Our contact maps (7.5Å)			CMOS contact maps (7Å)	
	LAGR	A_purva	A_purva_sse	LAGR	CMOS
Easy instances (164)	161	164	164	150	161
Hard instances (616)	0	338	444	0	0
Total (780)	161	502	608	150	161

Figure 4 compares the time needed by LAGR to that of A_purva on the set of instances solved by both algorithms. We observe that A_purva is significantly faster than LAGR (up to several hundred times in the majority of cases). The use of SSE information can push this even further, since A_purva_sse is about 2.5 times faster than A_purva.

We observed that the time for solving instances (without using SSE information) that align domains from the same family varies between 0.04s and 4.27s (except for two instances); this time varies respectively from 17.9s to more than 1800s when aligning domains from different classes. In this manner our results confirmed once more the property (also observed in [2,14]) that : instances, such that both domains belong to the same family, seem to be easily solvable; in contrast to instances that align domains from different families.

Our next observation concerns the quality of gaps obtained by LAGR and A_purva on the set of unsolved instances. Remember that when a Lagrangian algorithm stops because of time limit (1800 sec. in our case) it provides an upper bound (UB), and a lower bound (LB), which is a real advantage of a B&B type algorithm compared to any meta-heuristics. The relative gap value

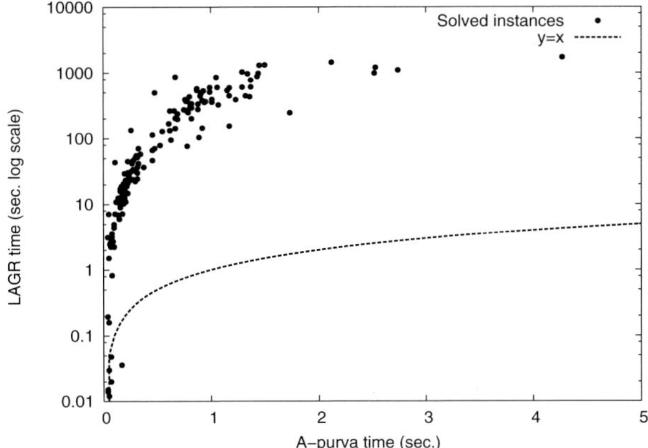

Fig. 4. A_purva versus LAGR running time comparison on the set of Skolnick instances solved by both algorithms. The A_purva time is presented on the x-axis, while the one of LAGR is on the y-axis. Often, A_purva is more than 100 times faster.

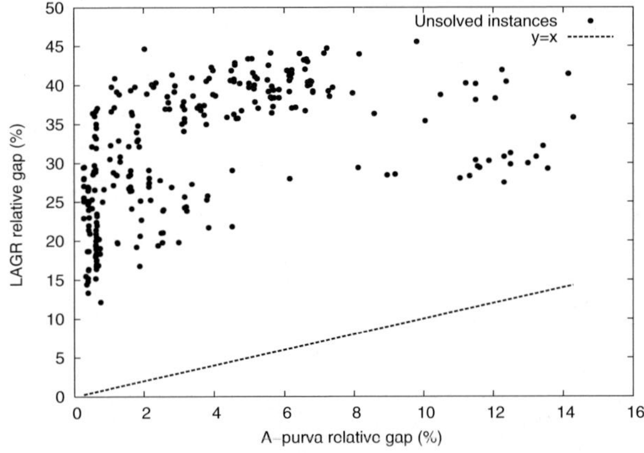

Fig. 5. Comparing relative gaps on the set of unsolved instances. The A_purva gaps (x-axis) are plotted against the LAGR gaps (y-axis). Relative gaps obtained by A_purva are substantially smaller.

$(UB - LB)/UB$, measures how far is the optimization process from finding the exact optimum. Fig. 5 shows the relative gaps of A_purva plotted against those of LAGR. The entire figure is very asymmetric to the advantage of our algorithm since the relative gaps of A_purva are always smaller than those of LAGR, meaning that the bounds of A_purva are always tighter.

4.2   A_purva as a Classifier

In this section we are interested in checking the ability of `A_purva_sse` to perform successfully as classifier in a given small lapse of time. We used the following protocol : we limited the runs of `A_purva_sse` to the root (i.e. B&B was not used), with a limit of 500 iterations for the subgradient descent. To measure the similarity between two proteins P_1 and P_2, we used the function defined in [14]: $Sim(P_1, P_2) = 2.LB/(|E_1| + |E_2|)$, where LB is the incumbent value found by `A_purva_sse`. These similarities were given to `Chavl` [20], a publicly available tool which proposes both a hierarchical ascendant classification and the cut corresponding to the best partition level (therefore, it does not require a similarity threshold). The obtained result was compared with the SCOP classification at a family level.

For the Skolnick set, the alignment of all couples was done in less than 810 seconds ($\simeq$ 1.04 sec/couple). The classification returned by `Chavl` was exactly the same as the classification at the family level in SCOP. To get a stronger confirmation of `A_purva` classifier capabilities, we performed the same operation on the Proteus_300 set. Aligning the 44850 couples required roughly 22 hours ($\simeq$ 1.82 sec/couple). The classification returned by `Chavl` contained 34 classes. The only difference between our classification and the one of SCOP at the family level is that 4 SCOP families were each split in two in our classification.

5   Conclusion

In this paper, we give an efficient exact algorithm for contact map overlap problem. The bounds are found by using Lagrangian relaxation, and the dual problem is solved by sub-gradient approach. The performance of the algorithm is demonstrated on a benchmark set of 40 domains and its superiority over the existing algorithms is obvious. We also propose a suitable filter based on secondary structures information which further accelerates the solution process. The capacity of the proposed algorithm to provide a convenient similarity measure was tested on a large data set of 300 protein domains. We were able to obtain in a short time a classification in very good agreement to the well known SCOP database.

Acknowledgment

Supported by ANR grant Calcul Intensif projet PROTEUS (ANR-06-CIS6-008) and by Hubert Curien French-Bulgarian partnership "RILA 2006" N^0 15071XF. N. Malod-Dognin is supported by Région Bretagne. We are thankful to Professor Giuseppe Lancia for numerous discussions and for kindly providing us with the source code of `LAGR`. All computations were done on the Ouest-genopole bioinformatics platform (http://genouest.org). Thanks to Jacques Nicolas and Basavanneppa Tallur for their suggestions.

References

1. Gibrat, J.-F., Madej, T., Bryant, S.H.: Surprising similarities in structure comparison. Curr. Opin. Struct. Biol. 6, 377–385 (1996)
2. Caprara, A., Carr, R., Israil, S., Lancia, G., Walenz, B.: 1001 Optimal PDB Structure Alignments: Integer Programming Methods for Finding the Maximum Contact Map Overlap. J. Comput. Biol. 11(1), 27–52 (2004)
3. Halperin, I., Ma, B., Wolfson, H., et al.: Principles of docking: An overview of search algorithms and a guide to scoring functions. Proteins Struct. Funct. Genet. 47, 409–443 (2002)
4. Godzik, A.: The Structural alignment between two proteins: is there a unique answer? Protein Science 5, 1325–1338 (1996)
5. Goldman, D., Israil, S., Papadimitriu, C.: Algorithmic aspects of protein structure similarity. In: IEEE Symp. Found. Comput. Sci., pp. 512–522 (1999)
6. Caprara, A., Lancia, G.: Structural Alignment of Large-Size Protein via Lagrangian Relaxation. In: RECOMB 2002, pp. 100–108 (2002)
7. Lancia, G., Istrail, S.: Protein Structure Comparison: Algorithms and Applications. Prot. Str. Anal. Des., 1–33 (2003)
8. Carr, R., Lancia, G.: Compact optimization can outperform separation: A case study in structural proteomics. 4OR 2, 221–233 (2004)
9. Agarwal, P.K., Mustafa, N.H., Wang, Y.: Fast Molecular Shape Matching Using Contact Maps. J. Comput. Biol. 14(2), 131–147 (2007)
10. Xu, J., Jiao, F., Berger, B.: A parametrized Algorithm for Protein Structure Alignment. In: Apostolico, A., Guerra, C., Istrail, S., Pevzner, P.A., Waterman, M. (eds.) RECOMB 2006. LNCS (LNBI), vol. 3909, pp. 488–499. Springer, Heidelberg (2006)
11. Garey, M., Johnson, D.: Computers and Intractability: A Guide to the Theory of NP-completness. Freeman and company, New York (1979)
12. Crescenzi, P., Kann, V.: A compendium of NP optimization problems, http://www.nada.kth.se/~viggo/problemlist/
13. Strickland, D.M., Barnes, E., Sokol, J.S.: Optimal Protein Structure Alignment Using Maximum Cliques. Oper. Res. 53, 389–402 (2005)
14. Xie, W., Sahinidis, N.: A Reduction-Based Exact Algorithm for the Contact Map Overlap Problem. J. Comput. Biol. 14, 637–654 (2007)
15. Bauer, M., Klau, G.W., Reinert, K.: Fast and Accurate Structural RNA Alignment by Progressive Lagrangian Optimization. In: R. Berthold, M., Glen, R.C., Diederichs, K., Kohlbacher, O., Fischer, I. (eds.) CompLife 2005. LNCS (LNBI), vol. 3695, pp. 217–228. Springer, Heidelberg (2005)
16. Andonov, R., Balev, S., Yanev, N.: Protein threading: From mathematical models to parallel implementations. INFORMS J. on Comput. 16(4) (2004)
17. Veber, P., Yanev, N., Andonov, R., Poirriez, V.: Optimal protein threading by cost-splitting. In: Casadio, R., Myers, G. (eds.) WABI 2005. LNCS (LNBI), vol. 3692, pp. 365–375. Springer, Heidelberg (2005)
18. Yanev, N., Veber, P., Andonov, R., Balev, S.: Lagrangian approaches for a class of matching problems. Comp. and Math. with Appl. 55(5), 1054–1067 (2008)
19. Andreeva, A., Howorth, D., Chandonia, J.-M., Brenner, S.E., Hubbard, T.J.P., Chothia, C., Murzin, A.G.: Data growth and its impact on the SCOP database: new developments. Nucl. Acid Res (2007)
20. Lerman, I.C.: Likelihood linkage analysis (LLA) classification method (Around an example treated by hand). Biochimie, Elsevier Editions 75, 379–397 (1993)

A Faster Algorithm for RNA Co-folding

Michal Ziv-Ukelson[1,*], Irit Gat-Viks[2,*], Ydo Wexler[3,*], and Ron Shamir[4]

[1] Computer Science Department, Ben Gurion University of the Negev, Beer-Sheva
[2] Computational Molecular Biology Department, Max Planck Institute for Molecular Genetics, Berlin, Germany
[3] Microsoft Research, Microsoft Corporation, Redmond, WA
[4] School of Computer Science, Tel Aviv University

Abstract. The current pairwise RNA (secondary) structural alignment algorithms are based on Sankoff's dynamic programming algorithm from 1985. Sankoff's algorithm requires $O(N^6)$ time and $O(N^4)$ space, where N denotes the length of the compared sequences, and thus its applicability is very limited. The current literature offers many heuristics for speeding up Sankoff's alignment process, some making restrictive assumptions on the length or the shape of the RNA substructures. We show how to speed up Sankoff's algorithm in practice via non-heuristic methods, without compromising optimality. Our analysis shows that the expected time complexity of the new algorithm is $O(N^4\zeta(N))$, where $\zeta(N)$ converges to $O(N)$, assuming a standard polymer folding model which was supported by experimental analysis. Hence our algorithm speeds up Sankoff's algorithm by a linear factor on average. In simulations, our algorithm speeds up computation by a factor of 3-12 for sequences of length 25-250.

Availability: Code and data sets are available, upon request.

1   Introduction

Within the last few years non-coding RNAs (ncRNAs) have been recognized as a highly abundant class of RNAs that do not code for proteins but nevertheless are functional in many biological processes, including localization, replication, translation, degradation, regulation and stabilization of biological macromolecules [19]. Thus, the computational identification of functional RNAs in genomes is a major, yet largely unsolved, problem. It is well known that structural conservation implies potential function and thus comparative structure analysis is the gold standard for the identification of functional RNAs and the determination of RNA secondary structures [22].

Many of the comparative genomics methods for the identification of functional RNA structures require a sequence alignment as input [16,13,23]. However, an alignment based on primary sequence alone is generally not sufficient for the identification of conserved secondary structure [6]. This is due to the fact that functional RNAs are not necessarily conserved on their primary sequence level. Instead, the stem-pairing regions of the functional RNA structures evolve such that substitutions that maintain the bonding between paired bases

---

* These authors contributed equally to the paper.

K.A. Crandall and J. Lagergren (Eds.): WABI 2008, LNBI 5251, pp. 174–185, 2008.

© Springer-Verlag Berlin Heidelberg 2008

are more likely to survive. The base-pair covariation in the structural stems includes multiple compensatory substitutions (*e.g. G:C → A:U*) and compatible single substitutions (*e.g. G:C → G:U*). In 1985 David Sankoff proposed an algorithm [4], which we shall call SA, designed for the simultaneous alignment and structure prediction of homologous structural RNA sequences (for brevity, we call that problem here the *co-folding* problem). SA merges the recursions of sequence alignment such as Smith and Waterman's algorithm [25] with those of RNA structure prediction algorithms [20,27]. SA takes as input two sequences and finds a local alignment (two substrings) such that the entire configuration of structure and alignment is optimal. To combine both structure and alignment scores, SA aims to maximize the weighted sum of the predicted structure's base-pairing interactions and the alignment cost. Base-pairing interactions scores (e.g., Nussinov et al. [20] or Zuker and Stiegler's energy-based methods [27]) guide homologous base pairs to align correctly and thus base-pair covariation can be naturally included as a factor in the alignment solution. The alignment cost aims to punish for insertions/deletions but encourage compensatory mutations among base-paired characters. SA preserves a common branching configuration for the aligned structures, but allows variations in the sizes of the stems and the loops (see Fig. 1).

Unfortunately, SA is not practical for most realistic applications due to its prohibitive computational cost: the algorithm requires $O(N^6)$ time and $O(N^4)$ space, for a pair of sequences of length N. The high complexity of this algorithm on one hand, and the need for practical solutions on the other hand, motivated attempts to reduce its complexity heuristically. In recent years, many studies have suggested practical heuristic methods to reduce SA's computational cost [1,3,15,14,7,21,11,10], highlighting the importance of this algorithm in current RNA comparative structure analysis. Here, in contrast to the above heuristic approaches, we obtain a non-heuristic speedup, which does not sacrifice the optimality of results.

Our algorithm extends the approach of Wexler, Zilberstein and Ziv-Ukelson [26], previously applied to speeding up the classical $O(N^3)$ *RNA secondary structure prediction* algorithms [20,27]. The classical algorithms for RNA secondary structure prediction are based on dynamic programming (DP) and their time complexities are $O(N^3)$, where the bottleneck is due to the fact that the recursion for computing the optimal folding includes a term that takes into account $O(N)$ possible branching points (see Figure 1) which "compete" for the optimal score. The speedup suggested in [26] to the classical $O(N^3)$ algorithm computes an exact optimal folding. This is done by pruning the number of branch points that need to be considered from $O(N)$ down to $\psi(N)$, where $\psi(N)$ is shown to be constant on average under standard polymer folding models. The accelerated algorithm uses a candidate-list approach to utilize two observations: (a) the main (2D) DP matrix computed by the classical algorithm for RNA secondary structure prediction obeys the triangle inequality; (b) a classical thermodynamic argument indicates that the probability for base-pair formation between two bases q indices apart is bounded by b/q^c for some constants $b, c > 0$ [17,18].

Similarly to the RNA folding algorithms, the time-complexity bottleneck of SA is also due to the computation of all the scores induced by competing branch points. However, in contrast to the RNA folding algorithms, in the SA case the branch point considerations scale up to two orders, as all possible combinations of pairs of branch point indices (*i.e.* all possible indices of sequence A × all possible indices of sequence B, where A and B are the two co-folded sequences) need to be considered. This adds a factor of $O(N^2)$ to the time complexity of SA which is due to branch point considerations. In this paper we extend the approach of [26] and apply it to speed up SA by reducing the number of branch points that need to be considered in the main recursion for the SA score computation. Although the method suggested in [26] is not directly applicable to the $4D$ DP matrix computed in the SA algorithm, we show how to interpret the terms in the SA algorithm so as to maintain a set of candidate lists that will reduce the amount of computations needed. The resulting algorithm is guaranteed to obtain the optimal solution to the problem. In order to do this, we first show that the DP table computed by the SA recursion also obeys a sort of triangle inequality, and that the main theorem of [26] regarding redundant candidate branch-point considerations extends to the branch-point pairs considered by SA. Based on these proofs, we give a new candidate-list variant of SA that exploits redundancies in the main SA DP table to speed it up.

Under the probabilistic model of self avoiding random walk [24], which has been verified experimentally for polymers [17,18], the expected time complexity of the new algorithm is $O(N^4\zeta(N))$, where $\zeta(N)$ is the expected maximal size of a candidate list. In contrast to [26], where $\zeta(N)$ converges to a constant, we show that in fact here $\zeta(N)$ converges to $O(N)$. Thus, our algorithm provides a theoretical speedup over the original SA by a linear factor on average. This behavior of the co-folding algorithm is verified in an experimental analysis, which shows a linear growth of the candidate list size with increasing sequence length. The faster new algorithm was implemented as a filter, denoted FASTCOFOLD, on top of a popular SA implementation of Havgaard et al. [15]. Our run-time benchmarks show that FASTCOFOLD is indeed faster in practice when compared to the standard SA version of the same code, by a factor of 3-12 for sequences of length 25-250.

The rest of this paper proceeds as follows. Section 2 gives the basic preliminaries and definitions including an algorithmic background. Section 3 examines properties of the folding and co-folding DP algorithms which are later exploited by our pruning method. The new algorithm is described and analyzed in Section 4, and a comparative performance analysis of the algorithm is given in Section 5.

Due to space restrictions, all proofs are omitted. Proofs, as well as a biological application of SA to the analysis of functional tandem repeats in *C. elegans*, will appear in an extended journal version of the paper.

2   Preliminaries and Definitions

RNA is typically produced as a single stranded molecule, which then folds upon itself to form a number of short base-paired stems (Fig 1). This base-paired structure is called the *secondary structure* of the RNA. Paired bases almost always occur in a nested fashion in RNA secondary structure. Under the assumption that the structure does not contain pseudoknots, a model was proposed by Tinoco et al. [12] to calculate the stability (in terms of free energy) of a folded

```
     Left Branch        ][    Right Branch
A-CGGCAAA---UUGGCCG-U   ][   GCUGCGUGCAAAGCGC
((((((.........))))))   ][   ...((((.....))))
AUCGCGAAAAAAUUGCGCGAU   ][   GCUG-GCGCAAAGC-C
```

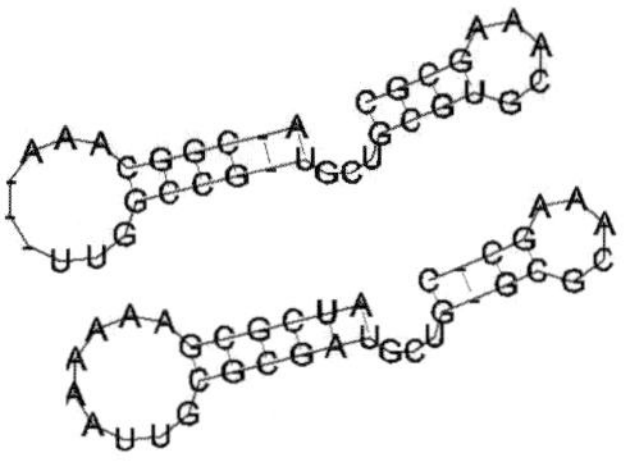

Fig. 1. Co-folding. Left: An example of a branching co-folding. Matched round parentheses indicate paired bases in the co-folding, and square brackets indicate a partition point. In the terminology of Section 2, the left branch is a co-terminus co-folding and the right side is a dangling co-folding. Right: The (aligned) secondary structures corresponding to the branching co-folding. Both branches consist of a stem (matched bases) and a loop.

RNA molecule by summing all contributions from the stabilizing, consecutive base pairs from the loop-destabilizing terms in the secondary structure. This model has been widely investigated since and parameters for this model were experimentally collected *e.g.* by the Turner group [5]. Based on this model, DP algorithms for computing the most stable structures were proposed [20,27].

The Sankoff co-folding algorithm [4] combines RNA folding with alignment, by taking into account both sequence and structure homology. This is formalized below in Definitions 1-4.

Definition 1. *An* **alignment** *of sequences R', T' is a pair of sequences R, T s.t. $|R| = |T|$, $R(T)$ is obtained from $R'(T')$ by adding '-'s such that for no index i, $r_i = t_i =$ '-'.*

The **alignment score** *is $\sum_i \sigma(R_i, T_i)$. Here, $\sigma(x, y)$ is the score of substituting x and y. The indel score $\sigma(x, -)$ is often indicated by $\gamma(x)$.*

Definition 2. *For a sequence F over the alphabet $\{(,), .\}$, where the parentheses are nested, define* $\mathbf{P(F)} = \{(\pi, \overline{\pi}) \mid \pi, \overline{\pi}$ *are paired parenthesis positions in* $F\}$.

Definition 3. *For a sequence T over the RNA alphabet $\{A, U, C, G, -\}$, a* **folding** *$F(T)$ is a series over the alphabet $\{(,), .\}$, annotated with the letters of T, such that (1)$|F(T)| = |T|$, (2) $F(T)$ is nested, and (3) for each paired parentheses positions $(\pi, \overline{\pi}) \in P(F)$, $\{T_\pi, T_{\overline{\pi}}\} \in \{\{A, U\}, \{G, C\}, \{G, U\}, \{-, -\}\}$.*

The **folding score** *is $\sum_{P(F)} \beta^T_{\pi\overline{\pi}}$, where β^T_{ij} denotes the score for the stability contribution of a base-pair between the characters in indices i and j of T.*

We note that some SA implementations relax requirement (3) in the above definition, and instead set β^T_{ij} to a score penalty in those cases when $\{T_i, T_j\} \notin \{\{A, U\}, \{G, C\}, \{G, U\}\}$. Furthermore, the implementation of the term β^T_{ij}

depends on the structure prediction algorithm employed by the specific SA variant. For example, in variants that are based on McCaskill's algorithm, such as PMCOMP [14], β_{ij}^T is the probability for the character at index i of T to form a base-pair with the character at index j of T. Alternatively, in SA variants that are based on Zuker's MFE approach, such as Foldalign [15], the score β_{ij}^T is in general computed dynamically depending on the structural context, making use of Turner parameters such as e.g. loop energy rules [27].

Definition 4. *A* **co-folding** *of two RNA sequences R', T' is the triplet R, T, F such that R, T is an alignment of R', T' and F is a folding of R and of T. The* **co-folding score** *is* $\sum_i \sigma(R_i, T_i) \; + \; \sum_{P(F)} (\beta_{\pi\overline{\pi}}^T + \beta_{\pi\overline{\pi}}^R + \tau(S_\pi, R_{\overline{\pi}}, T_\pi, T_{\overline{\pi}})),$ *where the term $\tau(R_\pi, R_{\overline{\pi}}, T_\pi, T_{\overline{\pi}})$ is a score that takes into account compensatory mutations and substitutions.*

Given a pair of RNA sequences A and B, the **local co-folding problem** *is to find two substrings $A[i \ldots j]$ and $B[k \ldots \ell]$ such that their co-folding has maximum score. The* **global co-folding problem** *is to find a co-folding of $A[1 \ldots N]$ and $B[1 \ldots N]$ of maximal score.*

Sankoff's algorithm solves the co-folding problem by DP. All of the current algorithms that are based on SA employ variants of the same basic DP recursion. We now demonstrate the recursions in FoldalignM [7] and Foldalign [15]. Given two RNA sequences A and B, let $S[i, j; k, \ell]$ hold the score of the best co-folding of the subsequences $A[i \ldots j]$ and $B[k \ldots \ell]$.

$$
S[i, j \,;\, k, l] = \max
\begin{cases}
(1) \; S[i+1, j; k, \ell] \; + \gamma(A_i), \\
(2) \; S[i, j; k+1, \ell] \; + \gamma(B_k), \\
(3) \; S[i, j-1; k, \ell] \; + \gamma(A_j), \\
(4) \; S[i, j; k, \ell-1] \; + \gamma(B_\ell), \\
(5) \; S[i+1, j; k+1, \ell] \; + \sigma(A_i, B_k), \\
(6) \; S[i, j-1; k, \ell-1] \; + \sigma(A_j, B_l), \\
(7) \; S[i+1, j-1; k, \ell] + \beta_{ij}^A \; + \gamma(A_i) \; + \gamma(A_j) \\
(8) \; S[i, j; k+1, \ell-1] + \beta_{kl}^B \; + \gamma(B_k) \; + \gamma(B_\ell) \\
(9) \; S[i+1, j-1; k+1, \ell-1] + \beta_{ij}^A + \beta_{kl}^B + \tau(A_i, A_j, B_k, B_l), \\
(10) \; \max\limits_{i<m<j, k<n<l} \{ S[i, m; k, n] + S[m+1, j; n+1, \ell] \}
\end{cases}
\tag{1}
$$

where all entries of S are initialized to 0. Terms $(1)-(4)$ account for gaps in one of the two sequences. Terms (5) and (6) describe the extension of both subsequences with an unpaired position. Terms (7) and (8) describe the extension of only one of the subsequences with a base-pair. These two terms are conditional and are only applied in certain contexts: term (7) can only be applied if, in the co-folding corresponding to $S[i+1, j-1; k, \ell]$, A_{i+1} is base paired with A_{j-1}, and term (8) can only be applied if in the co-folding corresponding to $S[i, j; k+1, \ell-1]$ B_{k+1} is base paired with $B_{\ell-1}$. The constraints on terms 7 and 8 are only necessary in order to preserve the common branching configuration for the two aligned structures. Term (9) describes the extension of both sequences by a base-pair match, and Term (10) describes a branching event, as demonstrated in Figure 1. The following two definitions are also exemplified in Figure 1.

Definition 5 (co-terminus folding and co-folding)

*A folding of a sequence $s_i \ldots s_j$ is a **co-terminus** folding if s_i pairs with s_j. Otherwise it is called a **dangling** folding.*

*Similarly, a **co-terminus** co-folding over a pair of sequences $A[i,j]$ and $B[k, \ell]$ is a co-folding in which A_i pairs with A_j and B_k pairs with B_ℓ. Otherwise it is called a **dangling** co-folding.*

Definition 6. *A **partition point** in a given co-folding of $A[i,j]$ versus $B[k, \ell]$ is an index pair (p_1, p_2), such that there is no co-terminus folding over $A_x \ldots A_y$ in this co-folding, where $i \leq x \leq p_1$ and $p_1 \leq y \leq j$, and in addition there is no co-terminus folding over $B_z \ldots B_w$ in this co-folding, where $k \leq z \leq p_2$ and $p_2 < w \leq \ell$.*

Time and Space Complexity Analysis of SA
Let $N = max\{|A|, |B|\}$. Computing the matrix S requires $O(N^4)$ space. For each entry $S[i, j, k, l]$ to be computed in this matrix, the algorithm employs the recursion of Eq. 1. The bottleneck of Eq. 1 is term (10), which considers $O(N^2)$ competing sums of pairs. Therefore, the time complexity is $O(N^6)$.

3   Properties of RNA Co-folding

In this section we generalize the triangle inequality and polymer zeta properties, which were used for speeding up the folding algorithm [26], to the RNA Co-Folding problem. We start with a short review of the quadrangle inequality and the triangle inequality in the context of speeding up dynamic programming. Let M be an $n \times n$ matrix in which each entry $M(i, j)$, such that $i \leq j$, is computed by the following formula:

$$M(i, j) = \min_{i < i' \leq j} \{M(i, i') + M(i' + 1, j)\}$$

The well-known inverse quadrangle inequality property [9] is defined as follows.

Definition 7. *A matrix M obeys the **inverse quadrangle inequality** condition iff*

$$\forall \ i < i' < j < j' \qquad M(i, j') \leq M(i, j) + M(i', j') - M(j', j)$$

Both the quadrangle and the inverse quadrangle inequalities have previously been used to speed up dynamic programming [2,9]. However, both the quadrangle inequality and the inverse quadrangle inequality are strong constraints on the input behavior, and do not apply to the matrix S computed by SA. However, a special weaker case of the classical inverse quadrangle inequality, the *triangle inequality* property, which is much more common in practice in various applications, will be extended in this paper and used to speed up RNA folding prediction.

Definition 8. *A matrix M obeys the **triangle inequality** property iff*

$$\forall \ i < j < j' \qquad M(i, j') \leq M(i, j) + M(j + 1, j').$$

The matrix S is four dimensional and therefore the standard triangle inequality does not apply. However, we consider the following "extended" inverse triangle inequality property of a 4D matrix.

Definition 9. *A four dimensional matrix M obeys the* **4D inverse triangle inequality** *property iff*

$$\forall \ i < j < j' \text{and} \ \forall \ k < \ell < \ell'$$

$$M[i, j'; k, \ell'] \geq M[i, j; k, \ell] + M[j + 1, j'; \ell + 1, \ell']$$

The next claim is immediate from Definition 9 and Eq. 1.

Claim 1. *The matrix S, as computed by Eq. 1, obeys the 4D inverse triangle inequality.*

We next turn to review the polymer zeta property in the context of RNA folding.

Definition 10. *Consider the space θ of all possible foldings for a given RNA string $s_i \ldots s_j$ under a given folding model Λ. Let $P(i, j)$ denote the probability for a folding in θ to be a co-terminus folding, and let $j - i = q$. We say that Λ has the* **polymer-zeta property** *if $P(i, j) \leq b/q^c$ for some constants $b, c > 0$.*

Experimental work has shown that RNA folding obeys the polymer-zeta property, namely, the probability that a co-terminus folding is formed over the subsequence, pairing two positions at distance q monomers apart, is $P(q) = b/q^c$ where $b = 1$ and $c > 1$ [17,18]. This fact is explained by modeling the folding of a polymer chain as a self-avoiding random walk (SAW) in a 2D lattice [24].

The theoretical exponent for the 2D SAW model is known to be $c = 1.5$ [8]. In this paper we assume that the secondary structures corresponding to SA cofoldings obey the polymer zeta property. This assumption is supported by our computational results on real data, described in Section 5.

4   A Candidate List Filter for Computing the Matrix S

In this section we describe an alternative approach to the computation of S, which prunes redundant computations in the bottleneck term (10) of Eq. 1 without sacrificing optimality of results. The algorithm saves operations by filling the $O(N^4)$ matrix S in a specific order, avoiding certain computations that are suboptimal. For each combination of subsequence start points and end points, the original algorithm requires $O(N^2)$ sums in term (10). Instead, our algorithm will identify certain endpoint combinations for which a fraction of the sums is already guaranteed not to yield the optimal score, thereby saving from the $O(N^2)$ time needed for computing all sums.

We start by describing the order in which we traverse and fill the matrix S. Recursion 1 requires the availability of both values $S[i+1, j, k, \ell]$ and $S[i, j, k+1, \ell]$ during the consideration of terms (1) and (2), correspondingly, in the computation of $S[i, j, k, \ell]$. Symmetrically, it also requires the availability of $S[i, j-1, k, \ell]$ during the computation of term (3) and of $S[i, j, k, \ell-1]$ during the computation of term (4). Theorem 1, which is given later in this section, shows that, for a given row in S, some partition points (with smaller m and n values) dominate others (with greater m and n values) in the computation of term (10), and that this dominance holds for all other entries in the row such that $j > m$ and $\ell > n$. In order to exploit this dominance property, as well as maintain the precedence order necessary for the application of Recursion 1, we will processes the entries of S in decreasing i index order and in decreasing k index order first (in other words: we compute S row-by-row, in decreasing row index order). For each index-pair (i, k), defining a specific row in S, we will compute the entry values in increasing ℓ index order first and then in increasing j index order (in other words: we maintain a left-to-right cell-traversal order within each row). Therefore, let *beam(i,k)* denote the ordered series of entries $S[i, j; k, \ell]$, for $j = i \ldots N, \ell = k \ldots N$, first in increasing ℓ index, and then in increasing j index. For simplicity of presentation, we will refer to each entry by its order of traversal within its beam, *i.e.* entry $S[i, j; k, \ell]$ will be denoted *entry (j, ℓ) of beam(i, k)*. Clearly, there are $O(N^2)$ possible beams in S and each beam covers $O(N^2)$ entries, left-to-right.

Note that each sum in term (10) is defined by its start points (i, k), its end points (j, ℓ) and its partition point (m, n). However, when considering all the $O(N^4)$ sums computed per beam, we note that a specific partition point (m, n) participates in the sums applied per computations for all end indices (m', n') such that $m \leq m' \leq j$ and $n \leq n' \leq \ell$. Therefore, in this section we view term (10) of Recursion 1 as a competition between partition points (m, n), $m = i + 1 \ldots j - 1, n = k + 1 \ldots \ell - 1$ for the branching event that yields the best score for $S[i, j; k, \ell]$. The term $S[i, m; k, n]$ of Recursion 1.(10) will be called the *left branch*, while the other term will be called the *right branch*. For each entry traversed by a beam, the naive SA computes the sums corresponding to $O(N^2)$ partition points in Recursion 1.(10). The following lemma and theorem show that some of these partition points are dominated by others (*i.e.* there are other partition points that yield equal or better score) and can thus be excluded from the computation.

Lemma 1. *Without loss of optimality, Recursion 1 can be constrained so that the left branch in term (10) is always a co-terminus co-folding.*

Naively, after constraining all left branches in term (10) to co-terminus co-foldings, there are still $O(N^2)$ partition points that compete for the optimal score in term (10), and thus altogether $O(N^4)$ sums of pairs computed per beam. However, the next theorem exposes a dominance relationship among the competing partition points, based on the 4D inverse triangle inequality property of S.

Theorem 1. *Suppose $S[i, j; k, \ell] \leq S[i, m; k, n] + S[m + 1, j; n + 1, \ell]$ for some $i < m < j$ and $k < n < \ell$. Then, $\forall\, j' > j, \ell' > \ell$ $S[i, j; k, \ell] + S[j+1, j'; \ell+1, \ell'] \leq S[i, m; k, n] + S[m + 1, j'; n + 1, \ell']$.*

Theorem 1 exposes redundancies in the repeated computation of term (10) throughout the beam entry traversal, redundancies which could be avoided by maintaining a list of only those candidate partition points that are not dominated by others.

Definition 11 (candidate). *A partition point (m, n) is a **candidate** during the computation of all entries $(j, \ell) \in beam(i, k)$ such that $m < j$, $n < \ell$, iff*
    1. $S[i, m; k, n]$ corresponds to a co-terminus co-folding.
    2. $S[i, m; k, n] > S[i, m'; k, n'] + S[m'+1, j; n'+1, \ell]\ \forall\, i < m' < m, k < n' < n$.

The above definition can be applied to speed up the computation of $S(i, j; k, \ell)$, as follows: rather than considering all possible $O(N^2)$ partition points, one could query the list that contains only partition points that satisfy the candidacy criteria above. This can be done by further reformulating term (10) as follows,

$$(10) \quad \max_{\substack{\forall (m,n) \in candidate\_list: \\ m<j \wedge n<l}} \{ S[i, m; k, n] + S[m+1, j; n+1, \ell] \} \quad (2)$$

The pseudocode for the new algorithm, denoted FASTCOFOLD, is given below. *Procedure ComputeBeam* in Algorithm FASTCOFOLD replaces the original term (10) in Recursion 1 with its constrained reformulation as Eq. 2. This is implemented as a candidate list that is empty at the start of each beam traversal, and is extended throughout the left-to-right computation of the entries of the beam, by appending to the list only those partition points that are candidates by Definition 11.

Each partition point (m, n) is considered for candidacy once per procedure call, when entry $S[i, m; k, n]$ is reached by the beam traversal, and will join the candidate list only if, at this point, the value of this entry dominates all its preceding partition points on the list. Each entry traversed by $beam(i, k)$ is computed by the new algorithm as before, with the only difference being that the maximum of term (10) is taken only over preceding pairs (m, n) from the candidate list, as formalized in Eq. 2.

In the following theorem we assume that the 2D structures corresponding to SA co-foldings follow the RNA 2D SAW model (see Section 3), and that therefore the probability for a co-terminus co-folding follows the polymer-zeta property with $c > 1$. This assumption is supported by the experimental results in Section 5.

Theorem 2. *Algorithm* FASTCOFOLD *improves* SA *by a linear factor on average.*

Algorithm FASTCOFOLD *:*
1 *for* each row $i := N$ to 1 *do*
2   *for* each row $k := N$ to 1 *do*
3   *call Procedure ComputeBeam(i,k);*

Procedure ComputeBeam(i,k):
1 *candidate_list* $\leftarrow NULL$
2 *for* each column $j := i$ to N *do*
3   *for* each column $l := k$ to N *do*
4          $S\_dang[i, j; k, l]$ $\leftarrow$
             maximal score among terms $(1) - (6)$ *of Eq. 1*
5          $S\_co\text{-}terminus[i, j; k, l]$ $\leftarrow$
             maximal score among terms $(7) - (9)$ *of Eq. 1*
6          $S\_branch[i, j; k, l]$ $\leftarrow$
                 $\max_{\substack{\forall (m,n) \in candidate\_list: \\ m \leq j \wedge n \leq l}} \{S[i, m; k, n] + S[m+1, j; n+1, l]\}$
7          $S[i, j; k, l] \leftarrow \max\{S\_dang[i, j; k, l], S\_branch[i, j; k, l]\}$
8   *if* $(S\_co\text{-}terminus[i, j; k, l] > S[i, j; k, l])$ *then*
9              $S[i, j; k, l] \leftarrow S\_co\text{-}terminus[i, j; k, l]$
10   Append (j, l) to the *candidate_list* for (i, k)

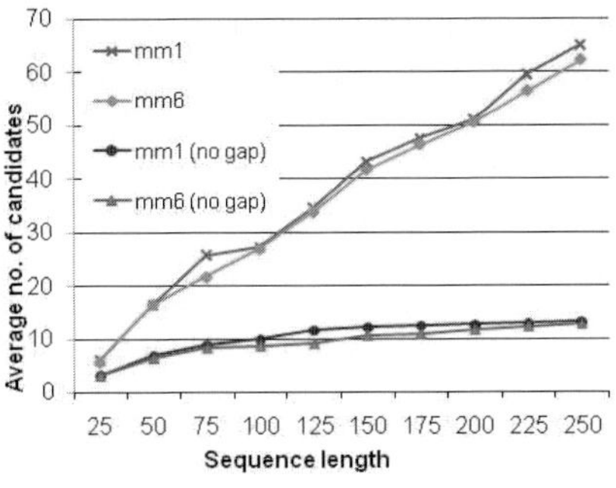
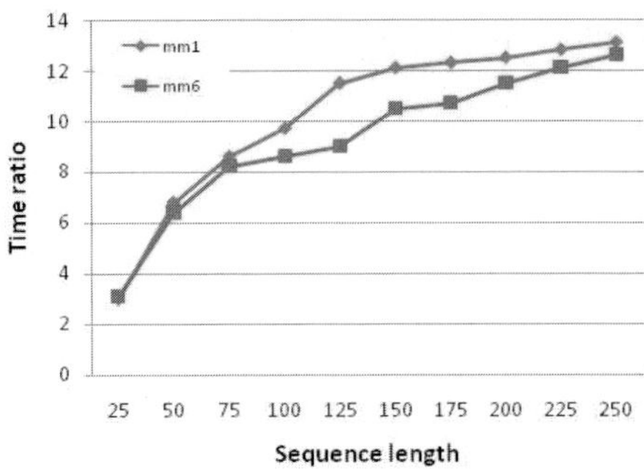

Fig. 2. Performance of FASTCOFOLD . (left) The average number of candidates in a list when running FASTCOFOLD. A "(no gap)" next to the benchmark name indicates that the parameter defining the allowed size difference between any two aligned subsequences was set to 0. (right) The average ratio between the run time of SA as implemented in Foldalign [15] and FASTCOFOLD (without any gap constraints), for different sequence lengths.

5   Performance Testing

To test the power of algorithm FASTCOFOLD in practice, we implemented it as a filter on top of the FoldAlign SA program [15] in its version that computes an optimal solution with no heuristic shortcuts. We then compared the performance of FASTCOFOLD with that of the original version of FoldAlign. For this, we generated sequences of length 25,50,75 ..., 250. Two datasets of 50 sequences each were generated for each length. Each pair of sequences of the same length in the same data set were co-folded. Altogether, we performed 24,500 co-foldings. The two sets were generated randomly according to a Markov model trained on RNA sequences randomly chosen from complete human mRNA sequences taken from the RefSeq database at NCBI `www.ncbi.nlm.nih.gov/RefSeq`. Sets "mm1" and "mm6" were generated using a Markov model of order 1 and 6, respectively. We ran FASTCOFOLD on each data set and measured running times and candidate list sizes. Two versions of FoldAlign and FASTCOFOLD were run: (1) allowing no gaps; (2) the full version allowing gaps of unbounded size. The results for FASTCOFOLD are shown in Figure 2(left). Reassuringly, in all runs the averages grow at most linearly with the length of the sequence. Figure 2(right) plots the average ratio between the run times of the two algorithms (both applied without any gap constraints) as a function of sequence length for both data sets. Runs were conducted on an Intel Xeon 2.8GHZ computer with 4GB RAM. The overall run-time for an all-against-all co-folding of a set of 50 sequences (including I/O time) varied from 2 seconds for the 25 bps sequences, to up to more than 12 hours for the 250 bps sequences. We suspect that the non-linear behavior of this graph is mostly due to exhausting memory resources in the benchmark computer.

Note that two heuristic constraints are used by Foldalign to reduce the time complexity. One constraint binds the total size of the allowed gaps, $(j - i) - (\ell - k) \leq \delta$. The other constraint, which can only be applied in the case of local alignments, binds the size of the compared subsequences $j - i \leq \lambda$ and $\ell - k \leq \lambda$. By applying both heuristics, a constrained version of SA is obtained with time complexities of $O(N^5)$ for global alignment with gaps bounded by a constant, and $O(N^4)$ for local alignment with both gaps and alignment sizes bounded by a constant. We observe that when δ is set to zero (no gaps allowed), ζ converges to a constant (see the "no gap" plot in Figure 2.a). Thus, using FASTCoFOLD, one can achieve an $O(N^4)$ SA that limits only the gaps and not the size of the alignment.

Acknowledgments. We thank Gary Benson for helpful advice and for elaborate demonstrations of TRDB. We are also grateful to the anonymous WABI2008 referees for their helpful comments. RS was supported in part by the Wolfson Foundation and by the Raymond and Beverley Sackler Chair in Bioinformatics. A preliminary version of this study was performed when MZU was supported by a fellowship from the Edmond J. Safra Program in Bioinformatics at Tel Aviv University.

References

1. Uzilov, A.V., Keegan, J.M., Mathews, D.H.: Detection of non-coding RNAs on the basis of predicted secondary structure formation free energy change. BMC Bioinformatics 7, 173 (2005)
2. Crochemore, M., Landau, G.M., Schieber, B., Ziv-Ukelson, M.: Re-use dynamic programming for sequence alignment: An algorithmic toolkit, pp. 19–60 (2005)
3. Mathews, D., Turner, D.: Dynalign: An algorithm for finding the secondary structure common to two RNA sequences. Journal of Molecular Biology 317, 191–203 (2002)
4. Sankoff, D.: Simultaneous solution of the RNA folding, alignment and protosequence problems. SIAM Journal on Applied Mathematics 45, 810–825 (1985)
5. Mathews, D.H., Burkard, M.E., Freier, S.M., Wyatt, J.R., Turner, D.H.: Predicting oligonucleotide affinity to nucleic acid target. RNA 5, 1458 (1999)
6. Rivas, E., Eddy, S.R.: Secondary structure alone is generally not statistically significant for the detection of non-coding RNAs. Bioinformatics 16, 583–605 (2000)
7. Torarinsson, E., Havgaard, J.H., Gorodkin, J.: Multiple structural alignment and clustering of RNA sequences. Bioinformatics 23(8), 926–932 (2007)
8. Fisher, M.E.: Shape of a self-avoiding walk or polymer chain. J.Chem. Phys. 44, 616–622 (1966)
9. Giancarlo, R.: Dynamic Programming: Special Cases. Oxford University Press, Oxford (1997)
10. Kiryu, H., Tabei, Y., Kin, T., Asai, K.: Murlet: a practical multiple alignment tool for structural RNA sequences. Bioinformatics 23, 1588–1598 (2007)
11. Holmes, I.: Accelerated probabilistic inference of RNA structure evolution. BMC Bioinformatics 6, 73 (2005)
12. Tinoco, I., Borer, P.N., Dengler, B., Levine, M.D., Uhlenbeck, O.C., Crothers, D.M., Gralla, J.: Improved estimation of secondary structure in ribonucleic acids. Nature New Biology 246, 40–41 (1973)
13. Hofacker, I.L., Fekete, M., Stadler, P.F.: Secondary structure prediction for aligned RNA sequences. Journal of Molecular Biology 319, 1059–1066 (2002)

14. Hofacker, I.L., Bernhart, S., Stadler, P.: Alignment of RNA base pairing probability matrices. Bioinformatics 20, 2222–2227 (2004)
15. Havgaard, J.H., Lyngso, R.B., Stormo, G.D., Gorodkin, J.: Pairwise local structural alignment of RNA sequences with sequence similarity less than 40%. Bioinformatics 21(9), 1815–1824 (2005)
16. Pederson, J., Bejerano, G., Siepel, A., Rosenbloom, K., Lindblad-Toh, K., Lander, E., Kent, J., Miller, W., Haussler, D.: Identification and classification of conserved RNA secondary structres in the human genome. PLOS Computational Biology 2, 33 (2006)
17. Kabakcioglu, A., Stella, A.L.: A scale-free network hidden in the collapsing polymer. ArXiv Condensed Matter e-prints (September 2004)
18. Kafri, Y., Mukamel, D., Peliti, L.: Why is the DNA denaturation transition first order? Physical Review Letters 85, 4988–4991 (2000)
19. Mandal, M., Breaker, R.R.: Gene regulation by riboswitches. Cell 6, 451–463 (2004)
20. Nussinov, R., Jacobson, A.B.: Fast algorithm for predicting the secondary structure of single-stranded RNA. Proc. Natl. Acad. Sci. 77(11), 6309–6313 (1980)
21. Dowell, R.D., Eddy, S.: Efficient pairwise RNA structure prediction and alignment using sequence alignment constraints. BMC Bioinformatics 7, 400 (2006)
22. Griffiths-Jones, S.: The microrna registry. Nucleic Acids Research 32, D109–D111 (2003)
23. Washietl, S., Hofacker, I.L.: Consensus folding of aligned sequences as a new measure for the detection of functional RNAs by comparative genomics. Journal of Molecular Biology 342, 19–30 (2004)
24. Vanderzande, C.: Lattice Models of Polymers (Cambridge Lecture Notes in Physics 11). Cambridge University Press, Cambridge (1998)
25. Waterman, M.S., Smith, T.F.: Rapid dynamic programming algorithms for RNA secondary structure. Adv. Appl. Math. 7, 455–464 (1986)
26. Wexler, Y., Zilberstein, C., Ziv-Ukelson, M.: A study of accessible motifs and the complexity of RNA folding. Journal of Computational Biology 14(6), 856–872 (2007)
27. Zuker, M., Stiegler, P.: Optimal computer folding of large RNA sequences using thermodynamics and auxiliary information. Nucleic Acids Research 9(1), 133–148 (1981)

An Automated Combination of Kernels for Predicting Protein Subcellular Localization

Cheng Soon Ong[1,2] and Alexander Zien[1,3]

[1] Friedrich Miescher Laboratory, Tübingen, Germany
[2] Max Planck Institute for Biological Cybernetics, Tübingen, Germany
[3] Fraunhofer Institute FIRST, Berlin, Germany

Abstract. Protein subcellular localization is a crucial ingredient to many important inferences about cellular processes, including prediction of protein function and protein interactions. While many predictive computational tools have been proposed, they tend to have complicated architectures and require many design decisions from the developer.

Here we utilize the multiclass support vector machine (m-SVM) method to directly solve protein subcellular localization without resorting to the common approach of splitting the problem into several binary classification problems. We further propose a general class of protein sequence kernels which considers all motifs, including motifs with gaps. Instead of heuristically selecting one or a few kernels from this family, we utilize a recent extension of SVMs that optimizes over multiple kernels simultaneously. This way, we automatically search over families of possible amino acid motifs.

We compare our automated approach to three other predictors on four different datasets, and show that we perform better than the current state of the art. Further, our method provides some insights as to which sequence motifs are most useful for determining subcellular localization, which are in agreement with biological reasoning. Data files, kernel matrices and open source software are available at
`http://www.fml.mpg.de/raetsch/projects/protsubloc`.

1   Introduction

Support vector machines (SVMs, e.g. [1]) are in widespread and highly successful use for bioinformatics tasks. One example is the prediction of the subcellular localization of proteins. SVMs exhibit very competitive classification performance, and they can conveniently be adapted to the problem at hand. This is done by designing appropriate kernel functions, which can be seen as problem-specific similarity functions between examples. The kernel function implicitly maps examples from their input space $\mathcal{X}$ to a space $\mathcal{H}$ of real-valued features (e.g. $\mathcal{H} = \mathbf{R}^d, d \in \mathbf{N} \cup \{\infty\}$) via an associated function $\Phi : \mathcal{X} \to \mathcal{H}$. The kernel function k provides an efficient method for implicitly computing dot products in the feature space $\mathcal{H}$ via $k(\mathbf{x}_i, \mathbf{x}_j) = \langle \Phi(\mathbf{x}_i), \Phi(\mathbf{x}_j) \rangle$.

Many different types of features have been used for SVM-based subcellular localization prediction. One popular class of features are compositions, i.e. histograms of subsequences. The most common choice of subsequences are single amino acids. One can also generalise this to pairs of adjacent amino acids, pairs of amino acids with

K.A. Crandall and J. Lagergren (Eds.): WABI 2008, LNBI 5251, pp. 186–197, 2008.
© Springer-Verlag Berlin Heidelberg 2008

one position gap between them, pairs separated by two positions, and also for longer patterns [2]. A more widespread idea is to capture the statistics of signal peptides by computing compositions on relevant parts of a protein separately [3,4]. In section 2, we define kernels that generalize these ideas. Apart from compositions, further features include the search for known motifs [5,6] or PFAM domains [3], the use of PSI-BLAST profiles [7], and the use of PSI-BLAST similarities to other sequences [8]. In some cases, even SVMs or other classifiers are employed for feature generation or for motif detection [5,6].

When more than one set of features have been defined and computed, the task is to combine the evidence they yield into a single final prediction. This is often done in complex, hand-crafted architectures that frequently consist of two (or even more) layers of learning machines or decision systems. Instead, we utilize the novel multiclass multiple kernel learning (MCMKL) method [9], which optimally selects kernels from a given set and combines them into an SVM classifier (Section 3). Both are jointly applied to protein subcellular localization prediction in Section 4.

2   Motif Composition Kernels

2.1   Amino Acid Kernel and Motif Kernel

Before we consider motifs consisting of several amino acids, we define a kernel on individual amino acids (AAs). This will be useful as an ingredient to the more complex motif kernel. The AA kernel takes into account pairwise similarity of amino acids.

Let $\mathcal{A}$ be the set of 20 amino acids. A substitution matrix M consists of a real-valued element m_{ab} for each pair of amino acids a and b. As substitution matrices are not in general valid kernel functions, we apply some transformations. It has been shown that every sensible substitution matrix M implies a matrix R of amino acid substitution probabilities via $m_{ab} = \frac{1}{\lambda} \log \frac{r_{ab}}{q_a q_b}$. Here q_a is the so-called background probability of a, its relative frequency of appearance in any protein sequence. Given the constraints $\sum_a \sum_b r_{ab} = 1$ and $q_a = \sum_b r_{ab}$ and the symmetry of both M and R, R can be computed from M. We do so for the popular BLOSUM62 matrix [10] serving as M.

The elements of the obtained R, being substitution probabilities, are positive, and thus R can be seen as a (complete) similarity graph between amino acids with weighted edges. From this we derive a positive definite kernel k_1^{AA} on the amino acids by taking the graph Laplacian:

$$k_1^{AA}(a, b) = \sum_c r_{ac} - r_{ab} \ . \tag{1}$$

Note that other choices of kernels are possible. One alternative is the diffusion kernel, which is computed by taking the matrix exponential of a scalar multiple of R. In this context we prefer the graph Laplacian since it does not have any parameters to be adjusted. We extend the AA-kernel to r-tuples of amino acids ("motifs") by simply adding kernel values over the components. For $s, t \in \mathcal{A}^r$ we define the motif kernel

$$k_r^{AA}(s, t) = \sum_{i=1}^{r} k_1^{AA}(s_i, t_i) \ . \tag{2}$$

Note that these kernels cannot be direclty applied to variable-length protein sequences.

2.2   Motif Compositions

Previous work has shown that the amio acid composition (AAC) of a sequence is a useful basis for classifying its subcellular localization [11]. An advantage of this set of features is that it is robust with respect to small errors in the sequences, as may be caused by automated determination from genomic DNA. In subsequent work the AAC has been generalized in two directions.

First, instead of just considering the AAC of the entire protein sequence, it was calculated on different subsequences [3,6,12]. This is motivated by the fact that important indications of localization are not global. For example, the targeting of a protein to the mitochondrion or to the chloroplast is indicated by an N-terminal signal peptide with specific properties (for example pH or hydrophobicity) that are reflected by the AAC.

Second, it was noted that features which represent dependencies between two or more amino acids can increase the prediction performance. This seems plausible since there exist a number of (rather short) known motifs that are important for subcellular targeting. Examples include the C-terminal targeting signal for microbodies (SKL), the C-terminal endoplasmatic reticulum targeting sequence (KDEL), and the bipartite nuclear targeting sequence (which consists of five basic amino acids, R or K, in a certain arrangement). Existing prediction methods that generalize the AAC to higher order compositions do so in at least two ways: [2] and [8] use composition of pairs of amino acids, possibly with fixed-length gaps between them; [4] consider distributions of consecutive subsequences of length r, where a reduced size alphabet is used to avoid combinatorial explosion of the feature space for large r.

Here we carry the generalization a bit further, by allowing for patterns consisting of any number r of amino acids in any (fixed) positional arrangement. For example, we could choose the frequencies of occurance of AA triplets with two positions gap between the first two and no gap between the second two, corresponding to a pattern ($\bullet,\circ,\circ,\bullet,\bullet$). For any given pattern, we can compute the empirical distribution of corresponding motifs from a given AA sequence. This is a histogram of occurrences of each possible r-mer sequence. The example above will result in a histogram of all possible 3-mers where each sequence is represented by the counts of the occurrences of each 3-mer with the specified gap. Note that the combinatorial explosion of possible motifs for increasing order r is not a real problem, because the number of motifs with positive probability is bounded by the protein length, and we employ sparse representations.

2.3   Motif Composition Kernels

The feature sets defined just above are histograms, and after normalization they are probability distributions over discrete sets. While we can use standard kernels (like the Gaussian RBF) on these data, this would neglect the fact that they are not arbitrary vectors, but in fact carry a special structure. Hence we use kernels that are specially designed for probability distributions [13]. These kernels have the added benefit of allowing us to model pairwise similarities between amino acids. To our knowledge, this is the first time such kernels have been applied to protein sequence analysis.

We use the Jensen-Shannon divergence kernel (corresponding to $\alpha = 1$ in [13]), which is based on a symmetric version of the Kullback-Liebler divergence of information theory. Applied to histograms on patterns of order r we have

$$k_r^{JS}(p, q) = \sum_{s \in \mathcal{A}^r} \sum_{t \in \mathcal{A}^r} k_r^{AA}(s, t) \left(p(s) \log \frac{p(s)}{p(s) + q(t)} + q(t) \log \frac{q(t)}{p(s) + q(t)} \right) ,$$

(3)

where p and q are the r-mer histograms obtained from two sequences, and s and t run over the amino acid motifs. For this paper, we define the kernels between amino acids $k^{AA}(s, t)$ using the summed graph Laplacian defined in Equations (1) and (2).

Even using these choices, we are still left with a large number of possible patterns (as defined in Section 2.2) to consider. As linear combinations of kernels are again valid kernels [1], we could fix coefficients (e.g., uniform weights) and work with the resulting weighted sum kernel. However, it can be difficult to find optimal weights.

3   Multiclass Multiple Kernel Learning

Multiple kernel learning (MKL) is a technique for optimizing kernel weights β_p in a linear combination of kernels, $k(\mathbf{x}, \mathbf{x}') = \sum_p \beta_p k_p(\mathbf{x}, \mathbf{x}')$. Thereby MKL is capable of detecting useless sets of features (noise) and eliminating the corresponding kernel (by giving it zero weight). Consequently MKL can be useful for identifying biologically relevant features [14,15]. In this paper we use the newly proposed multiclass extension (see Figure 1) of MKL, called multiclass multiple kernel learning, MCMKL [9].

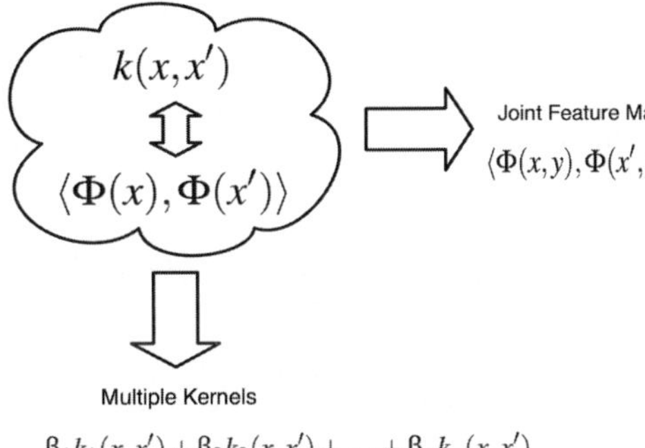

Multiclass multiple kernel learning combines two generalizations of SVMs. The first is to learn more complex output spaces (including multiclass) than the simple binary classification task by extending the kernel to also encode classes (the horizontal arrow). The second is to simultaneously optimize over the coefficients in a linear combination of kernels, hence automatically weighting different feature spaces (the vertical arrow).

Fig. 1. The approach in [9] generalizes the idea of kernel machines in two directions

While binary SVMs have a single hyperplane normal $\mathbf{w}$ in feature space, multiclass SVMs (as considered here) have a different hyperplane normal $\mathbf{w}_u$ for each class u. Thus a trained MCMKL classifier has a separate confidence function

$$f_u(\mathbf{x}) = \left\langle \mathbf{w}_u , \sum_p \beta_p \Phi_p(\mathbf{x}) \right\rangle = \sum_i \alpha_{iu} \sum_p \beta_p k_p(\mathbf{x}_i, \mathbf{x})$$

(4)

for each class u, where the latter equality derives from the expansion of the hyperplane normals $\mathbf{w}_u = \sum_i \alpha_{iu} \Phi(\mathbf{x}_i)$ (due to the Representer Theorem [1] or by Lagrange dualization). The predicted class for a given example $\mathbf{x}$ will be chosen to maximize the confidence, that is $\hat{y}(\mathbf{x}) = \arg\max_u f_u(\mathbf{x})$.

Using an approach similar to that in [15], we convert the dual into an equivalent semi-infinite linear program (SILP) formulation by a second (partial) dualization.

$$\max_{\boldsymbol{\beta}} \theta \quad \text{s.t.} \quad \forall \boldsymbol{\alpha} \in \mathcal{A} : \ \theta \le \frac{1}{2} \sum_k \beta_k \|\mathbf{w}_k(\boldsymbol{\alpha})\|^2 - \sum_i \alpha_{iy_i},$$

$$and \quad \sum_{k=1}^{p} \beta_k = 1, \quad \forall k : \ 0 \le \beta_k \ , \tag{5}$$

where

$$\mathcal{A} = \left\{ \boldsymbol{\alpha} \ \middle| \ \begin{array}{l} \forall i : 0 \le \alpha_{iy_i} \le C \\ \forall i : \forall u \ne y_i : \alpha_{iu} \le 0 \\ \forall i : \sum_{u \in \mathcal{Y}} \alpha_{iu} = 0 \\ \forall u \in \mathcal{Y} : \sum_i \alpha_{iu} = 0 \end{array} \right\}$$

is the set of admissable parametrizations for the first constraint. For details on the proof see the Supplement and [9]. For a fixed $\boldsymbol{\beta}$, Equation (5) is a quadratic program (QP) which is only slightly more complicated than a standard SVM: it solves a direct multi-class SVM. Furthermore, for fixed $\boldsymbol{\alpha}$ the optimization problem in $\boldsymbol{\beta}$ is a linear program (LP). However, the constraint on θ has to hold for every suitable $\boldsymbol{\alpha}$, hence the name (refering to the infinitely many constraints).

We follow [15] and use a column generation strategy to solve (5): Solving the QP given by the constraints for a fixed $\boldsymbol{\beta}$ results in a particular $\boldsymbol{\alpha}$, which gives rise to a constraint on θ which is linear in $\boldsymbol{\beta}$. We alternate generating new constraints and solving the LP with the constraints collected so far (Figure 2). This procedure is known to converge [16,15]. For more details on this model and how it can be trained, that is how the values of the parameters α_{iu} and β_p can be optimized, see [9]. Essentially the same model, though with a particular arrangement of kernels and a different optimization, is developed in [17]. Another related approach is described in [18].

4   Computational Experiments

To predict subcellular localization, we use motif kernels up to length 5 as defined in Section 2. Note that 20^5 (3.2 million) different motifs of the form ($\bullet,\bullet,\bullet,\bullet,\bullet$) exist; due to the Jensen-Shannon transformation (eq. 3) the feature spaces are even infinite dimensional. Apart from using the whole sequence, we compute the motif kernels on different sections of the protein sequences, namely the first 15 and 60 amino acids from the N-terminus and the 15 amino acids from the C-terminus (inspired by [6]). This results in $4 \times 2^{(5-1)} = 64$ motif kernels.

We augment the set of kernels available to the classifier by two small families based on features which have been shown to be useful for subcellular localization [19,20]. Using the pairwise E-value of BLAST as features, we compute a linear kernel, a Gaussian RBF kernel [1] with width 1000, and another Gaussian kernel with width 100000 from the logarithm of the E-value of BLAST. The second additional kernel family is derived from phylogenetic profiles [21]. Using the results from their webserver (`http://apropos.icmb.utexas.edu/plex/`) as features, we compute a linear kernel and a Gaussian kernel of width 300. The Gaussian kernel widths were selected from a coarse grid by running MCMKL separately on each of the two kernel families; the range of the grid was inspired by the distribution of pairwise Euclidean distances of the feature vectors.

In total, we thus consider 69 candidate kernels. This renders manual selection and weighting tedious or even impossible, and thus calls for MKL. As in standard binary single-kernel SVMs, there is a parameter "C" in the MCMKL method to tune the regularization. We normalize each kernel such that setting $C = 1$ will at least be a reasonable order of magnitude (refer to [9] for details).

The subsequent protocol for all our experiments is as follows:

- Ten random splits into 80% training and 20% test data are prepared.
- For each training set, the parameter C is chosen using 3-fold cross validation on the training set only. We search over a grid of values $C = \{1/27, 1/9, 1/3, 1, 3, 9, 27\}$. For all tasks, the best C is chosen by maximizing the average F1 score on the validation (hold out) part of the training set. The F1 score is the harmonic mean of precision p and recall r, $f1 = (2 * p * r)/(p + r)$. For more details, see the Supplement.
- Using the selected C, we train MCMKL on the full training set and predict the labels of the test set.

To compare with existing methods, we compute several different measures of performance. We assess the proposed an two different datasets, for which results with other methods are reported in the literature. The first is the data set used for training and evaluating TargetP [22]. The second dataset is a database of bacterial proteins, PSORTdb [5].

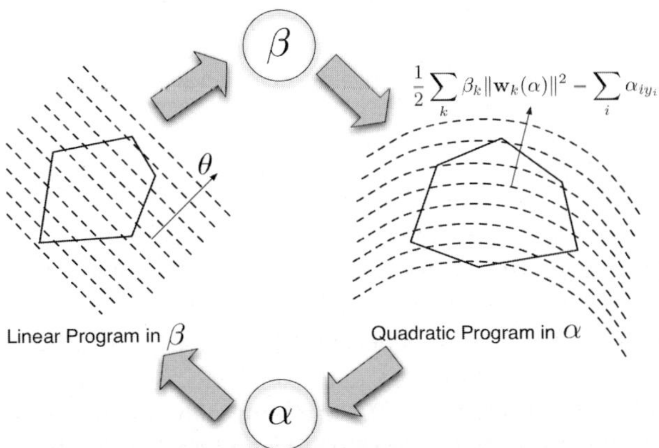

Fig. 2. The SILP approach to MKL training alternates between solving an LP for β and a QP for α until convergence [9,15]

Table 1. Summary of comparative Protein Subcellular Localization results

dataset name	performance measure	performance [%] of MCMKL	competitor	competing method, reference, year
TargetP plant	avg. MCC	89.9 ± 1.1	86.0	TargetLoc [6], 2006
TargetP nonplant	avg. MCC	89.7 ± 0.8	86.6	TargetLoc [6], 2006
PSORT+, 15.0% out	avg. Prec/Recall	95.5 / 94.7	95.9 / 81.3	PSORTb [5], 2004
PSORT-, 13.3% out	avg. Prec/Recall	96.4 / 96.3	95.8 / 82.6	PSORTb [5], 2004
PSORT-	avg. recall	91.3	90.0	CELLO II [20], 2006

A summary of the overall performance is shown in Table 1. Details of performance are available in the Supplement.[1]

4.1   Comparison on TargetP Dataset

The original plant dataset of TargetP [22] is divided into five classes: chloroplast (ch), mitochondria (mi), secretory pathway (SP), cytoplasm (cy), and nucleus (nuc). However, in many reported results, cy and nuc are fused into a single class "other" (OT), and hence we do the same to enable direct comparison. Non-plant is similar, but lacks the chloroplasts. Each of the 10 random splits contains 21 cross validation optimizations (3-fold cross-validation to select from 7 values of C), with an additional larger optimization at the end. For the 3-class problem plant, the computation time for each split is roughly 10 hours on a 2.4Ghz AMD64 machine.

The results in [22] are reported in terms of Matthew's correlation coefficient (MCC). As can be seen in Table 1, the MCC values obtained with our proposed method are significantly better. Details for each class are shown in the Supplement. The features most often selected for classification as seen in Figures 3(c) and 3(d) are the kernels computed from BLAST E-values as well as phylogenetic information. The lists of *all* kernels selected are in the Supplement. From Table 2, we see the motif kernels which are most frequently selected. The motif kernel with pattern (●,○,○,○,○) only measures the simple amino acid composition. However, this encodes important global properties like protein mass or charge; it is thus reassuring to see that it gets selected. However, observe that several long patterns (●,○,○,○,●) and (●,●,○,○,●) are selected in the N-terminus region, indicating the presence of long meaningful subsequences in that region.

4.2   Comparison on PSORTdb Dataset

We also run computations on sequences and localizations of singly localized proteins in bacteria obtained from PSORTdb [5] and compare the performance to the prediction tool PSORTb v2.0 [5]. PSORTb v2.0 can withhold a prediction when it is uncertainabout the localization. From their supplementary website we estimate the

---

[1] The Supplement is freely available for download as `protsubloc-wabi08-supp.pdf` at `http://www.fml.tuebingen.mpg.de/raetsch/projects/protsubloc`

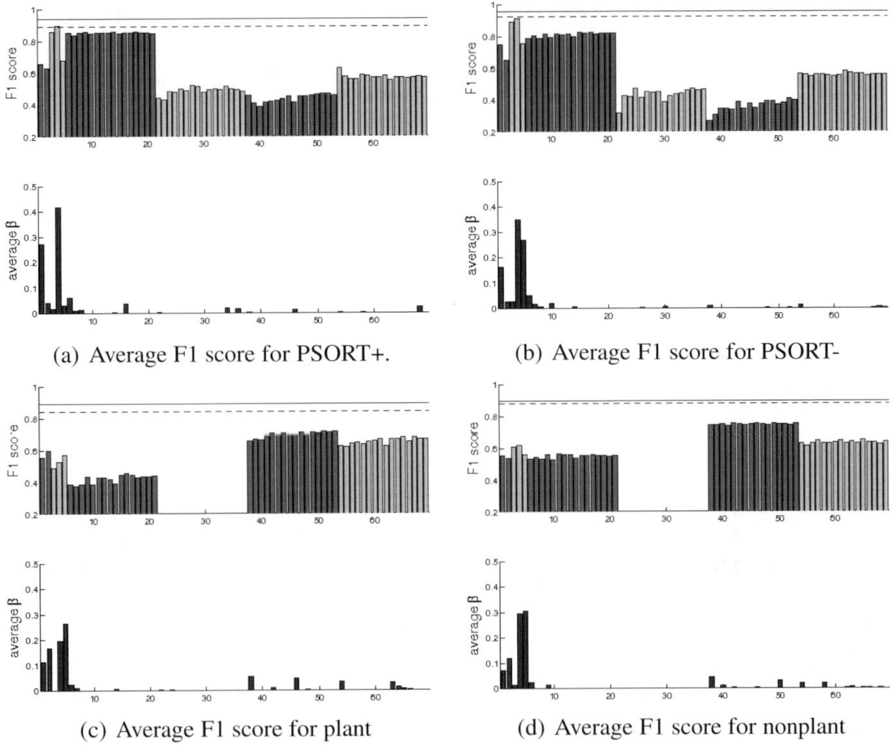

(a) Average F1 score for PSORT+.

(b) Average F1 score for PSORT-

(c) Average F1 score for plant

(d) Average F1 score for nonplant

Fig. 3. Average F1 score for various features. The horizontal axis indexes the 69 kernels. The two phylogenetic profile kernels and the three BLAST E-value kernels are on the left. The following four blocks are the motif kernels on: the whole sequence, the 15 AAs at the C-terminus, the 15 and 60 AAs at the N-terminus. In each subfigure, the bars in the lower panel display the average optimized kernel weight. The bars in the upper panels show the accuracy obtained from using each kernel just by itself. The dashed black line shows the performance when all the kernels are equally weighted (after appropriate normalization). The solid blue line shows the result of our method for comparison.

proportion of "unknown" predictions to be 15.0% for the Gram positive and 13.3% for Gram negative bacteria. We estimate probabilistic outputs from our method by applying the logistic transformation, that is $\hat{p}(y|\mathbf{x}) = \frac{\exp f_y(\mathbf{x})}{\sum_u \exp f_u(\mathbf{x})}$ to the SVM confidences. To obtain comparable figures, we discard the same fractions of most uncertain predictions, i.e. those with the lowest $\hat{p}(\hat{y}(\mathbf{x})|\mathbf{x})$. The mean and standard deviations on this reduced test set are reported in Table 1 in the comparison with PSORTb.

In 2004, PSORTb claimed to be the most precise bacterial localization prediction tool available [5]. However, as the results in Table 1 suggest, our method performs dramatically better. The performance assessments for various measures and for each class, which are reported in the Supplement, confirm this. The numbers show that our method in general matches the high precision of PSORTb and that we have extremely high recall levels, resulting in significantly better F1 scores for most localizations in

Table 2. Sequence motif kernels which have been selected at least 8 times out of the 10 splits for each dataset

times selected	mean β_k	kernel (PSORT+)
10	6.23%	motif (●,○,○,○,○) on $[1, \infty]$
10	3.75%	motif (●,○,●,○,●) on $[1, \infty]$
9	2.24%	motif (●,○,●,●,●) on $[1, 60]$
10	1.32%	motif (●,○,○,○,●) on $[1, 15]$
8	0.53%	motif (●,○,○,○,○) on $[1, 15]$

times selected	mean β_k	kernel (plant)
10	5.50%	motif (●,○,○,○,○) on $[1, 15]$
10	4.68%	motif (●,○,○,○,●) on $[1, 15]$
10	3.48%	motif (●,○,○,○,○) on $[1, 60]$
8	3.17%	motif (●,●,○,○,●) on $[1, 60]$
9	2.56%	motif (●,○,○,○,○) on $[1, \infty]$

times selected	mean β_k	kernel (PSORT-)
10	5.04%	motif (●,○,○,○,○) on $[1, \infty]$
10	1.97%	motif (●,○,○,●,○) on $[1, \infty]$
9	1.57%	motif (●,●,○,○,○) on $[1, \infty]$
10	1.51%	motif (●,○,○,○,○) on $[1, 60]$
10	1.14%	motif (●,○,○,○,○) on $[1, 15]$
10	0.82%	motif (●,○,○,○,●) on $[-15]$

times selected	mean β_k	kernel (nonplant)
9	4.48%	motif (●,○,○,○,○) on $[1, 15]$
10	3.23%	motif (●,○,○,●,●) on $[1, 15]$
9	2.32%	motif (●,○,○,○,○) on $[1, \infty]$
9	2.17%	motif (●,○,○,○,○) on $[1, 60]$
8	1.92%	motif (●,○,○,●,○) on $[1, 60]$
9	1.48%	motif (●,●,●,○,○) on $[1, \infty]$
8	0.94%	motif (●,○,●,○,○) on $[1, 15]$

both PSORT datasets. We also show in our Supplement that we are still competitive with PSORTb, even when predicting on all sequences (as opposed to withholding when unsure). Our method further compares favorably to CELLO II [20], for which results on the Gram negative bacterial protein set are published.

Similar to the motifs selected in the plant dataset, the BLAST E-values and phylogenetic profiles are important (Figure 3(a) and 3(b)). Note also that in all datasets, both the BLAST E-value as well as the log transformed version turn out to be useful for discrimination. This demonstrates one of the major dilemmas of using only one fixed kernel, as it may be possible that some transformation of the features may improve classification accuracy. Note that in Table 2 the motif (●,○,○,○,●) near the C-terminus, $[-15, \infty]$, has very little weight, and all other motifs are shorter. Indeed, in other experiments (results not shown), MCMKL with motifs up to length 4 performs equally well.

5   Discussion

First we note that our proposed method improves on established and state of the art methods for predicting protein subcellular localization. This is the case with respect to various figures of merit, which also demonstrates the robustness of our method. The success off our approach comes despite the fact that its design required little time and care: in contrast to complex competing methods, we only need to provide a sufficient set of candidate feature spaces (i.e. kernels) and do not have to worry about which one of them is best. In fact, Figures 3(a)-3(d) show that there is no single best kernel for all localization prediction tasks, and that kernel combinations can improve on each single kernel.

In our setting, the simple unweighted sum of all considered kernels reliably yields high accuracy. Note that this depends on the normalization of the kernels; while we use a heuristic, but justified scaling scheme, no theoretically optimal task-independent scaling method is known. However, we successfully use modern machine learning methods, specifically multiclass multiple kernel learning (MCMKL), to learn an optimal kernel weighting (the values β_p) for the given classification problem. Indeed the MCMKL reweighting consistently outperforms the plain normalization: it reduces the error (1 - score) by roughly 20%. We even used MCMKL to adjust real-valued kernel parameters by selecting them from a pre-specified coarse grid.

Note that the performance of the kernels taken by themselves is not a good indication of their weights in the optimized combination. For example in the plant experiments, motif kernels are best individually, but BLAST and phylogeny kernels obtain higher weights. We speculate that correlating information of the kernels is one reason for this: instead of choosing very similar kernels, MCMKL chooses a mixture of kernels that provide complementary information. Thus one should include as many diverse forms of information as possible. However, the weights of the kernels also depend on their prior scaling (normalization), and more machine learning research is necessary to fully understand this issue.

In addition to improving the accuracy, MCMKL also helps to understand the trained classifier. For example, in the plant data, the motif kernels on N-terminal subsequences (both length 15 and 60) provide the most informative feature spaces. The reason for this is most likely that they are best suited to detect the chloroplast and mitochondria transit peptides which are known to be in the N-terminal. For bacteria, which do not have organelles and corresponding signal peptides, the composition of the entire protein is more useful; probably because it conveys properties like hydrophobicity and charge. However, the BLAST kernels, which can pick up protein structure via remote homology, are assigned even higher weights. For the bacterial datasets, the BLAST kernels obtain more weight and perform better individually than the phylogenetic kernels. Phylogenetic profiles have only been shown to work well with organellar proteins [23], and the evidence in Figures 3(c) and 3(d) shows that they can help (slightly) for eukaryotes.

Since signal peptides are usually longer than 5 amino acids, our motifs may not capture all the information. However, one would expect that BLAST scores and phylogenetic profiles capture this information. This is reflected by the high weight given to these kernels (Figure 3). Note however that the localization signal may be distributed across the amino acid sequence, and only brought together by protein folding – the so called signal patches. If each component of the signal patches is relatively short, then this can be captured by our motif kernels.

While MKL did successfully identify relevant kernels (e.g., motif patterns), in this work we did not narrow it down to the specific features (i.e., the motifs). A promising goal for future work is to determine which particular motifs are most important; this can be modeled as an MKL task (cf. [15]). The idea of this approach is to represent the kernel by a sum of subkernels and to learn a weight (importance) for each of them. However, it seems that the imposed sparsity of the solution, while useful on the level of one kernel for each motif pattern, can be harmful at finer resolutions [15], and alternative approaches are more successful [24].

6 Summary and Outlook

We propose a general family of histogram-based motif kernels for amino acid sequences. We further propose to optimize over sets of kernels using a modern multiclass multiple kernel learning method, MCMKL [9]. We demonstrate that this approach outperforms the current state of the art in protein subcellular localization on four datasets. This high accuracy is already achieved with only using information from the amino acid sequence, while our method offers a principled way of integrating other data types. A promising example would be information mined from texts like paper abstracts [25]. Further, by selecting and weighting kernels, MCMKL yields interpretable results and may aid in getting insight into biological mechanisms.

Finally, the MCMKL framework [9] is very general and could be beneficial for a variety of (multiclass) bioinformatics prediction problems. For example, in this work we have only considered the case of singly located proteins, but in general proteins may exist in several possible locations in the cell. MCMKL does allow learning with multiple classes for each data point (each label y would be the corresponding subset of classes), and it will be interesting to see its performance on multiply located proteins. Application to more different prediction tasks is facilitated by the large and increasing set of existing sequence and structure kernels. MCMKL also allows to guide the learning process with different types of prior knowledge, including the relationships of classes to each other (by a kernel on the classes, c.f. [9]). These exciting opportunities remain to be explored.

Acknowledgement

We thank Alex Smola for stimulating discussions, Gunnar Rätsch for practical help with optimization with CPLEX, and Lydia Knüfing for proofreading the manuscript. This work was supported in part by the IST Programme of the European Community, under the PASCAL Network of Excellence, IST-2002-506778. This publication only reflects the authors' views.

References

1. Schölkopf, B., Smola, A.J.: Learning with Kernels. MIT Press, Cambridge (2002)
2. Park, K.J., Kanehisa, M.: Prediction of protein subcellular locations by support vector machines using compositions of amino acids and amino acid pairs. Bioinformatics 19(13), 1656–1663 (2003)
3. Guda, C., Subramaniam, S.: TARGET: a new method for predicting protein subcellular localization in eukaryotes. Bioinformatics 21(21), 3963–3969 (2005)
4. Yu, C.-S., Lin, C.-J., Hwang, J.-K.: Predicting subcellular localization of proteins for gram-negative bacteria by support vector machines based on n-peptide compositions. Protein Science 13, 1402–1406 (2004)
5. Gardy, J.L., Laird, M.R., Chen, F., Rey, S., Walsh, C.J., Ester, M., Brinkman, F.S.L.: PSORTb v.2.0: expanded prediction of bacterial protein subcellular localization and insights gained from comparative proteome analysis. Bioinfomatics 21, 617–623 (2004)

6. Höglund, A., Dönnes, P., Blum, T., Adolph, H.-W., Kohlbacher, O.: MultiLoc: prediction of protein subcellular localization using N-terminal targeting sequences, sequence motifs, and amino acid composition. Bioinfomatics (2006)

7. Xie, D., Li, A., Wang, M., Fan, Z., Feng, H.: LOCSVMPSI: a web server for subcellular localization of eukaryotic proteins using SVM and profile of PSI-BLAST. Nucleic Acids Research 33, W105–W110 (2005)

8. Garg, A., Bhasin, M., Raghava, G.P.S.: Support vector machine-based method for subcellular localization of human proteins using amino acid composition, their order, and similarity search. The Journal of Biological Chemistry 280(15), 14427–14432 (2005)

9. Zien, A., Ong, C.S.: Multiclass multiple kernel learning. In: International Conference on Machine Learning (2007)

10. Henikoff, S., Henikoff, J.G.: Amino acid substitution matrices from protein blocks. In: Proceedings of the National Academy of Sciences, pp. 10915–10919 (1992)

11. Reinhardt, A., Hubbard, T.: Using neural networks for prediction of the subcellular location of proteins. Nucleic Acids Research 26, 2230–2236 (1998)

12. Cui, Q., Jiang, T., Liu, B., Ma, S.: Esub8: A novel tool to predict protein subcellular localizations in eukaryotic organisms. BMC Bioinformatics 5(66) (2004)

13. Hein, M., Bousquet, O.: Hilbertian metrics and positive definite kernels on probability measures. In: Cowell, R., Ghahramani, Z. (eds.) Proceedings of AISTATS 2005, pp. 136–143 (2005)

14. Lanckriet, G., De Bie, T., Cristianini, N., Jordan, M.I., Stafford Noble, W.: A statistical framework for genomic data fusion. Bioinfomatics 20(16), 2626–2635 (2004)

15. Sonnenburg, S., Rätsch, G., Schäfer, C.: A general and efficient multiple kernel learning algorithm. In: Neural Information Processings Systems (2005)

16. Hettich, R., Kortanek, K.O.: Semi-Infinite Programming: Theory, Methods, and Applications. SIAM Review 35(3), 380–429 (1993)

17. Lee, Y., Kim, Y., Lee, S., Koo, J.-Y.: Structured multicategory support vector machines with analysis of variance decomposition. Biometrika 93(3), 555–571 (2006)

18. Roth, V., Fischer, B.: Improved functional prediction of proteins by learning kernel combinations in multilabel settings. BMC Bioinformatics 8 (suppl. 2), 12 (2007)

19. Nair, R., Rost, B.: Sequence conserved for subcellular localization. Protein Science 11, 2836–2847 (2002)

20. Yu, C.-S., Chen, Y.-C., Lu, C.-H., Hwang, J.-K.: Prediction of protein subcellular localization. Proteins: Structure, Function and Bioinformatics 64(3), 643–651 (2006)

21. Pellegrini, M., Marcotte, E.M., Thompson, M.J., Eisenberg, D., Yeates, T.O.: Assigning protein functions by comparative genome analysis: Protein phylogenetic profiles. Proceedings of the National Academy of Sciences 96(8), 4285–4288 (1999)

22. Emanuelsson, O., Nielsen, H., Brunak, S., von Heijne, G.: Predicting subcellular localization of proteins based on their N-terminal amino acid sequence. Journal of Molecular Biology 300, 1005–1016 (2000)

23. Marcotte, E.M., Xenarios, I., van der Bliek, A.M., Eisenberg, D.: Localizing proteins in the cell from their phylogenetic profiles. Proceedings of the National Academy of Sciences 97(22), 12115–12120 (2000)

24. Zien, A., Sonnenburg, S., Philips, P., Rätsch, G.: POIMS: Positional Oligomer Importance Matrices – Understanding Support Vector Machine Based Signal Detectors. In: Proceedings of the 16th International Conference on Intelligent Systems for Molecular Biology (2008)

25. Höglund, A., Blum, T., Brady, S., Dönnes, P., San Miguel, J., Rocheford, M., Kohlbacher, O., Shatkay, H.: Significantly improved prediction of subcellular localization by integrating text and protein sequence data. In: Pacific Symposium on Biocomputing, pp. 16–27 (2006)

Fast Target Set Reduction for Large-Scale Protein Function Prediction: A Multi-class Multi-label Machine Learning Approach

Thomas Lingner and Peter Meinicke

Department of Bioinformatics, Institute for Microbiology and Genetics,
University of Göttingen, Goldschmidtstr. 1, 37077 Göttingen, Germany
`thomas@gobics.de, pmeinic@gwdg.de`

Abstract. Large-scale sequencing projects have led to a vast amount
of protein sequences, which have to be assigned to functional categories.
Currently, profile hidden markov models and kernel-based machine learning methods provide the most accurate results for protein classification.
However, the prediction of new sequences with these approaches is computationally expensive. We present an approach for fast scoring of protein sequences by means of feature-based protein sequence representation
and multi-class multi-label machine learning techniques. Using the Pfam
database, we show that our method provides high computational efficiency and that the approach is well-suitable for pre-filtering of large
sequence sets.

Keywords: protein classification, large-scale, multi-class, multi-label,
Pfam, homology search, metagenomics, target set reduction, protein
function prediction, machine learning.

1 Introduction

In the last decade, the number of known protein sequences has rapidly increased, mainly due to several genome sequencing projects. Currently, large-scale
shotgun sequencing as used in metagenomics produces large amounts of protein
sequences with unknown phylogenetic origin [1]. For further analysis of these protein sequences, they have to be classified in terms of structural and functional
properties. Because experimental determination is time-consuming and expensive, several computational methods have been proposed for protein function
prediction and protein classification [2,3].

Widely-used methods for functional assignment are based on annotation
transfer from homologues by means of pairwise sequence alignments. Here, heuristic approaches such as BLAST [4] are used to search well-annotated databases
for similar protein sequences. However, these methods require the evaluation of
all pairwise alignments, which is computationally demanding for large sets of sequences. Furthermore, pairwise alignment methods often fail when sequence similarity is below 60% residue identity. Finally, this kind of annotation transfer may
be erroneous and may further lead to propagation of erroneous annotation [2,3].

K.A. Crandall and J. Lagergren (Eds.): WABI 2008, LNBI 5251, pp. 198–209, 2008.
© Springer-Verlag Berlin Heidelberg 2008

Another widely-used approach for protein classification is based on models that statistically represent properties of a set of related sequences, e.g. protein families. For example, the Pfam database [5] provides a comprehensive collection of manually curated multiple alignments of protein domain families. Each family is represented by a Profile Hidden Markov Model (PHMM, [6]) that has been constructed from the multiple alignment. The PHMMs can be used to classify new sequences into the Pfam categorization scheme, e.g. with the HMMER package (`http://hmmer.janelia.org/`). The threshold for assignment of a sequence to a particular model in Pfam is chosen to detect only verified family members. Because of the resulting highly specific models and the amount of manually verified annotation, Pfam and HMMER provide a highly valuable combination for automatic functional assignment of protein sequences. However, PHMMs may fail in the detection of remote homologues, i.e. homologues with a sequence similarity below 30% residue identity [7]. A major drawback of the PHMM approach is that for classification, every candidate sequence has to be aligned to all models. This is computationally demanding and several methods have been proposed to tackle this problem, e.g. by means of parallelization or hardware-acceleration [8]. Another approach is to use a "pre-filter", i.e. a computationally more efficient algorithm that reduces the set of candidate sequences or models (e.g. `http://www.microbesonline.org/fasthmm/`). For example, the Pfam website (`http://pfam.janelia.org/help`) provides a perl script that uses the BLAST method for pre-filtering. According to the author of the script, a speed-up factor of 10 may be achieved, accompanied by a slightly reduced sensitivity.

As an alternative, computationally more efficient alignment-free methods could be used as pre-filters. Recently, several feature-based machine learning approaches have been proposed for protein classification (for an overview see [9,10,11]) and detection of remote homologues [12,13,14,15]. Many of these discriminative methods have been shown to provide state-of-the-art classification performance. Therefore, feature-based approaches in general could be used for pre-filtering, too. However, so far the evaluation of these methods has been limited to detection of members of a particular protein family or problems with few exemplary categories and a relatively small number of sequences.

In that context, alignment-based kernel methods have been shown to outperform other approaches for remote homology detection [16,17]. In comparison with fast feature-based methods, kernel-based approaches are computationally expensive [15] and thus unsuitable for large-scale classification problems. As a consequence, the application of kernel-based methods has been limited to evaluation on small-scale setups, e.g. on the widely-used setup described in [7] including 54 classes (superfamilies).

If the classification problem does involve many functional categories, discriminative methods require a multi-class evaluation setup. The most common way to handle a multi-class problem is to split it up into isolated two-class one-against-all problems in order to utilize binary classification methods [18,19]. However, for large-scale multi-class problems, discriminative training and hyperparameter search for several thousand classifiers is computationally demanding. To our knowledge, the application of discriminative approaches in computational biology has been limited to small-scale setups so far.

In protein classification, an example may belong to several functional categories, e.g. multi-domain proteins consist of several functional regions. This gives rise to so-called "multi-label" learning problems. Multi-class multi-label machine learning methods have first been applied within the research field of text categorization [20]. Recently, multi-label methods have also been applied in bioinformatics, for example for analysis of gene expression profiles [21,22], protein subcellular localization prediction [23] and also for protein classification [24]. Usually, multi-label classification can be splitted into a ranking of examples which provides a correctly ordered list of categories and a set size prediction method, which determines the number of relevant classes.

In this work, we present an approach for ranking of protein sequences that is based on feature-based protein sequence representation methods and multi-class multi-label machine learning techniques. Based on an evaluation setup involving a large part of the Pfam database, we show that our approach can be used as a computationally efficient pre-filter for protein function prediction methods.

2   Methods

2.1   Machine Learning Approach

In order to realize a multi-label ranking scheme within a highly imbalanced large-scale setup with $M = 4423$ classes, we implemented an extension of the so-called "regularized least squares classifiers" [25]. For computational efficiency, we utilize a linear one-against-all approach with simultaneous training of all M discriminants.

For ranking of a sequence, we compute the corresponding feature vector $\mathbf{x} \in \mathbb{R}^d$ and the linear discriminant of class i based on the weight vector $\mathbf{w}_i$ to obtain a score from the corresponding scalar product:

$$s_i(\mathbf{x}) = \mathbf{w}_i^T \mathbf{x} \tag{1}$$

where the upper T indicates vector (matrix) transposition. The sorting of all M scores s_i in descending order then produces the ranks of the classes, i.e. the first element in the sorted list with the highest score indicates the highest rank (M).

For the training of M discriminants we use N feature vectors $\mathbf{x}_j$ and the corresponding binary indicator vectors $\mathbf{y}_j$ for representation of the labels. The i-th dimension of $\mathbf{y}_j$ only has a non-zero value $(= 1)$ if sequence j is associated with class i. The discriminant weight vectors $\mathbf{w}_i$ are represented as columns in the $d \times M$ weight matrix $\mathbf{W}$ and the inverse class sizes are collected in the balancing vector $\mathbf{b} = [1/n_1, \ldots, 1/n_M]^T$ with n_i counting the number of sequences associated with class i. For optimization of the weight matrix we minimize the regularized squared error criterion:

$$E(\mathbf{W}) = \sum_{j=1}^{N} \mathbf{b}^T \mathbf{y}_j \|\mathbf{W}^T \mathbf{x}_j - \mathbf{y}_j\|^2 + \lambda \|\mathbf{W}\|_F^2 \qquad (2)$$

where $\|\cdot\|_F$ denotes the Frobenius norm. Minimization of the above cost function tries to approximate the indicator vectors by means of the discriminant scores. Thus, the training forces a discriminant to produce relatively high scores only if examples are associated with the corresponding class. In addition, the λ-weighted regularization term bounds the squared norm of the weight vectors. A suitable value of λ then provides a compromise between a low approximation error and good generalization. The balancing vector $\mathbf{b}$ down-weights the contribution of examples, which are exclusively associated with overrepresented classes.

Now, consider the N feature vectors $\mathbf{x}_j$ as column vectors in a $d \times N$ training matrix $\mathbf{X}$ and the binary indicator vectors $\mathbf{y}_j$ as columns in an $M \times N$ label matrix $\mathbf{Y}$. Using the $N \times N$ diagonal matrix $\mathbf{D} = \mathrm{diag}(\mathbf{Y}^T \mathbf{b})$ of example-specific balancing weights, the minimizer of the regularized error can be written as:

$$\hat{\mathbf{W}} = (\mathbf{X}\mathbf{D}\mathbf{X}^T + \lambda \mathbf{I})^{-1} \mathbf{X}\mathbf{D}\mathbf{Y}^T \qquad (3)$$

where $\mathbf{I}$ is the $d \times d$ identity matrix. The above solution requires inversion of a $d \times d$ matrix, which practically limits the approach to feature spaces with a moderate dimensionality. However, on current computers, spaces with approximately 10^4 dimensions can be used without problems. For ranking of a set of test examples represented by columns of a matrix $\mathbf{X}_{\text{test}}$, the $M \times N$ score matrix $\mathbf{S}$ of all examples can be efficiently computed using the matrix product

$$\mathbf{S} = \hat{\mathbf{W}}^T \mathbf{X}_{\text{test}} . \qquad (4)$$

The j-th column of $\mathbf{S}$ contains all M discriminant scores of the j-th example.

2.2   Protein Sequence Representation

In order to apply our machine learning approach to the protein ranking problem, we have to represent all amino acid sequences in a common vector space. Here, we describe two exemplary protein sequence representation methods that have been used for protein classification and remote homology detection and which can be associated with an interpretable feature space of moderate dimensionality.

***k*-mer Spectrum.** The k-mer Spectrum of a sequence S counts the occurrences of all k-length words in S and can be represented as a vector in the k-mer Spectrum feature space. According to the number of different k-mers, for protein sequences this feature space comprises 20^k dimensions. The dimension associated with a particular k-mer K counts the occurrences of K in a sequence.

For large values of k, the k-mer Spectrum feature space becomes very large. For example, the trimer ($k = 3$) Spectrum feature space comprises a moderate number of 8000 dimensions, whereas the feature space for $k = 5$ consists of $3.2 * 10^6$ dimensions. On the other hand, the algorithmic complexity to calculate the feature space representative $\mathbf{x}$ of a sequence of length L is only $O(L)$. The k-mer Spectrum for small values of k is a time and memory efficient protein sequence representation method and thus provides a suitable vector space for large-scale machine learning problems. In order to provide comparability with respect to sequences of different length, the feature vectors are normalized to Euclidean unit length.

For $k = 1$ and $k = 2$, the k-mer Spectrum corresponds to the amino acid composition and dipeptide composition of a sequence, respectively. Both methods have been successfully applied to protein classification [10,11]. In comparative studies, the dipeptide composition has been shown to perform similarly to other feature-based sequence representation methods [9]. The k-mer Spectrum has also been used for remote homology detection in [12], where it has been shown to outperform unsupervised approaches.

Oligomer Distance Histograms. In [15], Oligomer Distance Histograms (ODH) have been introduced for protein sequence representation and remote homology detection, where they have been shown to outperform the k-mer Spectrum. In the distance-based feature space, for each pair of k-mers (oligomers) there exists a specific histogram counting the occurrences of that pair at certain distances. For the set of $M = 20^k$ different k-mers, the feature vector of a sequence S comprises M^2 distance histogram vectors. Considering a maximum sequence length L_{max}, each histogram vector contains the counts of two k-mers at distances $d = 0, \ldots, L_{max} - k + 1$. For $k > 1$ and large L_{max} the ODH representation gives rise to a high dimensional feature space. For example, the feature space for $k = 3$ (trimers) and a maximum sequence length of $L_{max} = 1000$ comprises about $6.4 * 10^{10}$ dimensions. In this case, our feature-based machine learning approach cannot be applied.

On the other hand, in [15] the best detection performance has been achieved for $k = 1$ (monomers). Furthermore, sequences with less than L_{max} residues do not contribute to feature space dimensions associated with this distance. Therefore, a restriction of the maximum distance provides a suitable means to reduce the feature space dimensionality of ODHs. As an example, the ODH feature space for monomer pairs with a maximum distance of $D_{max} = 10$ residues comprises only 4020 dimensions, which is about half the number of dimensions required for the feature space associated with the trimer Spectrum. In that way, the ODH representation allows a fine-grained control of the feature space size, which linearly depends on the maximum distance. Note that the distance-based feature space for monomers incorporates the amino acid composition ($D = 0$), the dipeptide composition ($D = 1$) and trimer counts according to a central mismatch ($D = 2$).

The systematic evaluation of all pairwise k-mers in a sequence of length L to calculate the feature space representative $\mathbf{x}$ is of algorithmic complexity $O(L^2)$. If the maximum distance is restricted to D_{max}, the algorithmic complexity reduces to $O(D_{max}L)$. As the Spectrum feature vectors, ODH feature vectors are normalized to unit length.

2.3 Performance Measures

For evaluation of our approach, we use different performance indices. First, we measure the so-called *coverage* to analyze the ranking performance [22]. We define the *minimal set size* of a test example as the cardinality of the minimal set that includes all ranks above or equal to the minimal rank of a true category. Then, the coverage is defined as the average minimal set size over all examples reduced by 1. This measure is well-suited to evaluate our approach in terms of a pre-filtering for target set reduction. In this case, the coverage reflects the number of functional categories that have to be considered in a subsequent function prediction stage. Therefore, a smaller coverage allows a higher possible speed-up of the prediction. The coverage measure is widely used in multi-label learning problems. Besides the "classical" mean coverage, we also consider the median coverage, i.e. the median of the above minimal set size over all test examples. In addition, we use the so-called *one-error* to measure the ranking performance. The one-error evaluates how often on average the top-ranked category is not a true class of a particular example. Therefore, a one-error value of 0 is desirable, which means that all examples would have a correct functional category assignment for the highest rank.

In addition, we measure the average area under ROC and ROC50 curve over all families to measure the detection performance w.r.t. potential family members. The ROC curve reflects the dependency of the the the false positive rate $(1-\text{specificity})$ on the true positive rate (sensitivity) w.r.t. to variation of the classification threshold. The ROC50 score is the area under curve up to 50 false positives. ROC scores are particularly useful for analysis of rank-ordered lists and imbalanced problems [26] and are widely used in the evaluation of remote homology detection performance [7,15,16].

3 Experimental Setup

In order to evaluate our approach, we developed a test setup based on the Pfam database [5]. Pfam is a widely-used, comprehensive and well-annotated collection of protein domain families. Pfam 22.0 (released in July 2007) consists of 9318 families in the Pfam-A section, i.e. the manually curated part of the database. Each family comprises a seed alignment, which contains selected representative sequences of the family. For our setup, we use these seed sequences, which in total add up to 217445 examples in Pfam 22.0.

The seed alignments of Pfam are based on protein domains, i.e. functional regions of protein sequences. When functional categories have to be assigned to unannotated sequences, the boundaries of these domains are unknown. Therefore, we consider the complete protein sequences associated with the domains. Furthermore, multi-domain proteins include several domains, which may realize different functions. Therefore, multi-domain proteins may be associated with several Pfam families. In our setup, the label indicator vector of a protein sequence refers to all its relevant classes, i.e. Pfam families that have a significant match to this sequence.

For training of the discriminant weight vectors we use the regularized least squares approach described in section "Methods". For this purpose, we calculated the feature vectors for the k-mer Spectrum method using $k = 1, 2, 3$ and for the ODH method using $D_{max} = 10, 20, 30$. Higher values for k and D_{max} result in feature spaces that complicate the training due to high dimensionality. We evaluated different values of the regularization parameter $\lambda = \{10^m | m = -4, -3, \ldots, 4\}$ by means of 5-fold cross-validation. In order to provide a sufficiently high number of training and test examples for each class, we only consider Pfam families with at least 10 different seed sequences. In total, our setup consists of 147003 protein sequences from 4423 Pfam families. The size of the families varies from a minimum of 10 to a maximum of 1670 example sequences and thus gives rise to a highly imbalanced classification problem.

All methods were implemented using the Matlab® programming language. Files containing sequences and labels associated with our evaluation setup can be downloaded from `http://www.gobics.de/thomas/data/pfam/`

4   Results and Discussion

The results of our performance evaluation are summarized in table 1. As a main result, the ODH method clearly outperforms the k-mer Spectrum in terms of all performance indices. In particular, the coverage (columns 3 and 4) and the one-error values (column 5) are substantially better for the ODH method. In terms of ROC/ROC50 values (columns 6 and 7), all methods except for the monomer Spectrum show a good performance. On the other hand, the coverage and one-error values are more suitable measures for the utility of a particular method for target set reduction on multi-label problems (see also section "Performance Measures"). Although not tested here, we expect the combination of Pfam PHMMs and HMMER to perfectly classify the sequences. This would implicate a value of 1 for ROC/ROC50 values and 0 for the one-error. The mean coverage is expected to correspond to the average number of true categories over all examples reduced by 1, which is approximately 0.6 in our test setup.

The results in table 1 also show that for the ODH method the highest maximum distance $D_{max} = 30$ produces better results than smaller maximum distances. Similarly, higher values of k for the Spectrum method provide better detection performance and less coverage. Note that the ODH method for $D_{max} = 10$ performs better than the Spectrum method for $k = 3$, although the feature space associated with the ODH method comprises only about half the number of dimensions as compared to that of the Spectrum method. This indicates that the choice of a suitable feature mapping is more important than just a high dimensionality of the feature space. In [15], the ODH method was introduced without restriction of the maximum distance and clearly outperformed the Spectrum method on a small-scale benchmark setup for protein remote homology detection. Our results in this work indicate that the introduction of a maximum distance for the ODH method is a suitable means for application of this method to large-scale problems. In column 8, the best regularization

Table 1. Performance results of different methods. The first column indicates the method, where "Spectrum" refers to the k-mer Spectrum and "ODH" refers to Oligomer Distance Histograms using $k = 1$ with maximal distance D_{max} (see also section "Methods"). Column 2 denotes the dimensionality of the feature space associated with a particular method. Columns 3-7 show the different performance indices as described in section "Methods" in terms of average values over 5 cross-validation test folds. "ROC" and "ROC50" refer to the average area under ROC/ROC50 curve over all families with at least 10 positive test examples in each fold. In column 8, the best regularization parameter λ associated with a particular method is shown.

Method	d	Coverage		One-error	ROC	ROC50	best λ
		mean	median				
Spectrum ($k = 1$)	20	452.4	243.8	0.95	0.925	0.046	0.001
Spectrum ($k = 2$)	420	214.9	65.0	0.85	0.978	0.563	0.001
Spectrum ($k = 3$)	8000	116.7	4.8	0.57	0.987	0.827	1
ODH ($D_{max} = 10$)	4020	65.3	3.8	0.55	0.993	0.840	0.001
ODH ($D_{max} = 20$)	8020	47.6	2.0	0.43	0.995	0.881	0.001
ODH ($D_{max} = 30$)	12020	41.6	1.2	0.37	0.995	0.894	0.01

parameter λ w.r.t the coverage over all 5 cross-validation test folds is shown. The values indicate that the regularization parameter has to be chosen larger for feature spaces with higher dimensionality. On the other hand, our results only showed a small variation of the performance for a broad range of λ-values.

Column 3 of table 1 shows that on average only about 42 of 4423 families are required to detect all functional categories of a particular test example with the ODH method for maximum distance $D_{max} = 30$. If the approach is used as a pre-filter for protein function prediction, this corresponds to a speed-up factor of 106. This is one order of magnitude higher than the speed-up that is usually achieved with alignment-based methods. Furthermore, the one-error for the ODH method with $D_{max} = 30$ indicates that for 63% of the test examples the class associated with the highest rank implies the correct assignment to a functional category.

However, the total speed-up also depends on the computational efficiency of the target set reduction method. Therefore, we measured the running time for classification with different methods on an AMD Opteron 870 workstation. Table 2 shows that the ranking with the monomer Spectrum ($k = 1$) took 585 seconds for all categories and 147003 examples. As another example, the ranking with the ODH method using $D_{max} = 30$ took 3380 seconds. On average, 23 ms are required for ranking of a single sequence with this more sensitive method as compared with the monomer Spectrum method. This is about 1270 times faster than functional assignment of a single sequence with the HMMER package using 4423 models on the same hardware (on average 33 seconds per sequence). In general, the feature extraction process and the calculation of the matrix product can easily be parallelized, which further reduces the required running time. Therefore, our approach provides a suitable means for target set reduction on huge sequence collections, which are routinely analyzed in metagenomics [1,27].

Table 2. Runtime comparison of different methods for sequence scoring. The first column denotes the method, whereby "Spectrum" refers to the k-mer Spectrum and "ODH" refers to Oligomer Distance Histograms using $k = 1$. Column 2 denotes the dimensionality of the feature space associated with a particular method. Columns 3 and 4 show the runtime in seconds for feature extraction and matrix multiplication during the scoring process for all 147003 sequences (see also section "Machine Learning Approach"). Columns 5 and 6 denote the total runtime in seconds for scoring and the average runtime in milliseconds for scoring of a single sequence, respectively.

Method	d	$t_{featExt}$ (s)	$t_{matMult}$ (s)	t_{total} (s)	t_{avg} (ms)
Spectrum ($k = 1$)	20	575	10	585	4.0
Spectrum ($k = 2$)	400	622	80	702	4.8
Spectrum ($k = 3$)	8000	643	1500	2143	14.6
ODH ($D_{max} = 10$)	4020	1027	800	1827	12.4
ODH ($D_{max} = 20$)	8020	1082	1525	2607	17.7
ODH ($D_{max} = 30$)	12020	1130	2250	3380	23.0

Table 1 shows that mean and median coverage of a particular method differ significantly. This results from the shape of the minimal set size distribution for the test examples. Figure 1 shows the dependency of the corresponding fraction of sequences on a given minimal set size in terms of curve plots for different methods. It is clearly visible that for the ODH method a large number of test examples show a low minimal set size, while only few examples show a very high minimal set size. If one is willing to accept a significant loss of sensitivity, e.g. for a coarse estimation of functional profiles in metagenomics, the target set reduction allows a considerable speed-up.

We also analyzed whether the performance of our approach critically depends on the family size or on the number of associated classes of an example. Therefore, we calculated the correlation coefficient of the minimal set size and the number of true categories of a particular example using the results from the ODH method with maximum distance $D_{max} = 30$. The low correlation coefficient of -0.0031 indicates that examples with many functional categories do not lead to higher minimal set sizes than examples with few categories. Since only multi-domain proteins have more than one assigned category, this result indicates that our approach is suitable for ranking of single-domain and multi-domain proteins, as well. Furthermore, we computed the correlation coefficients between ROC/ROC50 values and the family size. The low correlation values (ROC: 0.1288, ROC50: 0.046) indicate that the performance does not critically depend on family size. Note that we measured the ROC/ROC50 performance only for families with at least 10 positive test examples in each test fold.

In this work, we limited the evaluation setup to Pfam families with at least 10 different seed sequences. In practice, our approach is also suitable to be used with the complete Pfam database for learning of 9318 family-specific discriminants. As a draft study, we evaluated Pfam families with 2 to 9 seed sequences using 2-fold cross-validation. The corresponding data set consists of 4750 families with 22944 sequences in total. Here, the ODH method using $D_{max} = 30$ achieves a

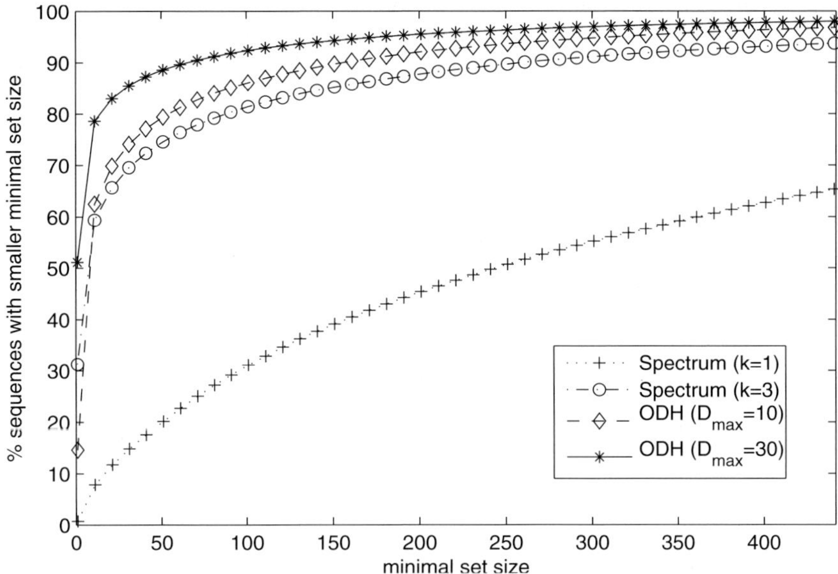

Fig. 1. Coverage curves for different methods. The coverage curve of a particular method plots the relative number of examples that have a minimal set size below a given threshold. Only minimal set size values up to 442 (10% of the classes) are shown.

mean coverage of 32.4 and a one-error of 0.26. The performance of the Spectrum method for $k = 3$ is inferior, resulting in a mean coverage of 94.6 and a one-error of 0.54. A direct comparison to the results in table 1 is not possible, because the average number of true categories over all example sequences is much lower for the small families (1.07 instead of 1.6). Furthermore, the results of the two cross-validation folds show large differences. One possible reason is that for the smallest families only one example sequence is used for training and testing. In practice, sequences from the Pfam full alignments [5] could be used to extend the training set of small families.

In contrast to PHMMs, feature-based approaches allow the interpretation of discriminative sequence features that have been learned from the data. Therefore, these methods are not only useful for target set reduction, but they could also be used for analysis of biologically meaningful properties of the sequences. In [15] we showed how discriminative features of the ODH method could be interpreted in a remote homology detection context.

5   Conclusion

In this work, we presented an approach for large-scale ranking of protein sequences for function prediction. Our method is based on explicit representation of sequences in feature spaces that allow fast scoring and efficient training of discriminants. We developed a setup for evaluation of multi-label machine learning techniques, which is based on a large part of the Pfam database. We showed that our approach can be used for fast target set reduction in terms of a ranking of functional categories. The reduced target set can be analyzed with highly accurate but computationally more expensive methods. As a main result, we showed that the ranking performance critically depends on the choice of a suitable feature space for representation of protein sequences.

Although our approach worked well for most of the Pfam families and sequences, for some categories the performance was very low. In this work, we showed that the detection performance does not critically depend on family size. First results also indicate that functional categories can be predicted for multi-domain proteins and single-domain proteins with similar accuracy. A detailled analysis of the reasons for the low performance in a few cases will be part of future work. In order to use our approach as a stand-alone protein classification method, our ranking has to be extended by a so-called "set size predictor" [21], which estimates the number of relevant functional categories from the discriminant scores. For this purpose, we are currently investigating several set size prediction methods.

Acknowledgments. This work was partially supported by the Federal Ministry of Research and Education project "MediGRID" (BMBF 01AK803G).

References

1. Yooseph, S., et al.: The Sorcerer II Global Ocean Sampling expedition: expanding the universe of protein families. PLoS Biol. 5, 16 (2007)
2. Friedberg, I.: Automated protein function prediction–the genomic challenge. Brief. Bioinformatics 7, 225–242 (2006)
3. Pandey, G., Kumar, V., Steinbach, M.: Computational approaches for protein function prediction. Technical Report TR 06-028, Department of Computer Science and Engineering, University of Minnesota, Twin Cities (2006)
4. Altschul, S.F., Gish, W., Miller, W., Myers, E.W., Lipman, D.J.: Basic local alignment search tool. J. Mol. Biol. 215(3), 403–410 (1990)
5. Finn, R., et al.: Pfam: clans, web tools and services. Nucleic Acids Res. 34, D247–251 (2006)
6. Eddy, S.R.: Profile hidden Markov models. Bioinformatics 14(9), 755–763 (1998)
7. Liao, L., Noble, W.S.: Combining pairwise sequence similarity and support vector machines for detecting remote protein evolutionary and structural relationships. J. Comput. Biol. 10(6), 857–868 (2003)
8. Walters, J.P., Meng, X., Chaudhary, V., Oliver, T.F., Yeow, L.Y., Schmidt, B., Nathan, D., Landman, J.I.: MPI-HMMER-Boost: Distributed FPGA Acceleration. VLSI Signal Processing 48(3), 223–238 (2007)
9. Ong, S., Lin, H., Chen, Y., Li, Z., Cao, Z.: Efficacy of different protein descriptors in predicting protein functional families. BMC Bioinformatics 8, 300 (2007)
10. Strope, P., Moriyama, E.: Simple alignment-free methods for protein classification: a case study from G-protein-coupled receptors. Genomics 89, 602–612 (2007)

11. Han, L., Cui, J., Lin, H., Ji, Z., Cao, Z., Li, Y., Chen, Y.: Recent progresses in the application of machine learning approach for predicting protein functional class independent of sequence similarity. Proteomics 6, 4023–4037 (2006)
12. Leslie, C., Eskin, E., Noble, W.S.: The spectrum kernel: a string kernel for SVM protein classification. In: Pac. Symp. Biocomput., pp. 564–575 (2002)
13. Ben-Hur, A., Brutlag, D.: Remote homology detection: a motif based approach. Bioinformatics 19 (suppl. 1), 26–33 (2003)
14. Leslie, C.S., Eskin, E., Cohen, A., Weston, J., Noble, W.S.: Mismatch string kernels for discriminative protein classification. Bioinformatics 20(4), 467–476 (2004)
15. Lingner, T., Meinicke, P.: Remote homology detection based on oligomer distances. Bioinformatics 22(18), 2224–2231 (2006)
16. Saigo, H., Vert, J.P., Ueda, N., Akutsu, T.: Protein homology detection using string alignment kernels. Bioinformatics 20(11), 1682–1689 (2004)
17. Rangwala, H., Karypis, G.: Profile-based direct kernels for remote homology detection and fold recognition. Bioinformatics 21(23), 4239–4247 (2005)
18. Rifkin, R., Klautau, A.: In Defense of One-Vs-All Classification. Journal of Machine Learning Research 5, 101–141 (2004)
19. Jensen, L.J., Gupta, R., Staerfeldt, H., Brunak, S.: Prediction of human protein function according to Gene Ontology categories. Bioinformatics 19, 635–642 (2003)
20. Schapire, R., Singer, Y.: Boostexter: A system for multiclass multi-label text categorization (1998)
21. Elisseeff, A., Weston, J.: A kernel method for multi-labelled classification. In: Dietterich, T.G., Becker, S., Ghahramani, Z. (eds.) NIPS, pp. 681–687. MIT Press, Cambridge (2001)
22. Zhang, M.L., Zhou, Z.H.: A k-nearest neighbor based algorithm for multi-label classification. The IEEE Computational Intelligence Society 2, 718–721 (2005)
23. Lee, K., Kim, D., Na, D., Lee, K., Lee, D.: PLPD: reliable protein localization prediction from imbalanced and overlapped datasets. Nucleic Acids Res. 34, 4655–4666 (2006)
24. Diplaris, S., Tsoumakas, G., Mitkas, P.A., Vlahavas, I.P.: Protein classification with multiple algorithms. In: Bozanis, P., Houstis, E.N. (eds.) PCI 2005. LNCS, vol. 3746, pp. 448–456. Springer, Heidelberg (2005)
25. Rifkin, R., Yeo, G., Poggio, T.: Regularized Least Squares Classification. In: Advances in Learning Theory: Methods, Model and Applications NATO Science Series III: Computer and Systems Sciences, vol. 190, pp. 131–153. IOS Press, Amsterdam (2003)
26. Cohen, G., Hilario, M., Sax, H., Hugonnet, S., Geissbuhler, A.: Learning from imbalanced data in surveillance of nosocomial infection. Artif. Intell. Med., 7–18 (2006)
27. Hoff, K., Tech, M., Lingner, T., Daniel, R., Morgenstern, B., Meinicke, P.: Gene prediction in metagenomic fragments: a large scale machine learning approach. BMC Bioinformatics 9, 217 (2008)

Multiple Instance Learning Allows MHC Class II Epitope Predictions Across Alleles

Nico Pfeifer and Oliver Kohlbacher

Division for Simulation of Biological Systems, Center for Bioinformatics Tübingen,
Eberhard Karls University Tübingen, 72076 Tübingen, Germany

Abstract. Human adaptive immune response relies on the recognition of short peptides through proteins of the major histocompatibility complex (MHC). MHC class II molecules are responsible for the recognition of antigens external to a cell. Understanding their specificity is an important step in the design of peptide-based vaccines. The high degree of polymorphism in MHC class II makes the prediction of peptides that bind (and then usually cause an immune response) a challenging task. Typically, these predictions rely on machine learning methods, thus a sufficient amount of data points is required. Due to the scarcity of data, currently there are reliable prediction models only for about 7% of all known alleles available.

We show how to transform the problem of MHC class II binding peptide prediction into a well-studied machine learning problem called multiple instance learning. For alleles with sufficient data, we show how to build a well-performing predictor using standard kernels for multiple instance learning. Furthermore, we introduce a new method for training a classifier of an allele without the necessity for binding allele data of the target allele. Instead, we use binding peptide data from other alleles and similarities between the structures of the MHC class II alleles to guide the learning process. This allows for the first time constructing predictors for about two thirds of all known MHC class II alleles. The average performance of these predictors on 14 test alleles is 0.71, measured as area under the ROC curve.

Availability: The methods are integrated into the EpiToolKit framework for which there exists a webserver at `http://www.epitoolkit.org/mhciimulti`

1   Introduction

The adaptive immune system is one of the most advanced and most important systems in humans. It can direct immune responses according to various kinds of invading microorgansims and even recognize and destroy tumor cells [1]. A very important part of this system are T-cells. These cells are involved in a cascade which results in a large production of very specific antibodies. Furthermore, activated T-killer cells are able to induce apoptosis in deficient cells. One of the first steps in activating T-cells is to present short peptides to them. There are two

K.A. Crandall and J. Lagergren (Eds.): WABI 2008, LNBI 5251, pp. 210–221, 2008.
© Springer-Verlag Berlin Heidelberg 2008

major complexes, which present peptides at the cell membrane. They are called major histocompatibility complex class I (MHCI) and major histocompatibility complex class II (MHCII). MHCI presents peptides which originate from proteins produced within cell. MHCII presents peptides originating from the outside of the cell. There are various different MHCII alleles which have very specific sets of peptides to which they can bind. At present there are more than 750 unique MHCII alleles known [2] (regarded on the protein sequence level), but for less than 3 % of them there exist sufficient experimental data from binding studies [3]. Since every human has at most eight different MHCII alleles, it is very important for vaccine design to know which peptides can bind to the particular alleles. A good predictor for MHC binding can reduce the number of possible peptides and therefore saves a lot of time – and money-consuming experiments.

In contrast to MHCI, the binding clefts of the MHCII are open. This is why the length of the binding peptides varies significantly (from 8 to more than 30 amino acids). Nevertheless, analyses of MHCII structures revealed that the part of the peptide responsible for binding to MHCII is usually nine amino acids long. This part is also called binding core of the peptide. For most of the experimental data it is unknown which part of the peptide actually is the binding core, which complicates the problem of MHCII peptide binding prediction compared to MHCI peptide binding prediction. The binding clefts of MHCI are closed and the binding peptides have a length between eight and ten. There are various methods for MHCII binding peptide prediction for alleles for which there exists sufficient experimental data. Some of these models are based on positional scoring matrices [4,5,6,7,8,9], others use Gibbs samplers [10] or hidden Markov models [11]. Further works have used the ant colony search strategy [12], artificial neural networks [13], partial least squares [14] or support vector machines with standard kernel functions [15,16,17]. Very recently Wang *et al.* [18] combined several of these predictors to build a new predictor. There have also been efforts to improve binding prediction by using structural information [19].

To the best of our knowledge all but two of the models for MHCII binding peptide prediction are based on experimental data for the particular alleles for which the predictions are for. The models of Singh *et al.* [8] and Zaitlen *et al.* [19] are the only methods which were shown to predict binding for alleles without training on them. However, the model by Singh *et al.* is only applicable for 51 alleles [18] which is about 7 % of all known alleles and Zaitlen *et al.* require three-dimensional structures of a similar allele to perform this kind of predictions which limits the number of alleles accessible through the method. Since there is very few experimental data, it is desirable to be able to predict peptide binding for alleles, for which few or no experimental data is available. This is what our method is able to do. Similar ideas have also recently been introduced for MHCI predictions, although based on different machine learning techniques and for a far simpler problem (MHCI peptides have more or less identical lengths) [20,21,22].

We use similarities of the binding pockets of the alleles to build classifiers for alleles, which do not need experimental data of the target allele to reach good prediction performance. The similarities are incorporated into the predictions using a specialized kernel function, which is based on the normalized set kernel by Gärtner *et al.* [23]. To do so, the problem is transformed into a multiple instance learning problem [24]. The classifier is trained using the kernel function and Support Vector Regression (SVR) [25]. Using this method we are for the first time able to build predictors for about two thirds of all MHCII alleles. Assessment of their quality in blind predictions for alleles with known data reveals that the predictions are of sufficient quality for use in vaccine design. Furthermore, we show that our transformation of the problem into the multiple instance learning problem enables us to build classifiers which are as good or even better as the best methods for MHCII binding peptide prediction.

2   Methods

2.1   Multiple Instance Learning

In standard supervised binary classification, the associated label for every instance out of the sets of training samples is known. Every instance can be represented as (x_i, y_i) where $x_i \in \mathcal{X}$ and $y_i \in \{-1, 1\}$. We define the set of positive training examples as $S_p = \{(x, y)|x \in \mathcal{X} \wedge y = 1\}$ and the set of negative training examples as $S_n = \{(x, y)|x \in \mathcal{X} \wedge y = -1\}$. In multiple instance learning [24] not every label y_i for every x_i is known. The positive label is only known for sets of instances which are called bags. For every bag X_i with label $+1$ it is only known that at least one instance of X_i is associated with label $+1$. Every instance in a negative bag is associated with label -1. More formally this means that the set of positive bags is $X_p = \{(X_i, 1)|\exists x_j \in X_i : (x_j, y_j) \in S_p\}$. The set of negative bags is $X_n = \{(X_i, -1)|\forall x_j \in X_i : (x_j, y_j) \in S_n\}$. The multiple instance learning problem is to find the best predictor for predicting the labels of bags.

Kernels for multiple instance learning were introduced by Gärtner *et al.* [23] in 2002. The normalized set kernel by Gärtner *et al.* [23] is the following:

$$k(X, X') := \frac{\sum\limits_{x \in X, x' \in X'} k_{\mathcal{X}}(x, x')}{f_{\text{norm}}(X) f_{\text{norm}}(X')} \tag{1}$$

with $k_{\mathcal{X}}$ being a kernel on $\mathcal{X}$. Gärtner *et al.* [23] evaluated different normalization functions f_{norm} and showed that averaging $(f_{\text{norm}}(X) = \#X)$ and feature-space normalization $(f_{\text{norm}}(X) = \sqrt{\sum\limits_{x \in X, x' \in X} k_{\mathcal{X}}(x, x')})$ perform equally well on the studied datasets. Preliminary results on our data also suggested that both methods perform equally well (data not shown). Therefore, in the following only normalization by feature space normalization was considered.

Gärtner *et al.* [23] hypothesized in their paper, that the kernel could also be used for multiple instance regression [26, 27]. In this setting every bag X_i has a label $y_i \in \mathbb{R}$.

2.2 Multiple Instance Learning for MHCII Prediction

Since for most of known MHCII binders the binding core is unknown, one cannot directly use the binding core for training a learning machine. Since previous work on MHCII prediction [28,9] suggests that only aliphatic (I, L, M, V) and aromatic (F, W, Y) amino acids in position one are common, we represent every putative binder by a bag containing all 9-mers (putative binding cores) with aromatic or aliphatic amino acid at position one. By this, we transformed the data directly into a multiple instance learning problem in which every positive bag has at least one positive binding core. All negative bags just contain false binding cores. Formally this means that every putative binder s_i of length m is represented by a bag (X_i, y_i).

$$X_i = \{x | x = s_{i_{k,k+1,\ldots,k+8}} \wedge k \geq 1 \wedge k + 8 \leq m \wedge x_1 \in \{F, I, L, M, V, W, Y\}\}$$

are all putative binding cores and y_i is the measured binding affinity of s_i. For less than 3% of the putative binders of the benchmark set X_i was empty (see also Section 3.1).

In this work we introduce two classifiers for MHCII binding peptide prediction. The first classifier is just trained on parts of the data of the allele for which the predictions should be made. This classifier is called C_{single} in the following. It will be shown that the performance of C_{single} is comparable to the best methods in the field. This classifier is particularly useful for alleles, for which sufficient binding data is available.

The second classifier is not trained on data of the allele for which the predictions should be made. Instead, data from other alleles is combined in a way which reflects the similarity of the binding pockets of the target allele to the binding pockets of the other alleles. This classifier will be called C_{multi} in the following. Because no data of the allele, for which the predictions should be made is needed, one can build classifiers for alleles with little or no experimentally determined binders.

In this work, we use the normalized set kernel with an RBF kernel for C_{single}. Furthermore, we introduce a new kernel based on the normalized set kernel for C_{multi}. Throughout the paper we use RBF kernel or k_{RBF} for the Gaussian RBF kernel with the kernel function $k(x, x') = \exp^{-\frac{\|x-x'\|^2}{2\sigma^2}}$.

2.3 Feature Encoding

Venkatarajan and Braun [29] evaluated in 2001 different physicochemical properties of amino acids. They performed a dimension reduction on a large set of features from the AAindex database [30] and showed that every amino acid can be represented adequately by a five-dimensional feature vector. This encoding was already used in a recent study on MHC binding by Hertz *et al.* [31] and will be called *PCA encoding* in the following.

2.4 MHCII Binding Peptide Prediction for Alleles with Sufficient Data

For alleles, for which enough experimental data is available, we build predictors which are just trained on binding peptide data for the particular allele. In this setting we use the *normalized set kernel* [23] with $k_\mathcal{X}$ being the RBF kernel. $\mathcal{X}$ is the set of all putative binding cores. This means that $\mathcal{X}$ is the set of every possible nine amino acid long peptide sequence in PCA encoding for which the first amino acid is aliphatic or aromatic. This means that every input vector has length 45. The classifier is trained using this kernel function together with ν-SVR [25].

2.5   Combining Allele Information with Peptide Information

Representation of MHCII Alleles. Sturniolo *et al.* [9] showed in 1999 that there is a correspondence between the structures of the binding pockets of the MHCII and the polymorphic residues in this region. They defined certain positions inside the amino acid sequence of the allele sequences and showed that alleles having the same residues at these positions also have similar binding pocket structures. This was done for several alleles and binding pockets for peptide positions 1, 4, 6, 7 and 9 because these positions are assumed to have the most influence on binding [9].

To represent each allele, we encoded every polymorphic residue of the pockets 1, 4, 6, 7, and 9 by PCA encoding and calculated a mean of the encoded vectors for every pocket position. This resulted in a 25×1 dimensional vector $p = \left(p_1^T, p_4^T, p_6^T, p_7^T, p_9^T\right)^T$ for every allele, which is called *pocket profile vector* in the following. To get the polymorphic residues for alleles that were not defined by Sturniolo *et al.* [9], we used the HLA-DRB1, HLA-DRB3, HLA-DRB4 and HLA-DRB5 alignments of the IMGT/HLA database [2] (release 2.18.0).

We computed the sequence logo [32] for an alignment of all HLA-DRB1, HLA-DRB3, HLA-DRB4 and HLA-DRB5 alleles. An extract of this can be seen in Fig. 1. The whole sequence logo can be found on our webpage (http:// www-bs.informatik.uni-tuebingen.de/Publications/Materials/WABI08MHCII). Since the alignments show very good conservation for alleles HLA-DRB1, HLA-DRB3, HLA-DRB4 and HLA-DRB5 at the non-pocket positions, we assume that this procedure is applicable at least for these HLA-DRB alleles which constitute 525 of all 765 unique MHCII alleles (on the protein sequence level), currently contained in the IMGT/HLA database [2].

Similarity Function of MHCII Binding Pockets. Our goal was to get a similarity measure between pocket positions of alleles between zero (totally different) and two (identical). Since we have the pocket profile vectors, a natural idea is to take the Pearson correlation between the corresponding positions of the pocket. To get a similarity measure we added one which means that the similarities are in the interval $[0, 2]$. The resulting similarity measure for each pocket $i = 1, 4, 6, 7, 9$ is then

$$\mathrm{sim}_i(p, p') := \mathrm{Pearson}(p_i, p_i') + 1. \tag{2}$$

This function was used in our work to measure similarity between the binding pockets corresponding to peptide position 1, 4, 6, 7 and 9.

Combining Allele Pocket Profiles with Peptide Information. For MHCI binding peptide prediction there have been two approaches for learning binding prediction models for alleles which do not need experimental data for the target allele [21, 22]. Both methods measure the similarity of the alleles only for the whole allele. This means that one allele would also have a large similarity if most of the binding pockets are similar and only one or two positions are not. This model is particularly dangerous if the position, which is very different is crucial for the binding process in the target allele. Therefore, we use similarities between the binding pockets directly in our kernel function to be able to account for these cases, too.

We now define the kernel $k_{\mathrm{pw-RBF}}$, which is defined on $\mathcal{A} \times \mathcal{X}$. $\mathcal{A}$ is the set of all possible pocket profile vectors and $\mathcal{X}$ is again the set of all possible nine amino acid long peptide sequences in PCA encoding. Let $p = \left(p_1^T, p_4^T, p_6^T, p_7^T, p_9^T\right)^T$ be the *pocket profile vector* of peptide sequence s. Let $x = (x_1^T, x_2^T, ..., x_9^T)^T$ be a putative binding core of sequence s, for which every x_i is the PCA encoding of the amino acid at position i in the putative binding core. Let p' and x' be defined analogously for peptide sequence s'. In C_{single} the inner kernel function of the normalized set kernel is a standard RBF kernel:

$$k_{\mathrm{RBF}}(x, x') = \exp^{-\frac{\|x-x'\|^2}{2\sigma^2}} . \tag{3}$$

As mentioned above, the kernel function should be able to weight positions according to the similarity of the alleles. Therefore, we designed a positionally-weighted RBF-kernel:

$$k_{\mathrm{pw-RBF}}((p, x), (p', x')) = \exp^{-\frac{w_1 \times \|x_1 - x_1'\|^2 + w_2 \times \|x_2 - x_2'\|^2 + ... + w_9 \times \|x_9 - x_9'\|^2}{2\sigma^2}} . \tag{4}$$

In our setting the weights are determined using the sim function, which was mentioned above:

$$w_i := \mathrm{sim}_i(p, p') \ \forall i = 1, 4, 6, 7, 9 \tag{5}$$

Since the other positions are not as important for binding, we set the weights w_2, w_3, w_5 and w_8 (which correspond to peptide positions 2, 3, 5 and 8) to 0.5.

In this work $k_{\mathrm{pw-RBF}}$ is used as the inner kernel function of the *normalized set kernel* [23] in conjunction with ν-SVR [25] for C_{multi}. Similar to [33] one can show that $k_{\mathrm{pw-RBF}}$ is positive definite using the Schoenberg Theorem [34] since

$$w_1 \times \|x_1 - x_1'\|^2 + w_2 \times \|x_2 - x_2'\|^2 + ... + w_9 \times \|x_9 - x_9'\|^2 =$$
$$w_1 \times \|x_1' - x_1\|^2 + w_2 \times \|x_2' - x_2\|^2 + ... + w_9 \times \|x_9' - x_9\|^2 \tag{6}$$

and

$$w_1 \times \|x_1 - x_1'\|^2 + w_2 \times \|x_2 - x_2'\|^2 + ... + w_9 \times \|x_9 - x_9'\|^2 = 0 \ \forall x = x'. \tag{7}$$

Training Choices for Leave-Allele-out Predictions. We designed a procedure to get the largest possible training set, in which the similarities of the target allele to the other alleles are reflected in the number of training samples

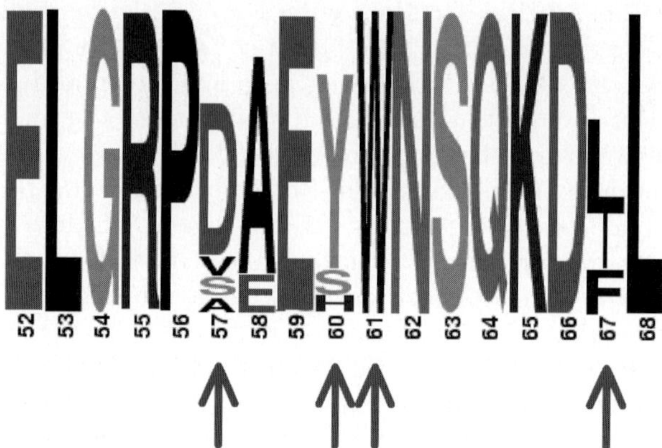

Fig. 1. This figure shows the conservation of an excerpt of the alignments of all HLA-DRB1, HLA-DRB3, HLA-DRB4 and HLA-DRB5 alleles from the IMGT/HLA database [2]. The arrows pointing at positions of the sequence logo [32] are polymorphic residues, which are used for the binding pocket profile vectors.

from the particular alleles. The idea is that training samples from more similar alleles should enable better predictions for the target allele than distant ones. To compute similarities between alleles, we calculated the Pearson correlation between alleles using the pocket profile vectors. Let p_i and p_j be the pocket profile vectors of alleles i and j. The similarity between these alleles is the Pearson correlation of p_i and p_j scaled linearly to $[0,1]$. This value is called $allelesim_{i,j}$ in the following. For a particular target allele, the procedure is the following:

Let j be the target allele. For every allele $i \neq j$ with n_i peptide sequences, we compute the maximal number t_i such that $allelesim_{i,j} \times t_i \leq n_i$. Then we choose the minimum of all t_i, which is now called t^*. For every allele $i \neq j$, we then choose $t^* \times allelesim_{i,j}$ peptide sequences from allele i and assign them to the training set.

Since for some alleles there exist more than $t^* \times allelesim_{i,j}$ peptides we wanted to make sure that we do not miss important peptides. Therefore, we chose the $t^* \times allelesim_{i,j}$ peptides which had the biggest kernel value to any of the target sequences. The intuition behind is that training samples with a very low kernel value to the target sequences will influence the prediction less than other samples since the output label is a linear combination of kernel functions [25].

3   Results

In this section we compare our classifiers to other state-of-the-art methods. In particular we compare our performance to the results of Wang *et al.* [18] who performed a large scale evaluation on MHC class II prediction methods. On their benchmark dataset we show that our predictor C_{single}, which is trained on parts of the target allele dataset, performs equally well or better than all other methods. Furthermore, we show that C_{multi} can predict binding for alleles without using any training data of the target allele and achieves performance which is comparable to the best predictors which are trained on binding data of the target allele.

3.1 Data

We show the performance of our classifiers on an MHCII benchmark dataset, introduced by Wang *et al.* [18]. This dataset contains peptide binding data for 14 human alleles as well as three mouse alleles which were measured in the laboratory of Wang *et al.* [18]. Binding affinities of the benchmark dataset were given in IC_{50} values. The smaller the IC_{50} value, the better is the binder. There are many peptides with very high IC_{50} values. Since the cutoff for binders is between 500 and 1000 nM there is not a big difference between a non-binder with IC_{50} value of 10,000 or 20,000 nM. Therefore, we transformed the IC_{50} values like Nielsen *et al.* [5]. Let a_i be the binding affinity of peptide i. The log-transformed binding affinity a_i' is defined as $a_i' := 1 - \log_{50000}(a_i)$. Like Nielsen *et al.* [5], we set the $a_i' < 0$ to zero. In the following, the dataset will be called $D_{\mathrm{benchmark}}$. Like Wang *et al.* [18], we consider peptides as binder, if their IC_{50} value is smaller than 1000 nM [18] ($a_i' > 0.3616$).

Peptide sequences, for which no binding core could be found (aliphatic or aromatic amino acid at position one) were excluded from all evaluations. This was the case for less than 3% of peptides (270 out of all 9478). Out of these peptides only 64 peptides are considered as binder ($a_i' > 0.3616$). Since the whole dataset contains 6475 binders in total, this means that our assumption that every binder has to have a binding core with an aliphatic or aromatic amino acid at position one just misses 64 out of 6475 binders which is under 1%.

3.2 Performance of C_{single}

Wang *et al.* [18] recently compared the performances of state-of-the-art predictors for MHCII binding. We show a comparison to the top four methods of their evaluation. Wang *et al.* [18] measured the performance of the ARB method [4] by ten-fold cross-validation. The performance of the other methods was evaluated using available webservers. The authors justified this procedure by the fact that they measured the performance on unpublished data, which had been measured in their labs. Therefore, it is unlikely that any of these methods (except the ARB method) was trained on parts of this dataset. To compare the performance of C_{single} to this evaluation, we performed a ten-fold cross-validation using parameter ranges $C \in \{0.001, 0.01, 0.1, 1, 10, 100, 1000\}$, $\nu \in \{0.2 * 1.4^i | i = 1, 2, ..., 5\}$ and $\sigma \in \{0.0625, 0.125, 0.25, ..., 16\}$. Table 1 shows that C_{single} outperforms all other single methods. The column "Consensus" corresponds to the consensus approach of Wang *et al.* [18] in which the best three classifiers for the particular allele are combined to achieve higher accuracy. One could assume that with C_{single} as one of these three predictors the accuracy will improve, since the performance of C_{single} is comparable to the consensus approach. All performances were measured in area under the ROC curve.

3.3   Performance of Leave-Allele-out C_{multi} and $C_{\text{supertype}}$

To show, that C_{multi} performs well although it is not trained on any data of the target allele, we conducted the following experiment. The training samples are chosen as described in the "Methods" section. We then performed a five-fold cross-validation on the training set to determine the best hyperparameters. For this $C \in \{0.01, 0.1, 1, 10\}$, $\nu \in \{0.2 * 1.4^i | i = 1, 2, ..., 5\}$ and $\sigma \in \{0.0625, 0.125, 0.25, ..., 4\}$. With the best hyperparameters of the cross-validation we trained our classifiers with the whole training set. We then measured the area under the ROC curve performance on the target allele. To show that the $k_{\text{pw--RBF}}$ kernel function of C_{multi} really improves the MHCII binding prediction if data from various alleles is combined we introduced $C_{\text{supertype}}$. This method is trained like C_{multi} but uses the same kernel function as C_{single}.

It can be seen in Table 1 that the classifier C_{multi} performs quite well on $D_{\text{benchmark}}$ although it was not trained on any binding data of the target allele. One can hypothesize that this performance could also be reached for other alleles, for which no binding data is available, since we did not use any data of the target allele. C_{multi} performs even better than C_{single} on some alleles which shows that the method is not just valuable for new alleles but also for predictions for alleles for which there exists binding data. The performance of $C_{\text{supertype}}$ is worse than the performance of C_{multi}. This suggests that our kernel function, which takes the similarities of the alleles into account, is very valuable for this kind of predictions. The fact that $C_{\text{supertype}}$ still has a good performance could be due to the fact that there exist many promiscuous peptides for MHCII [35].

3.4   Implementation

All methods were implemented in C++. We used LIBSVM [36] for support vector learning. The predictions for all alleles are integrated into EpiToolKit [37] for which there exists a webserver at http://www.epitoolkit.org/mhciimulti.

4   Discussion

The proposed method is a novel approach for predicting MHC class II binding peptides for alleles lacking experimental data and thus opens up new alleys for the design of peptide-based therapeutic or prophylactic vaccines. Obviously, a conclusive validation of predicitons for alleles without experimental data is difficult. The leave-one-allele out predictions presented here indicate, however, that the method performs very well. One could object that restricting the first amino acid of the binding core to aromatic and aliphatic amino acids is a strong assumption. Nevertheless, if one selects all putative binding peptides of the 9478 peptide sequences in $D_{\text{benchmark}}$ for which no binding core with an aromatic or aliphatic residue at position one can be found, a classifier which just predicts 0 (non-binder) would have 0.7630 classification rate on these peptides. In other words, only for 270 peptides or 2.85% out of the 9478 peptides no binding core

Table 1. Performance comparison of C_{single}, C_{multi} and $C_{\text{supertype}}$ to the best four methods presented in [18]. The C_{multi} and the $C_{\text{supertype}}$ classifier are just trained on data of different alleles than the target allele. The performance of C_{single} and ARB are measured by ten-fold cross validation. All other methods are trained on binding data of the target allele which was not contained in the benchmark dataset.

MHCII type	# peptides	ARB	PROPRED	SMM-align	Consensus	C_{single}	C_{multi}	$C_{\text{supertype}}$
DRB1*0101	3882	0.76	0.74	0.77	0.79	**0.81**	0.58	0.75
DRB1*0301	502	0.66	0.65	0.69	0.72	**0.73**	0.64	0.65
DRB1*0401	512	0.67	0.69	0.68	0.69	0.67	**0.76**	0.64
DRB1*0404	449	0.72	0.79	0.75	**0.80**	0.79	0.73	0.73
DRB1*0405	457	0.67	0.75	0.69	0.72	**0.82**	0.73	0.78
DRB1*0701	505	0.69	0.78	0.78	0.83	0.82	**0.85**	0.78
DRB1*0802	245	0.74	0.77	0.75	**0.82**	0.77	0.78	0.77
DRB1*0901	412	0.62	-	0.66	**0.68**	0.64	0.62	0.62
DRB1*1101	520	0.73	0.80	0.81	0.80	**0.85**	0.84	0.68
DRB1*1302	289	**0.79**	0.58	0.69	0.73	0.74	0.69	0.62
DRB1*1501	520	0.70	0.72	**0.74**	0.72	0.72	0.67	0.76
DRB3*0101	420	0.59	-	0.68	-	**0.73**	0.51	0.47
DRB4*0101	245	0.74	-	0.71	0.74	**0.80**	0.64	0.71
DRB5*0101	520	0.70	0.79	0.75	0.79	0.81	**0.87**	0.69
Mean		0.71	0.73	0.73	0.76	0.76	0.71	0.69

with aromatic or aliphatic residue at position one can be found and only 64 out of these are considered as binders (log-transformed binding affinity greater than 0.3616). This is why we think that the heuristic is very well applicable and reflects a general property for MHCII binding which is also supported by previous work [28, 9]. Moreover, the restriction to these binding cores is one of the key parts of this work because if one selected all 9-mers as binding cores this would add very much noise to the bags and the positional weighting of the binding core would not have a big effect since nearly every residue of a peptide (except the residues at the ends) would be at every position in one of the instances in the bag. Ultimately, only experimental testing or structure-based studies will reveal whether some of the rarer allels might deviate from this behavior on the first position.

Acknowledgements. We thank Andreas Kämper for proofreading the manuscript. This work was supported by Deutsche Forschungsgemeinschaft (SFB 685, project B1).

References

1. Topalian, S.L.: MHC class II restricted tumor antigens and the role of CD4+ T cells in cancer immunotherapy. Curr. Opin. Immunol. 6(5), 741–745 (1994)
2. Robinson, J., Waller, M.J., Parham, P., Groot, N.d., et al.: IMGT/HLA and IMGT/MHC: sequence databases for the study of the major histocompatibility complex. Nucleic Acids Res. 31(1), 311–314 (2003)

3. Peters, B., Sidney, J., Bourne, P., Bui, H.H., Buus, S., et al.: The immune epitope database and analysis resource: from vision to blueprint. PLoS Biol. 3(3), 91 (2005)
4. Bui, H.H., Sidney, J., Peters, B., Sathiamurthy, M., Asabe, S., et al.: Automated generation and evaluation of specific MHC binding predictive tools: ARB matrix applications. Immunogenetics 57(5), 304–314 (2005)
5. Nielsen, M., Lundegaard, C., Lund, O.: Prediction of MHC class II binding affinity using SMM-align, a novel stabilization matrix alignment method. BMC Bioinformatics 8, 238 (2007)
6. Rammensee, H.G., Friede, T., Stevanović, S.: MHC ligands and peptide motifs: first listing. Immunogenetics 41(4), 178–228 (1995)
7. Reche, P.A., Glutting, J.P., Zhang, H., Reinherz, E.L.: Enhancement to the RANKPEP resource for the prediction of peptide binding to MHC molecules using profiles. Immunogenetics 56(6), 405–419 (2004)
8. Singh, H., Raghava, G.P.: ProPred: prediction of HLA-DR binding sites. Bioinformatics 17(12), 1236–1237 (2001)
9. Sturniolo, T., Bono, E., Ding, J., Raddrizzani, L., Tuereci, O., et al.: Generation of tissue-specific and promiscuous HLA ligand databases using DNA microarrays and virtual HLA class II matrices. Nat. Biotechnol. 17(6), 555–561 (1999)
10. Nielsen, M., Lundegaard, C., Worning, P., Hvid, C.S., Lamberth, K., Buus, S., Brunak, S., Lund, O.: Improved prediction of MHC class I and class II epitopes using a novel Gibbs sampling approach. Bioinformatics 20(9), 1388–1397 (2004)
11. Noguchi, H., Kato, R., Hanai, T., Matsubara, Y., Honda, H., Brusic, V., Kobayashi, T.: Hidden markov model-based prediction of antigenic peptides that interact with MHC class II molecules. J. Biosci. Bioeng. 94(3), 264–270 (2002)
12. Karpenko, O., Shi, J., Dai, Y.: Prediction of MHC class II binders using the ant colony search strategy. Artif. Intell. Med. 35(1-2), 147–156 (2005)
13. Brusic, V., Rudy, G., Honeyman, G., Hammer, J., Harrison, L.: Prediction of MHC class II-binding peptides using an evolutionary algorithm and artificial neural network. Bioinformatics 14(2), 121–130 (1998)
14. Guan, P., Doytchinova, I.A., Zygouri, C., Flower, D.R.: MHCPred: A server for quantitative prediction of peptide-MHC binding. Nucleic Acids Res. 31(13), 3621–3624 (2003)
15. Dönnes, P., Kohlbacher, O.: SVMHC: a server for prediction of MHC-binding peptides. Nucleic Acids Res. 34, 194–197 (Web Server issue) (2006)
16. Salomon, J., Flower, D.: Predicting class II MHC-peptide binding: a kernel based approach using similarity scores. BMC Bioinformatics 7(1), 501 (2006)
17. Wan, J., Liu, W., Xu, Q., Ren, Y., Flower, D.R., Li, T.: SVRMHC prediction server for MHC-binding peptides. BMC Bioinformatics 7, 463 (2006)
18. Wang, P., Sidney, J., Dow, C., Mothé, B., Sette, A., Peters, B.: A systematic assessment of MHC class II peptide binding predictions and evaluation of a consensus approach. PLoS Comput. Biol. 4(4), 1000048 (2008)
19. Zaitlen, N., Reyes-Gomez, M., Heckerman, D., Jojic, N.: Shift-invariant adaptive double threading: Learning MHC II - peptide binding. In: Speed, T., Huang, H. (eds.) RECOMB 2007. LNCS (LNBI), vol. 4453, pp. 181–195. Springer, Heidelberg (2007)
20. DeLuca, D., Khattab, B., Blasczyk, R.: A modular concept of hla for comprehensive peptide binding prediction. Immunogenetics 59(1), 25–35 (2007)
21. Jacob, L., Vert, J.P.: Efficient peptide-MHC-I binding prediction for alleles with few known binders. Bioinformatics 24(3), 358–366 (2008)

22. Nielsen, M., Lundegaard, C., Blicher, T., Lamberth, K., Harndahl, M., Justesen, S., Røder, G., Peters, B., Sette, A., Lund, O., Buus, S.: NetMHCpan, a method for quantitative predictions of peptide binding to any HLA-A and -B locus protein of known sequence. PLoS ONE 2(8), 796 (2007)
23. Gärtner, T., Flach, P.A., Kowalczyk, A., Smola, A.J.: Multi-instance kernels. In: Sammut, C., Hoffmann, A.G. (eds.) ICML, pp. 179–186. Morgan Kaufmann, San Francisco (2002)
24. Dietterich, T.G., Lathrop, R.H., Lozano-Pérez, T.: Solving the multiple instance problem with axis-parallel rectangles. Artif. Intell. 89(1-2), 31–71 (1997)
25. Schölkopf, B., Smola, A.J., Williamson, R.C., Bartlett, P.L.: New support vector algorithms. Neural Comput. 12(5), 1207–1245 (2000)
26. Dooly, D.R., Zhang, Q., Goldman, S.A., Amar, R.A.: Multiple-instance learning of real-valued data. J. Machine Learn Res. 3, 651–678 (2002)
27. Ray, S., Page, D.: Multiple instance regression. In: ICML 2001: Proceedings of the Eighteenth International Conference on Machine Learning, pp. 425–432. Morgan Kaufmann Publishers Inc, San Francisco (2001)
28. Hammer, J., Belunis, C., Bolin, D., Papadopoulos, J., Walsky, R., Higelin, J., Danho, W., Sinigaglia, F., Nagy, Z.A.: High-affinity binding of short peptides to major histocompatibility complex class II molecules by anchor combinations. Proc. Natl. Acad. Sci. USA 91(10), 4456–4460 (1994)
29. Venkatarajan, M.S., Braun, W.: New quantitative descriptors of amino acids based on multidimensional scaling of a large number of physical-chemical properties. Journal of Molecular Modeling 7(12), 445–453 (2001)
30. Kawashima, S., Ogata, H., Kanehisa, M.: AAindex: Amino acid index database. Nucleic Acids Res. 27(1), 368–369 (1999)
31. Hertz, T., Yanover, C.: Pepdist: A new framework for protein-peptide binding prediction based on learning peptide distance functions. BMC Bioinformatics 7 (suppl. 1), S3 (2006)
32. Crooks, G.E., Hon, G., Chandonia, J.M., Brenner, S.E.: WebLogo: a sequence logo generator. Genome Res. 14(6), 1188–1190 (2004)
33. Li, H., Jiang, T.: A class of edit kernels for SVMs to predict translation initiation sites in eukaryotic mRNAs. In: RECOMB, pp. 262–271 (2004)
34. Schoenberg, I.J.: Metric spaces and positive definite functions. Trans. Amer. Math. Soc. 44(3), 522–536 (1938)
35. Consogno, G., Manici, S., Facchinetti, V., Bachi, A., Hammer, J., et al.: Identification of immunodominant regions among promiscuous HLA-DR-restricted CD4+ T-cell epitopes on the tumor antigen MAGE-3. Blood 101(3), 1038–1044 (2003)
36. Chang, C.C., Lin, C.J.: LIBSVM: a library for support vector machines. (2001), http://www.csie.ntu.edu.tw/~cjlin/libsvm
37. Feldhahn, M., Thiel, P., Schuler, M.M., Hillen, N., Stevanović, S., et al.: EpiToolKit–a web server for computational immunomics. Nucleic Acids Res. (2008) (advanced access, doi:10.1093/nar/gkn229)

An Algorithm for Orienting Graphs Based on Cause-Effect Pairs and Its Applications to Orienting Protein Networks

Alexander Medvedovsky[1], Vineet Bafna[2], Uri Zwick[1], and Roded Sharan[1]

[1] School of Computer Science, Tel Aviv University, Tel Aviv 69978, Israel
{medv,zwick,roded}@post.tau.ac.il
[2] Dept. of Computer Science, UC San Diego, USA
vbafna@cs.ucsd.edu

Abstract. We consider a graph orientation problem arising in the study of biological networks. Given an undirected graph and a list of ordered source-target pairs, the goal is to orient the graph so that a maximum number of pairs will admit a directed path from the source to the target. We show that the problem is NP-hard and hard to approximate to within a constant ratio. We then study restrictions of the problem to various graph classes, and provide an $O(\log n)$ approximation algorithm for the general case. We show that this algorithm achieves very tight approximation ratios in practice and is able to infer edge directions with high accuracy on both simulated and real network data.

1   Introduction

One of the major roles of protein-protein interaction (PPI) networks is to transmit signals within the cell in response to genetic and environmental cues. Technologies for measuring PPIs (see, e.g., [3]) do not provide information on the direction in which the signal flows. It is thus a great challenge to orient a given network by combining causal information on cellular events. One such source of information is perturbation experiments in which a gene is perturbed and as a result other genes change their expression levels.

In graph theoretic terms, one is given an undirected graph and a list of cause-effect pairs. The goal is to direct the edges of the graph, assigning a single direction to each edge, so that a maximum number of pairs admit a directed path from the cause to the effect. In fact, by contracting cycles in the graph one can easily reduce the problem to that of orienting a tree. Hakimi et al. [4] studied a restricted version of the problem where the list of vertex pairs includes all possible pairs, giving a quadratic time algorithm for it. Another variant of the problem was studied in [1] and [5], where rather than maximizing the total number of pairs, an algorithm was given to decide if one can satisfy *all* given pairs.

In this paper we study the resulting tree orientation problem. We prove that it is NP-hard and hard to approximate to within a constant ratio, study restrictions

K.A. Crandall and J. Lagergren (Eds.): WABI 2008, LNBI 5251, pp. 222–232, 2008.

© Springer-Verlag Berlin Heidelberg 2008

of the problem to various graph classes, and provide an $O(\log n)$ approximation algorithm for the general case, where n is the size of the tree. We show that this algorithm achieves tight approximation ratios in practice and is able to infer edge directions with high accuracy on both simulated and real network data.

The paper is organized as follows: In Section 2 the graph orientation problem is presented and its complexity is analyzed. Section 3 provides exact and approximate algorithms for restrictions of the problem, and an approximation algorithm for the general case. Biological applications of the latter algorithm are described in Section 4. For lack of space, some proofs are shortened or omitted.

2   Problem Definition

Let $G = (V, E)$ be an undirected graph. An *orientation* $\boldsymbol{G}$ of G is a directed graph obtained from G by orienting each edge $(u, v) \subset E$ either from u to v or from v to u. Let $P \subseteq V \times V$ be a set of ordered source-target pairs. A pair $(a, b) \in P$ is satisfied by a given orientation $\boldsymbol{G}$ of G if there is a *directed* path from a to b in $\boldsymbol{G}$. Our goal is to find an orientation $\boldsymbol{G}$ of G that simultaneously satisfies as many pairs from P as possible.

If the graph G contains a cycle C, then it is easy to see that, for any set P, there is an optimal orientation of G in which all the edges of C are oriented in the same direction and, consequently, all pairs that connect two vertices in C are satisfied. The original problem can therefore be solved by *contracting* the cycle C and then solving an equivalent problem on the contracted graph. Thus, the interesting case is when the graph G is a tree.

Definition 1 MAXIMUM TREE ORIENTATION *(MTO): Given an undirected tree T and a set P of ordered pairs of vertices, find an orientation of the edges of T that maximizes the number of pairs in P that are satisfied.*

In the decision version of the problem, the input includes T, P, and an integer $k \leq |P|$, and the question is whether the edges can be directed so that at least k pairs in P are satisfied. As we show next, the problem is NP-hard even when T is a star or a binary tree.

Theorem 1. *MTO is NP-complete.*

Proof. The problem is clearly in NP. We show NP-hardness by reduction from MAX DI-CUT [8], which is defined as follows: given a directed graph $G = (V, E)$ and an integer $k \leq |E|$, is there a cut $A \subset V$ such that there are at least k edges $e = (u, v)$, with $u \in A$ and $v \in V \backslash A$.

We map an instance (G, k) of MAX DI-CUT into an instance $(T = (V', E'), P, k)$ of MTO in the following way: $V' = V \cup \{O\}$, $E' = \{(v, O) : v \in V\}$ and $P = E$.

Given a cut $A \subset V$ with k crossing edges, it is easy to see that each pair corresponding to such an edge can be satisfied: for all $v \in A$ direct the edge (v, O) toward O, and direct all other edges away from O.

On the other hand, suppose that we have directed the edges of T so that k pairs are satisfied. Note that if (u, v) is satisfied then u is directed toward O,

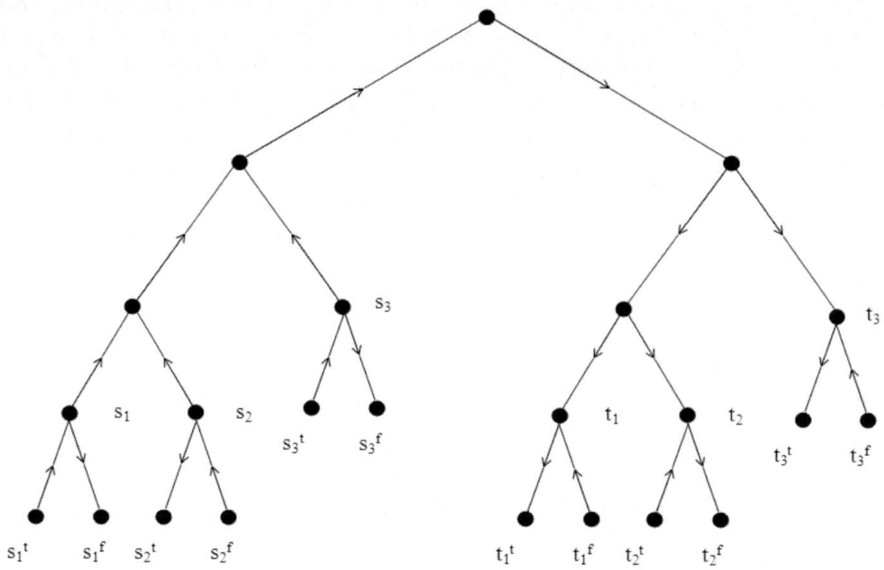

Fig. 1. An example of the reduction from MAX-2-SAT to MTO. The input 2-SAT formula is $(x_1 \vee \neg x_2) \wedge (x_3 \vee \neg x_2)$; x_1 and x_3 are assigned a *True* value, and x_2 is assigned a *False* value.

and no pair (v', u) can be satisfied. Therefore, the cut defined by $A = \{u \mid (u, O)$ is directed toward $O\}$, is of size k. ∎

Corollary 1. *MTO is NP-complete even on stars.*

As MAX DI-CUT is hard to approximate to within a factor of $\frac{11}{12} \simeq 0.9166$ (Håstad [6]), and the reduction is approximation preserving, we conclude:

Corollary 2. *It is NP-hard to approximate MTO to within a factor of $\frac{11}{12}$.*

Theorem 2. *MTO is NP-complete on binary trees.*

Proof. The problem is clearly in NP. We prove NP-hardness by a reduction from MAX 2-SAT, where each clause is assumed to contain exactly two literals. Suppose f is a 2-SAT formula with variables $x_1, ..., x_n$. Create a binary tree T with subtrees T_s and T_t, so that T_s has a leaf s_i, and T_t has a leaf t_i for each variable x_i. Create two child nodes s_i^t and s_i^f for each s_i, and t_i^t and t_i^f for each t_i (see Figure 1).

To complete the reduction we need to specify a set of pairs to be satisfied. This set will be composed of two subsets: P_1, forcing the choice of a truth value for each variable, and P_2, relating these truth values to the clauses in f.

The truth value of a variable will be set by forcing a directed path between s_i^t and s_i^f. If the path is directed from s_i^t to s_i^f we will interpret it as assigning the value *True* to x_i; if it is directed the other way, we will associate the value *False* with x_i. To this end, for every variable x_i participating in n_i clauses, we will add $3n_i$ pairs (s_i^t, s_i^f) and $3n_i$ pairs (s_i^f, s_i^t) to P_1. Similarly, we will force a path between t_i^t and t_i^f, indicating the truth value of x_i, but this time a path from t_i^t to t_i^f will indicate *False* and the opposite direction will indicate *True*. Again, this will be done by adding $3n_i$ pairs (t_i^t, t_i^f) and $3n_i$ pairs (t_i^f, t_i^t) to P_1. Finally, to force the consistency of truth association in T_s and T_t, we will force a directed path from s_i^t to t_i^t or from s_i^f to t_i^f by adding $3n_i$ pairs (s_i^t, t_i^t) and $3n_i$ pairs (s_i^f, t_i^f) to P_1.

The complementing subset of pairs is defined as follows:

$$P_2 = \{(s_i^t, t_j^t), (s_i^t, t_j^f), (s_i^f, t_j^t) | (x_i \vee x_j) \in f\} \cup$$
$$\{(s_i^f, t_j^t), (s_i^f, t_j^f), (s_i^t, t_j^t) | (\neg x_i \vee x_j) \in f\} \cup$$
$$\{(s_i^t, t_j^t), (s_i^t, t_j^f), (s_i^f, t_j^f) | (x_i \vee \neg x_j) \in f\} \cup$$
$$\{(s_i^f, t_j^t), (s_i^f, t_j^f), (s_i^t, t_j^f) | (\neg x_i \vee \neg x_j) \in f\}$$

Define $P = P_1 \cup P_2$. We claim that c clauses in f can be satisfied iff $18n + c$ pairs in P can be satisfied, where n is the total number of clauses in f.

Suppose that there is a truth assignment that satisfies c clauses in f. Direct the edges (s_i, s_i^t), (s_i, s_i^f), (t_i, t_i^t) and (t_i, t_i^f) according to the assignment. Direct all other edges in T_s upwards, and edges in T_t downwards. For each i, there are $9n_i$ satisfied pairs in P_1. Since $\sum_i n_i = 2n$, the number of satisfied pairs in P_1 is $18n$. Clearly, for every satisfied clause there is a satisfied pair from P_1. Thus, $18n + c$ pairs of P can be satisfied.

Conversely, suppose we have an orientation of T so that $18n + c$ pairs of P are satisfied. For each i there are at most $9n_i$ satisfied pairs in P_1. If the total number of satisfied pairs in P_1 is less than $18n$, then for some i there are less than $9n_i$ satisfied pairs (out of the ones associated with it). This implies that the directions of the edges (s_i, s_i^t), (s_i, s_i^f), (t_i, t_i^t), (t_i, t_i^f) are inconsistent. Thus, either $6n_i$, $3n_i$ or 0 of the corresponding pairs are satisfied. However, if we make these edge directions consistent, we add at least $3n_i$ satisfied pairs from P_1 and lose at most $3n_i$ pairs involving one of $s_i^t, s_i^f, t_i^t, t_i^f$ from P_2. Thus, w.l.o.g., we can assume that these edges are directed consistently, implying exactly $18n$ satisfied pairs from P_1. In addition, we have c satisfied pairs from P_2. Moreover, due to the consistency assumption, each clause can have at most one associated pair satisfied. It follows that c clauses can be satisfied in f. ∎

3   Exact and Approximation Algorithms for MTO

As we have shown that MTO is NP-hard, we describe polynomial time algorithms for special cases, and approximation algorithms for special cases and for the general case. We start by providing an integer programming (IP) formulation of the problem that will be useful for studying the practical performance of the algorithms we propose for MTO.

3.1   An Integer Program Formulation

Since every two vertices in a tree are connected by a unique path, MTO can be solved using the following integer program:

1. For each vertex pair $p \in P$ introduce a Boolean variable $y(p)$, indicating whether it is satisfied or not.
2. For each edge $e = (u, v) \in T$, where $u < v$, introduce a Boolean variable $x(e)$, indicating its direction (1 if it is directed from u to v, and 0 otherwise).
3. For each pair $p = (a, b) \in P$ and every tree edge $e = (u, v) \in T$, where $u < v$: if the path from a to b in T uses e in the direction from u to v, introduce a constraint $y(p) \leq x(e)$, and if it uses the edge in the direction from v to u, introduce a constraint $y(p) \leq 1 - x(e)$.
4. Maximize the objective function $\sum_{p \in P} y(p)$.

It is possible to consider an LP-relaxation of the above integer programming, but it is not very useful as a value of $|P|/2$ can always be obtained by setting $x(e) = y(p) = \frac{1}{2}$ for every $e \in T$ and $p \in P$.

3.2   Solving MTO on Paths

In this section we present a simple dynamic programming algorithm that solves MTO on a path in polynomial time.

Assume that the vertices on the path are numbered consecutively from 1 to n. The edges of the path are $(i, i + 1)$, for $1 \leq i < n$. We think of vertex i as lying to the *left* of vertex $i + 1$. We also let $[i, j] = \{i, i + 1, \ldots, j\}$.

Let P be the input set of pairs. For every $1 \leq i < j \leq n$, let $v_{ij}^+ = |\{(a, b) \in P \mid i \leq a < b \leq j\}|$ and $v_{ij}^- = |\{(b, a) \in P \mid i \leq a < b \leq j\}|$. In other words, v_{ij}^+ is the number of pairs of P with both endpoints in the interval $[i, j]$ that are satisfied when the edges $(i, i+1), \ldots, (j-1, j)$ are all oriented to the right, while v_{ij}^- is the number of such pairs satisfied when the edges are oriented to the left. Let v_{ij} be the maximal number of pairs of P with both endpoints in $[i, j]$ that can be simultaneously satisfied using *any* orientation of the edges in the interval $[i, j]$. We claim:

Lemma 1. *For every $1 \leq i < j \leq n$ we have $v_{ij} = \max\{\, v_{ij}^+, v_{ij}^-, \max_{i < k < j} v_{ik} + v_{kj}\}$.*

Proof. The proof that $v_{ij} \geq \max\{\, v_{ij}^+, v_{ij}^-, \max_{i<k<j} v_{ik} + v_{kj}\}$ is straightforward. We prove, therefore, the opposite inequality. Consider the orientation of $[i, j]$ that achieves the maximal value of v_{ij}. If all the edges in this orientation are oriented to the right, then $v_{ij} = v_{ij}^+$ and we are done. Similarly, if they are all oriented to the left, then $v_{ij} = v_{ij}^-$. Otherwise, there is a vertex $i < k < j$ for which the edges $(k - 1, k)$ and $(k, k + 1)$ have opposite orientations. It follows that no pair (a, b) with $a < k < b$ or $b < k < a$ can be satisfied by such an orientation. Hence, all edges satisfied by this orientation lie in either $[i, k]$ or $[k, j]$; thus, $v_{ij} = v_{ik} + v_{kj}$, as required.   ∎

As an immediate corollary we get:

Theorem 3. *MTO on paths of length n can be solved in $O(n^3)$ time.*

3.3 Approximating MTO on Stars

A *star* is a tree in which the root is directly connected to all the leaves (i.e., a tree with one level). In this section we describe an approximation algorithm for MTO on stars that will also serve as a building block in our approximation algorithms for general trees.

Lemma 2. *If T is a star then at least $1/4$ of all pairs can be satisfied.*

Proof. Choose a random orientation. Each pair is then satisfied with a probability of at least $1/4$.   ∎

It is easy to use the method of conditional expectations to obtain a deterministic linear time algorithm that produces an orientation of a star that satisfies at least $1/4$ of the pairs. This immediately gives us a $1/4$-approximation algorithm for the problem. As the MTO problem on stars is equivalent to the MAX-DI-CUT problem, an 0.874-approximation for the problem can be obtained using the semidefinite programming based approximation algorithms [2,9].

Approximating MTO on Caterpillars

Recall that a *caterpillar* is a graph in which all vertices are on a central path or at most one edge away from it. MTO is NP-complete even for caterpillars with maximum degree 3 (the proof is similar to the proof for binary trees, and is omitted for lack of space). We show the following:

Lemma 3. *Let T be a caterpillar. At least $1/8$ of all pairs can be satisfied.*

Proof. Partition edges into 'path' edges which lie on the caterpillar path, and 'bush' edges which "stick" from it. Direct the path edges in a single direction by choosing one of the two at random. Also, randomly direct each of the bush edges. Note that each pair of vertices (u, v) is connected by at most two bush edges and a sub-path. Therefore, the probability that (u, v) is satisfied by the random assignment is at least $1/8$. The claim follows.   ∎

3.4 Approximating MTO on Bounded-Depth Trees

In this section we present approximation algorithms for rooted, bounded-depth trees that make use of the approximation algorithm for stars. All the results can be extended to unrooted, bounded-diameter trees by rooting them at a "central" vertex so that their depth is bounded by roughly half the diameter.

Consider a tree T with d levels (and depth $d-1$). We denote the vertices at level i by L_i, starting from the root at level 1. For a node v, denote by T_v the subtree rooted at v. Two notions of separation will be useful to us in the approximation algorithms that we design in the sequel.

Definition 2. *A node w in a tree separates a pair (u, v), if w is on the path between u and v. w is called the* lowest common ancestor (LCA) *of u and v if in addition it lies on the lowest possible level in the tree.*

Lemma 4. *For any vertex v in T, at least $1/4$ of the pairs separated by v can be satisfied.*

Proof. Re-root the tree at v, and denote its subtrees by $T_1, \ldots, T_l$. We will direct all edges in T_i either toward v or away from v, consistent with the edge between v and T_i. To direct the edges between v and T_i, construct an instance of MTO on a star T' as follows: the root is v, and there are l leaves $v_1, \ldots, v_l$. For each (u, w) separated by v, where $u \in T_x, w \in T_y$, add (v_x, v_y) to P'. By Lemma 2, $1/4$ of the pairs in P' can be satisfied. The edge directions in T' are used to direct the edges of the tree. It is easy to see that any pair (u, w) separated by v, where $u \in T_x, w \in T_y$, is satisfied iff (v_x, v_y) is satisfied in T'. The claim follows. ∎

Corollary 3. *For any vertex v in T, at least $1/4$ of the pairs whose LCA is v can be satisfied.*

Definition 3. *For a tree T rooted at a vertex v, we denote by StarMTO(T, P, v) the star-based solution of the MTO problem on T, as described in the proof of Lemma 4.*

Lemma 5. *Let T be a rooted tree with d levels. At least $1/(4d)$ of the pairs in P can be satisfied.*

Proof. As there are d levels, there must be a level j that contains at least $|P|/d$ LCAs of the pairs in P. Compute StarMTO(T_v, P, v) for each node $v \in L_j$. By Corollary 3, at least $|P|/(4d)$ of the pairs are satisfied. ∎

The above lemma provides us with a lower bound on the number of pairs that can be satisfied. It implies an approximation algorithm to MTO with a ratio of $1/(4d)$, but the latter ratio can be improved as we show next:

Lemma 6. *Let T be a tree with d levels. MTO can be approximated to within a factor of $1/(2d)$ on T.*

Proof. We form a d-partite graph G_d, in which each node corresponds to a pair in P. The i-th layer is the set of pairs whose LCAs lie on L_i. We connect two vertices (in two layers) by an edge if the two pairs cannot be simultaneously satisfied. Clearly, the maximum number of pairs that can be satisfied is no more than a maximum sized independent set I in G_d.

For the algorithm, find an independent set I' in G_d. Next, solve StarMTO starting with the root level of the tree, and going down the tree. Specifically, for each vertex $v \in L_i$, solve StarMTO(T_v, I', v) on the pairs in the independent set I'. Note that some of the edge directions have been pre-set by previous levels. However, as I' is an independent set, the edge directions set by previous levels do not contradict any of the pairs in the current level. By Lemma 2, at least $|I'|/4$ pairs are satisfied. The approximation ratio is therefore

$$\frac{|I'|}{4|I|} = \frac{\alpha_d}{4}$$

where $\alpha_d = 2/d$ is the approximation ratio achievable for independent sets on a d-partite graph [7]. Overall we get an approximation ratio of $1/(2d)$. ∎

Lemma 6 implies a $1/(2\lg n)$ approximation algorithm for a complete binary tree, as it contains $\lg n$ levels.

Approximating MTO on General Trees

Lemma 7. *In every tree of size n there is a* centroid *node whose removal breaks the tree into components of size at most $n/2$.*

Our approximation algorithm for general trees is as follows:

MTO(T, P)

    1. Find a centroid v, with resulting subtrees $T_1, \ldots, T_l$.
    2. Let $A_s = \text{StarMTO}(T, P, v)$.
    3. For all $1 \le j \le l$, let $P_j - \text{MTO}(T_j, P)$.
    4. return $\max\{A_s, \sum_j P_j\}$.

Theorem 4. *Let T be a tree with n nodes. For any set P, MTO(T, P) finds an orientation that satisfies at least $1/(4\lg n)$ of the pairs.*

Proof. Let $R(n)$ be a lower bound on the fraction of the pairs satisfied by the orientation produced by the algorithm when run on a tree with n vertices. We show by induction that $R(n) \ge \frac{1}{4\lg n}$.

At the base of the induction $n = 2$. In this case, at least $1/2$ of the pairs are satisfied, and $R(n) \ge 1/2$. Suppose now that $R(k) \ge 1/(4\lg k)$ for any $k < n$. Let $P(n) = |P|R(n)$ denote the minimum number of pairs satisfied by running MTO on an input of size n. Also, let A be the subset of pairs separated by the centroid v. By the induction assumption, $\sum_j P(n_j) \ge \frac{|P|-|A|}{4\lg(n/2)}$. Therefore, applying to the recursion

$$P(n) = \max\left\{\frac{|A|}{4}, \sum_j P(n_j)\right\} \ge \max\left\{\frac{|A|}{4}, \frac{|P|-|A|}{4\lg(n/2)}\right\}$$

The two sub-expressions are balanced for $|A| = |P|/\lg n$, implying that

$$R(n) = \frac{P(n)}{|P|} \ge \frac{1}{4\lg n} \qquad\blacksquare$$

4   Applications to Simulated and Real Network Data

We implemented the general approximation algorithm described above and tested it on simulated and real network data. To evaluate its performance we also implemented the integer-program algorithm which provides an optimal solution to MTO.

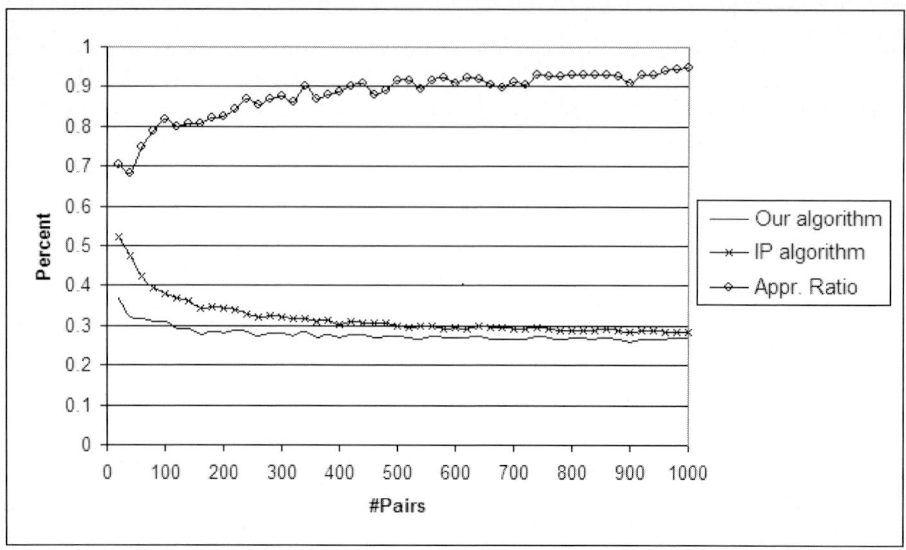

Fig. 2. Performance on simulated data. Percents of satisfied pairs are displayed for both an optimal solution (based on IP) and the approximation algorithm's solution. A third plot depicts the implied approximation ratio.

4.1   Performance on Simulated Data

As a first test of the algorithm, we measured its approximation ratio performance on random trees, by comparing the solutions it obtained to those obtained by the IP algorithm. The random trees contained 1,000 vertices and were generated by iteratively adding a vertex, and connecting it to one of the already existing vertices uniformly at random. The cause-effect pairs were generated by drawing vertices from the tree uniformly at random.

We tested the algorithm's performance when varying the number of cause-effect pairs from 20 to 1000. The percentage of satisfied pairs along with the implied approximation ratio are displayed in Figure 2. Evidently, the algorithm attains very high approximation ratios in practice reaching up to 0.9 and higher ratios on instances with 500 or more pairs. Notably, even on random instances for which at least 80% of the pairs could be satisfied (see a detailed description in the next paragraph) the average approximation ratio was 0.66 – much higher than the theoretical guarantee.

To test the utility of the algorithm in predicting edge directions, we simulated input with known edge directions as follows: we generated 20-1,000 pairs of vertices. 80% of the pairs were generated in a way that all could be satisfied simultaneously. The other 20% of the pairs were generated randomly, to simulate noise. We randomly chose 50 edges on the paths of the "correct" pairs, and tested the algorithm's accuracy in predicting their direction. The accuracy did not seem to depend on the number of pairs, and was 0.76 on average.

4.2   Biological Data

Next, we tested the performance of our algorithm on real data. To this end, we used a yeast protein-protein interaction (PPI) network consisting of 15,147 protein-protein interactions obtained from the Database of Interacting Proteins [11]. We complemented this network by additional 596 (kinase-substrate) PPIs from [10] for which the direction of signal flow is known (from the kinase to the substrate), represented as undirected edges in the constructed network. For cause-effect pairs, we used knockout data obtained from [12]. The data set contained 24,457 pairs of a knocked out gene (cause) and an affected gene (effect), out of which 14,295 are pairs of proteins from the network.

After removal of small disconnected components (of size ≤ 3) and cycle-contraction, we obtained a tree with 2,027 vertices and 3,370 cause-effect pairs. Interestingly, about 90% of the vertices in the contracted tree were aligned in a star form. Applying our approximation algorithm to the tree yielded an orientation that satisfied 3,262 of the 3,370 pairs. The optimal solution, obtained using integer programming, satisfied 3,295 pairs, implying a practical approximation ratio of 0.99. This tight ratio matches the ratios observed in the simulations (Figure 2).

The orientation produced by the algorithm provided predictions for 3,880 interaction directions. 148 of these interactions were from the kinase-substrate data set and, hence, their true directions were known. Remarkably, 147 of these 148 directions were predicted correctly. Notably, none of the kinase-substrate interactions were also cause-effect pairs, but rather lied on paths connecting such pairs.

5   Conclusions

In this paper we have studied the problem of orienting a graph so as to satisfy a maximum number of ordered pairs. We have given exact and approximate algorithm to certain restrictions of the problem, and an $O(\log n)$ approximation algorithm for the general case. The algorithm was shown to yield very tight approximation ratios in practice, and attained remarkable accuracy in predicting edge directions on a real protein network.

Several open problems that require further investigation include: (i) closing the gap between the guaranteed approximation ratio in the general case and the approximation hardness result; (ii) tackling the graph orientation problem when some of the edge directions are pre-set (in the biological context this happens when there is prior biological knowledge on directionality or when considering other types of interactions such as transcriptional regulatory ones); and (iii) improving the lower bound on the optimum number of pairs that can be satisfied.

Acknowledgments

We thank Richard Karp, Eran Halperin, Tomer Shlomi and Eytan Ruppin for stimulating discussions about the orientation problem. We thank Oved Ourfali for his help with the implementation. We thank Andreas Beyer and Silpa Suthram for providing us with the kinase-substrate data. This work was supported by a grant from the Israel Science Foundation (grant no. 385/06).

References

1. Arkin, E.M., Hassin, R.: A note on orientations of mixed graphs. Discrete Applied Mathematics 116(3), 271–278 (2002)
2. Feige, U., Goemans, M.X.: Aproximating the value of two prover proof systems, with applications to MAX 2-SAT and MAX DI-CUT. In: ISTCS, pp. 182–189 (1995)
3. Fields, S.: High-throughput two-hybrid analysis the promise and the peril. Febs. J. 272(21), 5391–5399 (2005)
4. Hakimi, S.L., Schmeichel, E., Young, N.E.: Orienting graphs to optimize reachability. Information Processing Letters 63(5), 229–235 (1997)
5. Hassin, R., Megiddo, N.: On orientations and shortest paths. Linear Algebra and its applications 114/115, 589–602 (1989)
6. Håstad, J.: Some optimal inapproximability results. Journal of the ACM 48(4), 798–859 (2001)
7. Hochbaum, D.S.: Efficient bounds for the stable set, vertex cover and set packing problems. Discrete Applied Mathematics 6, 243–254 (1983)
8. Kann, V., Lagergren, J., Panconesi, A.: Approximability of maximum splitting of k-sets and some other apx-complete problems. Information Processing Letters 58(3), 105–110 (1996)
9. Lewin, M., Livnat, D., Zwick, U.: Improved rounding techniques for the MAX 2-SAT and MAX DI-CUT problems. In: Proceedings of the 9th International IPCO Conference on Integer Programming and Combinatorial Optimization, London, UK, pp. 67–82. Springer, Heidelberg (2002)
10. Ptacek, J., Devgan, G., Michaud, G., Zhu, H., Zhu, X., Fasolo, J., Guo, H., Jona, G., Breitkreutz, A., Sopko, R., McCartney, R.R., Schmidt, M.C., Rachidi, N., Lee, S.J., Mah, A.S., Meng, L., Stark, M.J., Stern, D.F., De Virgilio, C., Tyers, M., Andrews, B., Gerstein, M., Schweitzer, B., Predki, P.F., Snyder, M.: Global analysis of protein phosphorylation in yeast. Nature 438, 679–684 (2005)
11. Salwinski, L., Miller, C.S., Smith, A.J., Pettit, F.K., Bowie, J.U., Eisenberg, D.: The database of interacting proteins: 2004 update. Nucleic Acids Research 32, D449 (2004)
12. Yeang, C.-H., Ideker, T., Jaakkola, T.: Physical network models. Journal of Computational Biology 11(2/3), 243–262 (2004)

Enumerating Precursor Sets of Target Metabolites in a Metabolic Network

L. Cottret[1,2], P.V. Milreu[4], V. Acuña[1,2], A. Marchetti-Spaccamela[3],
F. Viduani Martinez[4], M.-F. Sagot[1,2], and L. Stougie[5]

[1] Université de Lyon, F-69000, Lyon, Université Lyon 1, CNRS, UMR5558
[2] Projet Helix, INRIA Rhône-Alpes, France
{cottret,viacuna}@biomserv.univ-lyon1.fr, marie-france.sagot@inria.fr
[3] Sapienza University of Rome
alberto.marchetti@dis.uniroma1.it
[4] Universidade Federal de Mato Grosso do Sul
fhvm@dct.ufms.br, paulovieira@milreu.com.br
[5] Eindhoven University of Technology and CWI, Amsterdam
leen@win.tue.nl

Abstract. We present the first exact method based on the topology of a metabolic network to find minimal sets of metabolites (called precursors) sufficient to produce a set of target metabolites. In contrast with previous proposals, our model takes into account self-regenerating metabolites involved in cycles, which may be used to generate target metabolites from potential precursors. We analyse the complexity of the problem and we propose an algorithm to enumerate all minimal precursor sets for a set of target metabolites. The algorithm can be applied to identify a minimal medium necessary for a cell to ensure some metabolic functions. It can be used also to check inconsistencies caused by misannotations in a metabolic network. We present two illustrations of these applications.

1 Introduction

The metabolic capacities of an organism are directly defined by the set of its possible biochemical reactions. The links between reactions and compounds (or metabolites) that are used/produced by such reactions constitute the *metabolic network* of an organism. Once the metabolic network of an organism has been defined (see [2] for an overview of the metabolic data reconstruction process), the following important question arises: how are the essential metabolites for the organism produced? Equivalently, which are the metabolites that the organism needs to obtain from its environment to produce those essential metabolites? In the sequel, we call such metabolites *precursors*.

One way to answer this question is to manually inspect the metabolic pathways defined as present in the organism: the presence of any metabolic pathway is determined by comparing the set of reactions of a reconstructed metabolic network with the set of reactions in the reference metabolic pathways contained in metabolic databases such as METACYC from the BIOCYC database collection [1]

K.A. Crandall and J. Lagergren (Eds.): WABI 2008, LNBI 5251, pp. 233–244, 2008.
© Springer-Verlag Berlin Heidelberg 2008

or KEGG [5]. However, each reference metabolic pathway represents a very small part of the whole network and does not consider what occurs upstream of the pathway, nor whether some alternative organism-specific pathways exist.

With the goal of detecting inconsistencies in ECOCYC (the pathway-genome database dedicated to the bacterium *Escherichia coli*), Romero and Karp [8] used a whole-network approach to find precursors, while Handorf *et al.* [4] proposed a method to identify minimal metabolite sets required by an organism to produce all metabolites contained in a target set. In the method of Romero and Karp, the definition of potential precursors is very restrictive while it is very broad in Handorf *et al.*'s method. In this paper, we propose an exact method that may deal with any set of potential precursors defined by the user. Our method also takes into account the fact that most reactions are defined as reversible because of a lack of information on metabolite concentrations and enzyme kinetic properties. It can be used with a directed, undirected or mixed hypergraph representation of a metabolic network. It is not clear whether previous proposals could handle such cases.

In their paper, Romero and Karp did also not provide any details on how they dealt with cycles of reactions, a crucial issue when analysing metabolic networks. A similar consideration applies to the Handorf method, where a reaction can be fired only if all the metabolites are in the sub-network already produced by the process. The method is therefore not able to take into account metabolites which cannot be reached by such a process. The second main contribution of this paper is to address this problem by explicitly dealing with cycles when computing precursor sets. By a *cycle*, we refer to the concept of cycle in a hypergraph representation of a metabolic network which we describe in detail in Section 2.

This calls for the introduction of the new, biologically well-founded concept of "self-regenerating" metabolites that cannot be considered as available in infinite supply, e.g. provided by the environment, as is assumed to be the case for precursors. Such metabolites need to be continuously regenerated, but they have the ability to participate in their own regeneration, and in the subsequent generation of other metabolites. Self-regenerating metabolites will be part of at least one cycle. In [8], Romero and Karp very informally define what they call *bootstrapping* compounds that may be related to our self-regenerating metabolites but only partially and the list of such compounds needs furthermore to be provided as input by the user.

In Section 2, we give basic notations and definitions. In Section 3, we analyse the complexity of finding a minimal and a minimum precursor set. An exact method for enumerating all minimal precursor sets for a given set of target metabolites is described in Section 4. Finally, the method is illustrated with two applications in Section 5. For the sake of space most proofs are omitted.

2   Preliminaries

A metabolic network consists of a set of metabolites and a set of reactions. Each reaction transforms a subset of metabolites, the *substrates*, into another subset of metabolites, the *products* of the reaction. Such a network can be modeled as a directed hypergraph $G = (\mathcal{C}, \mathcal{R})$ with $\mathcal{C}$ the set of vertices corresponding to *metabolites* (also called *compounds*) and $\mathcal{R}$ the set of hyperedges corresponding to *reactions*. A hyperedge $r \in \mathcal{R}$ is directed away from a compound $c \in \mathcal{C}$ only if c is a substrate of r, and directed into c only if c is a product of r. Note that reactions can be reversible: each reversible reaction is modelled as two different reactions of opposite direction.

A solution to the problem of finding precursors for a specific set of target metabolites is a set of compounds that are in "infinite" supply (for instance, from the environment) and can be used as substrates of some reactions. These reactions will then produce new metabolites, thereby increasing the set of available metabolites. By iterating the process, we can check whether the target is produced.

This way of considering the dynamics of a network is not enough to model the real process. Indeed, the network could have cycles that regenerate their own metabolites: metabolites that are not available initially could still be used as substrates of reactions because they are part of a cycle in which they are both produced and consumed. We call such metabolites *self-regenerating* and we observe that they can be used to generate other metabolites that are not potential precursors. Self-regenerating metabolites and the metabolites they enable to be generated will be called the *continuously available* metabolites.

Let us consider the network of Figure 1 with A and C as potential precursors and E as target metabolite. In this case, B and D are self-regenerating metabolites that are *continuously available*. Note that G is also a *continuously available* metabolite.

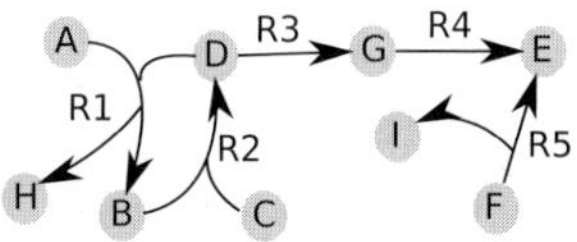

Fig. 1. A metabolic network G with set of metabolites $\mathcal{C} = \{A, B, C, D, E, F, G, H, I\}$, set of reactions $\mathcal{R} = \{R1, R2, R3, R4, R5\}$

For each reaction $r \in \mathcal{R}$, we call $Inp(r)$ the set of substrates of r and $Out(r)$ the set of products of r. In what follows, $\mathcal{P}(\mathcal{S})$ denotes the power set of a set $\mathcal{S}$.

Definition 1. *Given* $X \in \mathcal{P}(\mathcal{C})$, $Reach(X) \in \mathcal{P}(\mathcal{C})$ *is the set of compounds* y *for which there exists* $r \in \mathcal{R}$ *with* $Inp(r) \subseteq X$ *and* $Out(r) \ni y$.

In other words, $y \in Reach(X)$ if there exists a reaction in $\mathcal{R}$ producing y whose substrates are in X.

Given sets X, compounds in infinite supply, and Z, *continuously available* compounds, we wish to compute the total set of compounds that can be produced by the network.

Definition 2 (Reachability Function). *Let $X \in P(\mathcal{C})$ and $Z \in P(\mathcal{C})$ be two subsets of metabolites. The reachability function $f_Z : P(\mathcal{C}) \longrightarrow P(\mathcal{C})$ is defined as $f_Z(X) = X \cup Reach(X \cup Z)$.*

We define the functions $f_Z^k(X) = f_Z(f_Z^{k-1}(X))$, with $f_Z^1(X) = f_Z(X)$, as the function obtained by iterating k times the function f_Z.

Definition 3 (Scope Function). *Let $Z \in P(\mathcal{C})$ and $X \in P(\mathcal{C})$ be two subsets of metabolites. The scope function $f_Z^* : P(\mathcal{C}) \longrightarrow P(\mathcal{C})$ is $f_Z^*(X) = f_Z{}^k(X)$ for any k such that $f_Z{}^k(X) = f_Z{}^{k+1}(X)$.*

Note that f_Z is monotone, both in X and in Z and f_Z^* represents what may be produced from X with the help of Z and using reactions in $\mathcal{R}$. To define when a set of compounds X is a precursor set of a target T, we therefore need to impose that f_Z^* contains T, and f_Z^* can "regenerate" Z.

Definition 4 (Precursor Set). *A set of metabolites $X \subseteq P(\mathcal{C})$ is a precursor set of $T \subseteq P(\mathcal{C})$ if there exists a set $Z \subseteq P(\mathcal{C})$ such that $f_Z^*(X) \supseteq T \cup Z$.*

Note that in the above definition we are interested only in the existence of Z and not in characterising the set of available compounds that allows X to be a precursor set of T.

We now define a precursor set of a single target t in the hypergraph model.

Definition 5 (Hyperpath with a Set of *continuously available* Metabolites).

A set $H(X, Z, t) \in P(\mathcal{R})$ of reactions is a hyperpath from a set of metabolites X to t using another set of metabolites Z if it satisfies:

1. *The reactions in $H(X, Z, t)$ can be ordered $< r_1, r_2, \ldots, r_k >$ so that:*
   i) for all r_i, $Inp(r_i) \subset X \cup Z \cup Out(r_1) \cup \ldots \cup Out(r_{i-1})$;
   ii) $t \in Out(r_k)$;
   iii) for all $s \in Z$, there exists $j(s)$ such that $s \in Out(r_{j(s)})$,
2. *No proper subset of $H(X, Z, t)$ verifies the above.*

Clearly, if there is a hyperpath $H(X, Z, t)$ then X is a precursor set of t. For instance, in the Figure 1, if E is the target, and A and C are potential precursors, there is a hyperpath H(A, C, B, E) = r1, r2, r3, r4 such that the set A,C is a precursor set of E.

The reverse is shown in the following.

Lemma 6. *If X is a precursor set of t, then there exists a hyperpath $H(X, Z, t)$ for some $Z \in P(\mathcal{C})$.*

Sketch of the Proof
In the following we define recursively a sequence of reactions: starting from the target t, at each step i we choose a reaction r_i that produces a non-precursor and/or a substrate not yet produced by some of the previous reactions $r_1, \ldots,$ r_{i-1}. The obtained sequence may contain repetitions. By eliminating repetitions and inverting the list, we get an ordered set H that fullfills condition 1 of Definition 5.

In order to find reactions that can be reached from X, we only consider reactions in the set $W = \{r \in \mathcal{R} \mid Inp(r) \subseteq f_Z^*(X)\}$, *i.e.* reactions that takes as substrates only compounds available in the scope of X.

Let $N_0 = \{t\}$ and $A_0 = \emptyset$. At iteration i, $i \geq 1$, we define the sets $A_i = \cup_{j=1}^{i} Out(r_j) \setminus X$ and $N_i = \cup_{j=1}^{i} Inp(r_j) \setminus (A_i \cup X)$, i.e., A_i is the set of compounds produced in the first i reactions and N_i the set of compounds consumed but not yet made available in the first i reactions. In iteration i, select some $c_i \in N_{i-1}$. Since $c_i \notin X$, we know, by definition of W, that there exists a reaction $r_i \in W$ with $c_i \in Out(r_i)$. We update the set A_i by $A_{i-1} \cup (Out(r_i) \setminus X)$ and define the set $S_i = Inp(r_i) \cap A_i$, substrates of r_i that have been produced already. We update $N_i = (N_{i-1} \cup Inp(r_i)) \setminus (A_i \cup X)$.

This process is iterated until N_i is empty, which has to occur since the sequence of A_i is monotone. Let k be the first (and last) iteration such that $N_k = \emptyset$ and let $Z = \cup_{i=1}^{k} S_i$. The sequence $\omega = r_1, \ldots, r_k$ may have repetitions. We define $\bar{r}_1, \ldots, \bar{r}_\ell$ as the subsequence that contains only the first occurrence of each reaction and define $H = \{\bar{r}_1, \ldots, \bar{r}_\ell\}$ to be the set including all these reactions.

In the full version we shall show that the above defined set of reactions fullfills the conditions of Definition 5. $\qquad\square$

In the following we study the three problems below:

Problem MAL-PS(G, P, T): given a metabolic network $G = (\mathcal{C}, \mathcal{R})$ with $P \subset \mathcal{C}$ the set of all potential precursors and $T \subset \mathcal{C}$ the set of target metabolites, find a minimal precursor set $X \subset P$ of T in G.

Problem MIN-PS(G, P, T): given a metabolic network $G = (\mathcal{C}, \mathcal{R})$ with $P \subset \mathcal{C}$ the set of all potential precursors and $T \subset \mathcal{C}$ the set of target metabolites, find a minimum size precursor set $X \subset P$ of T in G.

Problem ALLMAL-PS(G, P, T): given a metabolic network $G = (\mathcal{C}, \mathcal{R})$ with $P \subset \mathcal{C}$ the set of all potential precursors and $T \subset \mathcal{C}$ the set of target metabolites, enumerate all minimal precursor sets $X \subset P$ of T in G.

Given the network of Figure 1, let E be the target and $\{\mathsf{A}, \mathsf{C}, \mathsf{F}\}$ the potential precursors. A solution to the MAL-PS(G, P, T) problem is $\{\mathsf{A}, \mathsf{C}\}$ or $\{\mathsf{F}\}$ whereas the precursor set $\{\mathsf{A}, \mathsf{C}, \mathsf{F}\}$ is not a solution because it is not minimal. The solution to the MIN-PS(G, P, T) problem is $\{\mathsf{F}\}$. Finally, the solution to the ALLMAL-PS(G, P, T) problem is given by $\{\mathsf{A}, \mathsf{C}\}$ and $\{\mathsf{F}\}$.

3   Minimal and Minimum Precursor Sets

The following useful Lemma is an interesting result in itself.

Lemma 7. *Given G, T, there exists a polynomial time algorithm to check whether a set X is a precursor set of T in G.* $\qquad\square$

The following algorithm for MAL-PS(G, P, T) resembles the one presented by Handorf *et al.* [4]. Given G, P and T, a simple algorithm to solve MAL-PS(G, P, T)

first sets $X = P$; at the beginning all compounds in X are unmarked. Then, the algorithm determines whether X is a precursor set of T in G using the algorithm in the proof of Lemma 7. If the answer is negative, then there is no precursor set and the algorithm stops. Otherwise, let u be an arbitrary unmarked compound of X and set $X' = X - \{u\}$; the algorithm determines whether X' is a precursor set of T in G. If so, then u is deleted from X: there exists at least one minimal solution that does not contain u. If not, then u remains in X and it is marked. The algorithm iterates this procedure until no unmarked compounds are left. Since all marked compounds are essential for a precursor set, they form a minimal precursor set of T in G. The following theorem states the correctness of the algorithm.

Theorem 8. *Given G, P, T, there exists a polynomial time algorithm that solves* MAL-PS(G, P, T). $\qquad\qquad\square$

The following is proved by a reduction from the Hitting Set problem [3].

Theorem 9. MIN-PS(G, P, T) *is NP-hard.* $\qquad\qquad\square$

4   Algorithm for Enumerating All Minimal Precursor Sets

To facilitate the exposition, we consider the case of a single target metabolite. The solution for several target metabolites is computed by adding an artificial node to the metabolic network and one irreversible reaction that has each target as substrate and the artificial node as only product, which then acts as the single target.

The algorithm is composed of two steps: the first one defines a special structure, called a *replacement tree*, that contains a representation of at least one hyperpath (see Section 2) for each precursor set of t. To achieve this, we proceed in a way that is analogous to the one adopted in the proof of Lemma 6; the main difference is that in this case X is unknown (in fact the algorithm is seeking all X that are precursor sets for t). Therefore, when the algorithm moves backwards in the network starting from t, it must consider all reactions and not only those in W. At the end of step 1, the replacement tree will contain a representation of at least one hyperpath for each minimal precursor set of t but also the representation of some hyperpaths that do not represent minimal precursor sets. In the second step, the replacement tree is used to enumerate all the precursor sets for the target metabolite, and sets of metabolites that are not precursor sets are removed.

The proof of the correctness of the algorithm uses Lemma 6 and is omitted. The time and space complexity of the algorithm are linear in the size of the replacement tree whose size can be exponential (note also that the number of solutions might in any case be exponential in the size of P). We finally observe that the two steps are described separately for the sake of clarity: in fact they can be executed simultaneously, thereby improving the time and space requirements.

Building the Replacement Tree
The replacement tree is rooted and directed from its root to the leaves. Nodes of the tree are labelled either by a metabolite or by a reaction. In the first case,

we have a *metabolite node* while in the second a *reaction node*. The children of a metabolite node are reaction nodes labelled by those reactions that produce the metabolite while the children of a reaction node are labelled by its substrates. In the tree, both metabolite and reaction nodes have only one parent whereas both can have several children. In the sequel, we use the terms *product node* for the parent of a reaction node and *substrate nodes* for the children of a reaction node.

The construction of the tree starts at the root t. For each reaction producing this metabolite, we create a reaction node, which has as parent the root, and as children new metabolite nodes corresponding to the substrates of the considered reaction. In this way, we obtain a tree (of depth 3) whose leaves are metabolites. This process is then iterated for each new metabolite node.

Let us call u a newly created metabolite node and c the corresponding metabolite of u. Along any branch of the tree, the process stops when one of these three conditions below is verified:

1. c corresponds also to an ancestor of u,
2. c corresponds to a child of one reaction node ancestor of u, or
3. c is not produced by any reaction (we cannot go further back in the network).

In the first case, we flag u as "continuously available". An example is metabolite node $u1 = $ H that is one of the children of N12 in the tree depicted in Figure 2. In this case, this metabolite node is flagged as continuously available (indicated by a star beside the node). Indeed, the metabolite is regenerated by the network.

In the second case, c is considered as a child of a reaction node of u; so it is not necessary to duplicate the search for a precursor by expanding the metabolite. An example is metabolite $u_2 = $ I in Figure 2. In this case, I is produced by reaction node N14 which is not ancestor of $u_2 = $ I in the tree. Since I is the child of the reaction node N13 ancestor of $u_2 = $ I, I has been already analysed. In the third case, c cannot be produced by any other reaction node so there is no need to expand it.

Therefore, when the process stops, all the leaves of the tree contain only metabolites not produced by any reaction or already visited metabolites.

The procedure implies that, for any solution X for target t, there exists a subtree whose set of reactions represents a hyperpath from X to t using some compounds that are regenerated (set Z in Lemma 6).

Lemma 10. *If X is a precursor set of t, then there exists a subtree of the replacement tree containing a hyperpath $H(X, Z, t)$ for some $Z \in \mathcal{P}(\mathcal{C})$.*

Enumerating the Solutions

We now use the replacement tree to enumerate all minimal precursor sets for t. This is done by successively processing subtrees that have a single reaction node r as root and one or more metabolite nodes as children that are all leaves in the tree. Depending on what those metabolite leaf nodes are (potential precursors, flagged metabolites or non-flagged metabolites), the subtree will either be eliminated, or it will be used to create a new subtree that will replace it. This effects a progressive compression of the original replacement tree until it has only three

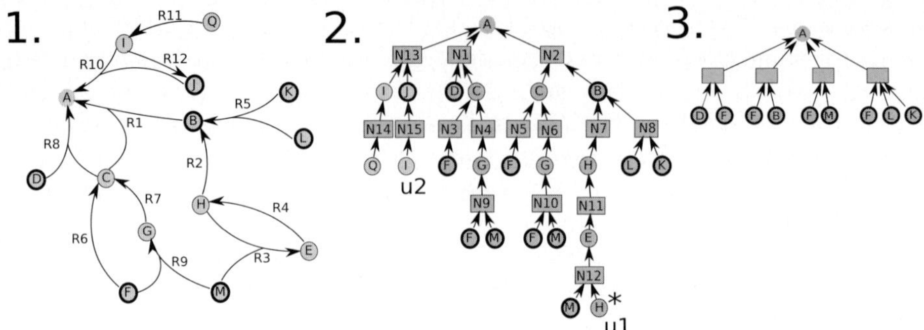

Fig. 2. An example of a metabolic network and of a replacement tree for the target metabolite A. **1.** The metabolic graph. The metabolite nodes surrounded in black are potential precursors. These are B, D, F, L, J, K and M. **2.** The replacement tree before compression. **3.** The replacement tree after compression. Each set of substrate nodes of a reaction node corresponds to a minimal precursor set.

levels composed of the root (level one), the reaction nodes producing the root (level two) and the minimal precursor sets (level three). The final compressed tree preserves the same properties as the initial tree with respect to the minimal precursor sets that produce the target t.

The compression algorithm starts by considering a reaction node whose substrate nodes are all leaves. Let r be the label of such a node, p be the parent of r, S the set of labels of its children and n the parent of p (note that S is a set of compounds and n is a reaction node). If r is such that either *i)* at least one of its substrate nodes is neither a potential precursor nor a flagged (*continuously available*) metabolite or *ii)* the potential precursors in S form a superset of the potential precursors that are substrate nodes of another reaction node having also p as parent, then the subtree rooted at r is simply eliminated from the tree.

If r is not eliminated, the subtree rooted at n is duplicated; namely all reaction and metabolite nodes are duplicated maintaining their label. Let n' be the root of this new subtree, p' the child of n' that corresponds to p, and r' the child of p' that corresponds to r. Furthermore, n' has the same parent as n.

We now modify the subtrees rooted at n and n' as follows:

- the metabolite nodes that are children of r in the subtree rooted at n are disconnected from r and r itself is eliminated;
- we replace node p' in the subtree rooted at n' with the set of children of r'. Both p' and r' are removed from n'.

We describe one example, using the replacement tree of Figure 2. Assume that, at some iteration of the compression algorithm, the subtree rooted at the reaction node N9 whose substrate nodes are leaves of the tree is considered. The children of N9 are potential precursors, and the parent of N9 has no other child. Therefore we cannot eliminate N9. Since the reaction node that is its immediate ancestor is N4 the subtree rooted at N4 is duplicated. Suppose the root of the duplicate subtree is labelled N4'. N4' is made the child of the parent of N4 labelled C. The metabolite nodes labelled F and M children of N9 are made the children of N4' in replacement of the copy of N4's only child labelled G. Reaction node N9 is removed. The parent of N4 has now 3 children: N3, N4 and N4'. If N4 is the new subtree considered at the next iteration of the algorithm, the algorithm would eliminate it since its only child is not a flagged leaf. If the subtree rooted at N4' is considered then its two children, labelled F and M, are potential precursors. However, since {F,M} is a superset of the set of potential precursors that are substrate nodes of N3 (only F), the subtree rooted at N4' can be eliminated and the next reaction node considered could be N3.

The process above described continues until the final compressed tree has only 3 levels: the root labelled by the target, the reaction nodes produced by the compression and the substrate nodes of these reaction nodes. The crucial property is that each step of the compression does not eliminate any minimal precursor set of t; it follows that labels of the children of a level 2 node directly correspond to a minimal precursor set of the target, as stated in the following lemma.

Lemma 11. *If X is a minimal precursor set of t, then there exists a child x of the root of the final compressed tree such the set of labels of the children of x coincides with X.*   □

5 Illustrations

We now briefly present two illustrations of our method. In the two cases, the set of potential precursors are defined in the same way. Since precursors are in general expected to be at the periphery of the network, we define as potential precursors all the metabolites not produced by any reaction and those that are involved in only one reaction which is reversible. This definition may not be sufficient for defining all potential precursors. We can therefore add some internal metabolites as further potential precursors. The final set and the procedure to get the metabolic data are given in the supplementary file available at `http://biomserv.univ-lyon1.fr/~cottret/WABI/annexes.pdf`.

5.1 Checking Inconsistencies in a Metabolic Database

The first example is very similar to the study done by Romero and Karp [8] whose goal was to check inconsistencies in the metabolic database ECoCYC [6].

There are two steps: the first proceeds exactly as in [8] on ECoCYC and is done to recover the target metabolites not reached by Romero and Karp's forward propagation algorithm. From the set of nutrients and of bootstrap metabolites given as input (these compounds are indicated by the user as being always present even though they may not be *continuously available* metabolites), the target metabolites not present in the scope of the nutrients are identified. In the second step, our method is applied on the network not produced by the forward propagation to find all the minimal precursor sets of the target metabolites not reached during the first step.

Data. The metabolic network we build contains 897 metabolites and 879 reactions, of which 104 are defined as irreversible.

The metabolites that are defined as bootstrapping during the first step, (which can be used during the forward propagation to fire a reaction), are those present in the minimal growth medium as indicated in [8], plus the first ten most connected metabolites and some metabolites whose presence in the cell seem obvious such as Coenzyme-A (see the supplementary file for a table of these bootstrapping metabolites). The only input metabolite for the forward propagation is glucose. The target metabolites are the 20 amino acids.

Results. In the first step, the forward propagation method in [8] returns a network with 513 reactions and 437 metabolites. This means that half of the network can be directly produced by injection of glucose, taking into account the presence of the bootstrapping metabolites selected in [8]. This subnetwork would even be bigger if our *continuously available* metabolites were also considered during the forward propagation. Among the 20 amino acids which are the building blocks for the synthesis of proteins, only 2 are not produced by the forward propagation: lysine and methionine.

Applying our method with lysine as target metabolite returns 9 sets of potentially missing precursors. Among them, one was noticeable: a set which contains only tetrahydrodipicolinate. Indeed, this metabolite is involved in the biosynthesis of lysine and, surprisingly, the pathway appears complete in *Escherichia coli*. Therefore the metabolite should not be a precursor. In fact, a closer look at the data reveals an error in EcoCyc. Indeed, tetrahydrodipicolinate is a substrate of reaction 1.3.1.26 which is indicated as irreversible, and furthermore in the "wrong" direction relatively to the known pathway. Once this reaction is made reversible, both lysine and methionine become present in the sub-network produced by the forward propagation.

5.2   Finding Nutrients Necessary to a Metabolic Function

The second example tests a case when the number of potential precursors is expected to be high. This is the case of the metabolic network of the endosymbiotic bacterium *Carsonella ruddii*. Indeed, this bacterium lives inside specialised cells of psyllids, phloem sap-feeding insects. *Carsonella ruddii* has the most reduced among known metabolic networks [7], hence the expected high number of potential precursors for its essential metabolites, such as the amino acids the bacterium provides to its host, who is not capable of producing them. Yet recently, the analysis of the bacterium genome [9] showed that half of the pathways involved in the biosynthesis of essential amino acids have been completely or partially lost. The bacterium therefore requires possibly many nutrients from the host to enable it to fill in these "holes". As an illustration, we chose to search for the precursors of one such essential amino acid, the arginine whose metabolic pathway appears to be complete in the bacterium.

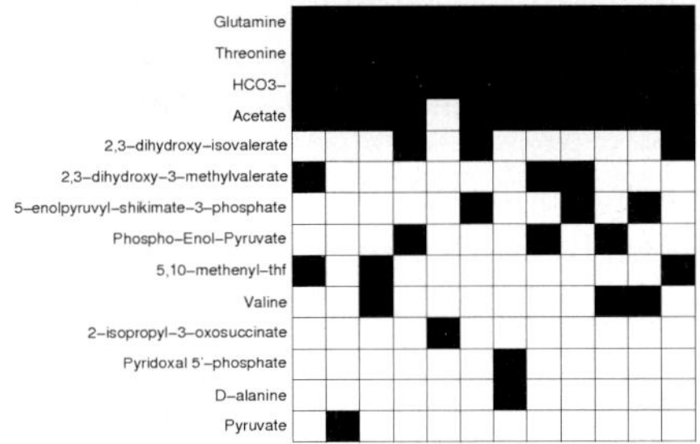

Fig. 3. The sets of precursors of the arginine in *Carsonella ruddii*

Data. We built a metabolic network containing 130 compounds and 71 reactions, 16 of which are defined as irreversible. As in the previous example, the boostrapping compounds defined in [8] were eliminated from the network.

Results. We found 12 minimal precursor sets for arginine. The results are presented in Figure 3. One interesting thing in these results is that glutamine, threonine, and ion bicarbonate are present in all solutions. Glutamine and ion bicarbonate are involved in reaction 6.3.5.5, which represents an essential step in the arginine biosynthesis pathway, as described in METACYC [1]. Threonine as precursor of arginine deserves to be discussed. The path between threonine and arginine goes through two different metabolic pathways by reversible reactions to produce aspartate, a key metabolite of the arginine biosynthesis. Interestingly, those two metabolic pathways (the "threonine biosynthesis from homoserine", and the "homoserine biosynthesis" pathways) are traversed in a direction inverse to the one classically indicated in the database for those specific pathways. Of course, this may be an artefact due to an imprecision concerning the direction of the reactions, but it may also mean that those reactions can be used in a direction inverse to the one indicated in the reference metabolic pathways.

6   Conclusion and Perspectives

We proposed the first topology-based exact method to find minimal precursor sets for a set of target metabolites. Despite the complexity of the problem, the method can be applied to genome-scale metabolic networks.

In contrast with previous methods, we deal in a formally clear way with the issue of cycles when searching for precursors in a given metabolic network. To this purpose, we defined the notion of "continuously available" compounds, which either are able to self-regenerate themselves once activated, or to be generated with the aid of self-regenerating metabolites. An implementation of the enumeration algorithm appears to allow finding all minimal precursor sets for networks of the sizes used in the examples in a time ranging from a few seconds to a few hours.

Our analyses show that some concepts would need further refinements. For instance, the assumption that all potential precursors are always in infinite supply from the environment may not be fully realistic from a biologicial point of view. Indeed, some nutrients are always available in some environmental conditions and not available in other conditions. The possibility to define different sets of potential precursors allows to search for the precursors in certain environments but needs some *a priori* biological information about the nutrients available in each condition.

Compressing the replacement tree while building it enables to reduce space while enumerating all solutions. Efficient though this is, we believe better results might be achievable. Both this and the previous problem require to define and deal well with cycles in hypergraphs representing metablic networks.

Acknowledgements. This work was funded by the ANR (Reglis project NT05-3_45205), the ANR and the BBSRC (MetNet4SysBio project ANR-07-BSYS 003 02), the INRIA, the CNRS and the French Ministry of Foreign and European Affairs (STIC-AmSud project) and the Dutch BSIK-BRICKS project. The authors would like also to thank the bioinformatic team at the Genoscope who provided us with the metabolic network data.

References

1. Caspi, R., et al.: MetaCyc: a multiorganism database of metabolic pathways and enzymes. Nucleic Acids Res. 34(Database issue), D511–D516 (2006)
2. Francke, C., Siezen, R.J., Teusink, B.: Reconstructing the metabolic network of a bacterium from its genome. Trends Microbiol. 13(11), 550–558 (2005)
3. Garey, M.R., Johnson, D.S.: Computers and Intractability (A guide to the theory of NP-completeness). W.H. Freeman and Company, New York (1979)
4. Handorf, T., Christian, N., Ebenhöh, O., Kahn, D.: An environmental perspective on metabolism. J. Theor. Biol (November 2007)
5. Kanehisa, M., et al.: From genomics to chemical genomics: new developments in KEGG. Nucleic Acids Res. 34(Database issue), D354–D357 (2006)
6. Keseler, I.M., et al.: EcoCyc: a comprehensive database resource for Escherichia coli. Nucleic Acids Res. 33(Database issue), D334–D337 (2005)
7. Nakabachi, A., et al.: The 160-kilobase genome of the bacterial endosymbiont carsonella. Science 314, 267 (2006)
8. Romero, P.R., Karp, P.: Nutrient-related analysis of pathway/genome databases. In: Pac. Symp. Biocomput., pp. 471–482 (2001)
9. Tamames, J., Gil, R., Latorre, A., Peretó, Silva, F., Moya, A.: The frontier between cell and organelle: genome analysis of candidatus carsonella ruddii. BMC Evol. Biol. 7, 181 (2007)

Boosting the Performance of Inference Algorithms for Transcriptional Regulatory Networks Using a Phylogenetic Approach

Xiuwei Zhang and Bernard M.E. Moret

Laboratory for Computational Biology and Bioinformatics
EPFL (Ecole Polytechnique Fédérale de Lausanne)
and Swiss Institute of Bioinformatics
Lausanne, Switzerland
{xiuwei.zhang,bernard.moret}@epfl.ch

Abstract. Inferring transcriptional regulatory networks from gene-expression data remains a challenging problem, in part because of the noisy nature of the data and the lack of strong network models. Time-series expression data have shown promise and recent work by Babu on the evolution of regulatory networks in *E. coli* and *S. cerevisiae* opened another avenue of investigation. In this paper we take the evolutionary approach one step further, by developing ML-based refinement algorithms that take advantage of established phylogenetic relationships among a group of related organisms and of a simple evolutionary model for regulatory networks to improve the inference of these networks for these organisms from expression data gathered under similar conditions.

We use simulations with different methods for generating gene-expression data, different phylogenies, and different evolutionary rates, and use different network inference algorithms, to study the performance of our algorithmic boosters. The results of simulations (including various tests to exclude confounding factors) demonstrate clear and significant improvements (in both specificity and sensitivity) on the performance of current inference algorithms. Thus gene-expression studies across a range of related organisms could yield significantly more accurate regulatory networks than single-organism studies.

1 Introduction

The widespread use in the life sciences of high-throughput gene-expression experiments has created a strong demand for suitable computational tools to analyze such data [19]. One of the most common analyses is the reconstruction of transcriptional (or gene) regulatory networks, models of the cellular regulatory system that governs transcription. In such networks, nodes are associated with genes or transcription factors, while arcs denote regulation. These networks can be inferred from gene-expression studies using various machine-learning or statistical inference methods. (Friedman [9] gave a survey of inference methods for probabilistic graphical models.) Other models have also been proposed, such as systems of differential equations [7].

Methods to reconstruct transcriptional regulatory networks from gene-expression data include Boolean networks and their generalizations [1,15], Bayesian networks [10]

K.A. Crandall and J. Lagergren (Eds.): WABI 2008, LNBI 5251, pp. 245–258, 2008.
© Springer-Verlag Berlin Heidelberg 2008

and dynamic Bayesian networks (DBNs) [14], differential equations [7,21], and many others. Results from these approaches, however, remain mixed: the high noise level in the data, along with the lack of well defined, realistic models for the regulatory networks, combine with other factors (such as the typically large number of genes tested vs. the small number of test samples—the so-called "tall dataset" problem) to make inference difficult. Researchers recognized early that additional information about gene expression could be used to good effect. Methods were developed to use time-series expression data [5,14,22,24]. Babu and his colleagues recently pioneered an evolutionary approach to the study of regulatory networks in *E. coli* and in *S. cerevisiae* [2,3,4,20]. They posit a simple evolutionary model for regulatory networks, which adds or removes edges to the network, and proceed to investigate how well such a model accounts for the dynamic evolution of the two most studied regulatory networks. Using a similar evolutionary model, Bourque and Sankoff [6] developed an algorithm to improve the inference of cross-species gene networks based on their phylogenetic relationships.

We use a phylogenetic approach to develop algorithms that boost the performance of any chosen network inference method by using phylogenetic information and a simple model of network evolution. We consider a scenario where orthologous regulatory networks have been (separately) inferred for a number of related organisms whose phylogenetic relationships are known. Our algorithms refine these networks by considering all of them at once, within the known phylogeny of the organisms, to produce networks with much higher specificity and sensitivity. Whereas Bourque and Sankoff used a parsimony approach and tightly integrated inference and refinement, our algorithms are formulated within a maximum likelihood (ML) framework and focus solely on refinement, thereby allowing one to use one's preferred network inference method. We test our algorithms on various types of simulated data with different standard network inference approaches, and for various specificity and sensitivity settings, and compare the results with a standard approach. The *receiver-operator characteristic (ROC)* curves for our algorithms consistently dominate those of the standard approaches; under comparable conditions, they also dominate the results from Bourque and Sankoff. We also investigate the source of these improvements to eliminate various confounding factors.

This paper is organized as follows: we provide some background in Sec. 2, present our algorithms in Secs. 3 and 4, describe our experimental design in Sec. 5, and discuss our results in Sec. 6.

2   Background

Our approach relies on placing inferred networks (obtained by one's favorite method) at the leaves of the known phylogenetic tree, reconstructing ancestral regulatory networks within this tree according to a model of network evolution, and propagating ancestral information back down to the leaves to improve the inferred networks. We use inference algorithms based on DBN and on differential equations. For the former, we use Murphy's Bayesian Network Toolbox [17]; for the latter, we use TRNinfer [21]. For ancestral sequence estimation, we use FastML [18].

2.1   DBNs for Network Inference

When DBNs are used to model regulatory networks, an associated structure learning algorithm is used to infer the networks from gene-expression data [8,11,14,15]. The implementation of this algorithm in the Bayesian Network Toolbox provides two optimization functions: an ML score and a Bayesian information criterion (BIC) score. Let D denote the dataset used in learning and G the (structure of the) network; the algorithm using ML scoring aims to return the structure $G^* = \arg\max_G \log Pr(D|G)$. However, transcriptional regulatory networks are typically sparse graphs, so that ML inferences often produce many false positive edges. The BIC score introduces a penalty on the complexity of G,

$$\log Pr(D|G, \hat{\Theta}_G) - 0.5\#G \log N \tag{1}$$

where $\hat{\Theta}_G$ is the ML estimate of parameters for G, N is the number of samples in D, and $\#G$ is the number of free parameters of G. This penalty makes inference more conservative, reducing the number of false positives in the networks inferred by maximizing the BIC score. In practice, heuristic search methods are used, as well as mild restrictions on the structure of the model, the latter aimed at reducing the number of possible network structures—such as a small bound on the maximum indegree of the nodes, a restriction that appears well supported by the data [1,15].

2.2   Differential Equations for Network Inference

Differential equations can describe causal relationships among components in a quantitative manner and are thus well suited to model transcriptional regulatory networks [7,21]. A regulatory system is represented by the equation $dx/dt = f(x(t)) - Kx(t)$, where $x(t) = (x_1(t), \cdots, x_n(t))$ denotes the expression levels of the n genes and K (a matrix) denotes the degradation rates of the genes. The regulatory relationships among genes are then characterized by $f(\cdot)$. [21] produced a tool, TRNinfer, that solves the differential equations by formulating them into linear programming problems.

2.3   ML-Based Reconstruction of Ancestral Nodes

Reconstructing ancestral information in phylogenetic work is typically in the nature of an anchoring step in the computation, particularly in parsimony-based approaches. When we have high confidence in the tree and the edge lengths are modest, however, an ML approach to ancestral inference can yield accurate results; FastML [18], using a user-specified character substitution matrix, infers labels for the internal nodes (on a site-by-site basis) that maximize the overall likelihood of the tree. The algorithm was initially designed for protein sequences, but can be used for any type of sequences with a suitable substitution matrix.

Fix a site, i.e., a character position in the sequence. Let i denote a node in the tree, l_i the length of the edge between node i and its parent, and a the value of a character at some node, chosen from a given set S of possible character values. For each character a at each node i, we maintain two variables:

- $L_i(a)$: the likelihood of the best reconstruction of the subtree with root i given that the parent of i is assigned character a.
- $C_i(a)$: the optimal character assigned to i given that its parent is assigned as a.

Finally, let π_a denote the initial distribution of character a and $p_{ab}(l)$ the probability of substitution of a with b along an edge of length l. For simplicity, assume that the given tree is binary; then our adaptation of the **FastML** algorithm carries out these steps:

1. If leaf i has character b, then, for each $a \in S$, set $C_i(a) = b$ and $L_i(a) = p_{ab}(l_i)$.
2. If i is an internal node and not the root, its children are j and k, and it has not yet been processed, then, for each $a \in S$, set
    - $L_i(a) = \max_{c \in S} p_{ac}(l_i) \times L_j(c) \times L_k(c)$
    - $C_i(a) = \arg\max_{c \in S} p_{ac}(l_i) \times L_j(c) \times L_k(c)$
3. If there remain unvisited nonroot nodes, return to Step 2.
4. If i is the root node, with children j and k, assign it the value $a \in S$ that maximizes $\pi_a \times L_j(a) \times L_k(a)$.
5. Traverse the tree from the root, assigning to each node its character by $C_i(a)$.

3   Approach

The input is a set of gene-expression data matrices, for several related organisms, along with a known phylogeny (with edge lengths) for this group of organisms. (Such phylogenies are typically well established though the edge lengths remain to be explored.) The first step is simply to run one's preferred algorithm for regulatory network inference, independently on each of the data matrices; in this study, we use two types of inference algorithms, respectively based on DBN and differential equations. The resulting networks are used to label the corresponding leaves of the phylogeny. We encode a network by the concatenation of the rows of its adjacency matrix—every code thus represents a valid network. Note that the initial networks themselves are the real inputs to our algorithm; we use the gene-expression data stage in our tests solely in order to enhance the verisimilitude of our simulations.

We then use our adaptation of the **FastML** algorithm to infer ancestral networks, which in turn are used to refine the sequences at the leaves. We present below two algorithms to carry out this refinement, both based on the intuition (verified in simulations) that ancestral sequences are more accurate than those at the leaves, but only up to some height in the tree—as distant ancestral sequences suffer from the inference errors of **FastML**. The two middle steps can be iterated: starting from the newly refined networks, we can once again infer ancestral networks and use the results to refine the leaves.

We realize that edge lengths obtained from an analysis of the sequences of (typically) a few genes need not reflect the amount of evolution in the regulatory networks—while both evolved on the same tree, their respective rates of evolution could differ considerably. As we still lack the knowledge required to formulate a more precise model of network evolution, using the same edge lengths is just the neutral choice.

4   Methods

4.1   Inferring the Initial Networks

In our study, we want to examine the ROC curves and so need to be able to trade off specificity and sensitivity. To this end, we modify the inference method based on DBN by generalizing Eq. 1 with a *penalty coefficient* k_p to adjust the penalty:

$$\log Pr(D|G, \hat{\Theta}_G) - k_p \# G \log N \qquad (2)$$

where k_p varies from 0 to 0.5. With $k_p = 0$, we have the ML score; with $k_p = 0.5$, Eq. 2 reduces to Eq. 1. For the TRNinfer algorithm, the parameter that it provides to adjust the sparseness of the networks does not afford sufficient control to generate sparse enough networks. We thus supplement it by applying different thresholds to the output connection matrix to choose final edges. We shall refer to these modified inference methods as *DBI* for that based on the DBN model and as *DEI* for that based on TRNinfer.

4.2   Inferring the Ancestral Sequences

In this study our adjacency matrices are binary, with a 1 in the (i, j) entry denoting an edge from node i to node j. We use binary matrices for simplicity's sake: generalization to weighted matrices is immediate and, indeed, the additional information present in a weighted matrix should further improve the results. The data used by FastML are thus:

- the proportions of 0s and 1s in the networks, $\Pi = \begin{pmatrix} \pi_0 & \pi_1 \end{pmatrix}$
- the topology of the phylogenetic tree;
- the *edge length* l_e of each edge e, i.e., the number of changes along this edge;
- for each edge length l_e, its corresponding substitution matrix, $P_s(l_e)$, which represents the mutation probability between 0 and 1.

The substitution matrices depend on edge length: the longer the edge, the higher the mutation probabilities. We choose a $P_s(1)$ for edge length 1 and calculate $P_s(l_e)$ for $l_e \geq 2$ using an exponential distribution, $P_s(l_e) = P_s^{l_e}(1)$.

4.3   Refining the Leaves

The underlying principle is simple: phylogenetically close organisms are likely to have similar regulatory networks; thus independent network inference errors at the leaves get corrected in the ancestral reconstruction process. Obviously, however, if too much evolution occurred, the ancestral reconstruction process itself generates errors. Thus a crucial aspect of our algorithm is how to use ancestral networks at various heights above the leaves to refine the leaves. We ran large series of experiments under various conditions (not shown); all showed an expected increase in accuracy when moving to the parents of the leaves, eventually replaced by a decrease when moving too far above the leaves. On the basis of our results, we chose to use only the immediate parents of the leaves for refinement—but note that these parents are themselves the product of a global ML inference and thus reflect the structure of the entire phylogeny.

4.4   A Fast Oblivious Refinement Algorithm

Our first algorithm, *RefineFast*, is designed to run quickly; it reposes complete trust in the networks associated with the parents of the leaves, using them to replace, rather than refine, the leaf networks.

1. From the current leaves, infer ancestral nodes using FastML.
2. For each leaf, pick its parent and evolve it (according to the length of the edge to the leaf and its substitution matrix) to generate a new child.
3. Use these new children to replace the old leaves.
4. Repeat Steps 1–3 until the total size of the leaf networks stabilizes.

We can use the same substitution matrices $P_s(l_e)$ in both Step 1 and Step 2, but choosing different substitution matrices can accelerate convergence.

The algorithm is deliberately oblivious: it uses the original networks only in the ancestral reconstruction, after which it replaces them with a sample network drawn from the distribution of possible children of the parent. When the original networks are noisy (a common occurrence), this simplistic procedure does quite well.

4.5   A Nonoblivious Refinement Algorithm

To use the information still present in the original leaf networks in the refinement step, we developed an ML-based refinement algorithm, *RefineML*. To use the existing leaf sequences, we assign each site of each leaf a belief (confidence) coefficient, k_b, which varies between 0.5 and 1. In the DBN framework, these coefficients can be calculated from the *conditional probability table* (CPT) parameters of the predicted networks. For each leaf i, we calculate the variables $L_i(a)$ and $C_i(a)$, as defined in Sec. 2.3, where a is the character value of the parent of leaf i inferred by FastML. Fix a site; then, using the notation introduced in Sec. 2.3, the *RefineML* algorithm can be described as follows:

1. Learn the CPT parameters for the leaf networks reconstructed by the base inference algorithm and calculate the belief coefficient k_b for every site.
2. From the current leaves, infer ancestral sequences using FastML.
3. For each leaf i with value b, let $Q_i(c) = k_b$ if $b = c$, $1 - k_b$ otherwise; then set
   - $L_i(a) = \max_{c \in S} p_{ac}(l_i) \times Q_i(c)$
   - $C_i(a) = \arg\max_{c \in S} p_{ac}(l_i) \times Q_i(c)$
4. For each leaf i, assign its most likely character from the variable $C_i(a)$.

5   Experimental Design

In order to evaluate the accuracy gains provided by our boosting algorithms, we need simulated data, as they allow us to control the parameters and, more importantly, to get an absolute assessment of accuracy. Ideally, such simulation should be complemented by applications to real data, but such data currently exist only for tiny examples— indeed, we hope that our positive results will incite researchers to collect such data on a larger scale. Simulations may lack realism and may also introduce a systematic bias in the results. We cannot assess the severity of the first problem, but we take specific precautions against systematic bias, in both the simulations and the analysis.

We generate test data from three pieces of information: the phylogenetic tree, the network at the root, and the substitution matrix (which is in turn influenced by the evolutionary rate). We use a wide variety of phylogenetic trees from the literature (of modest sizes: between 20 and 60 taxa) and several choices of root networks, the latter variations on part of the yeast network from the KEGG database [13], as also used by Kim *et al.* [14]; we also explore a wide range of evolutionary rates. Our networks are of modest size, with 16 genes each—this selection makes the gene-expression tables less "tall" and thus, at least in principle, less prone to generate errors in reconstruction, thus presenting a more challenging case for a boosting algorithm.

We need quantitative relationships in the networks in order to generate gene-expression data from each network. Therefore, in the data generation process, we use adjacency matrices with signed weights. Weight values are assigned to the root network, yielding a weighted adjacency matrix $A = (a_{ij})$. We can obtain the adjacency matrix for its child, $A' = (a'_{ij})$, by mutating A according to the substitution matrix and repeating as we traverse down the tree to obtain weighted adjacency matrices at the leaves. In other words, we evolve the weighted networks down the tree according to the model parameters—standard practice in the study of phylogenetic reconstruction [12,16].

To generate gene-expression data from the weighted networks, we use both Yu's GeneSim [23] and *DBNSim*, our own design based on the DBN model.

5.1   Gene-Expression Data Generated by DBNSim

For *DBNSim*, we follow [15], using binary gene-expression levels, where 1 and 0 indicate that the gene is, respectively, *on* and *off*. Denote the expression level of gene g_i by x_i, $x_i \in \{0,1\}$; if m_i nodes have arcs directed to g_i in the network, let the expression levels of these nodes be denoted by the vector $y = y_1 y_2 \cdots y_{m_i}$ and the weights of their arcs by the vector $w = w_1 w_2 \cdots w_{m_i}$. From y and w, we can get the conditional probability $Pr(x_i|y)$. Once we have the full parameters of the leaf networks, we generate simulated time-series gene-expression data. At the initial time point, the expression level of gene g_i is generated by the initial distribution $Pr(x_i)$; at time t, its expression level is generated based on y at time $t-1$ and the conditional probability $Pr(x_i|y)$.

5.2   Gene-Expression Data Generated by GeneSim

GeneSim [23] can produce simulated gene-expression values for a given weighted network as well as generate arbitrary network structures. In contrast to our *DBN-Sim* method, GeneSim gives continuous gene-expression levels. Denoting the gene-expression levels of the genes at time t by the vector $x(t)$, the values at time $t + 1$ are calculated according to $x(t + 1) = x(t) + (x(t) - z)C + \varepsilon$, where C is the weighted adjacency matrix of the network, the vector z represents *constitutive expression values* for each gene, and ε models noise in the data. The values of $x(0)$ and $x_i(t)$ for those genes without parents are chosen uniformly at random from the range $[0, 100]$, while the values of z are all set to 50. The term $(x(t) - z)C$ represents the effect of the regulators on the genes; this term needs to be amplified for the use of *DBI*, because of the required discretization. We use a factor k_e with the regulation term (set to 7 in our experiments), yielding the new equation $x(t + 1) = x(t) + k_e(x(t) - z)C + \varepsilon$.

5.3   Testing under Various Methods

With two data generation methods, *DBNSim* and GeneSim, and two network inference algorithms as our base algorithms, *DBI* and *DEI*, we can conduct experiments with different combinations of data generation methods and inference algorithms to verify that our boosting algorithms work under all circumstances. First, we use different data generation methods with the same inference algorithm. Since the binary gene-expression data generated by *DBNSim* does not fit *DEI*, we use *DBNSim* and GeneSim to generate data for *DBI*. We generate 200 time points for each gene-expression matrix, running the generation process 10 times to obtain mean and standard deviation. Second, we apply *DBI* and *DEI* to datasets generated by GeneSim to infer the networks. Since *DEI* does not accept large datasets (with many time points), here we used smaller datasets than the previous group of experiments with 75 time points, yielding expression level matrix of size 16×75. We run the generation process 20 times for each choice of tree structure and parameters and calculate mean and standard deviation. Finally, we conduct experiments with various evolutionary rates.

5.4   Comparing with the Bourque and Sankoff Approach

Bourque and Sankoff's algorithm [6], thereafter the B&S algorithm, also uses phylogenetic information to improve the inference of gene networks. We therefore conduct experiments, using continuous data, to compare our approach to theirs.

5.5   Where Is the Important Information?

Although we use only the direct parents to refine the leaves at each iteration, the leaves receive information from the whole tree, since the FastML algorithm assigns states to every internal node based on global information. We claim that the use of this global information is necessary. To verify this claim, we build a variation of our algorithms, that we call *RefineLocal*, where the ancestral reconstruction stops once the parents of leaves are reached. The resulting ancestral reconstruction, in other words, is now limited to exactly the parts of the tree used in the leaf refinement. *RefineLocal* works with both *RefineFast* and with *RefineML*, since it does not alter the refinement phase of the algorithm.

Part of the improvement is due to noise averaging, taking advantage of the independence in errors among the leaf networks. We claim that noise averaging not based on the correct phylogeny cannot produce the type of improvement we see. To verify this claim, we build a procedure that we call *RefineRandomTree*, which runs our full refinement procedures (either one), but does it on a tree where the initial inferred networks were randomly assigned to leaves. Since the tree topology is unchanged, the averaging effect over the data remains globally similar, but the phylogenetic relationships are destroyed. We run 100 such randomized tests and report the mean behavior.

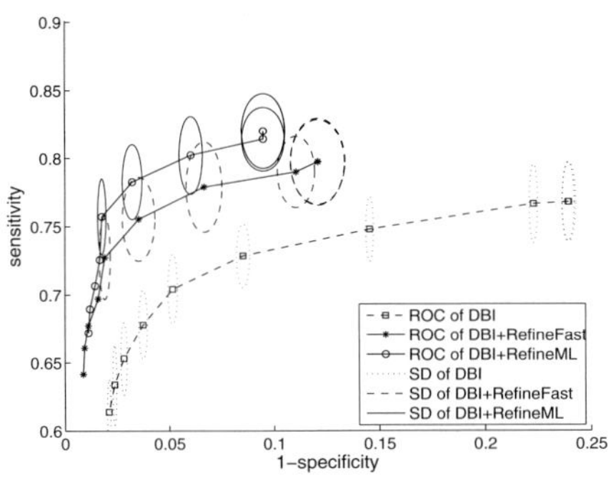

(a) *Results on* T1, *with standard deviations of sensitivity and specificity*

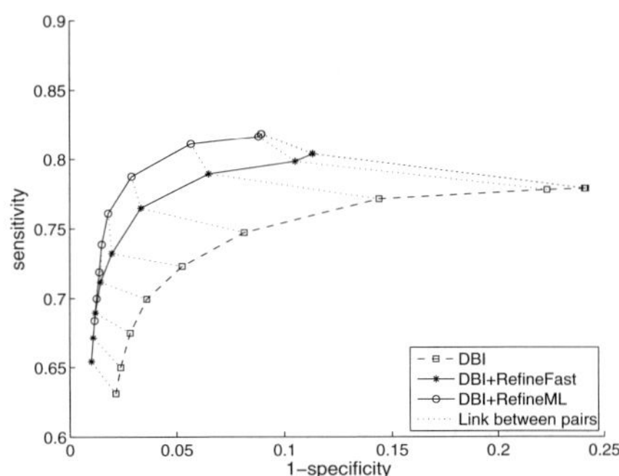

(b) *Results on* T2, *with dotted lines indicating matched penalty coefficients*

Fig. 1. ROC curves for DBI and boosting algorithms on the datasets generated by DBNSim

5.6   Measurements

We want to examine the predicted networks at different levels of sensitivity and specificity. For *DBI*, on each dataset, we apply different penalty coefficients to predict regulatory networks, from 0 to 0.5, with an interval of 0.05, which results in 11 discrete penalty coefficients. For each penalty coefficient, we apply *RefineFast*, *RefineML*, *RefineLocal*, and *RefineRandomTree* on the predicted networks. For *DEI*, we also choose 11 thresholds for each predicted weighted connection matrix to get networks on various sparseness levels. For each threshold, we apply *RefineFast*, *RefineLocal*, and *RefineRandomTree* on the predicted networks. We measure specificity and sensitivity to evaluate the performance of the algorithms and plot the values, as measured on the results for various penalty coefficients (for *DBI*) and thresholds (for *DEI*) to yield ROC curves. Recall that in such plots, the larger the area under the curve, the better the results.

6   Results and Analysis

Space constraints prevent us from showing more than a sample of our results. We show results on two representative trees: tree *T1* has 41 nodes on 6 levels and is better balanced than tree *T2*, which has 37 nodes on 7 levels. Both trees were generated with an expected evolutionary rate of 2.2 events (gain or loss of a regulatory arc in the network) per edge and resulting leaf networks have from 23 to 38 edges.

6.1   On Boosting under Different Experimental Settings

Different gene-expression data generation methods, same inference algorithm.
Fig. 1 shows the average performance of *RefineFast*, *RefineML*, and *DBI* on 10 noiseless datasets generated by *DBNSim* on trees *T1* (left) and *T2* (right). Throughout the range of parameters, our two algorithms clearly dominate *DBI*, with *RefineML* also dominating the simpler *RefineFast*: for every penalty coefficient, both sensitivity and specificity are improved from *DBI* to *RefineFast* and further improved from *RefineFast* to *RefineML*,

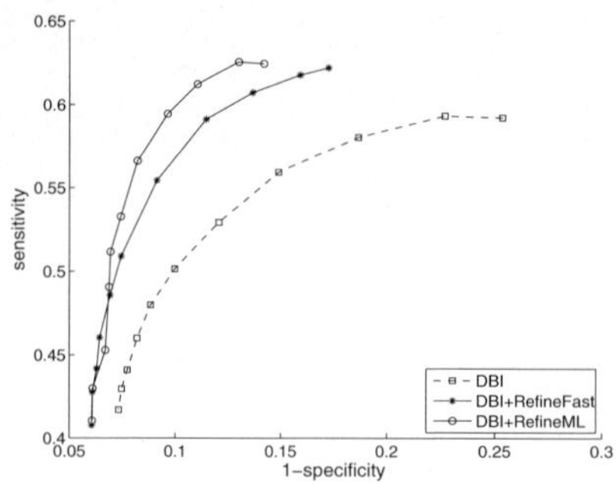

Fig. 2. ROC curves for DBI and boosting algorithms on the datasets generated by GeneSim

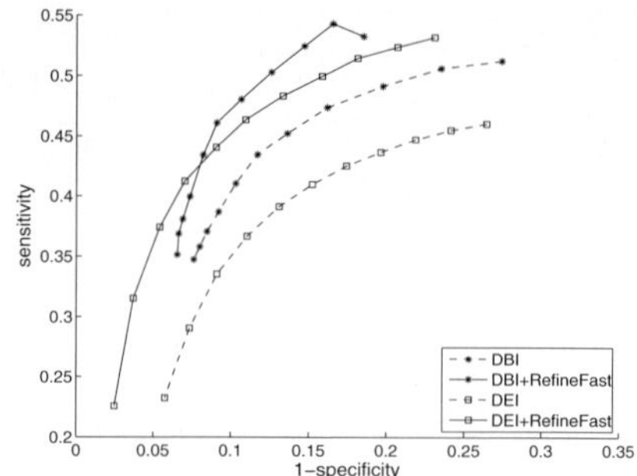

Fig. 3. Results for DBI and DEI with Refine-Fast boosting on GeneSim generated datasets

as easily seen on the right. Sample standard deviations of sensitivity and specificity for these three methods on the noiseless datasets on *T1* are shown as ellipses, the loci of one standard deviation around each point. The separation between the curves is almost always larger than the standard deviations, so that our assertions of dominance of one method over another hold, not only on average, but also in the vast majority of cases. Also, as this figure demonstrates, the boosting effect remains similar on different phylogenies—and so we present results only on *T1* hereafter.

All three algorithms behave on the noisy datasets much as on the noiseless ones. Our refinement algorithms yield more improvement on the noisy datasets, which are closer to the real data and thus cause more difficulties for *DBI* methods, yielding a larger margin for improvement. We thus show results for noiseless datasets only, as the level of improvement caused by our algorithms can only increase as the noise level in the data increases. Fig. 2 shows the results of the three algorithms on the noiseless datasets generated by GeneSim on *T1*. The boosting effects are much the same as seen in Fig. 1, but it is clear that the *DBI* base algorithm does worse on the datasets generated by GeneSim than on those generated by *DBNSim*, as might be expected.

Different inference algorithms, same gene-expression data generation method. The datasets used in this experiment are generated by GeneSim. Fig. 3 shows the ROC curves of *DBI* and *DEI*, along with *RefineFast* boosting, on the same datasets; the refinement algorithm clearly dominates the base algorithms.

Different evolutionary rates. The expected evolutionary rate (average edge length) was fixed in all experiments presented above. High rates of evolution cause various difficulties in phylogenetic reconstruction; we thus expect our method to become less effective as evolutionary rates increase. To study this problem, we conducted experiments on tree *T1* with a root network of 16 nodes and 24 edges, using different evolutionary rates to generate the leaf networks. Fig. 4 shows ROC curves for *RefineML* and *DBI* with evolutionary rates of 2.32, 4.76 and 6.67 on noiseless datasets. The loss in performance as the rate of evolution increases is clear for both methods; since *DBI* itself suffers (perhaps because some networks produced in the simulation violate implicit assumptions), the loss in performance of *RefineML* is a combination of worsened

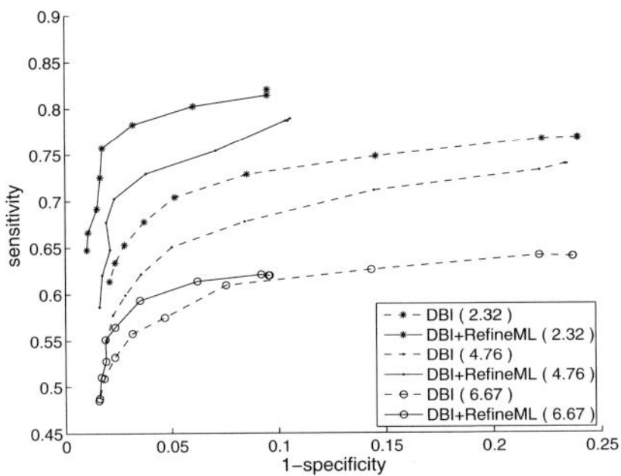

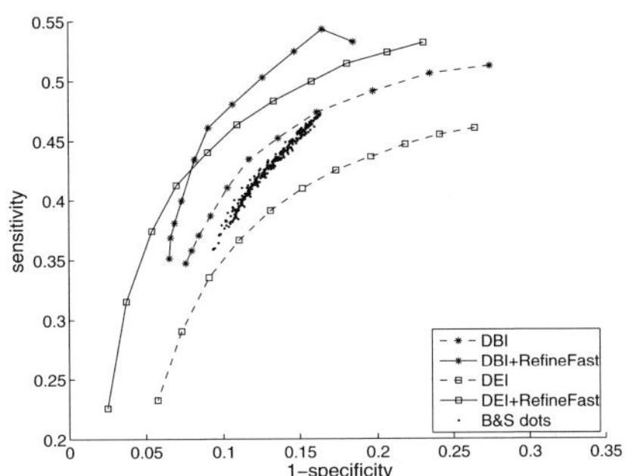

Fig. 4. ROC curves for DBI and RefineML under various evolutionary rates

Fig. 5. Performance for B&S and RefineFast based on DBI and DEI

leaf networks returned by *DBI* and worsened ancestral reconstruction by FastML. Yet boosting is evident in all cases and performance remains excellent at the very high evolutionary rate of 4.76: most paths from the root to a leaf in the tree have 5 edges and so, at that rate, have an expected length of 23.5, so that the expected number of changes from the root network almost equals the number of edges of that network—a very challenging problem and one that is remarkably well solved here.

6.2   On Performance with Respect to the B&S Algorithm

Since B&S requires continuous time-series gene-expression data, we use the same datasets, generated by GeneSim, as in Fig. 3. Fig. 5 presents the performance of B&S and *RefineFast* based on both *DBI* and *DEI*. The results of B&S are shown as a cloud of points, obtained under different parameter settings. B&S does better than plain *DEI*, but is clearly dominated by our *RefineFast* based on *DEI*, meaning that our refinement algorithm gains more improvement than B&S does.

6.3   On Applying ML Globally

We described earlier *RefineLocal*, a variant of our algorithms that infers ancestral networks only for the part of the tree that is used in the refinement phase. We use this algorithm to show that the improvement wrought in the leaves by our algorithms uses the phylogenetic information of the whole tree, not just the information present in the subforest induced by direct parents of leaves. Fig. 6(a) compares the performance of *RefineFast* with that of its localized version on noiseless datasets generated by *DBN-Sim* (the same datasets as in Sec. 6.1), while Fig. 6(b) does the same for *RefineML* on the same datasets. The plots are very similar: *RefineLocal* is clearly worse than the original algorithms, especially in terms of sensitivity. In fact, *RefineLocal* based on *RefineFast* does worse than *DBI*—due to the fact that the ancestral inference procedure introduces significant additional errors when limited to small subtrees. On the other hand, *RefineLocal* based on *RefineML* outperforms *DBI*—indicating that there is significant information present in the leaves, independent of the ancestral reconstruction.

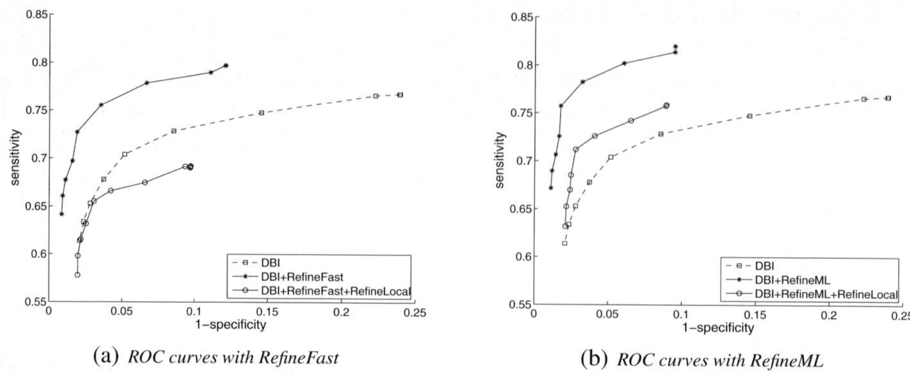

(a) *ROC curves with RefineFast*   (b) *ROC curves with RefineML*

Fig. 6. ROC curves for DBI and RefineLocal, showing RefineFast (left) and RefineML (right)

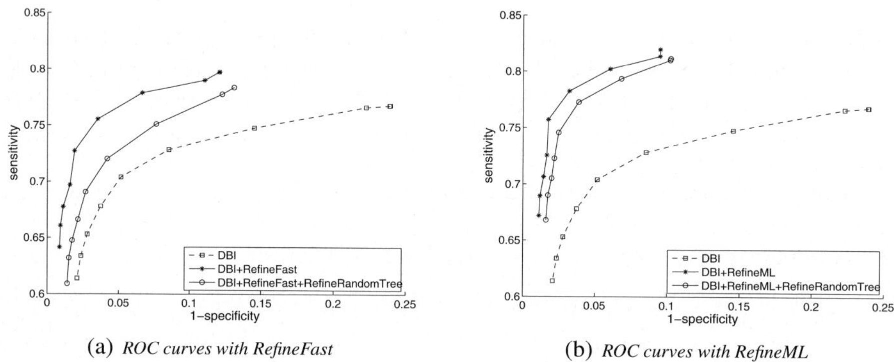

(a) *ROC curves with RefineFast*   (b) *ROC curves with RefineML*

Fig. 7. ROC curves for DBI and RefineRandomTree, with RefineFast (left) and RefineML (right)

6.4   On Phylogenetic Information

In Sec. 5.5 we introduced *RefineRandomTree*, which carries out our full algorithms, but on a tree where the leaves have been reshuffled randomly. Its purpose is to demonstrate that the improvements we observe are not due entirely to noise averaging among the leaf networks. Fig. 7 compares the performance of *RefineFast* (left) and *RefineML* (right) run on the correct phylogenetic tree with the average performance (over 100 runs) of the same algorithm run after randomly reshuffling leaf labels. Both *RefineFast* and *RefineML* show clearly worse performance on the reshuffled trees than on the correct one. The results on the shuffled trees are still better than the base algorithm *DBI*, which shows the error averaging effect of the trees. However, this improvement depends on the performance of the base algorithm: in other experiments (not shown) with larger gene-expression datasets, where *DBI* does better, *RefineFast* on the shuffled trees does not outperform *DBI*, while *RefineML* with shuffled trees does. Overall, the results demonstrate the value of correct phylogenetic data, the value of the information present in the original leaf networks, and the averaging effect of the trees.

7 Conclusions and Future Work

We described algorithms for boosting the accuracy of regulatory network inference on related organisms using a phylogenetic approach and gave experimental results demonstrating the effectiveness of these algorithms. Our algorithms yield significant gains in both sensitivity and specificity under all conditions and can, in principle, be used with any favorite network inference tool and any favorite phylogenetic reconstruction algorithm. Further assessments of the sensitivity of our methods to the accuracy of the branch length estimates and of their ability to correct for systematic errors in gene expression levels are planned.

Our approach requires comparable gene-expression data for orthologous regulatory networks, or the networks themselves, for a fair number of organisms; while cost and technical feasibility are no longer major obstacles to the accumulation of such data, ensuring comparability remains challenging, although we hope that our promising results will encourage researchers to consider collecting such data.

Many improvements to our approach are clearly possible, from refined models of network evolution to more precise handling of ancestral networks. Our approach to the use of phylogenetic information in network inference is not limited to transcriptional regulatory networks: it can be used, with suitable adaptations, for other types of signalling and metabolic networks.

References

1. Akutsu, T., Miyano, S., Kuhara, S.: Identification of genetic networks from a small number of gene expression patterns under the Boolean network model. In: Proc. 4th Pacific Symp. on Biocomputing (PSB 1999), pp. 17–28. World Scientific, Singapore (1999)
2. Babu, M.M., Luscombe, N.M., Aravind, L., Gerstein, M., Teichmann, S.A.: Structure and evolution of transcriptional regulatory networks. Curr. Opinion in Struct. Bio. 14(3), 283–291 (2004)
3. Babu, M.M., Teichmann, S.A.: Evolution of transcription factors and the gene regulatory network in Escherichia coli. Nucleic Acids Res. 31(4), 1234–1244 (2003)
4. Babu, M.M., Teichmann, S.A., Aravind, L.: Evolutionary dynamics of prokaryotic transcriptional regulatory networks. J. Mol. Bio. 358(2), 614–633 (2006)
5. Bar-Joseph, Z.: Analyzing time series gene expression data. Bioinformatics 20(16), 2493–2503 (2004)
6. Bourque, G., Sankoff, D.: Improving gene network inference by comparing expression time-series across species, developmental stages or tissues. J. Bioinform. Comput. Biol. 2(4), 765–783 (2004)
7. Chen, T., He, H.L., Church, G.M.: Modeling gene expression with differential equations. In: Proc. 4th Pacific Symp. on Biocomputing (PSB 1999), pp. 29–40. World Scientific, Singapore (1999)
8. Conant, R.C.: Extended dependency analysis of large systems. Int'l J. General Systems 14(2), 97–141 (1988)
9. Friedman, N.: Inferring cellular networks using probabilistic graph models. Science 303(5659), 799–805 (2004)
10. Friedman, N., Linial, M., Nachman, I., Pe'er, D.: Using Bayesian networks to analyze expression data. J. Comput. Bio. 7(3-4), 601–620 (2000)

11. Friedman, N., Murphy, K.P., Russell, S.: Learning the structure of dynamic probabilistic networks. In: Proc. 14th Conf. on Uncertainty in Art. Intell. UAI 1998, pp. 139–147 (1998)
12. Hillis, D.M.: Approaches for assessing phylogenetic accuracy. Syst. Bio. 44, 3–16 (1995)
13. Kanehisa, M., Goto, S., Hattori, M., Aoki-Kinoshita, K.F., Itoh, M., Kawashima, S., Katayama, T., Araki, M., Hirakawa, M.: From genomics to chemical genomics: new developments in KEGG. Nucleic Acids Res. 34, D354–D357 (2006)
14. Kim, S.Y., Imoto, S., Miyano, S.: Inferring gene networks from time series microarray data using dynamic Bayesian networks. Briefings in Bioinformatics 4(3), 228–235 (2003)
15. Liang, S., Fuhrman, S., Somogyi, R.: REVEAL, a general reverse engineering algorithm for inference of genetic network architectures. In: Proc. 3rd Pacific Symp. on Biocomputing (PSB 1998), pp. 18–29. World Scientific, Singapore (1998)
16. Moret, B.M.E., Warnow, T.: Reconstructing optimal phylogenetic trees: A challenge in experimental algorithmics. In: Fleischer, R., Moret, B.M.E., Schmidt, E.M. (eds.) Experimental Algorithmics. LNCS, vol. 2547, pp. 163–180. Springer, Heidelberg (2002)
17. Murphy, K.P.: The Bayes net toolbox for MATLAB. Computing Sci. and Statistics 33, 331–351 (2001)
18. Pupko, T., Pe'er, I., Shamir, R., Graur, D.: A fast algorithm for joint reconstruction of ancestral amino acid sequences. Mol. Bio. Evol. 17(6), 890–896 (2000)
19. Slonim, D.K.: From patterns to pathways: gene expression data analysis comes of age. Nature Genetics 32, 502–508 (2002)
20. Teichmann, S.A., Babu, M.M.: Gene regulatory network growth by duplication. Nature Genetics 36(5), 492–496 (2004)
21. Wang, R., Wang, Y., Zhang, X., Chen, L.: Inferring transcriptional regulatory networks from high-throughput data. Bioinformatics 23(22), 3056–3064 (2007)
22. Xu, R., Hu, X., Wunsch, D.C.: Inference of genetic regulatory networks from time series gene expression data. In: Proc. IEEE Int'l Joint Conf. on Neural Networks, vol. 2, pp. 1215–1220. IEEE Press, Piscataway (2004)
23. Yu, J., Smith, V.A., Wang, P.P., Hartemink, A.J., Jarvis, E.D.: Advances to Bayesian network inference for generating causal networks from observational biological data. Bioinformatics 20(18), 3594–3603 (2004)
24. Zhao, W., Serpedin, E., Dougherty, E.R.: Inferring gene regulatory networks from time series data using the minimum length description principle. Bioinformatics 22(17), 2129–2135 (2006)

Fast Bayesian Haplotype Inference
Via Context Tree Weighting

Pasi Rastas, Jussi Kollin, and Mikko Koivisto[*]

Department of Computer Science & HIIT Basic Research Unit
P.O. Box 68 (Gustaf Hällströmin katu 2b)
FIN-00014 University of Helsinki, Finland
`firstname.lastname@cs.helsinki.fi`

Abstract. We present a new, Bayesian method for inferring haplotypes
for unphased genotypes. The method can be viewed as a unification
of some ideas of variable-order Markov chain modelling and ensemble
learning that so far have been implemented only separately in some of
the state-of-the-art methods. Specifically, we make use of the Context
Tree Weighting algorithm to efficiently compute the posterior probability of any given haplotype assignment; we employ a simulated annealing scheme to rapidly find several local optima of the posterior; and we
sketch a full Bayesian analogue, in which a weighted sample of haplotype
assignments is drawn to summarize the posterior distribution. We also
show that one can minimize in linear time the average switch distance,
a popular measure of phasing accuracy, to a given (weighted) sample of
haplotype assignments. We demonstrate empirically that the presented
method typically performs as well as the leading fast haplotype inference
methods, and sometimes better. The methods are freely available in a
computer program BACH (Bayesian Context-based Haplotyping)

1 Introduction

Large-scale genotyping – that is, measuring the genomic variation at hundreds to
millions of marker loci for tens to thousands of subjects – has become a common
approach to the genetic mapping of complex traits and to the discovery of the
genomic structure and variation in general. The abundance of single-nucleotide
polymorphisms (SNPs) in the human genome, in particular, may enable powerful
association analyses as well as detecting larger chromosomal rearrangements,
such as deletion [1,2,3] and inversion [4] polymorphisms, using efficient statistical
and computational methods.

Crucial to methods that utilize multilocus genotypes is the modelling of
haplotypes, the maternal and paternal material that constitute a genotype. As haplotypes tend to be inherited as large blocks, broken only occasionally by recombina-
tions, they carry important information about ancient single-point mutations and
larger-scale structural changes. Unfortunately, the usual genotyping technologies

---

[*] Supported by the Academy of Finland under Grant 109101.

K.A. Crandall and J. Lagergren (Eds.): WABI 2008, LNBI 5251, pp. 259–270, 2008.
© Springer-Verlag Berlin Heidelberg 2008

cannot reliably measure the haplotypes *per se*, but only the unphased genotypes, that is, for any two loci it cannot be determined which of the measured alleles belong to the same haplotype, be it maternal or paternal. Thus, haplotypes appear as central latent variables in genotype analysis methods.

While the precise use of haplotypes should depend on the particularities of the data analysis problem at hand, it is plausible to test a model of haplotypes in the somewhat isolated *problem of haplotype inference:* given a set of multilocus genotypes, find good estimates of the underlying haplotype pairs, called a *haplotype assignment.* Indeed, this problem has attracted much attention in the recent years, and a variety methods have been proposed.

The state-of-the-art methods for haplotype inference are based on various ideas. Since Clark's [5] rule-based approach, several likelihood based methods that estimate "independent" haplotype frequencies over a short window of markers using an expectation–maximization (EM) algorithm or using related Gibbs sampling have been presented [6,7,8,9]; a sort of divide and conquer technique, called partition ligation, is commonly employed to handle larger numbers of markers. A more sophisticated method is implemented in HAP [10], based on fitting a perfect or almost perfect phylogeny model to short overlapping marker windows and solving conflicts to optimality using dynamic programming. Of the existing methods, the most accurate is perhaps PHASE [8]. It is a Bayesian Markov chain Monte Carlo method that samples haplotypes from a posterior distribution defined by a mosaic model, in which every haplotype is modelled as a concatenation of fragments extracted (fairly independently) from the other haplotypes. A drawback of PHASE is that it is impractically slow on large genotype samples. Three faster methods based on Hidden Markov models and sophisticated EM algorithms, HIT [11], GERBIL [12], and fastPHASE [13], were developed independently. Of these methods, fastPHASE, which uses an ensemble learning technique, seems the most accurate on average. In another direction, fast and accurate phasing methods have been built on variable-order Markov models that can be efficient in capturing rare higher-order dependencies. Such models are implemented in HaploRec [14] based on fast search for frequently occurring haplotype fragments, and more recently in Beagle [15] based on efficient estimation of related probabilistic automata and an EM-type search.

In this paper, we present a new haplotype inference method that can be viewed as a principled and unified treatment of certain key ideas underlying some of the state-of-the-art methods. Like in PHASE, we take a Bayesian approach in which haplotype inference relies on the posterior distribution of haplotype assignments. However, we model haplotypes using a variable order Markov model similar to those used in HaploRec and Beagle. Instead of fitting the model to given haplotypes, we show that the Bayesian approach of averaging over all the model parameters, including the context trees, can be done efficiently by using the celebrated Context Tree Weighting algorithm (CTW), originally developed for efficient data compression [16]. The remaining challenge is to efficiently sample haplotype assignments proportionally to the posterior, or, alternatively, to maximize the posterior probability. To this end, we consider two kinds local moves in the space of haplotype assignments: a forward sampling of a haplotype pair according to the given genotype and a simpler phase switch operation. Then, in the spirit of the ensemble EM technique of fastPHASE, we employ a simulated annealing procedure to rapidly find several local optima of the posterior distribution. We also sketch an analogous full-Bayesian method based on annealed importance sampling [17], which outputs a sample of independently drawn haplotype assignments, each associated with a real-valued weight. (An efficient implementation of this full-Bayesian method is work-in-progress.) Finally, we show that one can find in linear time a Bayes-optimal haplotype assignment, that is, one that minimizes the average switch distance, a popular measure of phasing accuracy, to a given (weighted) sample of haplotype assignments. This useful observation is technically rather simple and has appeared in other guises earlier: an unweighted, non-Bayesian setting is implemented in fastPHASE [13] and treated more formally recently [18]. An implementation of the methods is freely available in a computer program BACH (**B**ayesian **C**ontext-based **H**aplotyping).

Using the HapMap data [19] and simulated data, we have empirically compared the phasing accuracy of BACH against PHASE, fastPHASE, HAP, HIT, HaploRec, and Beagle. The results agree with earlier observations that PHASE, when computationally feasible, is generally the most accurate. Of the fast methods, HaploRec and BACH performed the best on average. Our study also contributes to mutual comparison of the aforementioned current leading methods, many of which were not included in some earlier comparison studies [20].

2   A Bayesian Variable Order Markov Model

Consider m SNPs labelled in their physical order by the numbers $1, 2, \ldots, m$. We assume that at each SNP in the population there occur two alleles, which we denote by 0 and 1. A haplotype is a sequence $h_1 \cdots h_m$ where $h_j \in \{0,1\}$ is an allele at SNP j. Two haplotypes h and h' determine an unphased genotype $g(h, h') = g = g_1 \cdots g_m$, where each single-SNP genotype g_j takes value 0 if $h_j = h'_j = 0$, value 1 if $h_j = h'_j = 1$, and value 2 otherwise.

We aim at a model that given n genotypes $\boldsymbol{g} - (g^1, \ldots, g^n)$ yields good estimates of the underlying $2n$ haplotypes $\boldsymbol{h} = (h^1, h^2, \ldots, h^{2n-1}, h^{2n})$, called a *haplotype assignment*; we denote the two haplotypes underlying g^i by h^{2i-1} and h^{2i}. We take a Bayesian approach and define a joint probability distribution of $\boldsymbol{g}$ and $\boldsymbol{h}$, from which the posterior distribution of the haplotypes is obtained by conditioning on $\boldsymbol{g}$. Based on the posterior distribution, "optimal" estimates of haplotype frequencies as well as individual haplotype pairs can be determined; see Sect. 5.

We begin with a basic decomposition:

$$p(\boldsymbol{h}|\boldsymbol{g}) \propto p(\boldsymbol{h}, \boldsymbol{g}) = p(\boldsymbol{g}|\boldsymbol{h})\, p(\boldsymbol{h})\,.$$

Here the first term is easily specified: it evaluates to 1 if for every i, the genotype g^i is the unique genotype determined by h^{2i-1} and h^{2i}, and to 0 otherwise.

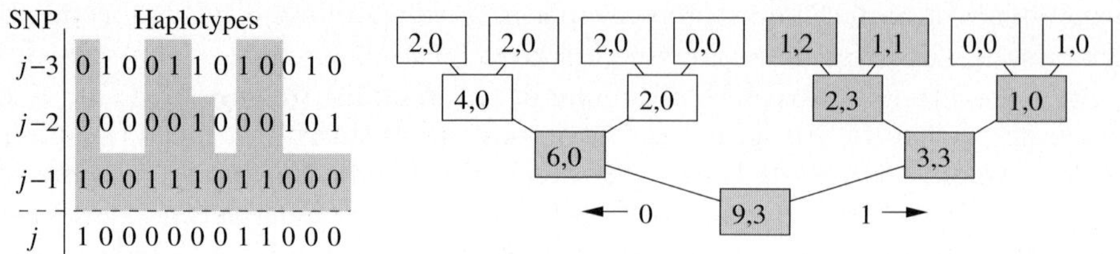

Fig. 1. The full context tree of marker j with maximum depth $D = 3$ for a sample of 12 haplotypes. Each node in the tree corresponds to a suffix. The numbers in each node are the empirical counts of 0s and 1s for marker j following the corresponding suffix. The gray nodes form a smaller context tree. The corresponding suffixes in the sample are shown on gray background.

The crucial part is the modelling of the latter term, the joint distribution of the $2n$ haplotypes. We next derive a variable-order Markov model, starting with the chain rule that holds for any probability distribution:

$$p(\boldsymbol{h}) = \prod_{j=1}^{m} p(\boldsymbol{h}_j | \boldsymbol{h}_1, \ldots, \boldsymbol{h}_{j-1}) \, ;$$

here and henceforth $\boldsymbol{h}_j$ denotes the alleles of the haplotypes at SNP j, that is, $\boldsymbol{h}_j = (h_j^1, h_j^2, \ldots, h_j^{2n-1}, h_j^{2n})$. The variable-order Markov model makes the restrictive, yet useful and biologically plausible assumption that $\boldsymbol{h}_j$ depends only on a relatively small context, i.e., the preceding markers. This assumption is reflected with linkage disequilibrium decreasing with marker distances.

We make the usual assumption that these contexts, or suffixes, form the leaves of a *context tree*. A context tree is a rooted tree where a node at depth d is a sequence $u = h_{j-d} \cdots h_{j-1} \in \{0, 1\}^d$; the node is either a leaf or the *parent* of its two *children*, $0u$ and $1u$. Thus, a context tree is uniquely determined by the leaves of the tree. See Fig. 1 for an illustration.

More formally, we specify a context tree by a function c_j that with each partial haplotype $h_1 \cdots h_{j-1}$ associates a suffix haplotype $h_{j-d} \cdots h_{j-1}$, a leaf of the context tree, with some context length d; we denote the set of these suffixes by $S_j = S_j(c_j)$. Let $c_j(\boldsymbol{h}_1 \cdots \boldsymbol{h}_{j-1})$ denote the composition of the suffixes over all the $2n$ partial haplotypes. By marginalizing over the unknown function c_j with respect to its prior distribution, we obtain

$$p(\boldsymbol{h}_j | \boldsymbol{h}_1, \ldots, \boldsymbol{h}_{j-1}) = \sum_{c_j} p(c_j) \, p\big(\boldsymbol{h}_j | c_j(\boldsymbol{h}_1 \cdots \boldsymbol{h}_{j-1})\big) \, . \tag{1}$$

We have implemented a simple prior over the context trees by restricting the maximum depth of a tree to D and assigning a probability $(1/2)^N$ to each tree with N nodes at depth less than D. This prior is the same as in the original CTW method and stems from the idea of a recursive construction of a context tree by either splitting or stopping at each node, independently, with equal probabilities.

The prior for stopping the tree construction could also have been set for each marker separately to reflect the effect of the marker distances.

To complete the model specification, it remains to specify the term $p(\boldsymbol{h}_j|c_j(\boldsymbol{h}_1\cdots\boldsymbol{h}_{j-1}))$. To this end, we associate with each leaf s of the context tree c_j a parameter θ_s that gives the probability that a haplotype has a 1 at marker j, given that its suffix up to marker $j-1$ is s. Considering these parameters independent a priori and assigning each a Beta$(1/2, 1/2)$ prior yields a closed-form expression [21]:

$$p(\boldsymbol{h}_j|c_j(\boldsymbol{h}_1\cdots\boldsymbol{h}_{j-1})) = \prod_{s\in S_j} \rho(a_s, b_s),$$

where a_s and b_s are the counts of haplotypes up to marker j with suffixes $s0$ and $s1$, respectively, and the *leaf score* $\rho(a_s, b_s)$ can be written [21] as

$$\rho(a_s, b_s) = \frac{\Gamma\left(\frac{1}{2} + a_s\right)\Gamma\left(\frac{1}{2} + b_s\right)}{\Gamma\left(\frac{1}{2}\right)\Gamma\left(\frac{1}{2}\right)\Gamma\left(1 + a_s + b_s\right)} = \frac{\frac{1}{2}\cdot\frac{3}{2}\cdot\frac{5}{2}\cdots\left(a_s - \frac{1}{2}\right)\cdot\frac{1}{2}\cdot\frac{3}{2}\cdots\left(b_s - \frac{1}{2}\right)}{(a_s + b_s)!}.$$

We have also experimented with a variant of the leaf score, defined as

$$\tilde{\rho}(a_s, b_s) = \left(\frac{a_s + \frac{1}{2}}{a_s + b_s + 1}\right)^{a_s}\left(\frac{b_s + \frac{1}{2}}{a_s + b_s + 1}\right)^{b_s}.$$

Like $\rho(a_s, b_s)$, this score can be interpreted as the probability of observing a_s suffixes $s0$ and b_s suffixes $s1$, but now with a fixed estimate of the θ_s parameter (rather than integrating θ_s out). Intuitively, $\tilde{\rho}(a_s, b_s)$ may perform better than $\rho(a_s, b_s)$ when the assumed Beta prior fits poorly with the observed data. Somewhat surprisingly, in our preliminary experiments $\tilde{\rho}(a_s, b_s)$ yielded consistently better results than $\rho(a_s, b_s)$, a phenomenon we currently do not fully understand. In Sect. 6 we report results only for $\tilde{\rho}(a_s, b_s)$.

3   The Context Tree Weighting Algorithm

We apply the CTW algorithm [16] – originally developed for compressing a single binary string – to efficiently evaluate the sum over context trees for haplotype (1). The idea is to compute for each possible haplotype suffix s the sum of scores of all possible subtrees rooted at s; denote this sum by ρ_s. If the length of s is the maximum D, then s must be a leaf and ρ_s is set to $\rho(a_s, b_s)$. Otherwise, ρ_s is obtained by averaging over the case that s is a leaf of a context tree and the case that it is has two children, $0s$ and $1s$:

$$\rho_s := \frac{1}{2}\rho(a_s, b_s) + \frac{1}{2}\rho_{0s}\rho_{1s}. \tag{2}$$

It is easy to show [16] that this recurrence yields

$$\rho_\lambda = \sum_{c_j} 2^{-N(c_j)} \prod_{s\in S_j(c_j)} \rho(a_s, b_s) = p(\boldsymbol{h}_j|\boldsymbol{h}_1,\ldots,\boldsymbol{h}_{j-1}),$$

where λ denotes the empty sequence, i.e., the root of a context tree, and $N(c_j)$ is the number of nodes at depth less than D in c_j.

The algorithm can be implemented to run in time $O(nD)$ for a single marker j. First, one builds a trie of substrings starting at $j-1$ backwards, as shown in Fig. 1. For each node s in this trie, the corresponding counts a_s and b_s can be computed according the values on marker j; this takes time $O(nD)$. Similarly, the values $\rho(a_s, b_s)$ for all nodes s can be computed in an incremental fashion in time $O(nD)$. Finally, the recurrence (2) can be solved along the trie in time $O(nD)$. Note that only the haplotype suffixes that are present in the data need to be considered (as $\rho(0,0) = 1$, the subtrees with zero counts will average to 1).

4   Finding Plausible Haplotypes Using Local Search

The posterior probability $p(\boldsymbol{h}|\boldsymbol{g})$ of a haplotype assignment $\boldsymbol{h}$, obtained by averaging over all possible context trees and the involved parameters, provides a Bayesian measure for the plausibility of $\boldsymbol{h}$. Ideally, we would like to explore the entire posterior distribution to fully take into account the uncertainty about each single assignment. Unfortunately, this seems computationally infeasible. What seems more practical is to quickly find several local optima, which then, when combined, provide a reasonable summary of the posterior distribution.

4.1   Local Search and Simulated Annealing

To maximize the posterior $p(\boldsymbol{h}|\boldsymbol{g})$, we have implemented a simulated annealing method that starts with an arbitrary haplotype assignment and then proceeds iteratively with simple *phase switch* moves that change the haplotype pair $(h, h') = (h^{2i-1}, h^{2i})$ for a randomly picked genotype g^i.

Phase Switch: Select the pair of haplotypes (h, h') for a random genotype g and a random marker j and switch the phase of the partial haplotypes of h and h' at markers $k \geq j$. That is, the new haplotype pair for g is $(h_1 \cdots h_{j-1} h'_j \cdots h'_m, h'_1 \cdots h'_{j-1} h_j \cdots h_m)$.

A proposed phase switch is accepted with probability $\min\{1, A^{1/\tau}\}$, where A is the ratio of the posterior probabilities of the proposed and the current haplotype assignment, and $\tau \geq 0$ is a decreasing temperature parameter. Each such an iteration can be computed in time $O(D^2)$ by storing the haplotype pair of each genotype as its switch sequence (see Sect. 5) and maintaining the tries used in the CTW algorithm for computing $p(\boldsymbol{h}|\boldsymbol{g})$. The posterior is evaluated with a large depth, $D = 40$; temperature τ is set to 1 and the best assignment found in mn steps is taken as the starting assignment in the next batch. Three more such batches are run with $\tau = 1/2, 1/4, 0$, each batch started from the best assignment found so far. The procedure is repeated for some number of times, 20 in our experiments, to obtain several local optima of the posterior, with the genotypes reversed in every other iteration as in Beagle [15].

We have found that the above described local search converges rapidly to a local optimum, but also that the quality of the optimum is highly sensitive to the initial haplotype assignment. We therefore designed a slower but otherwise more effective *forward sampling* heuristic for finding good initial assignments.

Forward Sampling:
1. Set the haplotypes $\boldsymbol{h}$ arbitrarily such that they match the genotypes $\boldsymbol{g}$.
2. For each genotype g^i, draw a new haplotype pair (h^{2i-1}, h^{2i}) from $p(h^{2i-1}, h^{2i}|\boldsymbol{h}^{-i}, \boldsymbol{g}) \propto p(\boldsymbol{h}|\boldsymbol{g})$, where $\boldsymbol{h}^{-i}$ consists of the remaining haplotype pairs in $\boldsymbol{h}$. Repeat this 10 times.
3. Return the centroid (see Sect. 5) of the last 5 samples.

Forward sampling can be implemented relatively efficiently using the idea of forward–backward sampling algorithm of hidden Markov models (see [15] and references therein). As that HMM algorithm is well-known, we here only mention some key features of the present application and omit most of the technical details. First, observe that given a genotype g^i, it suffices to consider drawing the haplotype h^{2i-1}, since h^{2i} is fully determined by g^i and h^{2i-1}. Second, notice that the jth allele of h^{2i-1} depends on the alleles at markers $j-1, j-2, \ldots, j-D$; the state of these D alleles corresponds to a hidden state in an analogous HMM. The time complexity of the sampling procedure essentially depends on the number of possible states, 2^D; the effect of the number of markers is linear, and the involved transition probabilities are easily obtained by statistics precomputed by the CTW algorithm for computing $p(\boldsymbol{h}|\boldsymbol{g})$. Thus, sampling a haplotype pair can be implemented in time $O(mD2^D)$ for a single genotype. We also note that the term 2^D can be improved to 2^k, where k is the maximum number of heterozygous sites of g in any D consecutive markers.

4.2   Sampling Haplotypes from the Posterior Distribution

We next sketch an annealed importance sampling (AIS) method [17] to draw haplotype assignments from the posterior distribution $p(\boldsymbol{h}|\boldsymbol{g})$ conditional on the given n genotypes, $\boldsymbol{g}$. In this simulated annealing type MCMC technique, the idea is to generate a sequence $\boldsymbol{h}^{(1)}, \boldsymbol{h}^{(2)}, \ldots, \boldsymbol{h}^{(T)}$ along a smooth chain of distributions $\pi^{(1)}, \pi^{(2)}, \ldots, \pi^{(T)}$. Only the last configuration, $\boldsymbol{h}^{(T)}$, will be retained with an appropriately defined weight $w(\boldsymbol{h}^{(T)})$; the procedure is repeated some number of times to obtain a collection $\mathcal{H}$ of independent weighted samples. For the posterior expectation of any function of interest, $\psi(\boldsymbol{h})$, an unbiased estimate is obtained as $\sum_{\boldsymbol{h}\in\mathcal{H}} w(\boldsymbol{h})\psi(\boldsymbol{h}) / \sum_{\boldsymbol{h}\in\mathcal{H}} w(\boldsymbol{h})$ [17].

For concreteness, consider a simple annealing scheme of the form $\pi^{(t)}(\boldsymbol{h}) \propto f^{(t)}(\boldsymbol{h}) = p(\boldsymbol{h}|\boldsymbol{g})^{t/T}$, with, say, $T = 1000$. First, $\boldsymbol{h}^{(1)}$ is drawn from $\pi^{(1)}$ by running a Metropolis algorithm, say, mn steps; suitable proposal distributions can be constructed based on the forward sampling and phase switch moves described above. Then, for $t = 1, \ldots, T-1$, an assignment $\boldsymbol{h}^{(t+1)}$ is drawn based on $\boldsymbol{h}^{(t)}$ using an analogous Metropolis kernel that now leaves $\pi^{(t+1)}$ invariant. Finally, the last configuration, $\boldsymbol{h}^{(T)}$, is stored together with a weight
$$w(\boldsymbol{h}^{(T)}) = \prod_{t=1}^{T-1} f^{(t+1)}(\boldsymbol{h}^{(t)}) / f^{(t)}(\boldsymbol{h}^{(t)}) = (p(\boldsymbol{h}^{(1)})p(\boldsymbol{h}^{(2)}) \cdots p(\boldsymbol{h}^{(T)}))^{1/T}.$$

5   Minimization of the Expected Switch Distance

In principle, the output of Bayesian inference of haplotypes is the posterior distribution. In practice, and especially for comparing different methods, one however needs to provide a single estimate for the haplotype pair underlying each observed genotype. How such an estimate should be constructed depends not only on the posterior but also on the cost assigned for mistakes in the estimate. The cost can usually be specified by a *loss function*, ℓ, that associates two haplotypes pairs, $\{h, h'\}$ and $\{\hat{h}, \hat{h}'\}$, a real-valued loss $\ell(\{h, h'\}, \{\hat{h}, \hat{h}'\})$; the total cost over all genotypes is assumed to be a sum of the individual losses. Given a loss function, a Bayes-optimal estimate is one that minimizes the posterior expected loss. When a posterior is summarized by a (weighted) sample of haplotype assignments, the expected loss is approximated by a (weighted) average over the sample.

While it is often not possible to reliably infer the correct haplotype pair, except for very short regions, it is usually possible to provide an estimate that is close to the true haplotypes in terms of the switch distance [22], a loss function commonly used in haplotype inference. The *switch distance* of two haplotype pairs $\{h, h'\}$ and $\{\hat{h}, \hat{h}'\}$, denoted by $\mathrm{sd}(\{h, h'\}, \{\hat{h}, \hat{h}'\})$, is the number of phase switches needed to turn $\{h, h'\}$ into $\{\hat{h}, \hat{h}'\}$. For example, if the true pair is $\{000000, 111111\}$ then the switch distance to $\{000111, 111000\}$ is 1 while the switch distance to $\{000110, 111001\}$ is 2.

Even though finding a Bayes-optimal estimate seems hard in general, it turns out that the special case of switch distance can be solved efficiently, in time linear in the sample size. To see this, first observe that any haplotype pair $\{h, h'\}$ compatible with a fixed genotype g is bijectively related to its switch sequence $\xi = (\xi_1, \ldots, \xi_{k-1})$ defined by $\xi_s = |\lambda_s - \lambda_{s+1}|$, where k is the number of heterozygous sites of g and $\lambda_s \in \{0, 1\}$ encodes (in an arbitrary way) the two possible phases of the sth heterozygous site of g; furthermore, the switch sequence can be constructed in linear time. Second, we observe that if $\{h, h'\}$ and $\{\hat{h}, \hat{h}'\}$ are two haplotype pairs compatible with g, and ξ and $\hat{\xi}$ their switch sequences, then $\mathrm{sd}(\{h, h'\}, \{\hat{h}, \hat{h}'\}) = \sum_{s=1}^{k} |\xi_s - \hat{\xi}_s| = ||\xi - \hat{\xi}||_1$, the 1-norm of the vector $\xi - \hat{\xi}$. Thus, it remains to find a switch sequence $\hat{\xi}$ that minimizes

$$\sum_{\xi \in \Xi} w(\xi) ||\xi - \hat{\xi}||_1 \,,$$

where Ξ denotes the multiset of switch sequences determined by the given sample of haplotype pairs, and $w(\xi)$ is the weight of the haplotype assignment that corresponds to ξ. Now, we know that for any set of binary sequences the centroid with respect to the metric induced by the 1-norm is obtained simply by coordinate-wise voting. This suffices for the unweighted case; see [18] for a somewhat more detailed argumentation. Having this, it is not difficult to see that the weighted case can be solved by weighted voting: set $\hat{\xi}_s$ to 1 if and only if

$$\sum_{\xi \in \Xi : \xi_s = 1} w(\xi) \;>\; \sum_{\xi \in \Xi : \xi_s = 0} w(\xi) \,.$$

Such a voting can obviously be carried out in time $O(k|\Xi|)$. Combining these observations gives

Theorem 1. *Let g be a genotype over m SNPs and $\mathcal{H}$ a set of haplotype pairs that are compatible with g, each pair associated with a real-valued weight. Then a centroid, that is, a haplotype pair that minimizes the sum of the weighted switch distances to the haplotype pairs in $\mathcal{H}$, can be found in time $O(m|\mathcal{H}|)$.*

6 Experimental Results

We have implemented the presented method in a prototype program BACH (Bayesian Context-based Haplotyping) and compared its performance against state-of-the-art software for haplotype inference on both real and simulated data.

6.1 Methods Compared

BACH is written in Java and the implementation is not optimized for speed. Currently, it contains only the simulating annealing method (with forward sampling and phase switch using tuning parameters $D = 8$ and $D = 40$, respectively); the annealed importance sampling analogue sketched in Sect. 4.2 is work-in-progress.

For comparison we included the latest versions of PHASE [23], fastPHASE [13], Beagle [15], HaploRec [14], HIT [11], and HAP [10]. We included PHASE as a reference, even though it took several hours to run on a single small (HapMap) dataset; we did not try to run it for the much larger, simulated datasets. The other tested methods scale well and take only some minutes or a few hours per large dataset. Specifically, the time complexity of BACH (with fixed maximum context length) is only linear, $O(mn)$; relatively large constant factors, however, render BACH the slowest of the fast methods. All methods except HAP were available as stand-alone applications, which we ran on a regular desktop PC. HAP was run on its web server, and due the amount of manual labour involved in sending the data and fetching the results, we tested HAP only on 24 of the the real datasets, selected at random.

We also tested deviating from the default parameters of the compared methods, and report the best results obtained. For example, we set "-nsamples=25" in Beagle for the real data, as suggested by its manual; this improved Beagle's performance slightly. For HaploRec we used option "-p S", here referred to as HaploRec-S, to use a segmentation model. By default, HaploRec uses a variable-order Markov model (option "-p VMM_AVG"), here referred to as HaploRec-V. For HIT, we used ten founders ($K = 10$).

6.2 Datasets

The real data, containing 132 datasets, were obtained from the HapMap database [19]. These datasets contain samples from human populations YRI (Yoruba) and CEU (Utah), from which there were 30 SNP trios available. The HapMap database contains 120 haplotypes inferred from these trios, which we took as the ground truth and converted them into 60 genotypes. From each chromosome we selected 100 SNPs in such a way that the average distance between adjacent SNPs is close to either 1, 3 or 9 kb. Thus, by CEU-3k (other combinations analogously) we refer to the datasets from the CEU population with an average SNP spacing of 3 kb.

To generate larger datasets, we followed the example of Browning and Browning [15] and used COSI [24] with the best-fit parameters to simulate 2,000 haplotypes of 1 Mb in length from a "European" population. We generated 100 sparse and dense datasets. The sparse datasets were filtered by first removing all SNPs with minimum allele frequency (MAF) ≤ 0.05, and then selecting at random 250 SNPs. We then sampled a subset of 120 haplotypes, and again eliminated SNPs with MAF ≤ 0.05. The SNPs were then selected to the final data set by selecting an informative subset by the method of Carlson et al. [25]. In the resulting set of SNPs, either the SNP was genotyped or there was at least one genotyped SNP that had the squared correlation coefficient $r^2 \geq 0.7$ with the omitted SNP. For dense sets the filtering was similar, except that we sampled 1,428 markers instead of 250, and the r^2 threshold for the tagging algorithm was 0.9. These datasets are referred to as Dense and Sparse. The former had median SNP count of 101, and the latter 367 markers.

6.3   Comparison of Phasing Accuracy

We evaluated the phasing accuracy of each method in terms of the switch error, that is, the switch distance between the predicted and the true haplotypes (see Sect. 5) divided by the maximum switch distance. We found that by average switch error (Table 1) BACH is among the most accurate methods on both the HapMap datasets and the synthetic datasets. On the HapMap datasets PHASE is superior, whereas HAP (on the selected 24 datasets) is consistently less accurate than the rest (details not shown).

We also compared the methods pairwise and examined the percentage of the HapMap datasets on which one method was more accurate than another method (Table 2). We observed that, after PHASE, the most accurate methods were fastPHASE, BACH, and HaploRec-S, with no clear order. Note that HAP and the synthetic datasets were not included in this comparison.

Table 1. Average switch errors of the tested methods

	PHASE	fastPHASE	BACH	Beagle	HaploRec-S	HaploRec-V	HIT
CEU-1k	0.0299	0.0343	0.0348	0.0405	0.0364	0.0376	0.0375
CEU-3k	0.0652	0.0692	0.0665	0.0764	0.0692	0.073	0.0745
CEU-9k	0.144	0.146	0.147	0.164	0.147	0.153	0.159
YRI-1k	0.0407	0.0579	0.0597	0.0645	0.0540	0.0601	0.0642
YRI-3k	0.0931	0.117	0.113	0.125	0.111	0.122	0.126
YRI-9k	0.189	0.204	0.198	0.223	0.193	0.204	0.220
Sparse	-	0.0398	0.0305	0.0317	0.0288	0.0316	0.0442
Dense	-	0.0169	0.0125	0.0116	0.0133	0.0145	0.0190

Table 2. Percentage of HapMap datasets won by a method on a column vs. a method on a row

	PHASE	fastPHASE	BACH	Beagle	HaploRec-S	HaploRec-V	HIT
PHASE	-	20.5	19.7	5.3	18.9	9.1	7.6
fastPHASE	78.0	-	48.5	17.4	55.3	34.8	15.2
BACH	79.5	45.5	-	18.2	53.8	28.0	15.2
Beagle	93.9	79.5	79.5	-	87.1	72.7	55.3
HaploRec-S	78.8	42.4	42.4	12.9	-	25.8	16.7
HaploRec-V	90.2	59.8	71.2	23.5	73.5	-	29.5
HIT	92.4	82.6	82.6	37.9	78.8	67.4	-

7 Concluding Remarks

We have presented a Bayesian implementation of variable-order Markov modeling of haplotypes. The promise of this approach is in its robustness, as it is not based of fitting a single model to the data. As we showed, the required sum over models can be efficiently computed using the context tree weighting algorithm [16]. We believe there is room for improving our heuristics for optimizing or exploring the resulting objective function (the posterior distribution). Yet, the techniques implemented in BACH were sufficient for demonstrating the potential of the approach: the phasing accuracy of BACH was competitive to the very best of the existing fast haplotype inference methods.

The context tree weighting algorithm has been previously successfully applied to protein classification [26]. In this application, a variable-order Markov model was extended to wild-card symbols that match to every alphabet symbol. Such wild-cards could be incorporated into our haplotype inference method as well.

References

1. Conrad, D.F., Andrews, T.D., Carter, N.P., Hurles, M.E., Pritchard, J.K.: A high-resolution survey of deletion polymorphism in the human genome. Nat. Genet. 38, 75–81 (2006)
2. Corona, E., Raphael, B.J., Eskin, E.: Identification of deletion polymorphisms from haplotypes. In: Speed, T., Huang, H. (eds.) RECOMB 2007. LNCS (LNBI), vol. 4453, pp. 354–365. Springer, Heidelberg (2007)
3. Kohler, J.E., Cutler, D.J.: Simultaneous discovery and testing of deletions for disease associations in SNP genotyping studies. Am. J. Hum. Genet. 81, 684–699 (2007)
4. Bansal, V., Bashir, A., Bafna, V.: Evidence for large inversion polymorphisms in the human genome from HapMap data. Genome Res. 17, 219–230 (2007)
5. Clark, A.G.: Inference of haplotypes from PCR-amplified samples of diploid populations. Mol. Biol. Evol. 7, 111–122 (1990)
6. Excoffier, L., Slatkin, M.: Maximum-likelihood estimation of molecular haplotype frequencies in a diploid population. Mol. Biol. Evol. 12, 921–927 (1995)
7. Long, J.C., Williams, R.C., Urbanek, M.: An E-M algorithm and testing strategy for multiple-locus haplotypes. Am. J. Hum. Genet. 56, 799–810 (1995)

8. Stephens, M., Smith, N., Donnelly, P.: A new statistical method for haplotype reconstruction from population data. Am. J. Hum. Genet. 68, 978–989 (2001)
9. Niu, T., Qin, Z., Xu, X., Liu, J.: Bayesian haplotype inference for multiple linked single-nucleotide polymorphisms. Am. J. Hum. Genet. 70, 157–169 (2002)
10. Halperin, E., Eskin, E.: Haplotype reconstruction from genotype data using imperfect phylogeny. Bioinformatics 20, 104–113 (2004)
11. Rastas, P., Koivisto, M., Mannila, H., Ukkonen, E.: A hidden Markov technique for haplotype reconstruction. In: Casadio, R., Myers, G. (eds.) WABI 2005. LNCS (LNBI), vol. 3692, pp. 140–151. Springer, Heidelberg (2005)
12. Kimmel, G., Shamir, R.: Genotype resolution and block identification using likelihood. In: Proceeding of the National Academy of Sciences of the United States of America (PNAS), vol. 102, pp. 158–162 (2005)
13. Scheet, P., Stephens, M.: A fast and flexible statistical model for large-scale population genotype data: Applications to inferring missing genotypes and haplotypic phase. Am. J. Hum. Genet. 78, 629–644 (2006)
14. Eronen, L., Geerts, F., Toivonen, H.: Haplorec: efficient and accurate large-scale reconstruction of haplotypes. BMC Bioinformatics 7, 542 (2006)
15. Browning, S., Browning, B.: Rapid and accurate haplotype phasing and missing-data inference for whole-genome association studies by use of localized haplotype clustering. Am. J. Hum. Genet. 81, 1084–1097 (2007)
16. Willems, F.M.J., Shtarkov, Y.M., Tjalkens, T.J.: The context-tree weighting method: Basic properties. IEEE Trans. Inform. Theory 41, 653–664 (1995)
17. Neal, R.M.: Annealed importance sampling. Statist. Comput. 11, 125–139 (2001)
18. Kääriäinen, M., Landwehr, N., Lappalainen, S., Mielikäinen, T.: Combining haplotypers. Technical Report C-2007-57, Department of Computer Science, University of Helsinki (2007)
19. The International HapMap Consortium: A haplotype map of the human genome. Nature 437, 1299–1320 (2005)
20. Marchini, J., Cutler, D., Patterson, N., et al.: A comparison of phasing algorithms for trios and unrelated individuals. Am. J. Hum. Genet. 78, 437–450 (2006)
21. Willems, F.M.J.: The context-tree weighting method: Extensions. IEEE Trans. Inform. Theory 44, 792–798 (1998)
22. Lin, S., Cutler, D.J., Zwick, M.E., Chakravarti, A.: Haplotype inference in random population samples. Am. J. Hum. Genet. 71, 1129–1137 (2002)
23. Stephens, M., Scheet, P.: Accounting for decay of linkage disequilibrium in haplotype inference and missing-data imputation. Am. J. Hum. Genet. 76, 449–462 (2005)
24. Schaffner, S.F., Foo, C., Gabriel, S., Reich, D., Daly, M.J., Altshuler, D.: Calibrating a coalescent simulation of human genome sequence variation. Genome Res. 15, 1576–1583 (2005)
25. Carlson, C.S., Eberle, M.A., Rieder, M.J., Yi, Q., Kruglyak, L., Nickerson, D.A.: Selecting a maximally informative set of single-nucleotide polymorphisms for association analyses using linkage disequilibrium. Am. J. Hum. Genet. 74, 105–120 (2004)
26. Eskin, E., Grundy, W.N., Singer, Y.: Protein family classification using sparse markov transducers. In: Proceedings of the Eighth International Conference on Intelligent Systems for Molecular Biology, pp. 134–145. AAAI Press, Menlo Park (2000)

Genotype Sequence Segmentation: Handling Constraints and Noise

Qi Zhang, Wei Wang, Leonard McMillan, Jan Prins,
Fernando Pardo-Manuel de Villena, and David Threadgill

UNC Chapel Hill

Abstract. Recombination plays an important role in shaping the genetic variations present in current-day populations. We consider populations evolved from a small number of founders, where each individual's genomic sequence is composed of segments from the founders. We study the problem of segmenting the genotype sequences into the minimum number of segments attributable to the founder sequences. The minimum segmentation can be used for inferring the relationship among sequences to identify the genetic basis of traits, which is important for disease association studies. We propose two dynamic programming algorithms which can solve the minimum segmentation problem in polynomial time. Our algorithms incorporate biological constraints to greatly reduce the computation, and guarantee that only minimum segmentation solutions with comparable numbers of segments on both haplotypes of the genotype sequence are computed. Our algorithms can also work on noisy data including genotyping errors, point mutations, gene conversions, and missing values.

1   Introduction

Recombination plays an important role in shaping the genetic variations present in current-day populations. Understanding the genetic variations and the genetic basis of traits is crucial for disease association studies. In this paper, we assume an evolution model (previously proposed and studied in [U, WG]) where a population is evolved from a small number of founder sequences. A real-world biological scenario is the Collaborative Cross (CC). The CC [THW, C] is a large panel of 1000 recombinant inbred (RI) mouse strains that were generated from a funnel breeding scheme initiated with a set of 8 founder strains followed by 20 generations of inbreeding. These 8 genetically diverse founder strains capture nearly 90% of the known variations present in the laboratory mouse. The resulting RI strains have a population structure that randomizes the known genetic variation, which will provide unparallel power for disease association studies.

Given a set of founder haplotype sequences, a sequence in the generated population is composed of segments from the founders. It is of interest to identify and label these segments according to their contributing founder. Although the segmentation for a haplotype sequence may be straightforward to compute, in many cases the sequence to be segmented is a genotype sequence for which the

K.A. Crandall and J. Lagergren (Eds.): WABI 2008, LNBI 5251, pp. 271–283, 2008.

© Springer-Verlag Berlin Heidelberg 2008

two haplotypes are not known completely and they may have different segmentations. For example, the genotype sequence for the strains generated during the intermediate generation in a 20 inbreeding generations of the CC contains two different haplotypes. In this paper, we study the segmentation problem of genotype sequences with the optimization for the minimum number of segments contained in the two associated haplotypes. Furthermore, we extend this basic model to include additional biologically-motivated constraints as well as noise. Since each autosome undergoes, on average, one recombination per meiosis, one expects that the number of founder switches per haplotype at a given generation of breeding to be comparable. Moreover, noise may exist in the founder sequences and the genotype sequence to be segmented. Sources of genotyping noise are both technical and biological. They include point mutations, gene conversions, genotyping errors, etc. Missing genotyping values are also very common in biological data sets.

Similar but different models were studied in [U, WG]. Ukkonen [U] first proposed the founder set reconstruction problem under the assumption that the sample set is evolved from a small set of founders. A dynamic programming algorithm was proposed which computes a minimum number of founders with a given set of sample haplotype sequences, where a segmentation of all the sequences in the sample set can be derived which contains the minimum number of founder switches. Wu and Gusfield [WG] proposed improved polynomial time algorithms for haplotype as well as genotype sample sequences for the special case where there are only two founders. Different from the problems considered in [WG, U], we study the problem where the set of founder sequences are already known, and compute the minimum segmentation for genotype sequences under biologically-relevant constraints and noise. A motivating biological example is the segmentation of the genotype sequences obtained from immediate generations of the CC to estimate the location of the recombination breakpoints. There is other related work on inferring the structure of the variation of the sequences, which include identifying haplotype blocks [DRSHL, GSNM, SHBCI], computing the phylogenies [GEL, G], etc.

In this paper, we propose two dynamic programming algorithms to compute the minimum segmentations for genotype sequences. Our algorithms run in polynomial time and consider biological constraints of the genotype segmentation problem, *i.e.*, the number of segments in both haplotypes are comparable. Moreover, our algorithms account for the potential noise sources in the data including point mutations, gene conversions, genotyping errors, and missing values.

2   The Minimum Segmentation Problem

Assume that we have a set of founding haplotypes $FS = \{F_1, \ldots, F_n, \ldots, F_N\}$. Each haplotype sequence is of length L: $F_n = f_1^n f_l^n \ldots f_L^n$, where $f_l^n \in \{0, 1\}$. Given an input sequence from a population which is derived exclusively from the founder set FS, we are interested in finding a possible segmentation of the sequence, where each segment is inherited from the corresponding region of one of the founders. We first consider the simple case where the input sequence is a haplotype, and then investigate the more interesting case where the input is a genotype sequence.

Given a haplotype sequence, $H = h_1 \ldots h_L$, $(h_l \in \{0,1\})$, a segment of H is denoted as $\overline{H}_k = h_{s_k} h_{s_k+1} h_{s_k+L_k-1}$, where s_k is the starting position of $\overline{H}_k$, and L_k is the length of $\overline{H}_k$. We consider a segmentation of H which divides the entire sequence into an ordered list of disjoint segments $Seg(H) = \{\overline{H}_1, \ldots, \overline{H}_k, \ldots, \overline{H}_K\}$, where each segment $\overline{H}_k$ is identical to the corresponding region of one of the founders and K is the number of segments in $Seg(H)$. In other words, for each segment $\overline{H}_k = h_{s_k} h_{s_k+1} h_{s_k+L_k-1}$, there exists a founder $F_n = f_1^n f_l^n \ldots f_L^n$ such that $h_{s_k+l_i} = f_{s_k+l_i}^n$, for $l_i = 0, 1, \ldots, L_k - 1$. Furthermore, a minimum segmentation is defined as the segmentation which contains the minimum number of segments. We denote the minimum segmentation as $MinSeg(H) = \{\overline{H}_1, \ldots, \overline{H}_{K_{min}}\}$, where $K_{min} = |MinSeg(H)|$ is the number of segments in $MinSeg(H)$.

If the input is a genotype sequence, we know that it represents two copies of different haplotype sequences, H_a and H_b. Assume that the genotype sequence is $G = g_1 \ldots g_L$, where $g_l \in \{0, 1, 2\}$. A site l is *homozygous* if $g_l = 0$ $(h_l^a = h_l^b = 0)$ or $g_l = 1$ $(h_l^a = h_l^b = 1)$; a site l is *heterozygous* if H_a and H_b take different alleles, in which case, $g_l = 2$. The process of determining whether $h_l^a = 0, h_l^b = 1$ or $h_l^a = 1, h_l^b = 0$ for a heterozygous site l is called *phasing*. The procedure of determining the two haplotype sequences from the genotype sequence by phasing all the heterozygous sites is called *Haplotype Inference*. For the genotype input case, a segmentation $Seg(G)$ consists of segmentations for both haplotype sequences: $Seg_a(H_a)$ and $Seg_b(H_b)$. The number of segments in $Seg(G)$ is the sum of the numbers of segments in $Seg_a(H_a)$ and $Seg_b(H_b)$: $|Seg(G)| = |Seg_a(H_a)| + |Seg_b(H_b)|$. The minimum segmentation is the segmentation which contains the minimum total number of segments: $|MinSeg(G)| = min\{|Seg(G)|\}$. Let $MinSeg(G) = \{Seg_a^*(H_a), Seg_b^*(H_b)\}$.

In this paper, we develop efficient algorithms for the minimum segmentation problem especially for the genotype input case. In addition to the basic models, there are other issues we may need to consider, such as genotyping errors, point mutations, missing values, the balance of the number of segments in both haplotypes, etc. We will explain later how we model these biological constraints and noise in our solutions.

Solutions for Haplotype Input: Computing the minimum segmentation for the haplotype input sequence is relatively easy and has been discussed in previous studies [WG, U]. A simple greedy algorithm can be applied to compute a minimum segmentation solution by scanning from left to right. Assume that the current site is i (initially it is site 1), and we have a minimum segmentation solution for the part of the input sequence from site 1 to site i. Starting from site i, we try to find the segment shared by the input sequence and one of the founders which extends furthest to the right. This greedy algorithm generates one of the minimum segmentation solutions.

A graph-based dynamic programming algorithm can be used to compute all minimum segmentation solutions given the input haplotype sequence and the founder set. At a high level, we first compute all maximal shared intervals between the input sequence and each founder sequence. The maximal shared interval between the input sequence and founder n is a region where the input sequence is exactly the same as founder n. We consider each shared interval as a node and connect two intervals with an edge if they overlap. In this way, a minimum segmentation solution corresponds to a shortest path from a node starting at the first site to a node ending at the last site. The complete set of the shortest paths can be computed, which are all possible minimum segmentation solutions.

3   Solutions for Genotype Input

The greedy algorithm and the graph-based algorithm for segmenting haplotype input sequences cannot be easily applied on genotype input. The major issue is that we do not know the exact sequences of the two haplotypes due to the multiple possible allele pairs at heterozygous sites. Second, the minimum segmentation solution for the genotype may not consist of the minimum segmentation solutions for each haplotype sequence.

In the following discussion, we describe two dynamic programming algorithms for solving the minimum segmentation problem for genotype input sequences. The first algorithm considers each site separately, the second algorithm considers a region of sites simultaneously, and is thus more efficient.

Site-based Dynamic Programming Algorithm: We consider for each site l, the possible founders for the two haplotype sequences H_a and H_b. If site l is a homozygous site, assuming $g_l = 0$ (without loss of generality), we have $h_l^a = h_l^b = 0$. Let $of^{a,l}$ be the original founder where h_l^a was inherited from at site l. Then $of^{a,l}$ must be one of the founders which also take 0 at site l: $of^{a,l} \in \{F_n | f_l^n = 0\}$. Similarly, we have the founder where h_l^b was inherited from as: $of^{b,l} \in \{F_n | f_l^n = 0\}$. Let $fp^l = \langle of^{a,l}, of^{b,l} \rangle$ denote the possible founder pair at site l, we have the set of all possible founder pairs as $FP^l = \{fp^l | fp^l \in \{F_n | f_l^n = 0\} \times \{F_n | f_l^n = 0\}\}$. If site l is a heterozygous site where $g_l = 2$, there are two possibilities: $h_l^a = 1 \wedge h_l^b = 0$ or $h_l^a = 0 \wedge h_l^b = 1$. Therefore, the possible founder pairs for heterozygous site l is $FP^l = \{fp^l | fp^l \in \{F_n | f_l^n = 0\} \times \{F_n | f_l^n = 1\} \cup \{F_n | f_l^n = 1\} \times \{F_n | f_l^n = 0\}\}$. We compute the founder pair set FP^l for each site l.

Assigning a founder pair from FP^l to each site l generates a segmentation of the input genotype sequence. The number of segments of both haplotypes (of the genotype) are the total number of *founder switches* between founder pairs of every consecutive sites plus 2. Consider two neighboring sites l and $l + 1$. If the corresponding founder pairs are $fp_{q_l}^l = \langle of_{q_l}^{a,l}, of_{q_l}^{b,l} \rangle$ $(1 \leq q_l \leq |FP^l|)$ and $fp_{q_{l+1}}^{l+1} = \langle of_{q_{l+1}}^{a,l}, of_{q_{l+1}}^{b,l} \rangle$ $(1 \leq q_{l+1} \leq |FP^{l+1}|)$, the number of founder switches between these two founder pairs $FounderSwitch(fp_{q_l}^l, fp_{q_{l+1}}^{l+1})$ can be computed as:

$$FounderSwitch(fp^l_{q_l}, fp^{l+1}_{q_{l+1}}) = \begin{cases} 0 : \text{if } of^{a,l}_{q_l} = of^{a,l+1}_{q_{l+1}} \wedge of^{b,l}_{q_l} = of^{b,l+1}_{q_{l+1}} \\ 1 : \text{if } of^{a,l}_{q_l} = of^{a,l+1}_{q_{l+1}} \wedge of^{b,l}_{q_l} \neq of^{b,l+1}_{q_{l+1}} \text{ or} \\ \quad\quad of^{a,l}_{q_l} \neq of^{a,l+1}_{q_{l+1}} \wedge of^{b,l}_{q_l} = of^{b,l+1}_{q_{l+1}} \\ 2 : \text{if } of^{a,l}_{q_l} \neq of^{a,l+1}_{q_{l+1}} \wedge of^{b,l}_{q_l} \neq of^{b,l+1}_{q_{l+1}} \end{cases} \quad (1)$$

Let $K_{min}(g_1 \ldots g_{l-1}|fp^l_{q_l})$ be the minimum number of segments in any segmentation solution over the subsequence $g_1 \ldots g_l$ which at site l takes the founder pair $fp^l_{q_l}$. The minimum number of segments over the entire genotype sequence $K_{min}(g_1 \ldots g_L)$ can be computed as:

$$K_{min}(g_1 \ldots g_L) = min_{fp^L_{q_L} \in FP^L} \{K_{min}(g_1 \ldots g_{L-1}|fp^L_{q_L})\} \quad (2)$$

The main recurrence of the dynamic programming algorithm is as follows:

$$K_{min}(g_1 \ldots g_{l-1}|fp^l_{q_l}) = min_{fp^{l-1}_{q_{l-1}} \in FP^{l-1}} \{K_{min}(g_1 \ldots g_{l-2}|fp^{l-1}_{q_{l-1}}) + \\ FounderSwitch(fp^{l-1}_{q_{l-1}}, fp^l_{q_l})\} \quad (3)$$

And initially,

$$K_{min}(\Phi|fp^1_{q_1}) = 2, \forall fp^1_{q_1} \in FP^1 \quad (4)$$

The solutions for this dynamic programming problem can be easily computed by populating a table T of L rows where row l has at most $|FP^l|$ entries. The entry $T(l, q_l)$, $1 \leq q_l \leq |FP^l|$ is filled with $K_{min}(g_1 \ldots g_{l-1}|fp^l_{q_l})$ during the computation. Row 1 is initialized according to Eq.(4), and row $i+1$ is computed after row i. During the computation of $T(l, q_l)$ according to Eq.(3), we keep the backtracking pointers from entry $T(l, q_l)$ to any $T(l-1, q_{l-1})$ where the minimum values are obtained. In this way, we are able to obtain all the minimum segmentation solutions.

There are at most N^2 founder pairs for each site l, i.e., $|FP^l| \leq N^2, \forall l$. Therefore, the table we populate is of size $O(LN^2)$. It takes constant time to compute $FounderSwitch(fp^l_{q_l}, fp^{l+1}_{q_{l+1}})$, then filling out a single entry in the table takes $O(N^2)$ time. Therefore, the computational complexity for the entire algorithm is $O(LN^4)$. The space complexity is $O(LN^2)$. For very long sequences and a small number of founders, i.e., $L \gg N^4$, the algorithm has linear time and space complexity in terms of the length of the input sequence. If we keep multiple backtrack pointers for each entry while populating the table, we are able to obtain all the minimum segmentation solutions.

Region-based Dynamic Programming Algorithm: For very long sequences, we propose a more efficient algorithm which considers a subregion of the entire sequence instead of a site at a time.

We first consider the homozygous regions, which are the regions of homozygous sites between any two consecutive heterozygous sites. Within a homozygous region, both copies of the haplotype sequences are the same and we know the exact allele at each site. Fig. 1 illustrates an example of a set of four founders $(F_1 - F_4)$ and a genotype input sequence G to be segmented. The length of

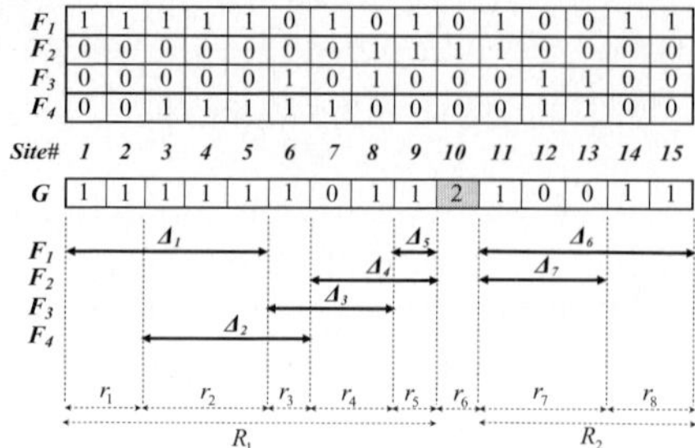

Fig. 1. An example subregions. F_1-F_4 are four founder sequences. G is the genotype sequence to be segmented. There are 15 sites in all sequences, where site 10 is the only heterozygous site. $R_1 : [1, 9]$ and $R_2 : [11, 15]$ are the homozygous regions. Δ_1-Δ_5 are the maximal shared intervals in R_1. Δ_6 and Δ_7 are the maximal shared intervals in R_2. $r_1 - r_8$ are the subregions for the entire sequence, out of which r_6 is the heterozygous subregions, and the remaining are the homozygous subregions.

each founder and the genotype sequence is 15, with 14 homozygous sites and 1 heterozygous site (site 10). The homozygous regions are $R_1 = [1, 9]$ and $R_2 = [11, 15]$. For each homozygous region, we compute all the maximal shared intervals between each founder and the haplotype sequences. A maximal shared interval Δ_i is an interval over which a haplotype and a founder shares the same allele at each site and the region cannot be extended further on either side. We represent each maximum shared interval as a triple, for example, $\Delta_i : (I_i, H_a, F_n)$ is a maximal shared interval between haplotype H_a and founder F_n over interval I_i. Since both haplotypes are the same, a maximal shared interval for haplotype H_a is also a maximal shared interval for haplotype H_b, therefore, the maximal shared interval for the homozygous regions can also be represented as $\Delta_i : (I_i, *, F_n)$. In Fig. 1, $\Delta_1 - \Delta_5$ are the maximal shared intervals within region R_1 for both haplotype sequences. We divide each homozygous region R_j into a set of subregions using the two end points of all maximal shared intervals inside R_j. For example, in Fig. 1, R_1 is divided into subregions r_1, r_2, r_3, r_4, and r_5. If we consider each heterozygous site as a 1-site subregion (e.g. r_6 in Fig. 1), together with all the subregions for the homozygous regions, we have a complete set of subregions $\{r_p\}$ which cover the entire sequence (e.g., $r_1 - r_8$ in Fig. 1).

For each homozygous subregion r_p, let $fp^{r_p} = \langle of^{a,r_p}, of^{b,r_p} \rangle$ be a possible founder pair for subregion r_p. We know that the set of possible founder pairs is $FP^{r_p} = \{\langle of^{a,r_p}, of^{b,r_p} \rangle | \exists \Delta_{i_1} = (I_{i_1}, *, of^{a,r_p}), \Delta_{i_2} = (I_{i_2}, *, of^{b,r_p}),$ where $I_{i_1} \supseteq r_p, I_{i_2} \supseteq r_p\}$. For example, the founder pair for the subregion r_2 in Fig. 1 could be $\langle F_1, F_1 \rangle$, or $\langle F_1, F_2 \rangle$, or $\langle F_2, F_1 \rangle$, or $\langle F_2, F_2 \rangle$. For each heterozygous subregion which is composed of a heterozygous site l, since h_a^l and h_b^l take different alleles, any possible founder pair should consist of a founder taking allele 1 and a founder taking allele 0. For example, in Fig. 1, the possible founder pairs for r_6 are $\langle F_1, F_2 \rangle$, $\langle F_2, F_1 \rangle$, $\langle F_2, F_3 \rangle$, $\langle F_2, F_3 \rangle$, $\langle F_2, F_4 \rangle$, and $\langle F_4, F_2 \rangle$.

Instead of considering each site, we consider each subregion in the dynamic programming solution. Assign $fp_{q_p}^{r_p} = \langle of_{q_p}^{a,r_p}, of_{q_p}^{b,r_p} \rangle$ to be the founder pair for subregion r_p, where $1 \leq q_p \leq |FP^{r_p}|$, and $fp_{q_{p+1}}^{r_{p+1}} = \langle of_{q_{p+1}}^{a,r_p}, of_{q_{p+1}}^{b,r_p} \rangle$ to be the founder pair for subregion r_{p+1}, where $1 \leq q_{p+1} \leq |FP^{r_{p+1}}|$. Similarly, we count the number of founder switches between $fp_{q_p}^{r_p}, fp_{q_{p+1}}^{r_{p+1}}$ as:

$$FounderSwitch(fp_{q_p}^{r_p}, fp_{q_{p+1}}^{r_{p+1}}) = \begin{cases} 0 : \text{if } of_{q_p}^{a,r_p} = of_{q_{p+1}}^{a,r_{p+1}} \wedge of_{q_p}^{b,r_p} = of_{q_{p+1}}^{b,r_{p+1}} \\ 1 : \text{if } of_{q_p}^{a,r_p} = of_{q_{p+1}}^{a,r_{p+1}} \wedge of_{q_p}^{b,r_p} \neq of_{q_{p+1}}^{b,r_{p+1}} \\ \phantom{1 : \text{if }} of_{q_p}^{a,r_p} \neq of_{q_{p+1}}^{a,r_{p+1}} \wedge of_{q_p}^{b,r_p} = of_{q_{p+1}}^{b,r_{p+1}} \\ 2 : \text{if } of_{q_p}^{a,r_p} \neq of_{q_{p+1}}^{a,r_{p+1}} \wedge of_{q_p}^{b,r_p} \neq of_{q_{p+1}}^{b,r_{p+1}} \end{cases} \quad (5)$$

Let $K_{min}(r_1 \ldots r_{p-1}|fp_{q_p}^{r_p})$ be the minimum number of segments in any segmentation solution over the subsequence covered by $r_1 \ldots r_p$ which takes the founder pair $fp_{q_p}^{r_p}$ at subregion r_p. The minimum number of segments over the entire genotype sequence $K_{min}(r_1 \ldots r_P)$ where r_P is the last subregion can be computed as:

$$K_{min}(r_1 \ldots r_P) = min_{fp_{q_P}^{r_P} \in FP^{r_P}}\{K_{min}(r_1 \ldots r_{P-1}|fp_{q_P}^{r_P})\} \quad (6)$$

The main recurrence of the dynamic-programming algorithm is as follows:

$$K_{min}(r_1 \ldots r_{p-1}|fp_{q_p}^{r_p}) = min_{fp_{q_{p-1}}^{r_{p-1}} \in FP^{P-1}}\{K_{min}(r_1 \ldots r_{p-2}|fp_{q_{p-1}}^{r_{p-1}}) + FounderSwitch(fp_{q_{p-1}}^{r_{p-1}}, fp_{q_p}^{r_p})\} \quad (7)$$

And initially,

$$K_{min}(\Phi|fp_{q_1}^{r_1}) = 2, \forall fp_{q_1}^{r_1} \in FP^{r_1} \quad (8)$$

We can also solve this dynamic programming problem by populating a table T which contains P rows where row p has at most $|FP^{r_p}|$ entries. Entry $T(p, q_p), 1 \leq q_p \leq FP^{r_p}$ is filled with $K_{min}(r_1 \ldots r_{p-1}|fp_{q_p}^{r_p})$ during the computation. There are at most N^2 founder pairs for each subregion r_p, i.e., $|FP^{r_p}| \leq N^2$. The table we populate is of size $O(PN^2)$. The computation of all the maximal shared intervals is $O(LN)$, and filling out each entry in the table costs $O(N^2)$. Thus the computational complexity of region-based dynamic programming algorithm is $O(LN + PN^4)$. Compared with the site-based algorithm which has a time complexity of $O(LN^4)$, if P is much smaller than L, we can greatly reduce the running time, especially for large L.

3.1   Enforcing the Constraints and Modeling Noise

Comparable Number of Founder Switches on Both Haplotypes: During each meiosis autosomes undergo one recombination on average. Thus, during the development of an recombinant inbred-line (RIL), one expects that the number of founder switches per haplotype at each generation to be comparable.

During each mating in the evolving history, each of the two haplotypes may be generated by a new recombination. Therefore, we may expect that for the given genotype to be segmented, the number of segments for the two haplotype sequences are comparable.

For a segmentation $Seg(G)$ of genotype sequence G, which is composed of a segmentation $Seg_a(H_a)$ on haplotype H_a and a segmentation $Seg_b(H_b)$ on haplotype H_b, we put an extra constraint on the minimum segmentation as follows: the difference of the numbers of the segments in the two haplotypes is no more than a threshold α: $||Seg_a(H_a)| - |Seg_b(H_b)|| < \alpha$, where α is a nonnegative integer. The definition of the minimum segmentation for G is then modified as: $|MinSeg(G)| = min\{|Seg(G)|\}$, where $MinSeg(G) = \{Seg_a^*(H_a), Seg_b^*(H_b)\}$ and $||Seg_a^*(H_a)| - |Seg_b^*(H_b)|| < \alpha$.

To compute the minimum segmentation solution with constraints, we propose an efficient heuristic which prunes out solutions that do not satisfy the constraint before they are fully computed during the table population process. This greatly reduces the computation, especially when there are a lot of minimum segmentation solutions that do not satisfy the constraint. We will explain how it works with the region-based algorithm. Assume that we have a founder set $\{F^1, \ldots, F^N\}$, and a genotype sequence G. For any minimum segmentation solution $MinSeg(G) = \{Seg_a^*(H_a), Seg_b^*(H_b)\}$, we have the following lemma:

Lemma 1. *For any homozygous region R on G, and any minimum segmentation solution $\{Seg_a^*(H_a), Seg_b^*(H_b)\}$, let the set of segments in $Seg_a^*(H_a)$ which completely fall inside R be $Seg_a^*(H_a) \cap R$, and the set of segments in $Seg_b^*(H_b)$ which completely fall inside R be $Seg_b^*(H_b) \cap R$, then we have*

$$||Seg_a^*(H_a) \cap R| - |Seg_b^*(H_b) \cap R|| \leq 2 \tag{9}$$

Proof. Details of the proof are presented in [ZWMPVT].

Lemma 2. *For a genotype sequence G containing Z heterozygous sites, any of its minimum segmentation $\{Seg_a^*(H_a), Seg_b^*(H_b)\}$ satisfies:*

$$||Seg_a^*(H_a)| - |Seg_b^*(H_b)|| \leq 3Z + 1 \tag{10}$$

Proof. For more details, please refer to [ZWMPVT].

We can use Lemma 2 in the dynamic programming algorithm when we are populating the table. Assume currently we are filling out the entry $T(p, q_p)$, *i.e.*, we are computing $K_{min}(r_1 \ldots r_{p-1} | fp_{q_p}^{r_p})$ according to Eq.(7), which is the minimum segmentation for the subsequence from r_1 to r_p with $fp_{q_p}^{r_p}$ as the founder pair for r_p. In addition to computing the minimum segmentation, we also keep track of the difference between the number of segments over two haplotypes in the minimum segmentation we have computed, $\delta(p, q_p) = ||Seg_a^*(H_a[r_1, r_p])| - |Seg_b^*(H_b[r_1, r_p])||$. Let the number of heterozygous sites in the remaining part of the sequence be $Z([r_{p+1}, r_P])$. Then if $\delta(p, q_p) - (3Z([r_{p+1}, r_P]) + 1) > \alpha$, according to Lemma 2, we know that the corresponding solution will not be able to generate a minimum segmentation solution for the entire sequence where the difference between the number of segments on both haplotypes is less than α. If the minimum segmentation solution we consider is from $T(p-1, q_{p-1})$ to $T(p, q_p)$, then if $\delta(p, q_p) - (3Z([r_{p+1}, r_P]) + 1) > \alpha$, we will not add the backtrack pointer from $T(p, q_p)$ to $T(p-1, q_{p-1})$.

Modeling Point Mutation, Genotyping Errors and Gene Conversions:
There are both biological and technical resources of noise in genotyping, which
include point mutations and gene conversions, and genotyping errors. We consider these potential noise sources in the data in our segmentation algorithms.

A point mutation or genotyping error can be treated as a single site mismatch
that falls within a shared interval between a copy of haplotype sequence and a
founder sequence. A gene conversion can be treated as a short sequence of mismatches that fall within a shared interval between a copy of haplotype sequence
and a founder sequence. In the following, we explain how we model these noise
sources in our region-based dynamic programming algorithm. We first consider
the point mutation, gene conversion, and genotyping error which happen side
a homogeneous range. During the computation of maximal shared intervals between sequence G and each founder F_n over a homozygous region R, assume that
we have two maximal intervals Δ_1, Δ_2 between G and F_n which are over the intervals $I_1 = [b_1, e_1]$ and $I_2 = [b_2, e_2]$. We know that I_1 and I_2 are not overlapping
or touching. Let I_1 be the interval on the left, $i.e.$, $e_1 < b_2 - 1$. If $e_1 = b_2 - 2$,
then there is a single mismatch at site $e_1 + 1$ within the combined region $[b_1, e_2]$.
Assume that both I_1 and I_2 are of enough length, then this single mismatch may
be a point mutation or genotyping error. Therefore, we create another shared
interval Δ_3 between G and F_n over the single-site interval $I_3 = [e_1 + 1, e_1 + 1]$.
This interval has a probability of $\beta < 1$ to be a shared interval between G and
F_n. β is defined as the probability of a single mismatch inside a shared interval being a point mutation or genotyping error. Similarly, if $e_1 < b_2 - 2$ but
$b_2 - 1 - e_1 < gc$, which means the gap between interval I_1 and I_2 is shorter
than a maximal possible length gc of a typical gene conversion, then this short
sequence of mismatches may be a gene conversion, assuming both I_1 and I_2 are
of enough length. We create another shared interval Δ_4 between G and F_n over
the short interval $I_4 = [e_1 + 1, b_2 - 1]$. This interval has a probability of $\gamma < 1$ to
be a shared interval between G and F_n. γ is defined as the probability of a short
sequence of mismatches inside a shared interval being a gene conversion. We can
consider the point mutation and genotyping error that happen at a heterozygous
site in a similar manner, where we check whether the heterozygous site is a single
mismatch falling into a shared interval, $i.e.$ the shared interval to the left of the
heterozygous site and to the right of the heterozygous site are from the same
founder. The maximal shared intervals computed without considering noise are
of probability 1. By modeling the noise using intervals with probability less than
1, we can compute the minimum segmentation solution with desired noise tolerance $\theta < 1$. When we compute each entry $T(p, q_p)$ in the table, we keep track
of the accumulated probability which is the multiplication of the probabilities
of the intervals in the founder pairs on the minimum segmentations solution so
far. We only keep the solution with the accumulated probability no less than θ.

Modeling Missing Values: Besides noise and incorrect values, there can also
be missing values in the data. Assume that the missing value is in founder F_n,
at site l, and the value at the same site on the sequence G is not missing. If l is a
homozygous site, we fill out f_l^n using g_l. If l is a heterozygous site, we consider f_l^n

can be either 0 or 1 when we create the founder pairs for this heterozygous site. If the missing value is on G at site l, we consider it to be either 0 or 1, which means this site can be in a maximal shared interval with any of the founder, no matter whether the founder has a missing value at the same site or not. In this way, we can generate the minimum segmentation solution with the smallest possible number of segments. The missing values in both founders and genotype sequences are filled with values in each solution (it may be filled with different values for different minimum segmentation solutions).

4   Experimental Results

We tested the performance of our region-based dynamic programming algorithm on the simulated data. As we presented earlier, the region-based algorithm has less computational complexity than the site-based algorithm. We only demonstrate the results on region-based algorithm.

The set of the founder sequences $\{F_n\}$ and the genotype sequence G (corresponding two haplotype sequences H_a, H_b) are generated so that: (a) The set of founders are generated randomly, except that, at each site, there is at least a founder taking 0 and at least a founder taking 1, (b) The number of the heterozygous sites in G is $h\_rate \times L$ where h_rate is a parameter representing the occurrence rate of the heterozygous sites, L is the total number of sites, and (c) H_a and H_b are generated by randomly patching up n_seg random segments from the founders.

Note that, the segmentation during the generation provides a lower bound on the number of segments in the minimum segmentation. The computed minimum segmentation may not have the same number of segments on both haplotypes. The code is implemented using Matlab, and the experiments are performed on an Intel Core 2 Duo 1.6GHz machine with 3GB memory.

Running Time. We evaluated the running time by varying the number of founders (N), the number of sites (L), the number of segments (n_seg), and the heterozygous sites occurrence rate (h_rate).

Fig. 2(a) highlights the running time by varying the number of founders from 2 to 10. Other parameters are fixed to be $L = 1000, n\_seg = 30, h\_rate = 0.01$. The complexity of the algorithm is $O(LN + PN^4)$, which is demonstrated by the superlinear increase in the running time with increasing N. Fig. 2(b) shows the running time with varying number of sites. All data sets contain 6 founders. n_seg for 10, 100, 1000, 5000, and 10000-site data sets are chosen to be 3, 8,

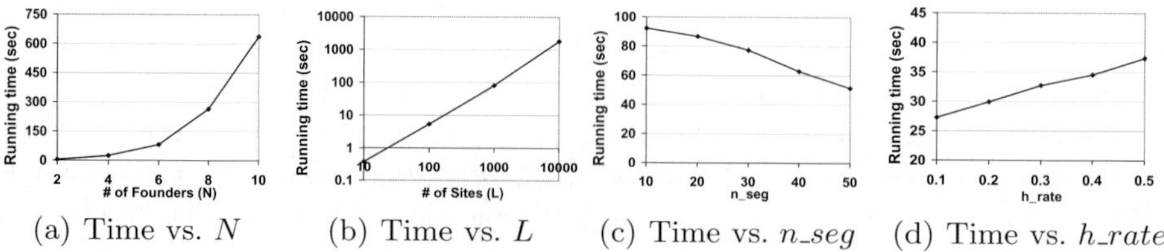

(a) Time vs. N (b) Time vs. L (c) Time vs. n_seg (d) Time vs. h_rate

Fig. 2. Running time with varying parameters

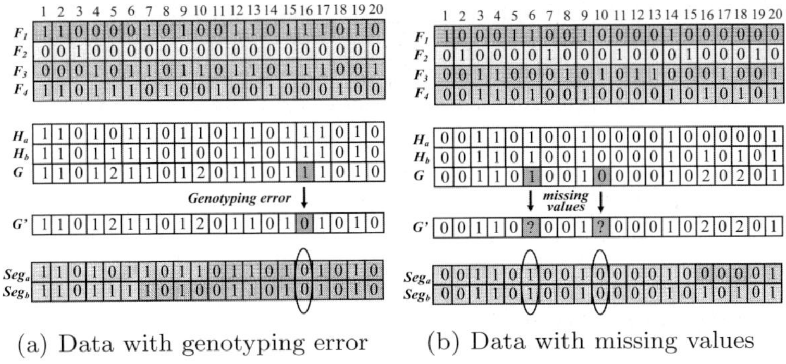

(a) Data with genotyping error   (b) Data with missing values

Fig. 3. Segmentation results (better viewed in color)

30, 150, 300, respectively. h_rate is 0.01 for all the data sets except for the 10-site set where the h_rate is set to be 0.1 to guarantee 1 heterozygous site. Both X-axis and Y-axis are in log scale, we can observe that the running time is linear to the number of sites L. In Fig. 2(c), the running time decreases as n_seg increases. The reason is due to the increased number of founder pairs generated from large shared intervals when n_seg is small. The data sets have 1000 sites, 6 founders, and h_rate is set to 0.01. Fig. 2(d) shows the increase of the running time with increasing h_rate. More heterozygous sites introduce more subregions which cause the running time to increase. The data sets contain 1000 sites, 4 founders and n_seg is set to 50.

Effect of Enforcing the Constraint. Table 1 shows the results on three datasets where different constraints are applied. As shown in the table, enforcing the constraint greatly reduces the number of minimum segmentation solutions generated. The running time also decreases with the application of the constraints. For dataset #1, there are 28 minimum segmentation solutions where both haplotypes take the same number of segments. However, the number of solutions increases to 40 when the number of segments over the two haplotypes can differ by 1. Similarly, for dataset #2, the number of solutions increased 5 times, and the run time also doubled. A similar trend is observed for dataset #3.

Error Tolerance. We tested our algorithms on simulated data with point mutation, genotyping errors, and missing values. Fig. 3 shows two example segmentation results. In Fig. 3(a), the data set contains four founders $\{F_1, F_2, F_3, F_4\}$ each of which has 20 sites. The two copies of the haplotypes H_a and H_b, and the corresponding genotype G is shown in the figure as the ground truth. A random site chosen to take a genotyping error. The resulting genotype G' has a genotyping error (1 is mistaken as 0) at site 16. Our segmentation solution on G' is shown at the bottom of Fig. 3(a). Although F_1 does not match G at site 16, F_1 is still chosen as the founder in Seg_a and Seg_b, since site 16 is a single mismatch inside a long shared interval with F_1. Fig. 3(b) shows the result on a data set with missing values. Two random sites (site 6 and site 10) are chosen

Table 1. Effect of Enforcing the Constraint

dataset	#1		#2			#3	
parameters	$N = 4, L = 20$		$N = 2, L = 50$			$N = 6, L = 20$	
	$n\_seg = 4, h = 0.1$		$n\_seg = 8, h = 0.05$			$n\_seg = 6, h = 0.05$	
α	0	1	0	1	2	0	1
# solutions	28	40	92	276	460	6	356
running time (sec)	0.515	0.718	0.453	0.656	0.812	1.09	1.437

to be the sites with missing values. As shown at the bottom of the figure, our algorithm still generates the correct minimum segmentation with the values at both sites filled in. Fig. 3 is better to be viewed in color.

5   Conclusions

In this paper, we studied the minimum segmentation problem for genotype sequence given a set of founders. We proposed dynamic programming algorithms which run in polynomial time. The algorithms can effectively handle the constraint which requires comparable number of founder switches on both haplotypes. Moreover, the algorithms can deal with the noise in the data such as genotyping errors, point mutations, missing values, etc.

References

[C]   Churchill, G.A., et al.: The Collaborative Cross, a community resource for the genetic analysis of complex traits. Nat. Genet. 36, 1133–1137 (2002)

[DRSHL]   Dally, M., Rioux, J., Schaffner, S.F., Hudson, T., Lander, E.: High-resolution haplotype structure in the human genome. Nat. Genet. 29, 229–232 (2001)

[G]   Gusfield, D.: Haplotyping as perfect phylogeny: conceptual framework and efficient solutions. In: RECOMB, pp. 166–175 (2002)

[GEL]   Gusfield, D., Eddhu, S., Langley, C.: Optimal, efficient reconstruction of phylogenetic networks with constrained recombination. J. Bioinf. Comput. Biol. 2, 173–213 (2004)

[GSNM]   Gabriel, S.B., Schaffner, S.F., Nguyen, H., Moore, et al.: The structure of haplotype blocks in the human genome. Science 296, 2225–2229 (2002)

[SHBCI]   Schwartz, R., Halldorson, B.V., Bafna, V., Clark, A.G., Istrail, S.: Robustness of inference of haplotyp block structure. J. Comput. Biol. 10, 13–19 (2003)

[THW]   Threadgill, D.W., Hunter, K.W., Williams, R.W.: Genetic dissection of complex and quantitative traits: from fantasy to reality via a community effort. Mamm Genome 13, 175–178 (2002)

[U]   Ukkonen, E.: Finding founder sequences from a set of recombinants. In: Guigó, R., Gusfield, D. (eds.) WABI 2002. LNCS, vol. 2452, pp. 277–286. Springer, Heidelberg (2002)

[VFM]   Valdar, W., Flint, J., Mott, R.: Simulating the Collaborative Cross: Power of Quantitative Trait Loci Detection and Mapping Resolution in Large Sets of Recombinant Inbred Strains of Mice. Genetics 172(3), 1783–1797 (2006)
[WG]   Wu, Y., Gusfield, D.: Improved algorithms for inferring the minimum mosaic of a set of recombinants. In: Ma, B., Zhang, K. (eds.) CPM 2007. LNCS, vol. 4580, pp. 150–161. Springer, Heidelberg (2007)
[ZWMPVT]   Zhang, Q., Wang, W., McMillan, L., Prins, J., Villena, F.P., Threadgill, D.: Genotype Sequence Segmentation: Handling Constraints and Noise. UNC Tech. Report (2008)

Constructing Phylogenetic Supernetworks from Quartets

Stefan Grünewald[1], Andreas Spillner[2], Kristoffer Forslund[3],
and Vincent Moulton[2]

[1] CAS-MPG Partner Institute for Computational Biology, Chinese Academy of
Sciences, Shanghai, China and Max-Planck-Institute for Mathematics in the Sciences,
Leipzig, Germany
stefan@picb.ac.cn
[2] School of Computing Sciences, University of East Anglia, Norwich, UK
aspillner@cmp.uea.ac.uk, vincent.moulton@cmp.uea.ac.uk
[3] Stockholm Bioinformatics Centre, Stockholm University, Sweden
krifo@sbc.su.se

Abstract. In phylogenetics it is common practice to summarize collections of partial phylogenetic trees in the form of supertrees. Recently it has been proposed to construct phylogenetic supernetworks as an alternative to supertrees as these allow the representation of conflicting information in the trees, information that may not be representable in a single tree. Here we introduce *SuperQ*, a new method for constructing such supernetworks. It works by breaking the input trees into quartet trees, and stitching together the resulting set to form a network. The stitching process is performed using an adaptation of the QNet method for phylogenetic network reconstruction. In addition to presenting the new method, we illustrate the applicability of SuperQ to three data sets and discuss future directions for testing and development.

1   Introduction

In phylogenetics it is common practice to summarize a collection of phylogenetic trees as a consensus tree [1] or, in case the trees are on different leaf sets (i.e. they are partial trees), as a supertree [2]. However, this may result in information loss due to conflicting groupings in the input trees that cannot be represented in any single tree. In an attempt to remedy this problem various methods have been recently proposed for constructing consensus/supernetworks rather than trees [3,4,5,6,7]. Such networks have the advantage that they allow for the representation of conflicting information, although at the price that a tree is no longer necessarily reconstructed. They have been used for applications such as visualizing large collections of trees (arising e.g. from Bayesian tree inference) [4], and multiple gene tree analyses [7].

In this paper, we describe a new method for constructing supernetworks from a collection of partial phylogenetic trees. Currently there are basically two methods available to construct supernetworks: *Z-closure* [7] and *Q-imputation* [3].

K.A. Crandall and J. Lagergren (Eds.): WABI 2008, LNBI 5251, pp. 284–295, 2008.

© Springer-Verlag Berlin Heidelberg 2008

Z-closure essentially works by converting the input trees into a set of partial bipartitions, corresponding to the edges of the input trees, and then iteratively applying a set-theoretical operation to produce a collection of bipartitions of the total leaf set. This collection of bipartitions is subsequently represented in the form of a split-network [8]. Q-imputation, on the other hand, first uses quartet information to complete the partial trees to trees on the total leaf set, and then constructs a consensus network for these trees. This is a split-network representing those bipartitions that are present in some large proportion of the completed trees.

Our new method, *SuperQ*, works in a different manner. Each of the partial trees is broken down into a collection of quartet trees, and the union of all of these quartet trees is then stitched together to form a split-network. To perform this stitching process we adapt the QNet algorithm for constructing phylogenetic networks from quartet trees [9,10]. One of the main technical difficulties that we have to overcome to do this is to find a way to assign weights to the edges in the resulting network. This difficulty arises as several quartet trees may be missing after the break-up process, and QNet requires all possible quartet trees. One of the main advantages of SuperQ over Z-closure and Q-imputation is that it is guaranteed to generate a planar network, whereas the latter methods can potentially produce rather complex (and therefore difficult to interpret) networks.

The rest of this paper is organized as follows. In the next section we describe our new method. In Section 3 we illustrate the applicability of SuperQ by using it on three datasets, and in Section 4 we discuss possible future directions for the testing and further development of our method.

2   Methods

In this section we describe our method for constructing supernetworks. We first introduce some terminology and notation – more details concerning these and related concepts may be found in e.g. [11].

2.1   Terminology and Notation

Let X be a finite set. A *quartet* on X is a bipartition $a_1 a_2 | b_1 b_2$ of a 4-element subset $\{a_1, a_2, b_1, b_2\}$ of X. By $\mathcal{Q}(X)$ we denote the set of all quartets on X. A *quartet system* on X is a subset of $\mathcal{Q}(X)$. A *quartet-weight function* on $\mathcal{Q}(X)$ is a map $\mu : \mathcal{Q}(X) \to \mathbb{R}_{\geq 0}$.

A *split* of X is a bipartition $A|B$ of X into two non-empty subsets A and B of X. We denote the set of all splits of X by $\Sigma(X)$. A *split system* on X is a subset of $\Sigma(X)$. A *split-weight function* on $\Sigma(X)$ is a map $\nu : \Sigma(X) \to \mathbb{R}_{\geq 0}$. The *support* of a split-weight function ν is the split system $\mathrm{supp}(\nu) = \{S \in \Sigma(X) : \nu(S) > 0\}$.

A split $S = A|B$ of X *extends* a quartet $q = a_1 a_2 | b_1 b_2$, denoted $S \succ q$, if $\{a_1, a_2\} \subseteq A$ and $\{b_1, b_2\} \subseteq B$, or $\{b_1, b_2\} \subseteq A$ and $\{a_1, a_2\} \subseteq B$. Every split-weight function ν on $\Sigma(X)$ induces a quartet-weight function μ_ν defined

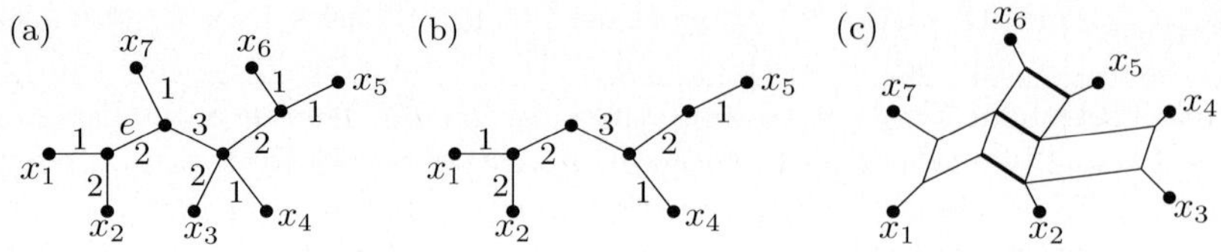

Fig. 1. (a) An X-tree with $X = \{x_1,\ldots,x_7\}$. (b) The subtree induced by $\{x_1, x_2, x_4, x_5\}$. (c) A split network on X. Splits are represented by parallel edges, e.g. the bold edges correspond to the split $x_1x_6x_7|x_2x_3x_4x_5$. The weight of the splits is usually represented by the length of the corresponding edges in the network.

by putting

$$\mu_\nu(q) = \sum_{S \in \Sigma(X), S \succ q} \nu(S)$$

for every quartet $q \in \mathcal{Q}(X)$.

As a formalization of phylogenetic trees we now recall the concept of an X-*tree* [11, ch. 2]. It is a graph theoretical tree with vertex set V and edge set E together with a map $\varphi : X \to V$ such that $\varphi^{-1}(v) \neq \emptyset$ for every vertex $v \in V$ of degree at most 2. Note that removing any edge e of an X-tree $T = (V, E, \varphi)$ breaks T into two subtrees with vertex sets V_1 and V_2, respectively, which induces a split $S_e = \varphi^{-1}(V_1)|\varphi^{-1}(V_2)$ of X associated to edge e. An *edge-weight function* on an X-tree $T = (V, E, \varphi)$ is a map $\omega : E \to \mathbb{R}_{\geq 0}$.

Note that an X-tree $T = (V, E, \varphi)$ with an edge-weight function ω induces a split-weight function ν_T on X given by

$$\nu_T(S) = \begin{cases} \omega(e) & \text{if } S = S_e \text{ for some edge } e \in E, \\ 0 & \text{else,} \end{cases}$$

for every split $S \in \Sigma(X)$, and also a quartet-weight function μ_T defined by setting $\mu_T = \mu_{\nu_T}$. Also note that the weight of the pendent edges of T does not influence μ_T. Formulated in terms of split-weight functions ν, μ_ν does not depend on the weight assigned by ν to the *trivial splits* of X, that is, splits $A|B \in \Sigma(X)$ with $\min\{|A|, |B|\} = 1$.

To illustrate some of the definitions given above, Figure 1(a) depicts an edge weighted X-tree T with $X = \{x_1,\ldots,x_7\}$. Associated to edge e of T is the split $S_e = x_1x_2|x_3x_4x_5x_6x_7$ of X. In Figure 1(b) we show the subtree T' of T spanned by the 4 elements in $\{x_1, x_2, x_4, x_5\}$. Note that $\mu_T(x_1x_2|x_4x_5) = 5$, that is, the induced weight of this quartet is the total weight of those edges in T' that are neither on the path from x_1 to x_2 nor on the path from x_4 to x_5.

2.2   Description of the Method

The input for our method is a collection $\mathcal{T} = \{T_1, T_2, \ldots, T_t\}$ of X_i-trees, $T_i = (V_i, E_i, \varphi_i)$, each with edge-weight function $\omega_i : E_i \to \mathbb{R}_{\geq 0}$. Define $X = \bigcup_{i=1}^{t} X_i$.

Our goal is to summarize the information contained in the collection $\mathcal{T}$ in the form of a split-weight function ν on $\Sigma(X)$ in such a way that the split system

SuperQ($\mathcal{T}$)

Input:	$\mathcal{T} = \{T_1, T_2, \ldots, T_t\}$ a collection that contains for every $i \in \{1, \ldots, t\}$ an X_i-tree T_i with an edge-weight function ω_i.
Output:	A split-weight function ν on $\Sigma(\bigcup_{i=1}^{t} X_i)$ with the property that $\mathrm{supp}(\nu)$ is circular.

---

1. Compute $X = \bigcup_{i=1}^{t} X_i$.
2. Compute the quartet-weight function μ^* on $\mathcal{Q}(X)$.
3. $\Sigma_{\mathcal{T}} = \mathrm{QNET}(X, \mu^*)$.
4. Compute the quartet system $\mathcal{Q}_{\mathcal{T}}$.
5. Fix an ordering $q_1, \ldots, q_r$ of the quartets in $\mathcal{Q}_{\mathcal{T}}$.
6. Fix an ordering $S_1, \ldots, S_s$ of the splits in $\Sigma_{\mathcal{T}}^*$.
7. Compute the vector $\mu_{\mathcal{T}}$.
8. Compute the reduced topological matrix M.
9. Find an initial solution y^* to the non-negative least squares problem (1).
10. For every $k \in \{1, \ldots, s\}$ set up the linear program Π_k and compute λ_k.
11. Define the split-weight function ν on $\Sigma(X)$ according to (2).
12. **return** ν.

Fig. 2. Pseudocode for an algorithm that computes a weighted circular split system from a collection of edge weighted partial trees

$\mathrm{supp}(\nu)$ has the following property: There exists an ordering $x_1, x_2, \ldots, x_n$ of X such that for every split $A|B \in \mathrm{supp}(\nu)$ there are $j, k \in \{1, \ldots, n\}$, $j \leq k$, such that $A = \{x_j, \ldots, x_k\}$ or $B = \{x_j, \ldots, x_k\}$. In general split systems with this property are known as *circular split systems* [12]. Any such split system can be represented in the form of an outer-labeled planar split network (see Figure 1(c) for a simple example, and cf. [8,13] for more details concerning these networks and how they can be used to represent split systems).

In Figure 2 we present pseudocode summarizing our algorithm SuperQ. After computing the set X (Line 1) we compute a quartet-weight function μ^* on $\mathcal{Q}(X)$ (Line 2) as follows. First we construct from the quartet-weight function μ_{T_i} on $\mathcal{Q}(X_i)$ a quartet-weight function μ_i on $\mathcal{Q}(X)$ by defining

$$\mu_i(q) = \begin{cases} \frac{1}{|\{j \in \{1, \ldots, t\}: q \in \mathcal{Q}(X_j)\}|} \cdot \mu_{T_i}(q) & \text{if } q \in \mathcal{Q}(X_i), \\ 0 & \text{else,} \end{cases}$$

for every quartet $q \in \mathcal{Q}(X)$. From this we then define a quartet-weight function μ^* on $\mathcal{Q}(X)$ by setting $\mu^*(q) = \sum_{i=1}^{t} \mu_i(q)$ for every $q \in \mathcal{Q}(X)$. Intuitively, μ^* assigns to a quartet $q = a_1 a_2 | b_1 b_2 \in \mathcal{Q}(X)$ the average weight that is induced on q by those input trees T_i with $\{a_1, a_2, b_1, b_2\} \subseteq X_i$.

Next (Line 3) the quartet-weight function μ^* is input into the QNet algorithm [9,10]. As an intermediate step this algorithm constructs a maximum circular split system $\Sigma_{\mathcal{T}} \subseteq \Sigma(X)$. Hence, the rest of the algorithm (Lines 4-12) is concerned with assigning non-negative weights to splits in $\Sigma_{\mathcal{T}}^*$, the set of non-trivial splits in $\Sigma_{\mathcal{T}}$.

To explain the difficulties that arise in the process, recall that the original QNet algorithm computes these weights (for details see [10]) using a non-negative least squares approach as described in e.g. [14]. However, in the context of our method we have to overcome the following problem: Let $\mathcal{Q}_0 = \{q \in \mathcal{Q}(X) : \forall i \in \{1, \ldots, t\}\ (q \notin \mathcal{Q}(X_i))\}$ be the set of quartets in $\mathcal{Q}(X)$ that are neither supported nor contradicted by any tree in the collection $\mathcal{T}$. Note that the quartet-weight function μ^* assigns weight 0 to the quartets in $\mathcal{Q}_0$ but the input trees do not naturally give rise to a weighting of these quartets. Therefore, when we compute a weighting ν of the splits in $\Sigma_{\mathcal{T}}^*$ such that the induced quartet-weight function μ_ν is as close as possible (in the least squares sense) to μ^* we need to ignore the quartets in $\mathcal{Q}_0$.

To describe our adaption of the non-negative least squares approach taken by the original QNet algorithm more formally, define $\mathcal{Q}_{\mathcal{T}} = \mathcal{Q}(X) \backslash \mathcal{Q}_0$, $r = |\mathcal{Q}_{\mathcal{T}}|$ and $s = |\Sigma_{\mathcal{T}}^*|$. Note that, since $\Sigma_{\mathcal{T}}$ is a maximum circular split system, $s = \binom{|X|}{2} - |X|$ [12]. Now, fix an arbitrary ordering $q_1, \ldots, q_r$ of the quartets in $\mathcal{Q}_{\mathcal{T}}$ and an arbitrary ordering $S_1, \ldots, S_s$ of the splits in $\Sigma_{\mathcal{T}}^*$. We define an $r \times s$-matrix $M = (m_{jk})$ by setting $m_{jk} = 1$ if $S_k \succ q_j$ and $m_{jk} = 0$ otherwise. We call M the *reduced topological matrix*.

We then solve the following non-negative least squares problem: compute a column vector $y = (y_1, \ldots, y_s)^\top$ with entries in $\mathbb{R}_{\geq 0}$ such that

$$\sum_{j=1}^{r} (\mu^*(q_j) - z_j)^2 \tag{1}$$

is minimum where $z = (z_1, \ldots, z_r)^\top = M \cdot y$. Note that the entries of vector y are the weights assigned to the splits in $\Sigma_{\mathcal{T}}^*$, and the entries of vector z are the weights induced on the quartets in $\mathcal{Q}_{\mathcal{T}}$. Also note that if the reduced topological matrix M has full rank then we obtain a unique solution vector y to this non-negative least squares problem. Otherwise we use the following strategy.

View the quartet-weight function μ^* restricted to $\mathcal{Q}_{\mathcal{T}}$ as a column vector $\mu_{\mathcal{T}} = (\mu^*(q_1), \ldots, \mu^*(q_r))^\top$, let $\|u\|$ denote $\sum_{j=1}^{r}(u_j)^2$ for every vector $u = (u_1, \ldots, u_r) \in \mathbb{R}^r$, and $\mathbf{0}$ denote the vector (of appropriate dimension) with all entries equal to zero. For two vectors $y = (y_1, \ldots, y_s)$ and $y' = (y_1', \ldots, y_s')$ we use $y \geq y'$ to express the fact that $y_k \geq y_k'$ for every $k \in \{1, \ldots, s\}$. Define $\delta_{opt} = \min\{\|\mu_{\mathcal{T}} - M \cdot y\| : y \in \mathbb{R}^s, y \geq \mathbf{0}\}$. Then the set of all solutions to the non-negative least squares problem (1) corresponds to the set of vectors $D_{opt} = \{y \in \mathbb{R}^s : \|\mu_{\mathcal{T}} - M \cdot y\| = \delta_{opt}, y \geq \mathbf{0}\}$. We then aim to compute for every split S_k, $1 \leq k \leq s$, a lower bound λ_k on the weight that this split is assigned over all solutions to the non-negative least squares problem (1), that is, $\lambda_k = \min\{y_k : (y_1, \ldots, y_s) \in D_{opt}\}$.

To this end, let $\ker(M) = \{y \in \mathbb{R}^s : M \cdot y = \mathbf{0}\}$ be the null-space or *kernel* of M. As a starting point we compute an arbitrary element y^* of D_{opt} using the algorithm NNLS by Lawson and Hanson [15, pp. 160-165]. Note that $y^* + y' \in D_{opt}$ for every $y' \in \ker(M)$ with the property that $y^* + y' \geq \mathbf{0}$. Moreover, every element $y \in D_{opt}$ can be written as $y = y^* + y'$ for some $y' \in \ker(M)$, since the non-negative least squares problem (1) restricted to the orthogonal complement of $\ker(M)$ has a unique solution. Hence, defining $K_{opt} = \{y' \in \ker(M) : y^* + y' \geq \mathbf{0}\}$, it follows that $D_{opt} = y^* + K_{opt}$. Consequently, for every $k \in \{1, \ldots, s\}$ the computation of the lower bound λ_k can be formulated as the following linear program Π_k:

objective: minimize $h(y) = h(y_1, \ldots, y_s) = y_k$
subject to: $y^* + y \geq \mathbf{0}$
$\qquad\qquad M \cdot y = \mathbf{0}$,

which can be solved using standard linear programming techniques. Putting $\overline{\lambda} = \sum_{k=1}^{s} \lambda_k / s$, the output split-weight function ν of our method is then defined by

$$\nu(S) = \begin{cases} \lambda_k & \text{if } S = S_k \in \Sigma_{\mathcal{T}}^*, \\ \overline{\lambda} & \text{if } S \text{ is a trivial split of } X, \\ 0 & \text{else}, \end{cases} \qquad (2)$$

for every split $S \in \Sigma(X)$. Clearly, $\text{supp}(\nu) \subseteq \Sigma_{\mathcal{T}}$ and, thus, $\text{supp}(\nu)$ is circular.

Note that ν provides a very conservative estimate of the weights of the splits in $\Sigma_{\mathcal{T}}^*$ – if a split S is assigned a positive weight $\nu(S)$ then S is assigned at least this weight in *every* solution to the non-negative least squares problem (1). To every trivial split of X we assign the average weight of a split in $\Sigma_{\mathcal{T}}^*$. This is done to improve the readability of the resulting split network.

2.3   Implementation

We implemented our algorithms for computing the quartet-weight function μ^* and subsequently applying QNet in JAVA. Our approach to estimate the weight of the splits in $\Sigma_{\mathcal{T}}^*$ was implemented in C using the GNU Linear Programming Kit (GLPK) version 4.21. The output is visualized using the current version of the SplitsTree program [16]. The examples we present below were processed on a 1.7 GHz desktop computer with 500MB RAM under Linux. As additional information we also computed the rank of the reduced topological matrix with MATLAB using the build-in function `rank`.

3   Examples

To illustrate the applicability of SuperQ, we present the output for 3 biological data sets. For the first data set, in which one of the input trees contains all taxa, this yields a reduced topological matrix M of full rank and, hence the non-negative least squares problem (1) has a unique solution. For the other two datasets M doesn't have full rank, therefore, (1) has no unique solution, and so the weighting of the splits was computed via the linear programs Π_k.

The first data set consists of seven gene trees for microgastrine wasps which originally appeared in [17] and was used as an example for Q-imputation in [3]. The total number of taxa is 45. In Figure 3 we depict the supernetwork constructed using SuperQ. It took approximately 15 minutes to compute this network. The taxa in capital letters form an outgroup which is clearly grouped together in the supernetwork produced by SuperQ.

The supernetwork for the wasp dataset obtained by Q-imputation presented in [3, p. 64] displays similar major groupings and is almost a tree. In contrast, the network constructed by SuperQ shown in Figure 3 displays much more conflicting signal. However, the smaller amount of conflicting signal in the supernetwork in [3] might be due to the fact that only those splits that are present in at least two of the trees imputed from the input trees are represented by the network.

For comparison purposes we also applied the Z-closure method [7] to the wasp data set. We used the implementation of this method available in the current version of SplitsTree [16], employing dimension filter 2 to obtain a network that can be easily visualized. The resulting split network is depicted in Figure 4. For dimension 3 or greater the network becomes very dense and hard to draw. A rough visual comparison between the supernetworks in Figures 3 and 4 reveals that major groupings are very similar although they are more pronounced in the network constructed by SuperQ. In particular, the split that separates the outgroup from the other taxa in Figure 4 has a relatively small weight compared to those splits that conflict with it.

The second data set consists of five gene trees for fungal species which was published in [18,19] and used as an example for Z-closure in [7]. The total number of taxa is 63. This is the largest dataset (with respect to the total number of taxa) that we analyzed with SuperQ. The resulting network is depicted in Figure 5. It took approximately 30 minutes to construct this network.

When we compare the supernetwork in Figure 5 with the output of Z-closure presented in [7, p. 156] we again find that both networks agree on many of the major splits. Note however that the Z-closure network in [7, p. 156] is unweighted (every split has weight 1) whereas we are able to obtain estimated weights for

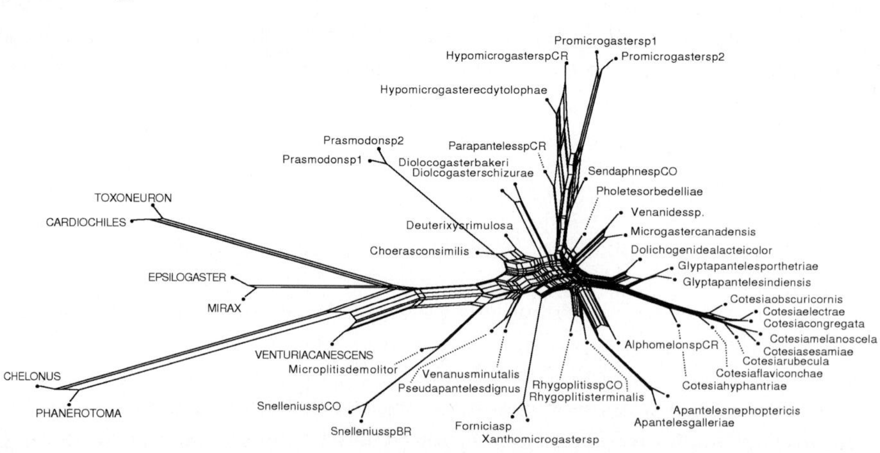

Fig. 3. The supernetwork produced by our method for the seven partial gene trees from [17] showing 45 species of microgastrine wasps

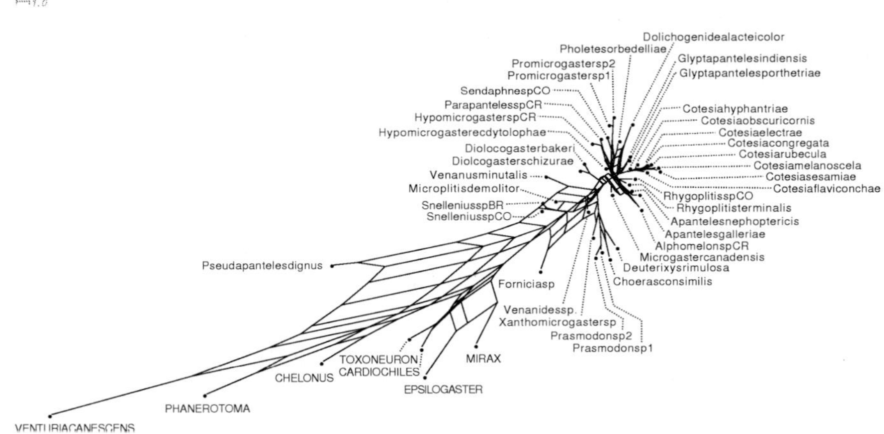

Fig. 4. The supernetwork produced by Z-closure [7] for the wasp dataset [17]. We used the implementation of Z-closure available in the current version of SplitsTree [16]. The settings we used were 100 runs, maximum dimension 2 and average relative weights for the splits.

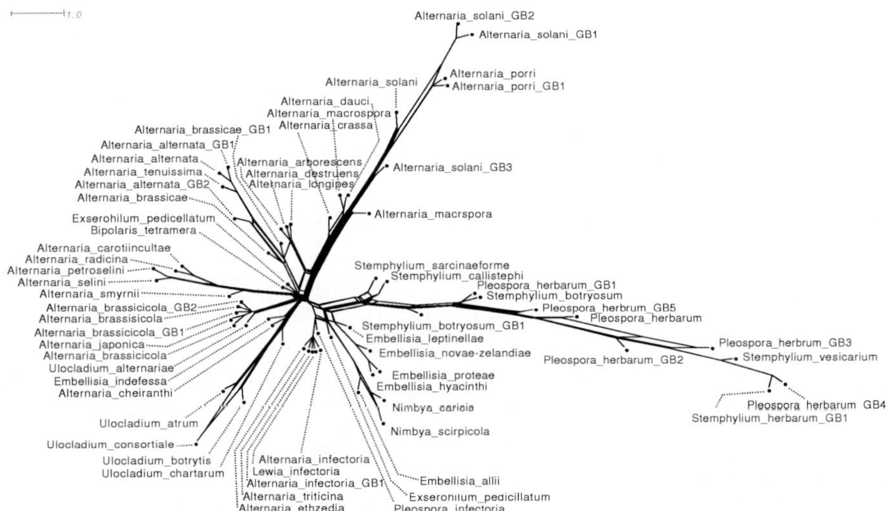

Fig. 5. The supernetwork constructed by our method for the five partial gene trees from [18,19] showing 63 species of funghi

the splits as described in Section 2.2. Curiously, even though in this example the rank of the reduced topological matrix was only 1531 (which is significantly less than 1952, i.e., the total number of splits in Σ_T^*), we found that the interval within which the weight of each split can vary in different solutions to the non-negative least squares problem (1) was very small for many of the splits. We will explore this phenomenon in future work.

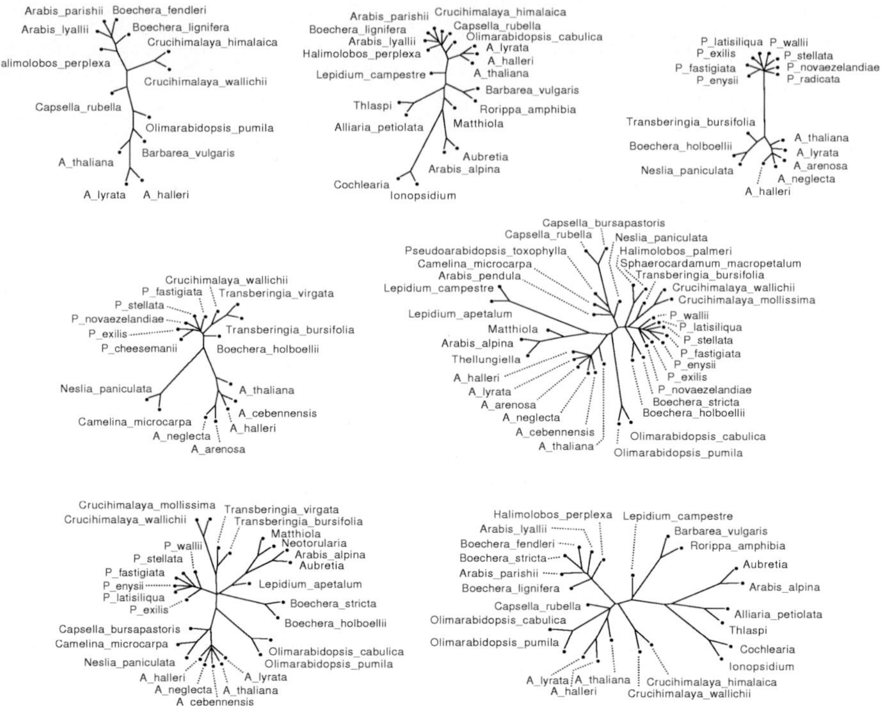

Fig. 6. Seven partial gene trees showing 50 species of Brassicaceae

The third dataset consists of 7 maximum likelihood trees estimated from independent genome loci of flowering plants from the family Brassicaceae. The gene trees, which are depicted in Figure 6, were reconstructed using nuclear and chloroplast nucleotide sequences obtained from GenBank, and others determined as part of a study on phylogenetic relationships among close relatives of *Arabidopsis* (McBreen et al., unpublished). The total number of taxa is 50. Note that in the trees shown in Figure 6 all pendent edges have the same weight. As these weights do not influence the quartet-weight function μ^* we only included these edges to make the trees easier to read. The network constructed by our method is shown in Figure 7. It took approximately 20 minutes to construct this network.

This example illustrates well a fundamental problem that every supernetwork construction method is confronted with. If certain taxa never appear together in an input tree then it can be difficult to decide how to group the taxa in the supernetwork. For example, the network in Figure 7 does not contain a split that separates all species of the genus *Pachycladon* (abbreviated by P) from the other taxa. The reason is that none of the input trees contains a *Pachycladon* species as well as *Rorippa amphibia* or *Barbarea vulgaris*. Therefore, the SuperQ algorithm cannot decide whether they all belong into one group or there should be two groups, one containing all *Pachycladon* species and one containing

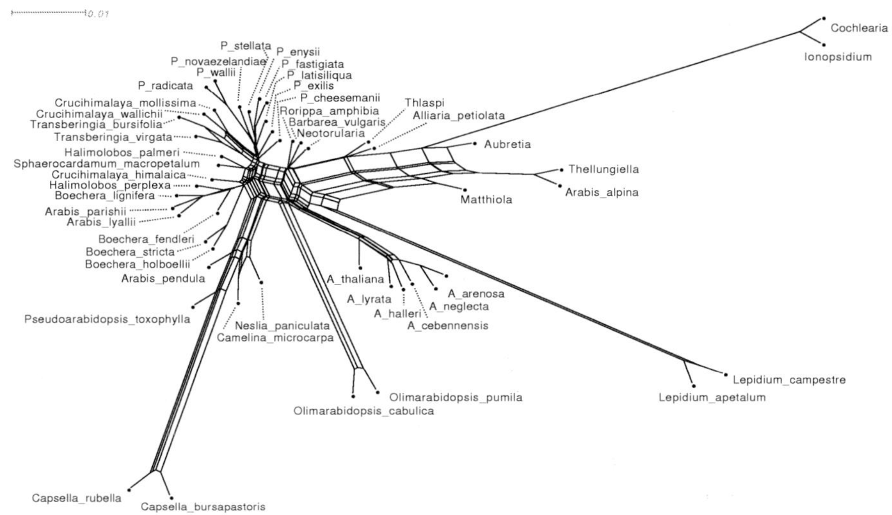

Fig. 7. The supernetwork constructed by SuperQ for the seven partial gene trees in Figure 6

Rorippa amphibia and *Barbarea vulgaris*. In the resulting supernetwork all of those potential splits are assigned weight zero.

4   Conclusions

In this paper we have presented SuperQ, a new method to construct supernetworks from partial trees. The examples presented in Section 3 indicate that our method works well in practice and, for these examples at least, generates networks that are similar in structure to the output of the Z-closure and Q-imputation methods.

Although further work needs to be done to compare the performance of SuperQ with the Z-closure and Q-imputation methods (and also supertree methods in general), one of the main advantages of SuperQ is that it is guaranteed to produce a planar network. Another useful property of SuperQ is that it does not depend on the order in which the quartets derived from the input trees are processed, except for ties in the score function used in the QNet algorithm to construct the split system $\Sigma_{\mathcal{T}}$ (such ties will occur only rarely in practice if the quartet weights are real numbers). Moreover, as with Z-closure (but in contrast to Q-imputation), the input to SuperQ is not restricted to partial trees but can clearly be applied more generally to any collection of partial split-networks.

Our experience from the data sets in Section 3 suggests that the computational time needed to process a data set with the current implementation of SuperQ is similar to Z-closure and Q-imputation. The split system $\Sigma_{\mathcal{T}}$ can be computed in $O(t \cdot n^4)$ time and space where $n = |X|$ and t is the number of input trees. The algorithm for estimating the weights of the splits has superpolynomial worst case run time but seems to work well in practice (as with QNet [9]). We expect that there is potential for a speed-up of SuperQ by taking advantage of the fact that the linear programs Π_k differ only in the objective function but not in the set of constraints.

In future work we will explore ways to pass more information on to the user about the different weightings of the split system Σ_T^* that result from solutions to the non-negative least squares problem (1). As mentioned above our current estimate of the weights is very conservative. As a first step we plan to compute also the maximum weight that a split receives in a solution of (1). Then for every split we have an interval within which the weight of the split must lie. If the length of the interval is zero (or very small), then the split has an (almost) unique weight in every solution of (1). Other issues we plan to address are ways to visualize the interdependence of the weights assigned to splits, and to tie the weight assigned to the trivial splits more directly with the input trees.

In conclusion there are now various methods available for computing phylogenetic supernetworks. Using such methods it should be possible to gain a greater understanding of the complexities hidden within collections of partial trees. As the number of fully sequenced genomes continues to expand and, correspondingly, the size of multiple gene tree analyses, these tools should become more and more useful.

Acknowledgment. We thank Peter Lockhart for providing one of the datasets and the anonymous referees for their helpful comments. In addition, Moulton and Spillner thank the Engineering and Physical Sciences Research Council (EP-SRC) (grant EP/D068800/1) for its support. Part of this work was done while the authors participated in the Phylogenetics Programme at the Isaac Newton Institute for the Mathematical Sciences.

References

1. Bryant, D.: A classification of consensus methods for phylogenies. In: Janowitz, M., Lapointe, F.J., McMorris, F., Mirkin, B., Roberts, F. (eds.) BioConsensus. DIMACS Series in Discrete Mathematics and Theoretical Computer Science, vol. 61, pp. 163–184. American Mathematical Society (2003)
2. Bininda-Emonds, O.R.P. (ed.): Phylogenetic supertrees: Combining information to reveal the tree of life. Kluwer Academic Publishers, Dordrecht (2004)
3. Holland, B., Conner, G., Huber, K., Moulton, V.: Imputing supertrees and supernetworks from quartets. Systematic Biology 56, 57–67 (2007)
4. Holland, B., Delsuc, F., Moulton, V.: Visualizing conflicting evolutionary hypotheses in large collections of trees: Using consensus networks to study the origins of placentals and hexapods. Systematic Biology 54, 66–76 (2005)
5. Holland, B., Jermiin, L., Moulton, V.: Improved consensus network techniques for genome-scale phylogeny. Molecular Biology and Evolution 23, 848–855 (2006)
6. Holland, B., Moulton, V.: Consensus networks: A method for visualising incompatibilities in collections of trees. In: Benson, G., Page, R.D.M. (eds.) WABI 2003. LNCS (LNBI), vol. 2812, pp. 165–176. Springer, Heidelberg (2003)

7. Huson, D., Dezulian, T., Klöpper, T., Steel, M.: Phylogenetic super-networks from partial trees. IEEE/ACM Transactions on Computational Biology and Bioinformatics 1, 151–158 (2004)
8. Huson, D., Bryant, D.: Application of phylogenetic networks in evolutionary studies. Molecular Biology and Evolution 23, 254–267 (2006)
9. Grünewald, S., Forslund, K., Dress, A., Moulton, V.: QNet: An agglomerative method for the construction of phylogenetic networks from weighted quartets. Molecular Biology and Evolution 24, 532–538 (2007)
10. Grünewald, S., Moulton, V., Spillner, A.: Consistency of the QNet algorithm for generating planar split graphs from weighted quartets. Discrete Applied Mathematics (in press, 2007)
11. Semple, C., Steel, M.: Phylogenetics. Oxford University Press, Oxford (2003)
12. Bandelt, H.J., Dress, A.: A canonical decomposition theory for metrics on a finite set. Advances in Mathematics 92, 47–105 (1992)
13. Dress, A., Huson, D.: Constructing split graphs. IEEE Transactions on Computational Biology and Bioinformatics 1, 109–115 (2004)
14. Winkworth, R., Bryant, D., Lockhart, P., Moulton, V.: Biogeographic interpretation of split graphs: studying quaternary plant diversification. Systematic Biology 54, 56–65 (2005)
15. Lawson, C.L., Hanson, R.J.: Solving least squares problems. Prentice-Hall, Englewood Cliffs (1974)
16. Huson, D.: SplitsTree: analyzing and visualizing evolutionary data. Bioinformatics 14, 68–73 (1998)
17. Banks, J.C., Whitfield, J.B.: Dissecting the ancient rapid radiation of microgastrine wasp genera using additional nuclear genes. Molecular Phylogenetics and Evolution 41, 690–703 (2006)
18. Pryor, B.M., Bigelow, D.M.: Molecular characterization of embellisia and nimbya species and their relationship to alternaria, ulocladium and stemphylium. Mycologia 95, 1141–1154 (2003)
19. Pryor, B.M., Gilbertson, R.L.: Phylogenetic relationships among alternaria and related funghi based upon analysis of nuclear internal transcribed sequences and mitochondrial small subunit ribosomal DNA sequences. Mycological Research 104, 1312–1321 (2000)

Summarizing Multiple Gene Trees Using Cluster Networks

Daniel H. Huson and Regula Rupp

Center for Bioinformatics ZBIT, Tübingen University, Sand 14, 72076 Tübingen,
Germany

Abstract. The result of a multiple gene tree analysis is usually a number of different tree topologies that are each supported by a significant proportion of the genes. We introduce the concept of a cluster network that can be used to combine such trees into a single rooted network, which can be drawn either as a cladogram or phylogram. In contrast to split networks, which can grow exponentially in the size of the input, cluster networks grow only quadratically. A cluster network is easily computed using a modification of the tree-popping algorithm, which we call network-popping. The approach has been implemented as part of the Dendroscope tree-drawing program and its application is illustrated using data and results from three recent studies on large numbers of gene trees.

1 Introduction

Increasingly, multiple gene analyses based on tens, hundreds or many thousands of genes are being used to investigate the evolutionary relationship between different species, e.g. [1,17,5]. These studies all confirm that *gene trees differ*, usually not just by random small amounts, but systematically, requiring mechanisms such as incomplete lineage sorting, or, more interestingly, hybridization, horizontal gene transfer or recombination, as an explanation.

The goal of a multiple gene analysis is usually to deduce a species tree from the computed gene trees, for example, by computing a consensus tree, or reporting a small number of possible trees. An alternative approach is to compute a phylogenetic network that represents a number of different gene trees simultaneously and thus stakes out the space of possible species trees.

The computer program SplitsTree [11,12] has helped to popularize the use of phylogenetic networks, in particular split networks [2]. A *split network* generalizes the concept of an unrooted phylogenetic tree and uses bands of parallel edges to represent incompatible "splits" (i.e., unrooted clusters) of the set of taxa under consideration.

Given a collection of trees (all on the same set of taxa), one approach is to select all splits that are contained in more than a fixed percentage of all trees and to represent these by a split network [10]. In the more general situation in which taxon sampling differs between trees, a distortion filter can be applied to obtain the main variants of the gene trees, in an unrooted and unweighted way, see [16].

K.A. Crandall and J. Lagergren (Eds.): WABI 2008, LNBI 5251, pp. 296–305, 2008.

© Springer-Verlag Berlin Heidelberg 2008

Such constructions give rise to split networks that summarize the main variants of the gene trees, in an unrooted and unweighted way, see Figure 1(a). In this simple example, the two networks are topologically equivalent. However, note that in general this will not be the case as the number of nodes and edges grows only quadratically in a cluster network and exponentially in a split network (in the worst case).

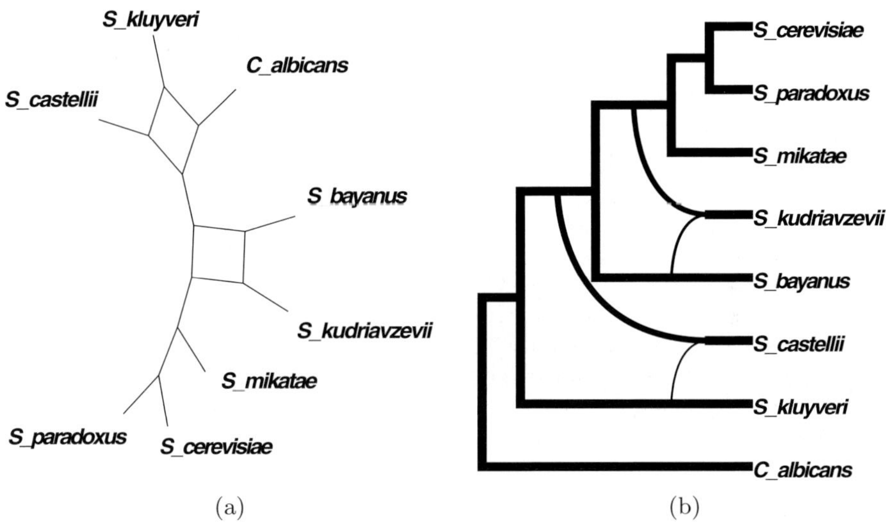

Fig. 1. Based on all splits contained in > 20% of 106 gene trees computed using maximum parsimony by Rokas [1], we show (a) the split network [10,12] and (b) a weighted cluster network. In the latter network, the edge width indicates the number of trees supporting a given edge.

Although split networks are increasingly being used to visualize the main competing phylogenetic signals in data sets, they have a number of drawbacks: (i) The bands of parallel lines used to facilitate incompatible splits can be confusing as they usually do not directly correspond to possible lineages. (ii) The networks are unrooted and thus cannot be drawn as cladograms or phylograms. (iii) In the worst case, the number of nodes and edges of the network can grow exponentially with the number of splits in the input.

To address these three problems, in this paper we introduce the concept of a *cluster network*, a rooted visualization of a set of clusters, which can be drawn as a cladogram or phylogram and has only a quadratic number of nodes and edges in the worst case, see Figure 1(b). After discussing how to compute a cluster network using a "network-popping" algorithm, we will address the problem of assigning weights to all edges in the network so as to show how strongly alternative phylogenies in the network are supported. To this end, we will introduce the concept of the *lowest single ancestor* of a node, which is related to the concept of the "lowest common ancestor".

We have implemented all algorithms discussed in this paper and will make them available in version 2.0 of our program Dendroscope [15] upon publication of this paper. (Referees can download a preliminary version from `www.dendroscope.org/internal`.) Below, we will illustrate the use of these new concepts on three examples: yeast data [1], plant chloroplast gene trees [17] and primate trees [5].

In cases where the topology of different gene trees varies quite substantially, for example, in prokaryotes due to horizontal gene transfer [4], see Section 4, it may be useful to use the term *species network* instead of cluster network to emphasize that the set of gene trees under consideration does not appear to support any one species tree.

2   Cluster Networks

Let $G = (V, E)$ be a connected, acyclic directed graph with node set V and edge set E that has exactly one node ρ of indegree 0, called the *root*. Nodes of outdegree 0 are called *leaves*, nodes of indegree ≤ 1 are called *tree nodes*, whereas nodes of indegree ≥ 2 are called *reticulate nodes*. An edge $e = (v, w)$ is called a *tree edge*, if its "target node" w is a tree node, and otherwise, it is called a *reticulate edge*.

Let X be a set of taxa. A *cluster network* N on X consists of such a graph G, together with a labeling of the nodes $\lambda : X \to V$ with the property that all leaves receive a label. Let N be a cluster network on X. Any tree edge $e = (v, w)$ defines a cluster $C \subseteq X$ on X, namely the set of all labels associated with nodes that can be reached from w. The set $\mathcal{C}(N)$ of all clusters obtainable in this way will be called the set of clusters *represented* by N. Note that we do not associate a cluster with any reticulate edge.

This definition is closely related to the concepts of a reticulate network [13], hybridization network [19,3] or recombination network [9,14]. However, there is a important difference in the interpretation of the networks. For example, in a hybridization network, a tree edge may represent more than one cluster, as any given reticulate node in such a network has precisely two incoming edges that represent alternative phylogenies which can be turned on or off at will. However, in a cluster network, all reticulate edges are always "on". It is useful to use the terms "softwired" and "hardwired" to emphasize this distinction.

In Figure 2, we list a collection of clusters that are incompatible and thus can not be represented by a tree, and show the corresponding cluster network.

A set of compatible splits, or compatible clusters, can be represented by an unrooted, or rooted, tree, respectively. Just as a collection of incompatible splits can be represented by a split network, we propose to represent a collection of incompatible clusters by a cluster network.

For this purpose, we will assume that all cluster networks considered have the following *reticulation separation* property: Every reticulate node has outdegree $= 1$. This property can easily be obtained by inserting a new edge before every

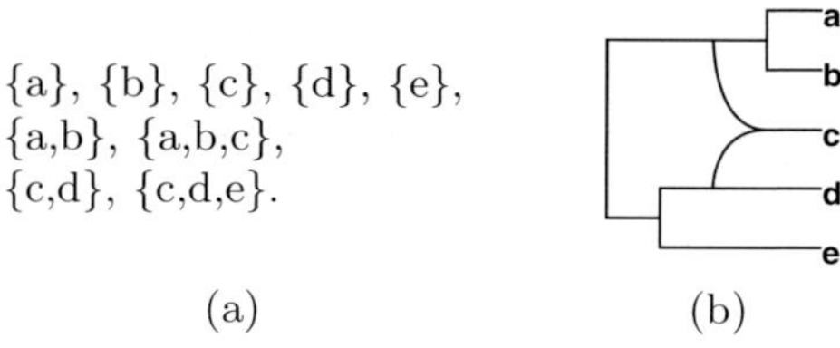

Fig. 2. (a) A collection of clusters and (b) a cluster network representing the clusters. "Tree" edges representing clusters are drawn in a rectangular fashion, whereas facilitating "reticulate" edges are curved.

node v that has indegree ≥ 2 and outdegree $\neq 1$. This new edge (v', v) connects a new node v' to v and all inedges of v are redirected to v'.

Let $\mathcal{C}$ be a set of k clusters on a taxon set X. We will now describe how to construct a cluster network that represents $\mathcal{C}$ (and only $\mathcal{C}$), using $O(k)$ nodes and $O(k^2)$ edges. As the algorithm is a generalization of the "tree-popping" method [18], we will refer to it as *network-popping*.

Algorithm 1 (Network-popping algorithm)
Input: Set of k clusters $\mathcal{C}$ on X
Output: Cluster network N representing $\mathcal{C}$

begin
 If $X \notin \mathcal{C}$ **then** *add X to $\mathcal{C}$*
 Create one node $v(C)$ per cluster $C \in \mathcal{C}$ and set $C(v) = C$
 Set the root $\rho = v(X)$
 for each $C \in \mathcal{C}, C \neq X$ *in order of decreasing cardinality* **do**
   Push ρ onto a stack S
   while S *is not empty* **do**
     Pop v off S
     Set $isBelow = false$
     for each *child w of v* **do**
       if $C \subset C(w)$ **then**
         Set $isBelow = true$
         if *w not yet visited* **then** *push w onto S*
     if $isBelow = false$ **then**
       Create a new edge $(w, v(C))$

 for each *node v with indegree ≥ 2 and outdegree $\neq 1$* **do**
   Create a new node v'
   Redirect all inedges of v to v'
   Create a new edge (v', v)

 for each *node v* **do**
   Set the label of v to $\lambda(v) = C(v) \setminus \bigcup_{w\ child\ of\ v} C(w)$
end

Fig. 3. The network-popping algorithm for constructing a cluster network

The network-popping algorithm proceeds in three steps, first constructing the "Hasse diagram" (as defined in graph theory), then inserting additional tree edges so as to ensure the reticulation separation property and then finally setting up the node labeling, see Figure 3. By construction, each node v of the original Hasse diagram corresponds to a cluster C and is the target of exactly one tree edge e, and so this edge can be used to represent C. As each node is incident to some tree edge and there are k tree edges, the number of nodes is at most $2k$ and the number of reticulate edges is at most $\binom{2k}{2}$. An upper bound for the running time of the algorithm is given by $O(nk^3)$, where n is the number of taxa and k is the number of clusters.

Note that the network-popping algorithm produces a tree, if and only if the set of input clusters is compatible.

Let $\mathcal{T}$ be a collection of rooted phylogenetic trees on X and let $\mathcal{C}$ be a subset of all clusters represented by trees in $\mathcal{T}$. If N is a cluster network representing $\mathcal{C}$ that was computed using the network-popping algorithm, then any tree $T \in \mathcal{T}$ is contained in N as a subtree, if all clusters of T are contained in $\mathcal{C}$. Otherwise, if N contains only a part of the clusters of T, then N contains a contraction of T, by properties of the algorithm.

Cluster networks can be drawn either as cladograms, in which edge lengths have no direct interpretation, or as phylograms, in which the tree edges are drawn to scale to represent evolutionary distances, using modifications of the standard algorithms used to draw trees [6], as we will describe in a forthcoming paper, see Figure 1(b).

3   Summarizing Gene Trees

Let $\mathcal{T}$ be a collection of gene trees on a taxon set X. Usually, the cluster network N of all clusters present in $\mathcal{T}$ will be too messy to be useful. One straight-forward idea is to consider only those clusters that are contained in a fixed percentage of the trees [10,16].

Additionally, to enhance the readability of the network N, one can use different line widths or a gray scale to visualize the *support* of each edge e in the network, indicating the percentage of trees in $\mathcal{T}$ that contain e. If e is a tree edge, then e represents some cluster $C \subset X$ and it is clear how to compute its support. However, if e is a reticulate edge, then it is not immediately obvious how the support of e is to be defined. We now address this issue in the following subsections.

3.1   The Lowest Single Ancestor

Let N be a cluster network. The *lowest single ancestor (lsa)* of a node v is defined as the *last* node $lsa(v)$ that lies on all paths from the root ρ of N to v, excluding v. The lsa of any node v is uniquely defined.

The lsa of all reticulate nodes of N can be computed in a single postorder traversal of the network: for each reticulate node v and every node w in the network, we keep track of which paths from v toward the root go through w. If, at any point, we encounter a node w such that the set of all paths through w equals the set of all paths currently in existence for v, then we have found the lca of v, namely w. [1]

---

[1] We can also define the lsa of a set of two or more nodes. For a tree, this coincides with the lowest common ancestor (lca). Using this fact and the lsa-tree of N one can show that the lookup of the lsa of any two nodes in a cluster network N can be performed in constant time, after a linear amount of processing, just as in the case of the lookup of lca in trees, which was described in [8], Chapter 8.

We obtain the *lsa-tree* $T_{lsa}(N)$ of a cluster network N as follows: for each reticulate node v in N, remove all reticulate edges leading to v and then add a new edge from $lsa(v)$ to v. This tree is well-defined. If the network was constructed from a collection of trees $\mathcal{T}$, then the lsa-tree $T_{lsa}(N)$ can be considered a new type of consensus tree that is different from all split-based consensus trees, such as the strict, loose or majority consensus. Another application of the lsa-tree is in drawing of cluster networks, as we will discuss in a forthcoming paper.

3.2   Estimating the Support of a Reticulate Edge

As stated above, we can define the support $s(e)$ of a tree edge e in N as the percentage of trees in the input set that contain the corresponding cluster. We will estimate the support of a reticulate edge by averaging the support of tree edges over appropriate sets of edges. In Figure 4, we show a reticulate node v with two reticulate edges e and f. Let A be the set of all tree edges between e and $lsa(v)$, let B be the set of all tree edges between f and $lsa(v)$, and let C be the set of all tree edges below v. Let $\bar{s}(A)$ denote the average support of all tree edges in A. In the case of a reticulation with two reticulation edges, we define the support of the reticulate edge e as:

$$s(e) = \bar{s}(C)\frac{\bar{s}(A \setminus B)}{\bar{s}(A \cup B)},$$

that is, as the average support of all tree edges below v, weighted by the ratio between the average support of tree edges only between e and $lsa(v)$ over the average support of all tree edges between v and $lsa(v)$. It is easy to adapt this definition for the case where a reticulation has more than two reticulation edges.

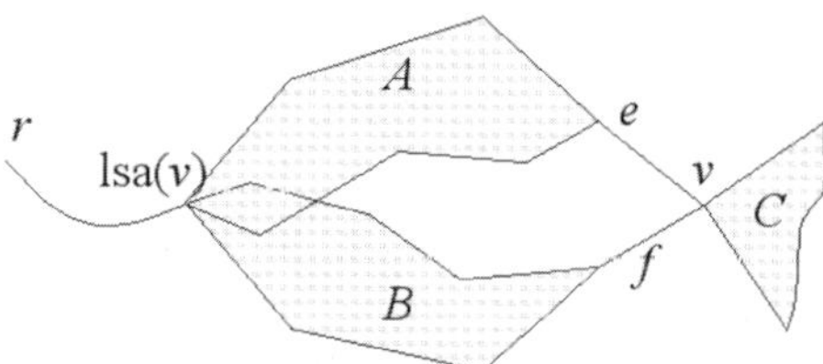

Fig. 4. The support for the reticulate edges e and f at the reticulate node v is computed by considering the average support of the tree edges in A, B and C. The network is drawn left-to-right, with the root r at the left.

4   Implementation and Examples

We have implemented all algorithms described in this paper within the framework of our program Dendroscope [15] and will make these features available in version 2.0 of the program, upon publication of this paper. Additionally, these algorithms will also be made available in a future release of SplitsTree4 [12].

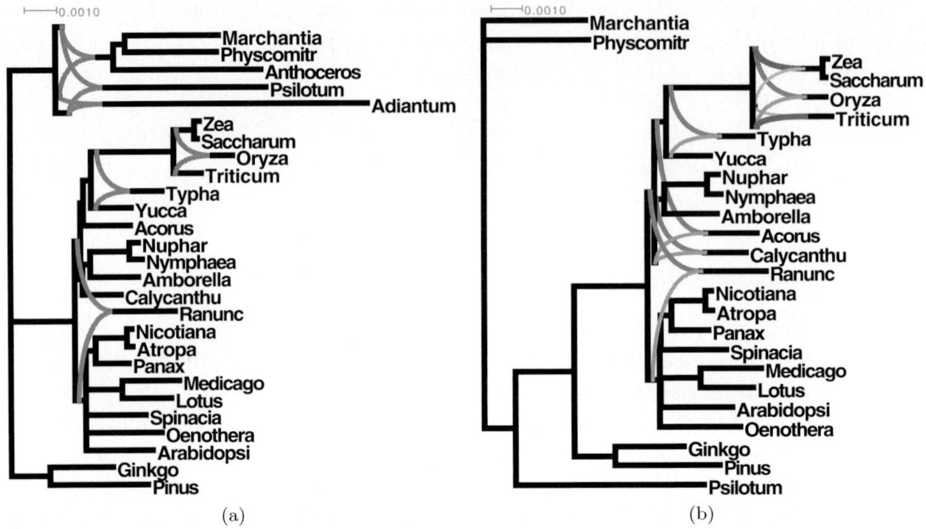

Fig. 5. Cluster networks computed from 59 gene trees compute from data reported in [17]. In (a), all 27 taxa are present. In (b), two very divergent taxa, *Adiantum* and *Anthoceros*, have been removed.

We will illustrate the application of these ideas using three examples. In [1], Rokas *et al.* investigate the evolutionary history of seven *Saccharomyces* species and a fungal outgroup *Candida albicans*, using alignments of 106 different gene sequences, totaling 127,026 nucleotides. The paper shows a number of different phylogenies obtained from the different genes, which are also compared with the tree obtained from the concatenation of all sequences. In [10], Holland *et al.* extract all splits contained in more than 20% of the trees and construct a split network from these, obtaining a network as show in Figure 1(a). In Figure 1(b), we show a cluster network constructed from the same data, in which different edge widths are used to indicate the support of the different reticulate edges, computed as described in the previous section. The cluster network clearly indicates which of the two placements of both *S. kudriavzevii* and *S. castellii* are favored by a majority of the trees.

In [17], Leebens-Mack *et al.* study the evolutionary history of plants using 61 chloroplast genes. We downloaded the gene sequence DNA alignments from the authors website and constructed trees using the logdet distance transformation [21] and the BioNJ tree construction algorithm [7], as implemented in SplitsTree4 [12]. For two genes, the logdet transformation was not applicable, due to singularities in the frequency matrix and these trees were discarded. In Figure 5, we show the cluster network for all clusters occurring in more than 20% of the computed trees. In (a), we show the network obtained from trees computed on the full dataset, whereas in (b) the network was obtained from trees computed on the subset of taxa obtained by removing the two taxa *Adiantum* and *Anthoceros*. Confirming the discussion in [17], in (a) we see that these two taxa are

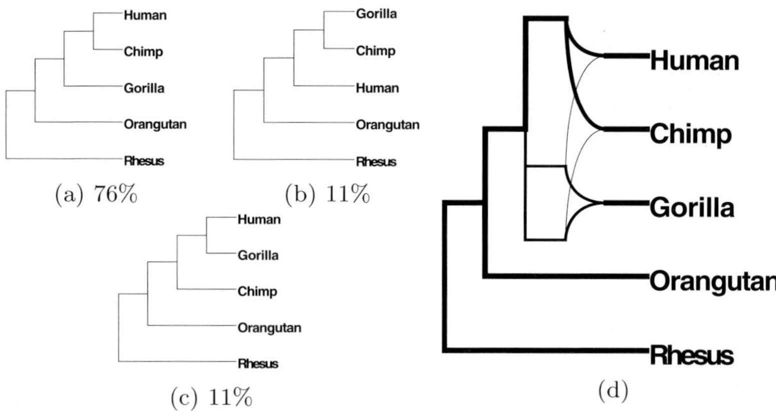

Fig. 6. A set of 11,945 "phylogenetically informative alignments" lead to the three main tree topologies shown in (a)–(c), with (a) occurring in $\approx 76\%$, and the other two each in $\approx 11\%$, of the cases [5]. The cluster network shown in (d) summarizes these three alternatives.

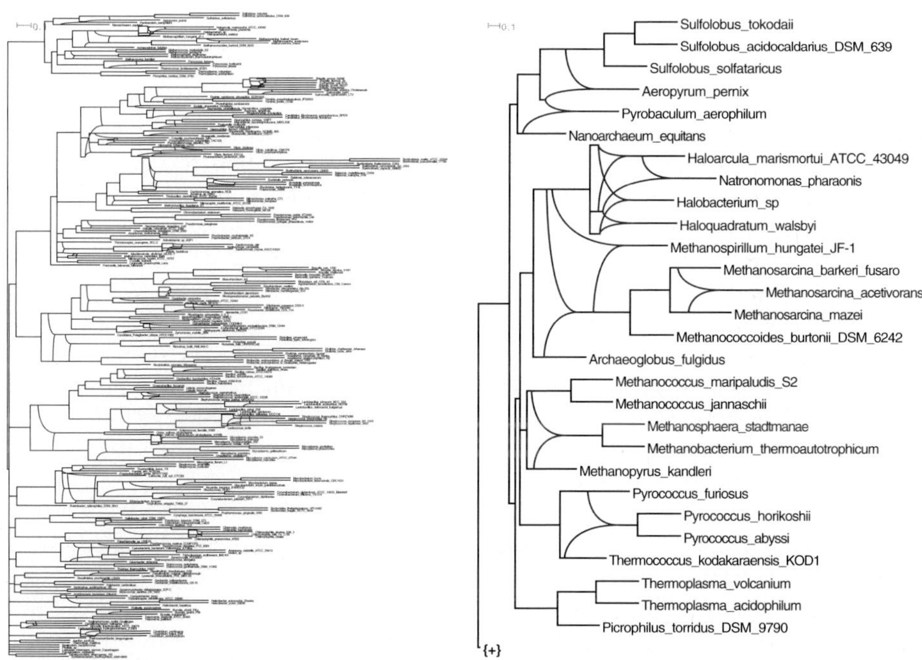

Fig. 7. (a) Overview of a cluster network based on 18 genes for 279 prokaryotes. (b) A detailed view of the archaea contained in the network.

very divergent and that they confound the two outgroup taxa *Marchantia* and *Physcomitr*. Also in line with the discussion in [17], in (b) we see that removal of the two divergent leads to a less certain placement of *Calycanthus*.

The evolutionary relationship between human, chimpanzee and gorilla, together with orangutan and rhesus monkey can now be investigated using genome-wide data [20]. In [5], strong support for three alternative phylogenies was reported, as shown in Figure 6 (a)-(c). Based on these three weighted trees, we obtain the cluster network N shown in (d).

There is much debate whether the evolutionary history of prokaryotes is best described by a tree, or whether a phylogenetic network is a more appropriate representation [4]. For a recently submitted manuscript, we built 18 different gene trees for 279 species of bacteria and archea with the aim of detecting possible cases of horizontal gene transfer. Using all clusters that occur in 20% or more of the trees, we obtain the cluster network depicted in Figure 7 which clearly shows the alternative placements of taxa suggested by the different genes. Computation of this network from the set of input trees took less than 1 second. Note that the alternative placements visible here are probably not due to horizontal gene transfer events but are probably caused by local rearrangements in the different trees based on lineage sorting effects or systematic and stochastic error.

Acknowledgements

We would like to thank Stephan Gruenewald, Pete Lockhart and Mike Steel for useful discussions. This work was undertaken at the Isaac Newton Institute of Cambridge University during the 2007 research programme on Phylogenetics and was also supported by DFG grant Hu 566/5-1.

References

1. King, N., Rokas, A., Williams, B.L., Carroll, S.B.: Genome-scale approaches to resolving incongruence in molecular phylogenies. Nature 425(6960), 798–804 (2003)
2. Bandelt, H.-J., Dress, A.W.M.: A canonical decomposition theory for metrics on a finite set. Advances in Mathematics 92, 47–105 (1992)
3. Bordewich, M., Semple, C.: Computing the minimum number of hybridization events for a consistent evolutionary history. Discrete Appl. Math. 155(8), 914–928 (2007)
4. Doolittle, W.F., Bapteste, E.: Pattern pluralism and the tree of life hypothesis. PNAS 104, 2043–2049 (2007)
5. Ebersberger, I., Galgoczy, P., Taudien, S., Taenzer, S., Platzer, M., von Haeseler, A.: Mapping Human Genetic Ancestry. Mol. Biol. Evol. 24(10), 2266–2276 (2007)
6. Felsenstein, J.: Inferring Phylogenies. Sinauer Associates, Inc. (2004)
7. Gascuel, O.: BIONJ: An improved version of the NJ algorithm based on a simple model of sequence data. Mol. Biol. Evol. 14, 685–695 (1997)
8. Gusfield, D.: Algorithms on Strings, Trees and Sequences. Cambridge University Press, Cambridge (1997)

9. Gusfield, D., Eddhu, S., Langley, C.: Efficient reconstruction of phylogenetic networks with constrained recombination. In: Proceedings of the IEEE Computer Society Conference on Bioinformatics, p. 363 (2003)

10. Holland, B., Huber, K., Moulton, V., Lockhart, P.J.: Using consensus networks to visualize contradictory evidence for species phylogeny. Molecular Biology and Evolution 21, 1459–1461 (2004)

11. Huson, D.H.: SplitsTree: A program for analyzing and visualizing evolutionary data. Bioinformatics 14(10), 68–73 (1998)

12. Huson, D.H., Bryant, D.: Application of phylogenetic networks in evolutionary studies. Molecular Biology and Evolution 23, 254–267 (2006), www.splitstree.org

13. Huson, D.H., Kloepper, T., Lockhart, P.J., Steel, M.A.: Reconstruction of reticulate networks from gene trees. In: Miyano, S., Mesirov, J., Kasif, S., Istrail, S., Pevzner, P.A., Waterman, M. (eds.) RECOMB 2005. LNCS (LNBI), vol. 3500, pp. 233–249. Springer, Heidelberg (2005)

14. Huson, D.H., Kloepper, T.H.: Computing recombination networks from binary sequences. Bioinformatics 21(suppl. 2), ii159–ii165 (2005)

15. Huson, D.H., Richter, D.C., Rausch, C., Dezulian, T., Franz, M., Rupp, R.: Dendroscope: An interactive viewer for large phylogenetic trees. BMC Bioinformatics 8, 460 (2007), www.dendroscope.org, doi:10.1186/1471-2105-8-460

16. Huson, D.H., Steel, M.A., Whitfield, J.: Reducing distortion in phylogenetic networks. In: Bücher, P., Moret, B.M.E. (eds.) WABI 2006. LNCS (LNBI), vol. 4175, pp. 150–161. Springer, Heidelberg (2006)

17. Leebens-Mack, J., Raubeson, L.A., Cui, L., Kuehl, J.V., Fourcade, M.H., Chumley, T.W., Boore, J.L., Jansen, R.K., de Pamphilis, C.W.: Identifying the Basal Angiosperm Node in Chloroplast Genome Phylogenies: Sampling One's Way Out of the Felsenstein Zone. Mol. Biol. Evol. 22(10), 1948–1963 (2005)

18. Meacham, C.A.: Theoretical and computational considerations of the compatibility of qualitative taxonomic characters. In: Felsenstein, J. (ed.) Numerical Taxonomy. NATO ASI Series, vol. G1, Springer, Berlin (1983)

19. Nakhleh, L., Warnow, T., Linder, C.R.: Reconstructing reticulate evolution in species - theory and practice. In: Proceedings of the Eighth International Conference on Research in Computational Molecular Biology (RECOMB), pp. 337–346 (2004)

20. Patterson, N., Richter, D.J., Gnerre, S., Lander, E.S., Reich, D.: Genetic evidence for complex speciation of humans and chimpanzees. Nature 441, 1103–1108 (2006)

21. Steel, M.A.: Recovering a tree from the leaf colorations it generates under a Markov model. Appl. Math. Lett. 7(2), 19–24 (1994)

Fast and Adaptive Variable Order Markov Chain Construction

Marcel H. Schulz[1,3], David Weese[2], Tobias Rausch[2,3], Andreas Döring[2], Knut Reinert[2], and Martin Vingron[1]

[1] Department of Computational Molecular Biology, Max Planck Institute for Molecular Genetics, Ihnestr. 73, 14195 Berlin, Germany
`{marcel.schulz,martin.vingron}@molgen.mpg.de`
[2] Department of Computer Science, Free University of Berlin, Takustr. 9, 14195 Berlin, Germany
`{weese,rausch,doering,reinert}@inf.fu-berlin.de`
[3] International Max Planck Research School for Computational Biology and Scientific Computing

Abstract. Variable order Markov chains (VOMCs) are a flexible class of models that extend the well-known Markov chains. They have been applied to a variety of problems in computational biology, e.g. protein family classification. A linear time and space construction algorithm has been published in 2000 by Apostolico and Bejerano. However, neither a report of the actual running time nor an implementation of it have been published since. In this paper we use the lazy suffix tree and the enhanced suffix array to improve upon the algorithm of Apostolico and Bejerano. We introduce a new software which is orders of magnitude faster than current tools for building VOMCs, and is suitable for large scale sequence analysis.

1 Introduction

Markov chains are often used in computational biology to learn representative models for sequences, as they can capture short-term dependencies exhibited in the data. The fixed order L of a Markov chain determines the length of the preceding context, i.e. the number of dependent positions, which is taken into account to predict the next symbol. A severe drawback of fixed order Markov chains is that the number of free parameters grows exponentially with L, s.t. training higher order Markov models becomes noisy due to overfitting. These considerations have led to the proposal of a structurally richer class of models called variable order Markov chains, which can vary their context length. There are two prominent methods in use by the community. One is the tree structured context algorithm of Rissanen [1], and the other is the probabilistic suffix tree of Ron et al. [2].

VOMCs have been used for different applications like classification of transcription factor binding sites, splice sites, and protein families [3–5]. Dalevi and co-workers applied VOMCs to the detection of horizontal gene transfer in bacterial genomes [6]. VOMCs have been used by Bejerano et al. for the segmentation

K.A. Crandall and J. Lagergren (Eds.): WABI 2008, LNBI 5251, pp. 306–317, 2008.
© Springer-Verlag Berlin Heidelberg 2008

of proteins into functional domains [7], and Slonim and others have shown how to exploit their structure for feature selection [8]. VOMCs do not require a sequence alignment, an advantage over many other sequence models. They dynamically adapt to the data and have been shown to be competitive to HMMs for the task of protein family classification [9]. For the task of DNA binding site classification they outperform commonly used Position Weight Matrices [3, 4, 10].

There are theoretical results on linear time and space construction of VOMCs [11]. However, currently no linear time implementation has been published and the available tools [6, 12] are suboptimal in theory and slow in practice. For example the pst software [12] implements the $\mathcal{O}(\|S\|^2 L)$ algorithm due to Ron et al. [2], which is used in the studies [5, 9, 13]. We believe that three major reasons account for the absence of a linear time implementation: (i) the linear time algorithm proposed by Apostolico and Bejerano appears to be quite complex [11], (ii) implementations of VOMC construction algorithms have always been explained separately for the methods of Ron et al. [2] and Rissanen [1], which led to the misconception that the first "has the advantage of being computationally more economical than the" second [13]; (iii) it is based on suffix trees which are known to be space-consuming and slow for large sequences due to bad cache locality behaviour.

A lot of research has focused on improving suffix tree construction algorithms by reducing the space requirements [14–17]. An index called enhanced suffix array can be an efficient replacement for a suffix tree [18]. In this work we will employ some of these strategies and devise and implement improved versions of the Apostolico-Bejerano (AB) algorithm.

In Section 2 we introduce necessary data structures and formulate a general approach to VOMC learning. Both methods [1] and [2] are implemented with the same construction algorithm in Section 3, where the AB algorithm and our improvements are explained. The superiority compared to former algorithms is demonstrated in Section 4.

2   Preliminaries

2.1   Definitions

We consider a collection S of strings over the finite ordered alphabet Σ. Without loss of generality choose $\sigma_i \in \Sigma$, s.t. $\sigma_1 < \ldots < \sigma_{|\Sigma|}$. Σ^L is defined as the set of all words of length L over Σ. ϵ is the empty string and $\Sigma^0 = \{\epsilon\}$. The length of a string $s \in S$ is denoted by $|s|$ and $\|S\|$ is defined as the concatenated length of all strings $s \in S$. $|S|$ is the number of strings in the collection, s^R is the reverse of s, and S^R is the collection of the reverse sequences of S. $u[i]$ denotes the i-th character of u. We denote as ur the concatenated string of strings u and r. The *empirical probability* $\widetilde{P}(r)$ of a subsequence $r \in S$ is defined as the ratio

$$\widetilde{P}(r) = \frac{N_S(r)}{\|S\| - (|r| - 1)|S|} \, , \tag{1}$$

where $N_S(r)$ is the number of all, possibly overlapping, occurrences of the subsequence r in strings $s \in S$. The denominator is the number of all possible positions of a word of length $|r|$ in S. This defines a probability distribution over words of length l, with $\sum_{r \in \Sigma^l} \widetilde{P}(r) = 1$. We define the *conditional empirical probability* $\widetilde{P}(\sigma|r)$ of observing the symbol σ right after the subsequence $r \in S$ as

$$\widetilde{P}(\sigma|r) = \frac{N_S(r\sigma)}{N_S(r*)} \, , \tag{2}$$

with $N_S(r*) = \sum_{\sigma \in \Sigma} N_S(r\sigma)$. This can be seen as the relative frequency of σ preceded by r.

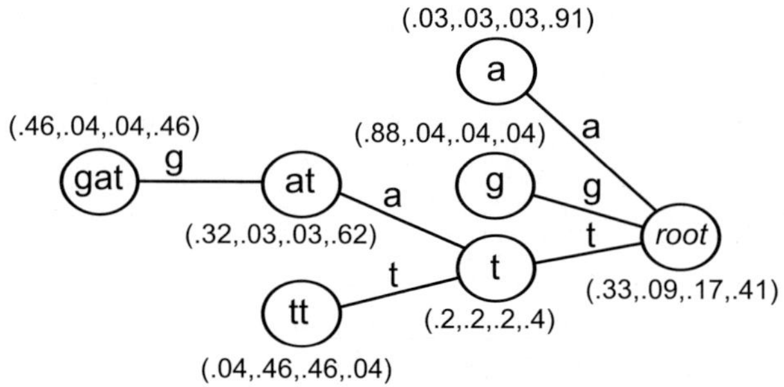

Fig. 1. A PST over the DNA alphabet {a,c,g,t}. The vector close to each node is the probability distribution for the next symbol, e.g., the symbol t occurs with probability 0.46 after all substrings ending with gat. The prediction of the string gattaca assign a probability to each character: $P(g)P(a)P(t|a)P(t|gat)P(a|tt)P(c|a)P(a) = 0.17 \cdot 0.33 \cdot 0.91 \cdot 0.46 \cdot 0.04 \cdot 0.03 \cdot 0.33$.

2.2   Variable Order Markov Chains

We now introduce the fundamental data structure, needed for learning a VOMC and for prediction of sequences with a VOMC. A convenient representation of a VOMC is the probabilistic suffix tree (PST) [2], sometimes called Context Tree [1]. A PST $\mathcal{T}$ is a rooted tree with edge labels from Σ, and no two child edges have the same character. A node with the concatenated string u, spelled from *node to root*, is denoted $\overleftarrow{u}$. A node $\overleftarrow{u}$ is called *maximal* if it has less than $|\Sigma|$ children. Each maximal node $\overleftarrow{u}$ in a PST has attached a probability vector $v^u \in [0,1]^{|\Sigma|}$, with $v_i^u = P(\sigma_i|u)$. Given $r[1]r[2] \ldots r[m]$, the next transition is determined as follows: walk down in the PST by following the edges labeled $r[m]r[m-1] \ldots$ as long as possible; use the transition probabilities associated to the last node visited.

Example 1. Consider the PST in Fig. 1. The underlying VOMC will map each context ending with character c to $v^\epsilon = (0.33, 0.09, 0.17, 0.41)$, the probability vector of the root. If the context is tat, the transitions correspond to the longest path node $\overleftarrow{at}$ and are given by $v^{at} = (0.32, 0.03, 0.03, 0.62)$.

2.3   A General Approach to Construct VOMCs

We divide VOMC learning algorithms [3, 4, 9, 1, 2] into two conceptional phases, *Support-Pruning* and *Similarity-Pruning*. In the *Support-Pruning* phase the PST is restricted to representative nodes. In the *Similarity-Pruning* phase redundant nodes of the PST are removed. Tuning the parameters of the *Support-Pruning* and *Similarity-Pruning* phases allows the control of the trade-off between bias and variance for the final VOMC. A strict pruning in each phase shrinks the PST and results in an underfitted VOMC, whereas limited pruning results in a PST that overfits the data. We want to solve the following problem:

Problem 1. Given a string collection S generated by a VOMC, the aim is to find the underlying PST $\mathcal{T}$ using only S.

We will introduce two widely-used solutions to Problem 1 by Rissanen [1] and Ron et al. [2] for learning VOMCs, represented by PSTs, according to our general approach.

Solution 1. Context Tree (S, t, L, K)
Support-Pruning. $\mathcal{T}$ is defined to contain the following nodes:

$$\mathcal{T} \leftarrow \left\{ \overleftarrow{r} \mid r \in \bigcup_{i=0}^{L} \Sigma^i \text{ and } N_S(r) \geq t \right\} \ . \tag{3}$$

We denote the first part of the condition in (3) as L-Pruning and the second part as t-Pruning.
Similarity-Pruning.   Prune recursively in bottom-up fashion all leaves $\overleftarrow{ur}$, with $|u| = 1$ and r possibly empty if:

$$\sum_{\sigma \in \Sigma} \left(\widetilde{P}(\sigma|ur) \cdot \ln \frac{\widetilde{P}(\sigma|ur)}{\widetilde{P}(\sigma|r)} \right) \cdot N_S(ur) < K \ . \tag{4}$$

The original solution proposed by Rissanen [1] set $t = 2$, and it was shown later to be a consistent estimator for any finite t by Bühlmann and Wyner [19].

Example 2. Consider the string set $S = \{\text{gattc}, \text{attgata}\}$ and set $t = 2$, $L = 3$, and $K = 0.04$. The *Context Tree* algorithm will build the PST depicted in Fig. 1. All probability vectors of nodes $\overleftarrow{r}$ have been smoothed by adding 0.01 to all $N_S(r\sigma)$, $\sigma \in \Sigma$, before the conditional probabilities are calculated with (2).

Solution 2. Growing PST $(S, P_{\min}, L, \alpha, \gamma_{\min}, k)$
Support-Pruning.   Let X be a string collection. Initially $\mathcal{T}$ is empty and $X = \{\epsilon\}$. While $X \neq \emptyset$, repeat the following: (i) pick and remove any $r \in X$; (ii) add $\overleftarrow{r}$ to $\mathcal{T}$; (iii) If $|r| < L$, extend X as follows:

$$X \leftarrow X \cup \left\{ \sigma r \mid \sigma \in \Sigma \text{ and } \widetilde{P}(\sigma r) \geq P_{\min} \right\} \ . \tag{5}$$

Similarity-Pruning.   Prune recursively in bottom-up fashion all leaves $\overleftarrow{ur}$, with $|u| = 1$ and r possibly empty if there exists no symbol $\sigma \in \Sigma$ such that

$$\widetilde{P}(\sigma|ur) \; \geq \; (1 + \alpha)\gamma_{\min} \tag{6}$$

and

$$\frac{\widetilde{P}(\sigma|ur)}{\widetilde{P}(\sigma|r)} \geq k \quad \text{or} \quad \frac{\widetilde{P}(\sigma|ur)}{\widetilde{P}(\sigma|r)} \leq \frac{1}{k} \; . \tag{7}$$

2.4   Suffix Trees

We will use artificial string markers $\$_j$ at the end of each string s_j to distinguish the suffixes of strings in a collection $S = \{s_1, \ldots, s_{|S|}\}$.

A suffix tree for a collection S over Σ is a rooted tree with edge labels from $(\Sigma \cup \{\$_1, \ldots, \$_{|S|}\})^*$, s.t. every concatenation of symbols from the root to a leaf node yields a suffix of $s_j\$_j$ for a string $s_j \in S$. Each internal node has at least two children and no two child edges of a node start with the same character. By this definition, each node can be mapped one-to-one to the concatenation of symbols from the *root to itself*. If that concatenated string is r the node is denoted $\overrightarrow{r}$. Notice that $\overleftarrow{u}$ is the node with the concatenated string u read from *node to root*. If $u = r^R$, $\overrightarrow{r}$ and $\overleftarrow{u}$ denote the same node. An important concept of suffix trees are *suffix links* [14]. They are used to reach node $\overrightarrow{r}$ from a node $\overrightarrow{\sigma r}$, $\sigma \in \Sigma$. Figure 2.1 shows a suffix tree built for Example 2.

3   Algorithms for VOMC Construction

3.1   The AB Algorithm

The algorithm of Apostolico and Bejerano builds a PST in optimal $O(\|S\|)$ time and space. We will explain the AB algorithm according to our general approach from Section 2.3. The *Support-Pruning* phase is implemented with a suffix tree built on S in linear time and space. Based upon that tree, all contexts which do not fulfill the conditions of the *Support-Pruning* phase are pruned, see Fig. 2.1 and (3),(5). A bottom-up traversal of the suffix tree yields the absolute frequencies and thus the probability vectors for each internal node. However, as the suffix tree is built over S and not S^R, going down in the suffix tree means to extend a subsequence to the right, whereas context extensions are subsequence extensions to the left. In other words, the father-son relation of the PST is unknown, an essential information for the *Similarity-Pruning* phase. Hence, so-called reverse suffix links (rsufs) are introduced, which are the opposite of suffix links, see Fig. 2.3. Via rsufs all existent nodes $\overrightarrow{\sigma r}$, $\sigma \in \Sigma$, can be reached from each node $\overrightarrow{r}$. Walking the reverse suffix link from $\overrightarrow{r}$ to $\overrightarrow{\sigma r}$ is equivalent to walking from father $\overleftarrow{r}$ to son $\overleftarrow{\sigma r}$ in the PST.

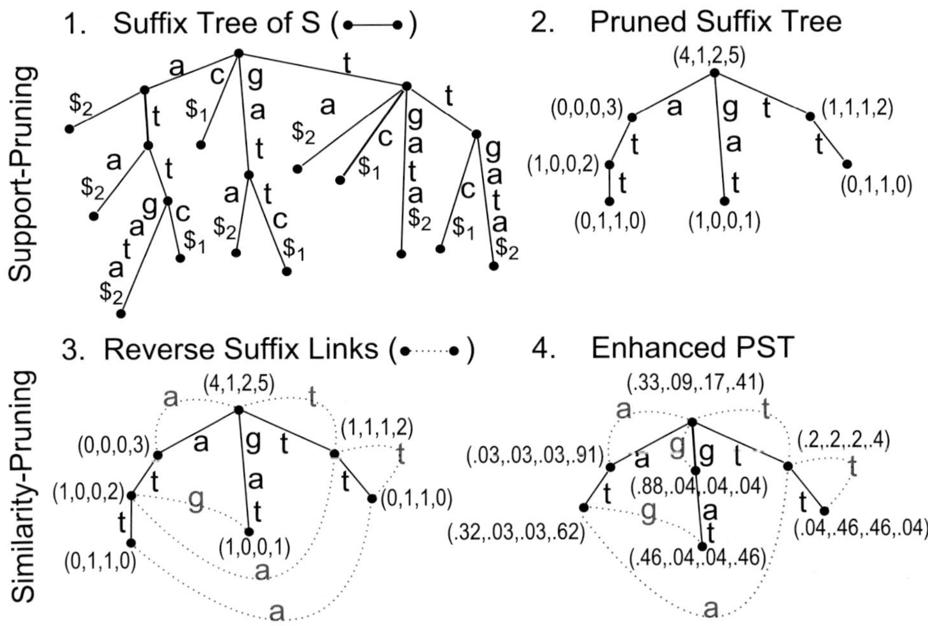

Fig. 2. Overview of our implementation of the AB algorithm [11] for PST construction. It can be divided into two pruning phases: Support-Pruning and Similarity-Pruning. The figure shows the detailed construction for the parameters given in Example 2. A detailed description can be found in Example 3.

Reverse suffix links, just as suffix links, can be constructed for a suffix tree in linear time [20]. Using rsufs and after adding missing nodes,[1] it can be shown that the *Similarity-Pruning* phase takes only $\mathcal{O}(\|S\|)$ time.

Example 3. Figure 2 shows how the AB algorithm constructs the PST for the parameters given in Example 2. In Fig. 2.1 the complete suffix tree of S is constructed. In step 2, a bottom-up traversal determines for all nodes the absolute frequencies and nodes $\vec{r}$, with $N_S(r) < 2$, are pruned. Reverse suffix links are added in Fig. 2.3, the start of the *Similarity-Pruning* phase. The node $\overrightarrow{att}$ is pruned, as (4) is satisfied, and the missing node $\vec{g}$ is added in step 4. Finally all vector entries are smoothed, see Example 2. The same PST as in Fig. 1 is obtained when only the dotted lines, the rsufs, are considered.

We call the resulting automaton *enhanced PST*, as it allows a traversal from node $\vec{r}$ to nodes $\overrightarrow{\sigma r}$ via rsufs and to nodes $\overrightarrow{r\sigma}$ via suffix tree edges, for all $\sigma \in \Sigma$. The enhanced PST can be modified to predict a new sequence of length m in $\mathcal{O}(m)$ time instead of $\mathcal{O}(mL)$ for a PST [11], which is another advantage of the

---

[1] It can happen that new auxiliary nodes need to be added to the suffix tree. Branching nodes in a PST are not necessarily branching in the suffix tree. However, the missing nodes, each a target of a reverse suffix link, can be created in $\mathcal{O}(\|S\|)$ time [11].

algorithm. We have implemented the AB algorithm using the *eager* write-only top-down (WOTD)-algorithm for suffix tree construction, one of the fastest and most space efficient suffix tree construction algorithms [15].

3.2   Adaptive Algorithms for VOMCs

VOMCs select the abundant contexts in the data, the carriers of information. They account for a small subset of suffix tree nodes, the pruned suffix tree, which is determined by a traversal of the suffix tree. In the following we suggest two different approaches to attain the pruned suffix tree more efficiently than constructing the complete suffix tree. $N_S(r)$ and the absolute frequencies can be obtained in a top-down traversal. After attaining the pruned suffix tree, we proceed with steps 3 and 4 of the AB algorithm; add reverse suffix links and prune similar nodes. The second approach is depicted for Solution 1 in Algorithm 1.

An Adaptive VOMC Algorithm with Enhanced Suffix Arrays. The suffix array stores the starting positions of all lexicographically ordered suffixes of a string [16]. We use the deep shallow algorithm by Manzini and Ferragina for suffix array construction [21]. An enhanced suffix array (ESA) is a suffix array which has two additional arrays, the lcp-table and the child-table which are used to simulate a top-down traversal on a suffix tree of S in linear time and space [18]. The ESA is an efficient replacement for the suffix tree in the first step of the *Support-Pruning* phase, see Fig. 2.1. It can be used to traverse a suffix tree top-down using $9 \cdot \|S\|$ bytes of memory [18], assuming that L is always smaller than 255. The traversed nodes are added to a new graph, the pruned suffix tree. This tree is used in the next steps, depicted in Fig. 2.2–2.4. After the traversal, the ESA is discarded. We refer to this algorithm as **AV-1**.

An Adaptive VOMC Algorithm with Lazy Suffix Trees. An approach more appealing than the previous one is to avoid building the complete ESA, but build only the parts of the suffix tree to be traversed. That means to entirely skip the first step of the *Support-Pruning* phase. The lazy WOTD-algorithm is perfectly suited to restrict the buildup of the suffix tree as it expands nodes top-down. A lazy suffix tree is a suffix tree whose nodes are created on demand, i.e. when they are visited the first time. Giegerich et al. introduced a lazy suffix tree data structure [22] that utilizes the WOTD-algorithm [15, 22] for the on-demand node creation. In the beginning, T contains only the root node and is iteratively extended by suffix tree nodes, to at most the entire suffix tree.

After creating a node $\overrightarrow{r}$ in T the values $|r|$ and $N_S(r)$, relevant for the *Support-Pruning* phase, are known. Thus the restriction for length L and $N_S(r)$ can be included to constrain the top-down construction, see lines 1–9 of Algorithm 1. This does not only save construction time, but also reduces the amount of space needed. Steps 3 and 4 of the AB algorithm are realized in lines 10–13. The adaptive algorithm with lazy suffix trees is referred to as **AV-2**.

Algorithm 1. createEnhancedPST(S, t, L, K)

Input   : string set S over Σ, min. support t, max. length L, pruning value K
Output : enhanced PST $\mathcal{T}$ for Solution 1
// Support-Pruning
1 Let X be a string collection and $\mathcal{T}$ be a suffix tree containing only the root node.
2 $X \leftarrow \{\epsilon\}$
3 **foreach** $x \in X$ **do**
4     $X \leftarrow X \setminus \{x\}$
5   **if** $|x| < L$ **then**
6   **foreach** $\sigma \in \Sigma$ *with* $N_S(x\sigma) \geq t$ **do**
7   *Let* $y \in \Sigma^*$ *be the longest string, s.t.* $N_S(x\sigma y) = N_S(x\sigma)$ *and* $|x\sigma y| \leq L$
8   Insert $\overrightarrow{x\sigma y}$ into $\mathcal{T}$.
9             $X \leftarrow X \cup \{x\sigma y\}$

// Similarity-Pruning
10 Insert auxiliary nodes into $\mathcal{T}$ [11] and add probability vectors to all nodes.
11 **foreach** $\overrightarrow{x}, \overrightarrow{\sigma x} \in \mathcal{T}$, $\sigma \in \Sigma$ **do**
12   Add a reverse suffix link from $\overrightarrow{x}$ to $\overrightarrow{\sigma x}$.
13 In a bottom-up traversal using only the reverse suffix links of T remove all nodes satisfying (4).
14 **return** T

---

4   Results

We compared the runtime of our algorithms within two experiments. In the first we investigated the improvement of the AV-1 and AV-2 algorithm compared to the AB algorithm. In the second, we compared our new tool with two existing software tools. For the *Growing PST* solution we used the **RST** algorithm [12] and for the *Context Tree* solution the **Dal** algorithm [6]. The algorithms were applied to data from applications mentioned in the introduction: three protein families from the classification experiment of Bejerano et al. [9], two bacterial genomes as in [6] retrieved from GenBank [23], a set of human promoter sequences [24], and the complete human proteome from Uniprot [25], see Table 2. Experiments were conducted under Linux on an Intel Xeon 3.2 GHz with 3 GB of RAM. Output of all programs was suppressed. For all tests of the *Context Tree* algorithm we fixed $K = 1.2$, and for the *Growing PST* algorithm we fixed $\alpha = 0$, $\gamma_{\min} = 0.01$, $k = 1.2$.

First we compared the runtime of the different *Support-Pruning* phases of our implemented algorithms AB, AV-1, and AV-2. The results are shown for four data sets with varying L, and three values for $P_{\min}$ or t in Fig. 3. The AB algorithm performs worse than the two adaptive algorithms for all sequences. The AV-2 algorithm outperforms the AB and AV-1 algorithms for restrictive parameters, but for small values of t the AV-2 implementation of the *Context Tree* algorithm becomes less efficient and is outperformed by the AV-1 algorithm.

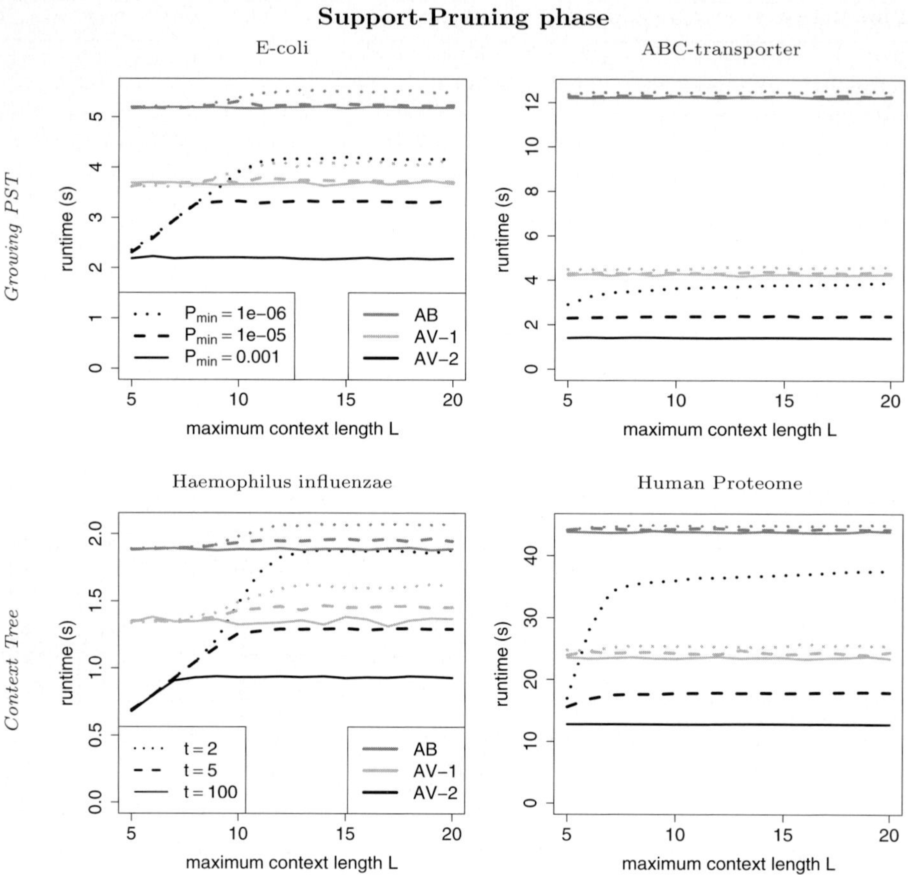

Fig. 3. Runtime results for different *Support-Pruning* phases of our implemented algorithms AB, AV-1, and AV-2, averaged over three runs

For the values of t in the H.inf. plot in Fig. 3 the runtime of the AV-2 algorithm depends on the L-Pruning up to context length 7, whereas for long contexts the t-Pruning dominates the L-Pruning. On the proteome data set a similar behaviour is already visible for smaller values of L, because of the larger alphabet size.

Second we compared the runtime for the complete VOMC construction of our implementations with the Dal algorithm and the RST algorithm. Table 1 reveals the superiority of our approach. The Dal algorithm is up to 11600 times slower than the AV algorithms and crashes for some instances due to insufficient memory. See Table 1 for the memory consumption for one such instance. The RST algorithm is up to 2700 times slower than our algorithms not including the parameter settings, where the RST algorithm was terminated after 100 minutes. However, our algorithms are efficient for all parameter settings and alphabets.

Table 1. Runtimes in seconds of our implementations for *Context Tree* and *Growing PST* solutions on various data sets and parameter settings

Runtime Context Tree Algorithms												
(L, t)	(5,2)				(9,5)				(15,100)			
Name	Dal	AB	AV-1	AV-2	Dal	AB	AV-1	AV-2	Dal	AB	AV-1	AV-2
H.inf.	6.56	2.04	1.72	**0.74**	28.9	3.98	3.78	**3.12**	485	2.16	1.75	**1.12**
E.coli	15.6	5.58	4.60	**2.54**	50.6	8.46	7.85	**6.89**	†	5.91	5.00	**3.75**
Promoters	76.5	50.4	31.1	**19.3**	179	53.9	**34.9**	36.1	†	52.3	**32.8**	34.4
7tm	83.1	1.62	1.61	**1.49**	339	0.82	**0.71**	**0.71**	812	0.21	0.14	**0.07**
ig	111	3.82	**3.33**	3.35	489	2.41	1.93	**1.89**	1200	0.98	0.46	**0.20**
ABC	618	41.2	31.3	**30.8**	†	30.3	20.5	**18.5**	†	16.2	5.97	**2.75**
Proteome	1810	59.2	51.0	**36.8**	†	60.3	51.6	**39.7**	†	39.1	27.4	**14.8**

Runtime Growing PST Algorithms												
$(L, P_{\min})$	$(5, 10^{-6})$				$(9, 10^{-5})$				(15, 0.001)			
Name	RST	AB	AV-1	AV-2	RST	AB	AV-1	AV-2	RST	AB	AV-1	AV-2
H.inf.	119	2.07	1.60	**0.71**	–	2.71	2.13	**1.68**	102	2.00	1.53	**0.70**
E.coli	305	5.42	4.29	**2.41**	–	6.08	5.03	**4.10**	252	5.46	4.19	**2.26**
Promoters	1240	48.7	29.3	**18.0**	–	49.4	29.7	**29.1**	1020	48.8	28.4	**15.4**
7tm	1710	4.40	4.39	**4.34**	2400	2.42	2.37	**2.36**	4.20	0.21	0.13	**0.06**
ig	3260	7.77	**7.21**	7.28	4900	2.51	1.91	**1.81**	11.3	0.93	0.39	**0.12**
ABC	–	22.2	11.8	**9.47**	–	16.7	6.17	**3.43**	85.5	15.9	5.20	**1.52**
Proteome	–	43.6	31.6	**20.2**	–	37.7	25.0	**12.7**	392	36.7	24.0	**7.39**

†   Program terminated due to insufficient internal memory ($> 3\,$GB).

–   Program run for more than 6000 s.

5   Discussion and Conclusions

We achieved an order of magnitude improvement for VOMC construction algorithms compared to two previous algorithms. In addition, we further improved the overall time and space consumption of [11] by replacing the suffix tree with more efficient and problem-oriented index data structures. All algorithms are implemented with SeqAn [26] in the *Pisa* tool, which is publicly available at http://www.seqan.de/projects/pisa.html. To our knowledge it is the most efficient tool for VOMC learning in the bioinformatics community.

In our experiments the lazy suffix tree approach [15] is faster than the ESA approach [18] and consumes roughly half of the memory. Improving the construction time of the suffix array, one of three tables of the ESA, is a topic of current research and an expanding field, outside the scope of this paper.

Other possible improvements of the *Support-Pruning* phase like a q-gram index or length limited suffix tree construction in linear time, e.g. [27], might be considered, but from our experience the lazy suffix tree is the most robust data structure for our general implementation, as it works for large alphabets and deep contexts as well.

Table 2. Data sets used in the experiments and memory usage of the *Context Tree* algorithms

Name	Description	$\|S\|$ (mb)	$\|\Sigma\|$	Memory Usage in MB ($L = 15, t = 100$)			
				Dal	AB	AV-1	AV-2
H.inf.	Haemophilus influenzae	1.9	5	2010	59.9	65.4	**31.6**
E.coli	Escherichia coli	4.6	5	†	129	159	**76.2**
Promoters	Human promoters	22.5	5	†	633	550	**373**
7tm	transmembrane family	0.2	24	1160	6.57	7.38	**3.93**
ig	Immunoglobulin family	0.5	24	1540	16.4	18.7	**11.8**
ABC	ABC transporter family	4.3	24	†	118	156	**81.1**
Proteome	Human proteome	17.3	24	†	456	503	**306**

In the future we will investigate extensions to the approach presented here to build more advanced types of VOMCs, which can learn also subset relations e.g. the approach presented by Leonardi [13].

References

1. Rissanen, J.: A universal data compression system. IEEE Transactions on Information Theory 29, 656–664 (1983)
2. Ron, D., Singer, Y., Tishby, N.: The power of amnesia: Learning probabilistic automata with variable memory length. Machine Learning 25, 117–149 (1996)
3. Ben-Gal, I., Shani, A., Gohr, A., Grau, J., Arviv, S., Shmilovici, A., Posch, S., Grosse, I.: Identification of transcription factor binding sites with variable-order Bayesian networks. Bioinformatics 21(11), 2657–2666 (2005)
4. Zhao, X., Huang, H., Speed, T.P.: Finding short DNA motifs using permuted Markov models. J. Comput. Biol. 12(6), 894–906 (2005)
5. Ogul, H., Mumcuoglu, E.U.: SVM-based detection of distant protein structural relationships using pairwise probabilistic suffix trees. Comput. Biol. Chem. 30(4), 292–299 (2006)
6. Dalevi, D., Dubhashi, D., Hermansson, M.: Bayesian classifiers for detecting HGT using fixed and variable order markov models of genomic signatures. Bioinformatics 22(5), 517–522 (2006)
7. Bejerano, G., Seldin, Y., Margalit, H., Tishby, N.: Markovian domain fingerprinting: statistical segmentation of protein sequences. Bioinformatics 17(10), 927–934 (2001)
8. Slonim, N., Bejerano, G., Fine, S., Tishby, N.: Discriminative feature selection via multiclass variable memory Markov model. EURASIP J. Appl. Signal Process 2003(1), 93–102 (2003)
9. Bejerano, G., Yona, G.: Variations on probabilistic suffix trees: statistical modeling and prediction of protein families. Bioinformatics 17(1), 23–43 (2001)
10. Posch, S., Grau, J., Gohr, A., Ben-Gal, I., Kel, A.E., Grosse, I.: Recognition of cis-regulatory elements with vombat. J. Bioinform. Comput. Biol. 5(2B), 561–577 (2007)
11. Apostolico, A., Bejerano, G.: Optimal amnesic probabilistic automata or how to learn and classify proteins in linear time and space. J. Comput. Biol. 7(3-4), 381–393 (2000)

12. Bejerano, G.: Algorithms for variable length Markov chain modeling. Bioinformatics 20(5), 788–789 (2004)
13. Leonardi, F.G.: A generalization of the PST algorithm: modeling the sparse nature of protein sequences. Bioinformatics 22(11), 1302–1307 (2006)
14. Kurtz, S.: Reducing the space requirement of suffix trees. Software Pract. Exper. 29(13), 1149–1171 (1999)
15. Giegerich, R., Kurtz, S., Stoye, J.: Efficient implementation of lazy suffix trees. Software Pract. Exper. 33(11), 1035–1049 (2003)
16. Manber, U., Myers, E.: Suffix arrays: A new method for on-line string searches. SIAM J. Comput. 22(5), 935–948 (1993)
17. Ferragina, P., Manzini, G., Mäkinen, V., Navarro, G.: Compressed representations of sequences and full-text indexes. ACM Trans. Algorithms 3(2), 20 (2007)
18. Abouelhoda, M., Kurtz, S., Ohlebusch, E.: Replacing suffix trees with enhanced suffix arrays. Journal of Discrete Algorithms 2, 53–86 (2004)
19. Bühlmann, P., Wyner, A.J.: Variable length Markov chains. Ann. Statist. 27(2), 480–513 (1999)
20. Maaß, M.G.: Computing suffix links for suffix trees and arrays. Inf. Process. Lett. 101(6), 250–254 (2007)
21. Manzini, G., Ferragina, P.: Engineering a lightweight suffix array construction algorithm. Algorithmica 40(1), 33–50 (2004)
22. Giegerich, R., Kurtz, S.: A comparison of imperative and purely functional suffix tree constructions. Sci. Comput. Program. 25, 187–218 (1995)
23. Benson, D.A., Karsch-Mizrachi, I., Lipman, D.J., Ostell, J., Wheeler, D.L.: GenBank. Nucleic Acids Res. 36(Database issue), D25–D30 (2008)
24. Fitzgerald, P.C., Sturgill, D., Shyakhtenko, A., Oliver, B., Vinson, C.: Comparative genomics of drosophila and human core promoters. Genome Biol. 7, R53 (2006)
25. The UniProt Consortium: The Universal Protein Resource (UniProt). Nucl. Acids Res. 36(suppl.1), D190–195 (2008)
26. Döring, A., Weese, D., Rausch, T., Reinert, K.: SeqAn an efficient, generic C++ library for sequence analysis. BMC Bioinformatics 9, 11 (2008)
27. Schulz, M.H., Bauer, S., Robinson, P.N.: The generalised k-Truncated Suffix Tree for time- and space- efficient searches in multiple DNA or protein sequences. Int. J. Bioinform. Res. Appl. 4(1), 81–95 (2008)

Computing Alignment Seed Sensitivity with Probabilistic Arithmetic Automata

Inke Herms[1] and Sven Rahmann[2]

[1] Genome Informatics, Faculty of Technology, Bielefeld University,
D-33501 Bielefeld, Germany
[2] Bioinformatics for High-Throughput Technologies,
Computer Science 11, TU Dortmund,
D-44221 Dortmund, Germany
`ihildebr@cebitec.uni-bielefeld.de`, `Sven.Rahmann@tu-dortmund.de`

Abstract. Heuristic sequence alignment and database search algorithms, such as PatternHunter and BLAST, are based on the initial discovery of so-called alignment *seeds* of well-conserved alignment patterns, which are subsequently extended to full local alignments. In recent years, the theory of classical seeds (matching contiguous q-grams) has been extended to *spaced seeds*, which allow mismatches within a seed, and subsequently to *indel seeds*, which allow gaps in the underlying alignment.

Different seeds within a given class of seeds are usually compared by their *sensitivity*, that is, the probability to match an alignment generated from a particular probabilistic alignment model.

We present a flexible, exact, unifying framework called *probabilistic arithmetic automaton* for seed sensitivity computation that includes all previous results on spaced and indel seeds. In addition, we can easily incorporate sets of arbitrary seeds. Instead of only computing the probability of at least one hit (the standard definition of sensitivity), we can optionally provide the entire distribution of overlapping or non-overlapping seed hits, which yields a different characterization of a seed. A symbolic representation allows fast computation for any set of parameters.

Keywords: Homology search, Seed sensitivity, Spaced seed, Indel seed, Multiple seeds, Probabilistic arithmetic automaton.

1 Introduction

Most heuristic homology search algorithms [1, 2, 3, 4] and some repeat detection algorithms [5] are based on a filtration technique. In the filtration step, one selects candidate sequences that share a common pattern of matching characters (the *seed*) with the query. These candidates are then further investigated by exact local alignment methods, such as the Smith-Waterman [6] algorithm. Initially, only contiguous perfect matches (e.g., DNA 11-mers in the initial BLAST implementation) were used as seeds. The PatternHunter (PH) tool by Ma et al. [7] was the first system to systematically advocate and investigate *spaced seeds*: PH looks for 18-mers with at least 11 matching positions distributed as 111*1**1*1**11*111,

K.A. Crandall and J. Lagergren (Eds.): WABI 2008, LNBI 5251, pp. 318–329, 2008.
© Springer-Verlag Berlin Heidelberg 2008

where 1 denotes a necessary match and $*$ denotes a don't care position (match or mismatch). Over time, various seed models have been proposed in the literature, including consecutive seeds [1, 2], spaced seeds [7, 8, 9, 10], subset seeds [11], vector seeds [12], and indel seeds [13].

Given a probabilistic model for alignments (a so-called *homology model*), different seeds in a class (e.g., all seeds with 11 match positions and length 18) can be compared according to their *sensitivity*, i.e., to match an alignment of given length. A good seed exhibits high sensitivity for alignment models that model evolutionarily related biosequences, and low sensitivity values for models that represent unrelated sequences. The latter property ensures that the seed does not detect too many *random hits*. Random hits decrease the efficiency of the filtration phase, since they are checked in vain for a significant alignment. An interesting finding was that the PatternHunter approach led to an increase in both sensitivity and filtration efficiency, compared to seeds of contiguous matches. Based on the observations in [7], the advantages of spaced seeds over consecutive seeds have been subsequently evaluated by many authors [8, 14, 15].

An extension to single seed models is the design of a multiple seed. This is a set of spaced seeds to be used simultaneously, such that a similarity is detected when it is found by (at least) one of the seeds. The idea to use a family of spaced seeds for BLAST-type DNA local alignment has been suggested by Ma et al. [7] and was implemented in PatternHunter II [16]. It has also been applied to local protein alignment in [17]. Since finding optimal multiple seeds (seed sets) is challenging, most authors concentrate on the design of efficient sets of seeds, leading to higher sensitivity than optimal single seeds [16, 18, 19]. Kucherov et al. [18] characterize a set of seeds only by its selectivity. Recent approaches [20, 21] approximate the sensitivity of multiple spaced seeds by means of correlation functions. Moreover, Kong [20] discussed that sensitivity should rather be measured via an averaging criterion.

When searching optimal seeds, one faces the following problems to evaluate candidate seeds (the second more general one has not yet been extensively considered in the literature).

Problem 1 (Sensitivity computation). Given a homology model (Sect. 2.2), a target length t of alignments, and a set of seeds (see Sect. 2.3 for a formal description of different seed models), what is the probability that an alignment of length t is matched by the seed (at least once)?

Problem 2 (Hit distribution). Given a homology model, a target length t of alignments, a set of seeds, and a maximal match number K, what is the probability that an alignment of length t is matched by the seed exactly (at least) k times, for $k = 0, \ldots, K$, when counting (a) overlapping matches, (b) non-overlapping matches only?

Related Work and Our Contributions. We present an exact method, called *Probabilistic Arithmetic Automaton* (PAA), to compute the sensitivity as well as the entire distribution of the number of overlapping or non-overlapping hits for a given seed or set of seeds. The model is very flexible as it comprises the majority of recent models. We generalize the Markov chain approach of Choi and Zhang [14], following the idea to characterize a seed by an averaging criterion rather than its sensitivity only. The obtained recurrences also allow a symbolic calculation as performed by Mak and Benson [22]. Moreover, the PAA can be designed for indel seeds Mak et al. [13]. Similar to [11], our model provides a unifying framework. However, we characterize multiple seeds by their sensitivity, whereas Kucherov et al. [11] design seed families with perfect sensitivity minimizing the number of false positives.

Outline. The rest of the paper is organized as follows: in Sect. 2 we first describe different homology and seed models before establishing the Probabilistic Arithmetic Automaton in Sect. 3. We derive recurrence relations to compute hit distributions and in particular the sensitivity of a given seed. Sect. 4 compares our results to recent approaches. We divide our findings regarding Problem 1 in three categories, namely for i) a single spaced seed, ii) a single indel seed and iii) a set of spaced seeds. Considerations about Problem 2 complete the paper.

2   Background

2.1   Notation

Let Σ denote an alphabet and let the set Σ^* include all finite words over Σ. A string $s = s[0]s[1]\ldots s[l-1] \in \Sigma^l$ has length $|s| = l$, $s[i]$ denotes the character at position i (where indexing starts with 0) and $s[i,j]$ refers to the substring of s ranging from position i to j. As usual, ϵ is the empty string. The concatenation of r and s is written as rs. Moreover, for two words $r, s \in \Sigma^*$, $r \lhd s$ indicates that r is a prefix of s, while $r \blacktriangleleft s$ says that r is a suffix of s. Finally, $\mathcal{L}(X)$ denotes the probability distribution ($\mathcal{L}$aw) of a random variable X with corresponding generic probability measure $\mathbb{P}$.

2.2   Homology Model

To be able to compute seed sensitivity, we need to describe random alignments with known degree of similarity (e.g., a certain per cent identity value). In the context of seed sensitivity computations, it is customary *not* to model random sequences and derive alignment properties from those, but to directly model alignments *representative strings* $\mathcal{A}$ over an alphabet Σ indicating the status of the alignment columns. In the simplest and most frequently studied case [7, 8, 9, 10, 14, 19] only substitution mutations are considered; that is $\Sigma = \{0, 1\}$, referring to matches (1) and mismatches (0). Indel seeds use the alignment alphabet $\Sigma = \{0, 1, 2, 3\}$, where additionally 2 denotes an insertion in the database sequence, and 3 indicates an insertion in the query sequence. There are various other alignment alphabets, e.g. the ternary alphabet representing match / transition / transversion in DNA [23], or even larger alphabets to distinguish different pairs of amino acids in the case of proteins [17]. The sequence states of alignment columns is modeled as a Markov chain (Σ, P, p^0) with *homology parameters* P, a transition matrix on the alphabet Σ, and p^0, an initial distribution on Σ.

If we consider $\Sigma = \{0, 1\}$ and an i.i.d. homology model, where the transition probability P_{ij} does not depend on i, then P takes the form $\begin{pmatrix} 1-p & p \\ 1-p & p \end{pmatrix}$ for a *match probability* $p \in [0, 1]$, which is the only homology parameter in this case and quantifies the average per cent identity of such alignments. For indel seeds, a first-order Markov chain is appropriate, since a substitution is more plausible than two consecutive indels (one in each sequence). Hence, the homology model should prohibit the pairs '23' and '32' in a representative string. With the intention to compare our results later on, we work with the transition matrix P proposed in [13]:

$$
\begin{array}{c}
 \\
0 \\
1 \\
2 \\
3
\end{array}
\begin{array}{cccc}
0 & 1 & 2 & 3
\end{array}
\begin{pmatrix}
p_0 & p_1 & p_g & p_g \\
p_0 & p_1 & p_g & p_g \\
p_0^* & p_1^* & p_g & 0 \\
p_0^* & p_1^* & 0 & p_g
\end{pmatrix},
\tag{1}
$$

where p_0 is the probability of a mismatch, p_1 is the probability of a match, and p_g refers to the probability of a gap in the alignment. In order to ensure a stochastic transition matrix, $p_i^* = p_i + p_g p_i/(p_0 + p_1)$ for $i = 0, 1$ redistributes p_g to match and mismatch characters. The initial distribution is given by $p^0 = (p_0, p_1, p_g, p_g)$. Other transition probabilities are possible, e.g. if alignments with affine gap costs should be modeled.

2.3   Seed Models

A seed $\pi = \pi[0]\pi[1]\ldots\pi[L-1]$ is a string over an alphabet of "care" and "don't care" characters. It represents alignment regions that indicate matches at the "care" positions. A seed is classified (L, ω) by its *length* $L = |\pi|$ and its *weight* ω, which refers to the number of "care" positions.

A *spaced seed* is a string over the alphabet $\Xi = \{1, *\}$. The "care" positions are indicated by '1', while '*' An *indel seed* according to Mak et al. [13] is a string over the alphabet $\Xi = \{1, *, ?\}$, where '1' and '*' are as above and '?' stands for zero or one character from the alignment alphabet $\Sigma = \{0, 1, 2, 3\}$. Consecutive '?' symbols may represent any character pair except '23' or '32'. By means of this interpretation, the model explicitly allows for indels of variable size. For example, 1??1 may detect indels of size 0, 1, or 2. It hence tolerates 2, 1, or 0 match/mismatch positions. To understand how indel seeds are used in the filtering step of homology search, we refer the reader to the original article.

In order to obtain the set of patterns defined by π, let $\Psi = \{[1], [01], [\epsilon 0123]\}$ contain the character sets induced by Ξ, where we write $[xy]$ as shorthand for $\{x, y\}$. The generalized string that refers to π is $G(\pi) = g(\pi[0])\ldots g(\pi[L-1])$, where $g : \Xi \to \Psi$ is a mapping such that '1' $\mapsto [1]$, '*' $\mapsto [01]$, and '?' $\mapsto [\epsilon 0123]$.

Definition 1 (Pattern set). *Let $\pi = \pi[0]\pi[1]\ldots\pi[L-1]$ be a seed of length L over the alphabet Ξ. The* pattern set

$$\mathcal{PS}(\pi) = \{s = s[0]\ldots s[L-1] \mid s[i] \in g(\pi[i])\}$$

contains all words satisfying the seed, this is all strings that match $G(\pi)$.

Example 1. For the spaced seed $\pi = 1 * 1 * 1$ of length 5 and weight 3, $G(\pi) = [1][01][1][01][1]$ and $\mathcal{PS}(\pi) = \{10101, 10111, 11101, 11111\}$. The patterns of the indel seed $\pi = 1 * 1?1$ with $G(\pi) = [1][01][1][\epsilon0123][1]$ are given by $\mathcal{PS}(\pi) = \{1011, 1111, 10101, 10111, 10121, 10131, 11101, 11111, 11121, 11131\}$.

Definition 2 (Hit position). *A seed π hits the representative string $\mathcal{A}$ ending at position n iff*

$$\exists\, M \in \mathcal{PS}(\pi) \text{ s.t. } \mathcal{A}[n - |M| + 1, n] = M \,.$$

Example 2. For the representative string 1011011110 for instance, the seed $\pi = 1 * 11 * 1$ hits at positions 6 and 9, the indel seed $\pi = 11 * 1?1$ hits at positions 7 and 8, respectively.

The idea to use a finite, non-empty set $\Pi = \{\pi_1, \ldots, \pi_m\}$ of spaced seeds, also called *multiple spaced seed*, turned out to further improve the quality of filtering [8, 9, 16, 19]. The patterns are collected in $\mathcal{PS}(\Pi) = \cup_{i=1}^m \mathcal{PS}(\pi_i)$. A multiple seed hits $\mathcal{A}$, if (at least) one of its components does (in contrast to Pevzner and Waterman [24] where all seeds are required to match).

3   Probabilistic Arithmetic Automata

We show that the framework of *Probabilistic Arithmetic Automata* (PAA), recently introduced by Marschall and Rahmann [25], provides a unified approach to seed sensitivity computation. While including previous related approaches [8, 11, 14], the PAA framework can handle both ungapped and gapped alignments. Moreover, it allows the investigation of overlapping and non-overlapping hits of a single or a multiple seed. It yields recurrence relations to compute the entire hit distribution and in particular seed sensitivity. From these recurrences we derive a polynomial which allows fast computation for any parameter set.

A PAA consists of three components, namely i) a Markov chain, ii) emissions associated with the states of this chain, and iii) a series of arithmetic operations performed on the emissions. In the context of seed sensitivity the Markov chain generates a random sequence alignment, emissions correspond to hit counts, and the accumulation of such counts yields the distribution of seed hits. The following formal definition is adopted from Marschall and Rahmann [25]:

Definition 3 (Probabilistic Arithmetic Automaton). *A* Probabilistic Arithmetic Automaton *is a tuple $\big(Q, T, q_0, E, \mu = (\mu_q)_{q \in Q}, N, n_0, \theta = (\theta_q)_{q \in Q}\big)$, where*

- Q *is a finite set of states,*
- $T = (T_{uv})_{u,v \in Q}$ *is a stochastic state transition matrix,*
- $q_0 \in Q$ *is called* start *state,*
- E *is a finite set called* emission *set,*
- *each* $\mu_q : E \to [0,1]$ *is a weight distribution associated with state* q,
- N *is a finite set called* value *set,*
- $n_0 \in N$ *is called* start *value,*
- *each* $\theta_q : N \times E \to N$ *is an operation associated with state* q.

The tuple (Q, T, δ_{q_0}) [1] defines a Markov chain on state space Q starting in q_0 with transitions according to T. Let $(Y_i)_{i \in \mathbb{N}_0}$ denote its associated state process. Accordingly, $(Z_i)_{i \in \mathbb{N}_0}$ represents the sequence of emissions and $(V_i)_{i \in \mathbb{N}_0}$ denotes the sequence of values $V_l = \theta_{Y_l}(V_{l-1}, Z_l)$, with $V_0 = n_0$, resulting from the performed operations. In order to compute seed sensitivity, the PAA should count the number of seed hits to a randomly generated alignment. Therefore, we add the emitted counts, so the operation is $\theta_q = +$ in each state. Thus, $V_l = V_{l-1} + Z_l$ with $V_0 = 0$, reducing the free parameters of the PAA to $(Q, T, q_0, E, \mu = (\mu_q)_{q \in Q})$.

3.1 PAA Construction

Our approach is related to [14]. Given the homology model (Σ, P, p^0), we aim to construct a Markov chain (Q, T, δ_{q_0}) that generates a proper random sequence alignment. Similar to the structure of an Aho-Corasick tree used in (approximate) string matching, the states of the underlying Markov chain have to incorporate all prefixes of the patterns of the considered seed π or multiple seed Π, respectively. Supplementary states are represented by the characters from the alignment alphabet, and an additional *start* state ensures the initial character distribution. Thus, for a given seed π of length L, the state space is

$$Q(\pi) = \{start\} \cup \{\sigma \in \Sigma\} \cup \{x \in \Sigma^* \mid 1 \leq |x| \leq L, \exists M \in \mathcal{PS}(\pi) : x \vartriangleleft M\}. \quad (2)$$

We mark a finite set $\mathcal{F}$ of final states in order to distinguish states that contribute to the number of seed hits. In our case, these states are those hit by π, that is $\mathcal{F} = \{q \in Q \mid \exists M \in \mathcal{PS}(\pi) : M \blacktriangleleft q\}$.

There is a non-zero outgoing transition from state u to state v if v is the maximal suffix of $u\sigma$, that is if there exists a $\sigma \in \Sigma$ s.t. $v = \mathrm{argmax}_{x \in \{y \in Q \mid y \blacktriangleleft u\sigma\}} |x|$. The corresponding transition probability is $P_{u[|u|-1]\sigma} = \mathbb{P}(\sigma \mid u[|u| - 1])$, where the conditional probabilities are given by the homology model. Thus, in the i.i.d. case it is just the character frequency p_σ. The stochastic state transition matrix of the PAA with $u, v \in Q$ is given by

$$T_{uv} = \begin{cases} p_\sigma^0 & \text{if } u = start, v = \sigma \in \Sigma, \\ P_{u[|u|-1]v[|v|-1]} & \text{if } u \neq start, \exists \sigma \in \Sigma : v = \mathrm{argmax}_{\{x \in Q \mid x \blacktriangleleft u\sigma\}} |x|, \\ 0 & \text{otherwise.} \end{cases}$$

$$(3)$$

---

[1] The initial state distribution given by δ_{q_0} is the Dirac distribution assigning probability 1 to $\{q_0\}$.

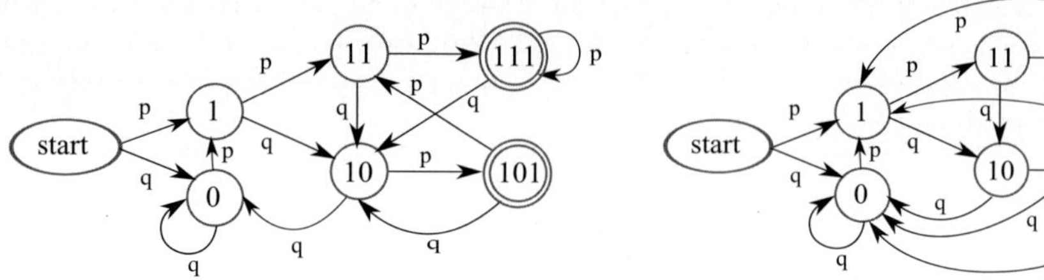

(a) PAA counting overlapping hits of $\pi = 1 * 1$.

(b) PAA counting non-overlapping hits of $\pi = 1 * 1$.

Fig. 1. Markov chain underlying the PAA for $\pi = 1 * 1$. 1(a) shows the transitions in order to count all occurrences of π in a random alignment over $\Sigma = \{0, 1\}$ (either i.i.d. with match probability p or Markov with $P_{\cdot 1} = p$ and $P_{\cdot 0} = q$). 1(b) is designed to count non-overlapping hits.

If we are interested in counting non-overlapping hits, the transitions outgoing a final state become

$$T_{uv} = \begin{cases} P_{1\sigma} & \text{if } u \in \mathcal{F}, v = \sigma \in \Sigma, \\ 0 & \text{if } u \in \mathcal{F}, v \notin \Sigma, \end{cases}$$

since the last character read was a 1 (by definition). Therewith, the Markov chain becomes reducible, and we can remove states from Q, that are not reachable from *start*.

The so constructed Markov chain (Q, T, δ_{start}) generates a random sequence alignment whose similarity level is specified by the homology parameters. It is adapted to a given seed through the choice of states. In order to count the number of seed hits, we still need to assign the family $\mu = (\mu_q)_{q \in Q}$ of weight output distributions. Let $C(q)$ denote the number of hit counts in state $q \in Q$. Then,

$$C(q) = |\{M \in \mathcal{PS}(\pi) : M \blacktriangleleft q\}|. \tag{4}$$

In the case of non-overlapping hits, $|\{M \in \mathcal{PS}(\pi) : M \blacktriangleleft q\}| = 1$ for all $q \in Q$, because of the reduced state space. Consequently, we deal with the emission set $E = \{0, 1, \ldots, c_{\max}\}$, $c_{\max} = \max_{q \in Q} C(q)$, and $|E|$-dimensional Dirac measures μ_q assigning probability 1 to $C(q)$.

Conclusion. To sum up, the PAA $(Q, T, q_0 = start, E = \{0, 1, \ldots, c_{\max}\}, \mu = (\mu_q)_{q \in Q})$ with Q and T given in (2), (3), and $\mu_q = \delta_{C(q)}$ with $C(q)$ given by (4), generates a random sequence alignment according to the specified homology model in order to count the number of hits of a given seed π or Π, respectively. The inherent *Markov Additive Chain* $(V_l)_{l \in \mathbb{N}_0}$ with

$$V_l = V_{l-1} + C(Y_l), \quad V_0 \equiv 0,$$

yields the number V_t of accumulated hits in a random alignment of length t.

Remark 1. We use an $\mathcal{O}(|\Sigma|\,|Q|\log|Q|)$ algorithm by Hopcroft [26] to minimize the size of the state space. Here, the initial partition is induced by grouping states with the same emitted value $C(q)$, the same ingoing and the same outgoing transition probability.

3.2 Hit Distribution and Seed Sensitivity

In order to compute the sensitivity of a given seed π or Π (Problem 1) and its entire hit distribution (Problem 2) for a target alignment length t, we seek the distribution $\mathcal{L}(V_t)$ of accumulated seed hits. This is obtained by marginalization from the joint state-value distribution $\mathcal{L}(Y_t, V_t)$ via $\mathbb{P}(V_t = k) = \sum_{q \in Q} \mathbb{P}(Y_t = q, V_t = k)$. For the sake of readability, we define $h_q^l(k) := \mathbb{P}(Y_l = q, V_l = k)$ and thus, $\mathbb{P}(V_l = k) =: h^l(k) = \sum_{q \in Q} h_q^l(k)$. Now, we can reformulate the mentioned problems. Seed sensitivity is commonly defined as

$$S(\pi, t) = \mathbb{P}(V_t \geq 1) = 1 - \mathbb{P}(V_t = 0) = 1 - h^t(0)\,. \tag{5}$$

It is related to the hit distribution

$$\mathbb{P}\big(\{\mathcal{A}\,|\,|\mathcal{A}| = t, V_t = k\}\big) = \mathbb{P}(V_t = k) = h^t(k) \quad \text{for } k \geq 0\,. \tag{6}$$

The PAA approach proposed in 3.1 yields the following system of recurrence relations:

$$h_q^l(k) = \begin{cases} \sum_{q' \in Q} h_{q'}^{l-1}(k)T_{q'q} & \text{if } q \notin \mathcal{F} \\ \sum_{q' \in Q} h_{q'}^{l-1}(k - C(q))T_{q'q} & \text{if } q \in \mathcal{F} \end{cases} \quad \text{for } q \in Q, l \geq 1, \tag{7}$$

with initial condition $h_{start}^0(0) = 1$.

In order to efficiently implement the computation of seed sensitivity, we make use of the vector $H^l(0) = \big(h_q^l(0)\big)_{q \in Q}$ and the update formula $H^l(0) = H^{l-1}(0)T'$. By T', we denote the transition matrix projected to the columns representing q with $C(q) = 0$. This means, all entries within a column corresponding to a state hit by π are set to 0 in order to ensure $h_q^l(0) = 0$ for q with $C(q) > 0$. The computation of $S(\pi, t)$ requires $\mathcal{O}(|Q|^2)$ space and $\mathcal{O}(t\,|Q|^2)$ time.

Parameter Free Calculation. Another advantage of our approach is the possibility of a parameter-free calculation as presented in Mak and Benson [22]. Without setting the homology parameters in advance, a polynomial (in these parameters) can be computed from (7) by means of a computer algebra system like Mathematica. Hence, one can quickly assess the sensitivity of a seed under different parameter values.

4 Results

We have computed the sensitivity of a given seed or set of seeds by means of the PAA approach. The analysis includes different seed and homology models.

4.1 Spaced Seeds

We considered amongst others the PH seed class $(18, 11)$ and the class $(11, 7)$ with a target alignment length of $t = 64$ as has been done in Choi and Zhang [14]. The computed sensitivities (data not shown) agree with those determined in the original article, except for class $(11, 7)$ for low similarity levels. Instead of sensitivity 0.05398 for $p = 0.1$ we compute 5.39821×10^{-6} and instead of sensitivity 0.1140 for $p = 0.3$ our result is 0.01140. However, our results are consistent with the polynomial obtained from the recurrences. They further agree with the following plausibility argument: Let h_i be the probability of the event that the seed from $(11, 7)$ hits the random alignment at position i. If we assume independence of positions, $h_i = p^7(1-p)^4 + 4p^8(1-p)^3 + \binom{4}{2}p^9(1-p)^2 + \binom{4}{3}p^{10}(1-p) + p^{11}$ for all $11 \leq i \leq 64$. For $p = 0.1$ this rough calculation yields a sensitivity of 5.40021×10^{-6}.

4.2 Indel Seeds

To show that our approach also works for indel seeds, we compare our results to sensitivities provided by Mak et al. [13]. Therefore, we investigate different seeds under a Markov homology model with $\Sigma = \{0, 1, 2, 3\}$ and transitions according to 1. Note that in the original article the authors work with *normalized positions*. This means, only positions in the representative string that refer to a character in the query are counted. Thus, any 2 in $\mathcal{A}$ does not contribute to the target alignment length t. We use the expected number $\bar{t}$ of characters to read up to the t'th normalized position and compute $S(\pi, \bar{t})$ as well as $S(\pi, t)$ accordant to (5). The results in Table 1 show that sensitivity is overestimated when using normalized positions, since some alignments are ignored, although sensitivity is defined as fraction of sequence alignments that are matched by a seed. Compare spaced and indel seeds of equivalent random hit rates, Mak et al. [13] observe that with increasing indel to mismatch ratio, indel seeds outperform spaced seeds. Our method yields the same results (compare Table 1), even if the "winning seed" changes with target length.

Table 1. Sensitivity of pairs of spaced and indel seeds with equivalent random hit rates for different homology parameters (t, p_1, p_0, p_g). Winning seeds are shown in bold.

seed	homology parameters	sensitivity by [13]	$S(\pi, \bar{t})$	$S(\pi, t)$
1111 ∗ 111111	$(64, 0.7, 0.2, 0.05)$	0.488697	0.494082	0.470369
1111 ∗ 111?1111	$(64, 0.7, 0.2, 0.05)$	0.487703	0.492503	0.468351
1111 ∗ 111111	$(100, 0.7, 0.2, 0.05)$	0.669123	0.668821	0.649304
1111 ∗ 111?1111	$(100, 0.7, 0.2, 0.05)$	0.670198	0.669884	0.650131
111 ∗ 11 ∗ 1111	$(64, 0.8, 0.15, 0.025)$	0.943899	0.943609	0.937750
111 ∗ 11?11 ∗ 111	$(64, 0.8, 0.15, 0.025)$	0.943214	0.942985	0.936985
111 ∗ 11 ∗ 1111	$(100, 0.8, 0.15, 0.025)$	0.991157	0.990945	0.989497
111 ∗ 11?11 ∗ 111	$(100, 0.8, 0.15, 0.025)$	0.991239	0.991044	0.989594

4.3   Multiple Seeds

It is already NP-hard [16] to find a single optimal seed by testing all seeds of a given class and selecting the best. Additionally, for a multiple seed there are exponentially many sets of seeds of a given weight. Hence, there is no exact algorithm known that computes optimal multiple seeds. However, "good" multiple seeds are computed by heuristic algorithms presented in [20] and [21]. These approaches use different quality measures correlated with sensitivity, while we compute exact sensitivity values even for non-i.i.d. homology models. Our results agree with the approximations.

The PAA framework can also be applied to design efficient sets of seeds. Similar to [19] we can successively find seeds that locally maximize the conditional probability to hit a random alignment, given that the seeds in the set do not match. This can be achieved by introducing absorbing states.

4.4   Alternative Criteria

In addition to recent work, our approach yields the entire distribution of seed hits. Therefore, we can rank seeds e.g. according to their probability to produce few overlapping hits or subject to the probability of at least two non-overlapping hits. The latter corresponds to the FASTA approach. There, candidates are required to contain at least two non-overlapping seed hits, which afterwards can probably be combined to a similarity.

We have calculated the distributions (6) of overlapping and non-overlapping hits for seed classes $(11, 7)$ and $(18, 11)$ (data not shown). Figure 2 shows the parameter ranges for best performing seeds from the PH seed class according to i) sensitivity and ii) the probability of at least two non-overlapping hits at target alignment length 64.

For some parameters, the optimal seed has only a slightly higher sensitivity than its competitors. In such cases, another criterion might be of interest. For instance, for $p = 0.4$, where the four top-ranking seeds have almost the same

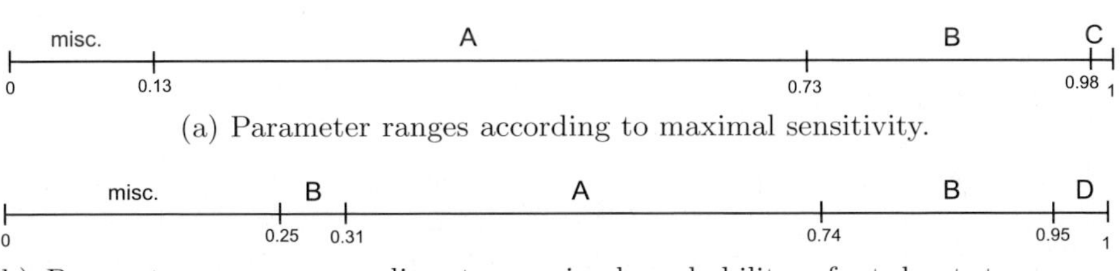

(a) Parameter ranges according to maximal sensitivity.

(b) Parameter ranges according to maximal probability of at least two non-overlapping hits.

Fig. 2. Parameter ranges for best performing seeds from PH seed class $(18, 11)$ at target length 64. 2(a) shows seeds with highest sensitivity. 2(b) displays winning seeds regarding the probability of at least two non-overlapping hits. A (PH):$111*1**1*1*$ $*11*111$, B:$111**1*11**1*1*111$, C:$11**111*1**1*111*1$, D:$111*1**11*1*1**111$.

sensitivity, we observe the PH seed A to maximize the non-overlapping hits criterion clearly.

5   Conclusions

We have presented a Probabilistic Arithmetic Automaton to compute seed sensitivity. It also provides the entire distribution of the number of overlapping or non-overlapping seed hits. The flexibility of the PAA framework allows various homology and seed models. In especially, we can handle sequence alignments with or without gaps and build the automaton for an appropriate seed. For that reason we have presented general definitions of the pattern set and the hit position. Further, symbolic representation of (7) yields a polynomial in the homology parameters, which gives the solution to Problem 1 or Problem 2 for any set of parameters. This enables us to differentiate between seeds that have very similar sensitivities. It remains to prove to what extent alternative criteria are reasonable. One idea is to investigate the hit distribution under two background models: one alignment model representing unrelated sequences, one model describing homologous sequences. Then, we would call a seed optimal that has a given high sensitivity (e.g. 95%) and maximizes selectivity.

Further, we show that one and the same approach is suitable for single and multiple seeds. In particular, we can compute the sensitivity of a multiple seed under non-i.i.d. homology models (in contrast to the DP algorithm by [16]). Clearly, despite minimization, the size of the automaton is exponential in the number of wildcards. Thus, the method should rather serve as verification for heuristic algorithms.

Acknowledgments. We thank T. Wittkop and E. Fritzilas for helpful comments on the manuscript. I. Herms is supported by the NRW Graduate School in Bioinformatics and Genome Research.

References

[1] Pearson, W., Lipman, D.: Improved tools for biological sequence comparison. Proc. Natl. Acad. Sci. USA 85, 2444–2448 (1988)

[2] Altschul, S.F., Gish, W., Miller, W., Myers, E., Lipman, D.: Basic local alignment search tool. J. Mol. Biol 215, 403–410 (1990)

[3] Altschul, S.F., Madden, T.L., Schäffer, A.A., Zhang, J., Zhang, Z., Miller, W., Lipman, D.J.: Gapped BLAST and PSI-BLAST: a new generation of protein database search programs. Nucleic Acids Res. 25(17), 3389–3402 (1997)

[4] Kent, W.J.: BLAT–the blast-like alignment tool. Genome Res. 12(4), 656–664 (2002)

[5] Gelfand, Y., Rodriguez, A., Benson, G.: TRDB–the tandem repeats database. Nucleic Acids Res. 35 (2007)

[6] Smith, T.F., Waterman, M.S.: Identification of common molecular subsequences. J. Mol. Biol. 147(1), 195–197 (1981)

[7] Ma, B., Tromp, J., Li, M.: Patternhunter - faster and more sensitive homology search. Bioinformatics 18, 440–445 (2002)

[8] Buhler, J., Keich, U., Sun, Y.: Designing seeds for similarity search in genomic DNA. In: Proceedings of the 7th annual international conference on Research in computational molecular biology, pp. 67–75 (2003)

[9] Brejová, B., Brown, D.G., Vinar, T.: Optimal spaced seeds for homologous coding regions. J. Bioinform. Comput. Biol. 1(4), 595–610 (2004)

[10] Choi, K.P., Zeng, F., Zhang, L.: Good spaced seeds for homology search. Bioinformatics 20(7), 1053–1059 (2004)

[11] Kucherov, G., Noé, L., Roytberg, M.: A unifying framework for seed sensitivity and its application to subset seeds. J. Bioinform. Comput. Biol. 4(2), 553–569 (2006)

[12] Brejová, B., Brown, D.G., Vinar, T.: Vector seeds: an extension to spaced seeds. J. Computer System Sci. 70(3), 364–380 (2005)

[13] Mak, D., Gelfand, Y., Benson, G.: Indel seeds for homology search. Bioinformatics 22(14), e341–e349 (2006)

[14] Choi, K.P., Zhang, L.: Sensitivity analysis and efficient method for identifying optimal spaced seeds. J. Computer System Sci. 68, 22–40 (2004)

[15] Li, M., Ma, B., Zhang, L.: Superiority and complexity of the spaced seeds. In: Proceedings of SODA 2006, pp. 444–453. SIAM, Philadelphia (2006)

[16] Li, M., Ma, B., Kisman, D., Tromp, J.: Patternhunter II: Highly sensitive and fast homology search. J. Bioinform. Comput. Biol. 2(3), 417–439 (2004)

[17] Brown, D.G.: Optimizing multiple seeds for protein homology search. IEEE/ACM Trans. Comput. Biol. Bioinform. 2(1), 29–38 (2005)

[18] Kucherov, G., Noé, L., Roytberg, M.: Multiseed lossless filtration. IEEE/ACM Trans. Comput. Biol. Bioinform. 2(1), 51–61 (2005)

[19] Sun, Y., Buhler, J.: Designing multiple simultaneous seeds for DNA similarity search. J. Comput. Biol. 12(6), 847–861 (2005)

[20] Kong, Y.: Generalized correlation functions and their applications in selection of optimal multiple spaced seeds for homology search. J. Comput. Biol. 14(2), 238–254 (2007)

[21] Ilie, L., Ilie, S.: Multiple spaced seeds for homology search. Bioinformatics 23(22), 2969–2977 (2007)

[22] Mak, D.Y.F., Benson, G.: All hits all the time: Parameter free calculation of seed sensitivity. In: APBC. Advances in Bioinformatics and Computational Biology, vol. 5, pp. 327–340. Imperial College Press (2007)

[23] Noé, L., Kucherov, G.: Improved hit criteria for DNA local alignment. BMC Bioinformatics 5, 149 (2004)

[24] Pevzner, P.A., Waterman, M.S.: Multiple filtration and approximate pattern matching. Algorithmica 13(1/2), 135–154 (1995)

[25] Marschall, T., Rahmann, S.: Probabilistic arithmetic automata and their application to pattern matching statistics. In: 19th Annual Symposium on Combinatorial Pattern Matching (accepted for publication, 2008)

[26] Hopcroft, J.E.: An n log n algorithm for minimizing states in a finite automaton. Technical report, Stanford, CA, USA (1971)

The Relation between Indel Length and Functional Divergence: A Formal Study

Raheleh Salari[1,3], Alexander Schönhuth[1,3], Fereydoun Hormozdiari[1],
Artem Cherkasov[2], and S. Cenk Sahinalp[1]

[1] School of Computing Science, Simon Fraser University, 8888 University Drive
Burnaby, BC, V5A 1S6, Canada
[2] Division of Infectious Diseases, Faculty of Medicine, University of British Columbia
2733 Heather Street, Vancouver, BC, V5Z 3J5, Canada
[3] The authors contributed equally

Abstract. Although insertions and deletions (*indels*) are a common type of evolutionary sequence variation, their origins and their functional consequences have not been comprehensively understood. There is evidence that, on one hand, classical alignment procedures only roughly reflect the evolutionary processes and, on the other hand, that they cause structural changes in the proteins' surfaces.

We first demonstrate how to identify alignment gaps that have been introduced by evolution to a statistical significant degree, by means of a novel, sound statistical framework, based on pair hidden Markov models (HMMs). Second, we examine paralogous protein pairs in E. coli, obtained by computation of classical global alignments. Distinguishing between indel and non-indel pairs, according to our novel statistics, revealed that, despite having the same sequence identity, indel pairs are significantly less functionally similar than non-indel pairs, as measured by recently suggested GO based functional distances. This suggests that indels cause more severe functional changes than other types of sequence variation and that indel statistics should be taken into additional account to assess functional similarity between paralogous protein pairs.

Keywords: Alignment statistics, Deletions, Insertions, GO, Pair Hidden Markov Models.

1   Introduction

Assessment of functional similarity of two proteins based on their amino acid composition is routinely done by alignment procedures. The reasoning behind this is that homology can be reliably inferred from alignment scores, if significantly high. Homology, in turn, is correlated to functional similarity [12]. However, it is well known that in the *twilight zone*, i.e. protein alignments between roughly 20% to 40% sequence identity, structural hence functional similarity cannot be reliably inferred [30]. Further criteria are needed to distinguish functionally similar proteins from functionally different proteins.

A couple of findings point out that indels, not substitutions, are the predominant evolutionary factor when it comes to functional changes. For example, indels happen to occur predominantly in loop regions [11] which strongly indicates that their occurrence is

K.A. Crandall and J. Lagergren (Eds.): WABI 2008, LNBI 5251, pp. 330–341, 2008.
© Springer-Verlag Berlin Heidelberg 2008

related to structural changes on the proteins' surfaces, hence functional changes. More-over, recent studies show that indels can be of particular interest with regard to disease. Indels are substantially involved in disease-causing mutational hot spots in the human genome [17]. Moreover, promising approaches to the design of novel antibacterial drugs based on the identification of indels have been recently described [6,5,21].

A more formal investigation of the relationship between indels and the functional changes they imply may employ the following general approach.

1. Compute all paralogous protein pairs in an organism.
2. Collect "Indel (I)" and "Non-Indel (NI)" pairs.
3. Compare functional similarity of "Indel" and "Non-Indel".

Obviously, there are several issues that need to be resolved in the above approach.

(i) First of all one needs to use a good definition of paralogous proteins (paralogs); this is a relatively easy task as most studies define two proteins as paralogs if their sequence identity and similarity, as given by a classical alignment, are above a threshold value and are statistically significant.

(ii) Then the organism on which the study will be based must be chosen. In this paper we have opted to examine paralogous protein pairs in E. coli K12. This choice of organism is motivated by our particular interest in studying bacterial organisms. [1]

(iii) It is then of key importance to provide a formal definition of an indel. Note that, while amino acid substitution is an evolutionary process that is reliably covered by the classical dynamic programming alignment procedures, realistic computational models for insertions and deletions (indels) have remained an insufficiently solved problem. There are sound indel studies both for DNA, [34,14,20] and proteins [2,28,4,23]. How-ever, models have usually been inferred for highly specific datasets and slightly contra-dict each other such that they are not generally applicable. Given the unclear actual situation, we have employed the classical Needleman-Wunsch alignment procedure with affine gap penalties to compute paralogous protein pairs, in order to account both for computational efficiency and soundness.

The main problem we deal with in this context was to reliably distinguish between alignment gaps that have been introduced by point mutations and those that have not. This is analogous to that of identifying alignment scores that indicate alignments that re-flect truly evolutionary relationships. The corresponding statistical frameworks [16,8] allow for reliable statistics even if empirical statistics, due to insufficient amounts of data, are not applicable. In analogy to this, we have developed a statistical framework, based on pair HMMs, with which to identify alignment gaps that are statistically sig-nificant, hence more likely refer to insertions and deletions, introduced by evolution. In order to measure the functional impact of indels compared to that of substitutions, "Indel" (I) and "Non-Indel" (NI) pairs that are compared with each other will refer to

----

[1] Note that, in prokaryotes, in addition to the classical transfer of genetic material from par-ent to offspring, other phenomena can be responsible for mutational changes. For example, horizontal gene transfer, that is, transfer of genetic material between arbitrary prokaryotic cells, sometimes of different species, has been made responsible for the rapid development of drug resistance [18,15]. Therefore, to study the paralogous proteins of prokaryotes is also of biomedical interest.

the same levels of sequence identity. Hence, the same amount of non-identical positions either contain significant indels (I) or a significant amounts of substitutions (NI).

(iv) The final challenge is the definition of functional similarity between protein pairs. One possible approach is to measure functional similarity by the use of GO based distances, as recently suggested [19,31] (see section 2.2 for further details).

According to the suggested strategy from above, we found that "Indel" protein pairs are **significantly** less functionally similar than "Non-Indel" pairs, although being of similar alignment scores. Perhaps this finding is quite intuitive. However, the main goal of this paper is a sound, formal framework for quantifying what an indel is with respect to sequence similarity and length and what particular indel lengths imply changes in functional similarity measured in terms of GO based distances. Our results do not only provide a rigorous framework for understanding the relationship between indels and functional similarity but also aim to lay some of the computational foundations of a novel large-scale approach to drug target search by extending the previous works of [6,5,21]. In sum, our major goals in this study are:

(1): To provide, based on pair HMMs, reliable formulas to assess the resulting from alignment methods with affine gap penalties.

(2):To carry out a large-scale study on the correlation of indel size and functional similarity for paralogous proteins (made in the context of E. coli). Note that, thanks to our arrangements, differences in functional similarity cannot be due to differences in alignment scores. Therefore, this study strongly suggests that indels, as evolutionary sequence variation, contribute significantly more to functional changes in the affected proteins than substitutions only.

To the best of our knowledge, these points have not been addressed in a formal sense before. In general, this study provides novel ideas on how to use classical alignment procedures more reliably in order to account for functional similarity, by incorporating indel statistics.

2 Methods

2.1 Indel Length Statistics

Problem Description. The driving problem is described by the following scenario. Let $T = \{(x^t, y^t) \mid t = 1, ..., |T|\}$ be a set of pairs of sequences over a common alphabet Σ. The alphabet Σ will later be identified with the set of amino acids and sets T will contain pairs of proteins of interest. Let $\mathcal{A}$ be an alignment procedure and

$$L_{\mathcal{A}}(x, y) \quad \text{resp.} \quad I_{\mathcal{A}}(x, y)$$

be the length of the alignment of $x = x_1...x_m, y = y_1...y_n$ resp. the length of the largest indel that can be found in the alignment, as computed by $\mathcal{A}$. If k is an integer, we are interested in the probabilities

$$P_{n,T}(I_{\mathcal{A}(x,y)\geq k}) := P(I_{\mathcal{A}}(x, y) \geq k \mid L_{\mathcal{A}}(x, y) = n, (x, y) \in T) \qquad (1)$$

that the largest indel in the alignment of x and y is greater than k or, equivalently, that the alignment contains an indel of length at least k, given that x and y have been

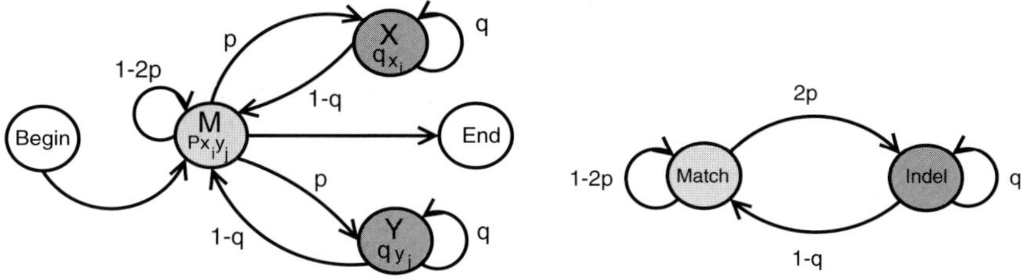

Fig. 1. Pair HMM (left) and the Markov chain resulting from it (right)

drawn from T such that the alignment of x and y is of length n. Clearly, k depends on n and also on T if, for example, T contains all protein pairs that yield a similar alignment score. An analytical treatment of the problem is highly desirable, as there usually are insufficient amounts of data for statistically reliable empirical distributions. Note that, in our case, we have only on the order of tens of proteins, sampled from the pools of interest, which is by far too little to infer reliable statistical estimates. This usually is also the justification of an analytical treatment of score statistics. Note that the score statistics problem refers to replacing $I_{\mathcal{A}}(x, y)$ by $S_{\mathcal{A}}(x, y)$, the alignment score attributed to an alignment of x and y computed by $\mathcal{A}$. In the case of database searches, pools T then may refer to all alignments where $x = x_0$ in all pairs (x, y) , for a single protein x_o. We will refer to the problem of computing probabilities of the type (1 as the indel length probability (ILP) problem.

Pair HMMs. We consider the ILP problem for $\mathcal{A}$ being the Needleman-Wunsch procedure for global alignments with affine gap penalties [22,13], as we are interested in statistics for paralogs which have been inferred by globally aligning all proteins in E. coli. Pair HMMs provide an approach to computing global alignments with affine gap penalties which is equivalent to the dynamic programming approach of Needleman-Wunsch. As pair HMMs are standard, we only give a brief description here and refer the reader to [10] for details.

Consider the pair HMM in Fig. 1. It generates alignments by traversing the hidden states according to state transition probabilities (parameterized by p and q) and emitting pairs of symbols according to emission probability distributions attached to the "hidden" states, i. e. matches/mismatches from the 'M' state by producing a pair of symbols x_i, y_j with probability $p_{x_i y_j}$ and symbols paired with gaps from states 'X' and 'Y'. Upon termination of the run through the hidden states, one has obtained an alignment of two sequences $x = x_1...x_m, y = y_1...y_n$, according to the sequence of symbol/gap pairs that had been generated along the run. As is well known [10], a Needleman-Wunsch alignment with affine gap penalties of two sequences $x = x_1...x_m, y = y_1...y_n$ can be obtained by computing the most likely sequence of hidden states (the *Viterbi path*) that yields an alignment of the two sequences by emitting suitable combinations of symbols along the run. Note that we can neglect transition probabilities referring to start and end states as they do not affect our further considerations.

Owing to the formulation of the ILP problem, we are only interested in statistics on sequences of hidden states, namely in probabilities referring to lengths of consecutive runs with either the 'X' or the 'Y' state. Therefore, we can direct our attention to the Markov chain of Fig. 1 which has been obtained from the pair HMM by collapsing the hidden states 'X' and 'Y' of the pair HMM into the 'Indel' state of the Markov chain. The Markov chain generates sequences over the alphabet $\tilde{\S} := \{'M', 'I'\}$ where $'M'$ and $'I'$ are shorthand for $'Match'$ and $'Indel'$.

Consider now $C_{n,k}$, the set of sequences over the alphabet $\tilde{\S}$ of length n that contain a consecutive $'I'$ stretch of length at least k. We suggest the following procedure in order to provide good approximations of $P_{n,T}(I_{\mathcal{A}(x,y)\geq k})$ from (1).

COMPUTATION OF $P_{n,T}(I_{\mathcal{A}}(x,y) \geq k)$:
1: Align all sequence pairs from T.
2: Infer p and q by training the pair HMM with the alignments.
3: $n \leftarrow$ length of the alignment of x and y
4: Compute $P(C_{n,k})$, the probability that the Markov chain generates a sequence from $C_{n,k}$.
5: **Output** $P(C_{n,k})$ as an approximation for $P_{n,T}(I_{\mathcal{A}}(x,y) \geq k)$.

The idea of step 1 and 2 is to design a pair HMM that generates alignments from the pool T. This is straightforwardly achieved by a standard Baum-Welch training with the Needleman-Wunsch alignments of the protein pairs from T. In our setting, this reduces to simply counting 'Match'-to-'Match' and 'Indel'-to-'Indel' transitions in the alignments under consideration to provide maximum likelihood estimates for the derived Markov chain, as the pair HMM is only of virtual interest. Clearly, any statistics on the pair HMM, thanks to the relationship with the Needleman-Wunsch procedure, will yield good approximations to alignment statistics. Note that this basic strategy has recently been successfully employed to model certain alignment features in a comparative genomics study on human-mouse DNA alignments [20]. Therefore, computing $P_{n,T}(I_{\mathcal{A}}(x,y) \geq k)$ has been translated to computing the probability that the Markov chain generates a sequence from $C_{n,k}$. However, surprisingly, efficient computation and/or closed formulas for such probabilities had been an unsolved mathematical problem so far. Related work only refers to respective formulas for i.i.d. processes on a two-letter alphabet [27].

Efficient Computation of $P_T(C_{n,k})$. Imagine that, as usual, we have collected the Markov chain parameters, as given by Fig. 1, into a state transition probability matrix and an initial probability distribution

$$A = (a_{ij})_{i,j=1,2} = \begin{bmatrix} q & 2p \\ 1-q & 1-2p \end{bmatrix} \quad \text{and} \quad \pi = e_2 = (0.0, 1.0)^T.$$

That is, state 1 corresponds to the 'Indel' state and state 2 corresponds to the 'Match' state. The initial distribution reflects that we assume that an alignment always starts from a 'Match' state (by adding a match to the alignment at the artificial position 0), in order to take into account that starting a global alignment with a gap is scored with a gap opening penalty just as if coming from a 'Match' state. For example, according to the laws that govern a Markov chain, the probability of being in state 1 at position t

in a sequence generated by the Markov chain is $P(X_t = 1) = e_1^T A^t \pi = e_1^T A^t e_2 = (1.0, 0.0) A^t (0.0, 1.0)^T$. We first would like to point out that naive approaches to computing $P(C_{n,k})$ fail. For example, let $B_{t,k}$ be the set of sequences that contain a consecutive 'Indel' run of length k, stretching over positions t to $t + k - 1$. Realizing that $C_{n,k} = \cup_{t=1}^{n-k+1} B_{t,k}$ and proceeding by inclusion-exclusion for computing probabilities (i.e. $P(C_{n,k}) = \sum_{m=1}^{n-k+1} (-1)^{m+1} \sum_{1 \le t_1 < ... < t_m \le n-k+1} P(B_{t_1,k} \cap ... \cap B_{t_m,k})$) results in a procedure that is exponential in n, i.e. in the length of the alignment.

Efficient computation of these probabilities is facilitated by the following trick which we have adopted from [27]. Instead of $B_{t,k}$, consider $D_{t,k}$, the set of sequences that have a run of state 1 of length k that not only stretches from positions t to $t + k - 1$ as before, but also ends at position $t + k - 1$, that is, the run is followed by a visit of state 2 at position $t + k$. According to elementary Markov chain theory, one obtains

$$P(D_{t,k}) = \begin{cases} P(B_{t,k}) \cdot a_{21} = (e_1^T A^t \pi) \cdot a_{11}^{k-1} a_{21} & t < n - k + 1 \\ P(B_{n-k+1,k}) = (e_1^T A^{n-k+1} \pi) \cdot a_{11}^{k-1} & t = n - k + 1 \end{cases} \tag{2}$$

where, in the second case, $D_{n-k+1,k} = B_{n-k+1,k}$ is the event that the last k positions in the run correspond to visits of state 1. Clearly, we have $C_{n,k} = \cup_{t=1}^{n-k+1} D_{t,k}$ as well as for the $B_{t,k}$. For technical convenience, we further define $r(s) := P(X_s = 1) = e_1^T A^s e_2$ for $s \ge 1$ and $Q_{l,m} := \sum_{\substack{1 \le s_1, ..., s_m \le l \\ s_1 + ... + s_m = l}} \prod_{i=1}^m r(s_i)$ for $1 \le m \le l \le n$ where the sum reflects summing over partitions of l into m positive, not necessarily different, integers s_i. Note further that the intersection of $D_{t_1,k}$ and $D_{t_2,k}$ is empty if $t_2 - t_1 \le k$ and that, after a run which is terminated by a visit of state 2, the probability distribution coincides with the initial probability distribution $\pi = (0.0, 1.0)^T$ which corresponds to being in a 'Match' state. This implies $P(D_{t_j,k} \cap D_{t_{j+1},k}) = P(D_{t_j,k}) P(D_{t_{j+1}-t_j-k-1,k})$ for $t_{j+1} - t_j > k$ $(*)$. Proceeding by inclusion-exclusion, we obtain $(a^* := (a_{11}^{k-1} a_{21}^m))$

$$P(C_{n,k}) \stackrel{(*)}{=} \sum_{m=1}^{n-k+1} (-1)^{m+1} \sum_{1 \le t_1 < ... < t_m \le n-k+1} P(D_{t_1,k}) \prod_{j=1}^{m-1} P(D_{t_{j+1}-t_j-k-1,k})$$

$$\stackrel{(2),(**)}{=} \sum_{m=1}^{n-k+1} (-1)^{m+1} \sum_{1 \le t_1 < ... < t_m \le n-k+1} r(t_1) \prod_{j=1}^{m-1} r(t_{j+1} - t_j - k - 1) \cdot (a^*)^m$$

$$\stackrel{(**)}{=} \sum_{m=1}^{n-k+1} (-1)^{m+1} \left[(\sum_{l=m}^{n-mk} Q_{l,m})(a^*)^m + Q_{n+1-mk,m}(a^*)^{m-1}(a_{11}^{k-1}) \right].$$

where, in the third line, a_{21} has to be replaced by 1 in a factor with $t_m = n - k + 1$

$(**)$, referring to the the special event D_{n-k+1} (see (2)).

Finally observe that the recursive relationship $Q_{l,m} = \sum_{s=1}^{l-m+1} r(s) Q_{l-s,m-1}$ yields a dynamic programming procedure to efficiently compute all of the needed $Q_{l,m}$, hence all of the needed $P(C_{n,k})$.

Table 1. Markov chain parameters

Similarity (%)	50 - 55	55 - 60	60 - 65	65 - 70	70 - 75	75 - 80	80 - 85	85 - 90	90 - 95	95 - 100
No. Alignments	20317	1793	709	345	186	102	62	49	40	26
$1 - 2p$	0.9349	0.9559	0.9687	0.9754	0.9828	0.9867	0.9893	0.9941	0.9959	0.9997
q	0.6383	0.6585	0.6666	0.6415	0.6188	0.6319	0.6516	0.6135	0.6651	0.8462

2.2 Functional Similarity

Gene Ontology (GO) [33] provides a well defined structured description of functional annotation and it has been reliably used in innumerable studies. Thanks to its organization, GO can be easily described. It consists of three "orthogonal" taxonomies (aspects) that contain terms to describe attributes related to molecular function, biological process and cellular component. The taxonomies are formally designed as Directed Acyclic Graphs (DAGs). A gene product is associated with a term if the term applies to the product. As a consequence, a protein can be identified with a subset of terms, hence a sub-DAG.

A variety of studies have been concerned with quantification of functional similarity based on comparison of GO terms. Among these approaches, Lord *et al.* [19] have proposed semantic similarity measures for GO terms that can be combined in different ways [19,32,31] to measure protein similarity. In parallel, studies on the correlation of semantic similarity and biological measures of similarity have confirmed the validity of these approaches [32,7]. Pesquita *et al.* [25] present a study that shows that a semantic similarity measure described by [29] is, relative to a variety of aspects, the best measure overall. We have opted to define functional similarity between two proteins as an extension of the semantic similarity measure given by [29] to protein similarity, as described by [31].

3   Results

3.1   Data, Alignment Method and Markov Chain Parameters

We downloaded the full set of 4342 proteins of E.coli K12 from the Uniprot database [35]. To calculate pairwise global alignments we used the "GGSEARCH" tool from the FASTA sequence comparison package [24]. As a substitution matrix, BLOSUM50 (default) was used. GGSEARCH implements the classical Needleman-Wunsch alignment algorithm with affine gap penalties.

To ensure high quality of the alignments in a first step, we discarded all alignments below 50% alignment similarity [2] which resulted in about 23000 aligned protein pairs. Note that alignment similarity is obviously correlated to the occurrence of gaps. In order to decorrelate our results from alignment similarity to a maximum degree (otherwise results might be due to differences in alignment similarity, not the occurrence of

---

[2] Alignment similarity, as defined by GGSEARCH, is the number of alignment positions of identical or highly similar amino acids, divided by the length of the alignment. Therefore, low similarity is either due to indels or badly matching amino acids.

indels), we grouped the alignments into ten different pools, according to their similarity scores, and inferred the parameters of the corresponding Markov chains, according to the procedure described in subsection 2.1. See table 1 for Markov chain parameters of the different pools.

After computation of these parameters, we discarded alignments of less than 20% identity and an E-value of greater than 10^{-6} (as computed by GGSEARCH) to ensure homology to a maximum degree. However, we had kept possibly non-homologous pairs for training the Markov chains to take into account the "alignment noise" we later would like to get rid of and to increase the amount of training data which enhances the reliability of the inference process.

3.2   Determining Indel and Non-indel Alignments

We would like to distinguish between two kinds of alignments, indel resp. non-indel alignments. To be more precise, we are interested in alignments of length n and maximal indel length k such that

$$P(C_{n,k}) \leq \theta_I \quad \text{resp.} \quad 1 - P(C_{n,k+1}) \leq \theta_{NI}$$

where θ_I, θ_{NI} are appropriate significance levels. Note that the quantity $1 - P(C_{n,k+1})$ is just the probability that an alignment does not contain an indel of size larger than k. In order to determine the significance levels θ_I, θ_{NI} we separated indel pairs (I) and non-indel pairs (NI) for varying θ_I, θ_{NI} and computed the average functional similarity in the corresponding groups for the three GO categories "Function", "Process" and "Component", according to the procedures described in section 2.2. We found that alignments with $P(C_{n,k}) \leq 10^{-6}$ contain large gaps that, rather than corresponding to insertions and deletions in the evolutionary sense we are interested in, reflect regions that separate domains. In order not to falsify our indel study, we discarded about 40 putative multi-domain alignments. [3]

Paralogous Protein Pairs: In the following, we will refer to the remaining protein pairs (sequence identity $\geq$ 20%, alignment similarity $\geq$ 50%, $P(C_{n,k}) > 10^{-6}$ as paralogous protein pairs. We observed that functional similarity, for all categories, decreases for the indel pairs while lowering θ_I and increases while lowering θ_{NI} (data available upon request). We have opted to choose $\theta_I = 0.045, \theta_{NI} = 0.25$ as it establishes an optimal trade-off between differences in GO similarity and sufficient amounts of alignments for statistical examinations.

3.3   The Twilight Zone

We sorted the set of paralogous protein pairs, resulting from the filtering procedure described above, according to their sequence identity. We computed the average GO similarity for indel pairs (I), non-indel pairs (NI) and overall (O) above a given identity threshold where indel and non-indel pairs are defined, as described in the above

---

[3] Note that this points out that our method can also be potentially used to get aware of multi-domain alignments that are a classical source of disturbances while screening databases for homologous counterparts [26].

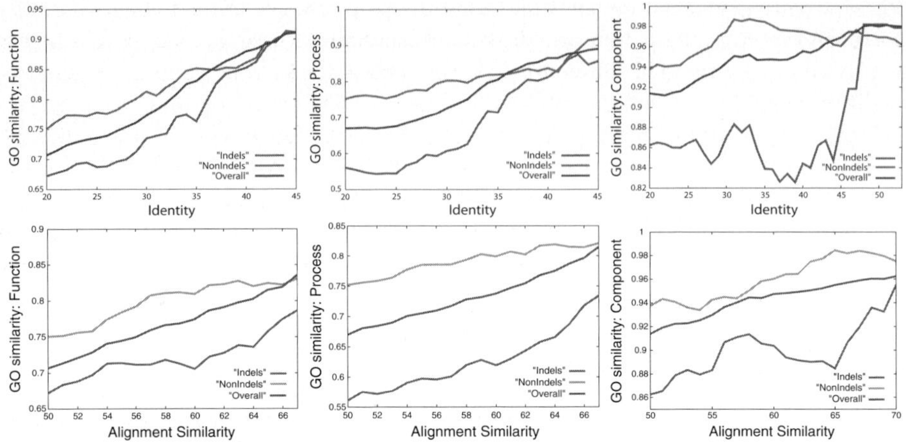

Fig. 2. GO similarity vs. identity (top row) resp. alignment similarity (bottom row) levels, for fixed p-values of 0.045 (indels I) and 0.25 (non-indels NI) for paralogous E.coli protein pairs (definitions see subs. 3.2)

subsection 3.2. Moreover, we carried out t-tests to determine the statistical significance of the differences. Results are displayed in the top row of Fig. 2 and Table 2. Clearly, GO similarity is lower for indel pairs and higher for non-indel pairs in all three GO categories in the region between 20% and 35% identity. T-tests confirmed statistical significance for all of the differences (see Table 2) in that region. As outlined before, given that the amount of non-identical, substitution or indel, positions in the compared alignments are on equal levels, this gives statistical evidence of that insertions and deletions highly likely are the more significant evolutionary cause of functional changes than substitutions.

Above $\approx 37.5\%$ identity for "Function" and "Process" and above $\approx 42.5\%$ identity for "Component" differences become insignificant which confirms that beyond the twilight zone, structural, hence functional, similarity can safely be assumed. In the area of 35% to 40% differences only remain significant for the category "Component", a phenomenon that remains to be explained.

We also display differences between the three groups when ordered according to alignment similarity (Fig. 2, bottom). Here as well, there are obvious differences in functional similarity depending on significant occurrence of either gaps or no gaps in the alignments. In sum, the results strongly suggest that an assessment of functional similarity should not only be based on common alignment quality measures (identity, similarity, score significance), but also on the significance of the occurrence of insertions and deletions.

4   Discussion and Outlook

We have conducted a large-scale study on the correlation of the occurrence of indels in the alignments of paralogous proteins in E. coli and their functional similarity. We have demonstrated, based on a sound statistical framework, that the occurrence of indels is

Table 2. Twilight Zone Statistics for paralogous E.coli protein pairs (definition see subs. 3.2)

GO category	Identity	No. Alignments			GO Similarity			T-test		
		I	NI	O	I	NI	O	I vs. O	NI vs. O	I vs. NI
Function	$\geq$ 20.0	841	647	4413	0.6724	0.7507	0.7069	3.6844	3.9313	4.9547
	$\geq$ 25.0	468	314	2219	0.6871	0.7783	0.7356	4.1953	2.9257	4.4796
	$\geq$ 30.0	183	159	1030	0.7347	0.8132	0.7736	2.0633	2.0135	2.6321
	$\geq$ 35.0	75	98	546	0.7629	0.8515	0.8304	2.2001	0.9201	2.1015
	$\geq$ 40.0	42	63	328	0.8522	0.8610	0.8796	0.7731	-0.7300	0.1808
Process	$\geq$ 20.0	813	670	4500	0.5609	0.7521	0.6699	8.4619	6.2435	9.5150
	$\geq$ 25.0	446	326	2235	0.5443	0.7602	0.6756	7.5987	4.4952	7.6959
	$\geq$ 30.0	177	160	1030	0.6069	0.8030	0.7261	4.3747	3.1777	4.9147
	$\geq$ 35.0	76	97	545	0.7140	0.8193	0.8053	2.3570	0.4757	1.9635
	$\geq$ 40.0	45	62	333	0.8122	0.8372	0.8653	1.3118	-0.8248	0.4209
Component	$\geq$ 20.0	542	381	2986	0.8625	0.9376	0.9139	4.7200	2.2012	4.4590
	$\geq$ 25.0	281	182	1479	0.8644	0.9483	0.9208	3.9673	2.0667	3.9138
	$\geq$ 30.0	123	106	750	0.8698	0.9746	0.9438	3.2578	2.6746	3.6899
	$\geq$ 35.0	60	63	411	0.8390	0.9843	0.9472	2.9245	3.2622	3.3570
	$\geq$ 40.0	36	39	257	0.8443	0.9746	0.9554	2.2478	1.2842	2.2405
	$\geq$ 45.0	19	25	175	0.8832	0.9604	0.9689	1.3316	-0.4529	1.0406

correlated to lower functional similarity of the aligned protein pairs, although the compared protein pairs had similar alignment scores. Functional similarity was measured by recently suggested GO based functional distance measures.

For an analytical treatment of indel significance, we have developed a statistically sound and computationally efficient strategy, based on pair HMMs. With it, we can separate true indel alignments from "indel noise" introduced by the classical dynamic programming procedures. This problem is analogous to that of computation of significance levels for alignment scores which can be obtained according to the well-known Dembo-Altschul-Karlin statistics [16,8]. However, to the best of our knowledge, the problem of indel significance had not been tackled before.

Future work will be to finetune our statistical model, as was done for alignment scores [1]. Models for indels in local alignments will also be developed. Interesting truly biomedical issues will also be addressed. For example, indel studies can analogously be conducted for orthologous protein pairs. A large-scale comparison of orthologous pairs of proteins from human and pathogens potentially will reveal new drug targets, as has already been demonstrated [6,5,21].

Acknowledgements

We would like to greatly thank Francisco Couto and Daniel Faria for computing the GO similarity values and Michael Coons for helpful discussions and proofreading the mathematical parts. AC, AS and SC are members of the Bioinformatics for Combating Infectious Diseases (BCID) project, funded by the SFU Community Trust Endowment Fund. AS is also funded by a postdoctoral stipend from the Pacific Institute for the Mathematical Sciences.

References

1. Altschul, S.F., Gish, W.: Local alignment statistics. Methods in Enzymology 266, 460–480 (1996)
2. Benner, S.A., Cohen, M.A., Gonnet, G.H.: Empirical and structural models for insertions and deletions in the divergent evolution of proteins. Journal of Molecular Biology 229, 1065–1082 (1993)
3. Chan, S.K., Hsing, M., Hormozdiari, F., Cherkasov, A.: Relationship between insertion/deletion (indel) frequency of proteins and essentiality. BMC Bioinformatics 8, 227 (2007)
4. Chang, M.S.S., Benner, S.A.: Empirical analysis of protein insertions and deletions determining parameters for the correct placement of gaps in protein sequence alignments. Journal of Molecular Biology 341, 617–631 (2004)
5. Cherkasov, A., Lee, S.J., Nandan, D., Reiner, N.E.: Large-scale survey for potentially targetable indels in bacterial and protozoan proteins. Proteins 62, 371–380 (2005)
6. Cherkasov, A., Nandan, D., Reiner, N.E.: Selective targetting of indel-inferred differences in 3D structures of highly homologous proteins. Proteins: Structure, Function and Bioinformatics 58, 950–954 (2005)
7. Couto, F.M., Silva, M.J., Coutinho, P.M.: Measuring semantic similarity between Gene Ontology terms. Data & Knowledge Engineering 61, 137–152 (2007)
8. Dembo, A., Karlin, S.: Strong limit theorem of empirical functions for large exceedances of partial sums of i.i.d. variables. Annals of Probability 19, 1737–1755 (1991)
9. Denver, D.R., Morris, K., Lynch, M., Thomas, W.K.: High mutation rate and predominance of insertions in the Caenorhabditis elegans nuclear genome. Nature 430, 679–682 (2004)
10. Durbin, R., Eddy, S., Krogh, A., Mitchison, G.: Biological sequence analysis. Cambridge University Press, Cambridge (1998)
11. Fechteler, T., Dengler, U., Schomburg, D.: Prediction of protein three-dimensional structures in insertion and deletion regions: a procedure for searching data bases of representative protein fragments using geometric scoring criteria. Journal of Molecular Biology 253, 114–131 (1995)
12. Gerlt, J.A., Babbitt, P.C.: Can sequence determine function? Genome Biology 1(5), reviews0005.1-0005.10 (2000)
13. Gotoh, O.: An improved algorithm for matching biological sequences. Journal of Molecular Biology 162, 705–708 (1982)
14. Gu, X., Li, W.-H.: The size distribution of insertions and deletions in human and rodent pseudogenes suggests the logarithmic gap penalty for sequence alignment. Journal of Molecular Evolution 40, 464–473 (1995)
15. Hsiao, W.W.L., Ung, K., Aeschliman, D., Bryan, J., Finlay, B.B., Brinkman, F.S.L.: Evidence of a large novel gene pool associated with prokaryotic genomic islands. PLoS Genetics 1, e62 (2005)
16. Karlin, S., Altschul, S.F.: Methods for assessing the statistic significance of molecular sequence features by using general scoring schemes. Proceedings of the National Academy of Sciences of the USA 87, 2264–2268 (1990)
17. Kondrashov, A.S., Rogozin, I.B.: Context of Deletions and Insertions in Human Coding Sequences. Human Mutation 23, 177–185 (2004)
18. Lake, J.A., Riveral, M.C.: Horizontal gene transfer among genomes: The complexity hypothesis. Proceedings of the National Academy of Science 96(7), 3801–3806 (1999)
19. Lord, P.W., Stevens, R.D., Brass, A., Goble, C.A.: Investigating semantic similarity measures across the Gene Ontology: the relationship between sequence and annotation. Bioinformatics 19, 1275–1283 (2003)

20. Lunter, G., Rocco, A., Mimouni, N., Heger, A., Caldeira, A., Hein, J.: Uncertainty in homology inferences: Assessing and improving genomic sequence alignment. Genome Research 18 (2007), doi:10.1101/gr.6725608
21. Nandan, D., Lopez, M., Ban, F., Huang, M., Li, Y., Reiner, N.E., Cherkasov, A.: Indel-based targeting of essential proteins in human pathogens that have close host orthologue(s): Discovery of selective inhibitors for Leishmania donovani elongation factor-$1-\alpha$. Proteins: Structure, Function and Bioinformatics 67, 53–67 (2007)
22. Needleman, S.B., Wunsch, C.D.: A general method applicable to the search for similarities in the amino acid sequence of two proteins. Journal of Molecular Biology 48, 443–453 (1970)
23. Pang, A., Smith, A.D., Nuin, P.A.S., Tillier, E.T.M.: SIMPROT: Using an empirically determined indel distribution in simulations of protein evolution. BMC Bioinformatics 6, 236 (2005)
24. Pearson, W.R., Lipman, D.J.: Improved tools for biological sequence comparison. Proc. Natl. Acad. Sci. USA 85, 2444–2448 (1988)
25. Pesquita, C., Faria, D., Bastos, H., Falco, A.O., Couto, F.M.: Evaluating GO-based semantic similarity measures. In: Proceedings of the 10th Annual Bio-Ontologies Meeting (Bio-Ontologies 2007) (2007)
26. Pipenbacher, P., Schliep, A., Schneckener, S., Schönhuth, A., Schomburg, D., Schrader, R.: ProClust: improved clustering of protein sequences with an extended graph-based approach. Bioinformatics 18(Supp.2), 182–191 (2002)
27. Peköz, E.A., Ross, S.M.: A simple derivation of exact reliability formulas for linear and circular consecutive-k-of-n F systems. Journal of Applied Probability 32, 554–557 (1995)
28. Qian, B., Goldstein, R.A.: Distribution of indel lengths. Proteins: Structure, Function and Bioinformatics 45, 102–104 (2001)
29. Resnik, P.: Semantic similarity in a taxonomy: an information- based measure and its application to problems of ambiguity in natural language. Artificial Intelligence Research 11, 95–130 (1999)
30. Rost, B.: Twilight zone of protein sequence alignments. Protein Engineering 12(2), 85–94 (1999)
31. Schlicker, A., Domingues, F.S., Rahnenführer, J., Lengauer, T.: A new measure for functional similarity of gene products based on gene ontology. BMC Bioinformatics 7, 302 (2006)
32. Sevilla, J.L., Segura, V., Podhorski, A., Guruceaga, E., Mato, J.M., Martnez-Cruz, L.A., Corrales, F.J., Rubio, A.: Correlation between gene expression and GO semantic similarity. IEEE/ACM Transactions on Computational Biology and Bioinformatics 2(4), 330–338 (2005)
33. The Gene Ontology Consortium. Gene Ontology: tool for the unification of biology. Nature Genetics 25, 25–29 (2000)
34. Thorne, J.L., Kishino, H., Felsenstein, J.: Inching toward reality: An improved likelihood model of sequence evolution. Journal of Molecular Evolution 34, 3–16 (1992)
35. The UniProt Consortium. The Universal Protein Resource (UniProt). Nucleic Acids Res. 35, D193-D197 (2007)

Detecting Repeat Families
in Incompletely Sequenced Genomes

José Augusto Amgarten Quitzau[1,2] and Jens Stoye[1]

[1] AG Genominformatik, Technische Fakultät
[2] International NRW Graduate School in Bioinformatics and Genome Research
Bielefeld University, Germany

Abstract. Repeats form a major class of sequence in genomes with implications for functional genomics and practical problems. Their detection and analysis pose a number of challenges in genomic sequence analysis, especially if the genome is not completely sequenced. The most abundant and evolutionary active forms of repeats are found in the form of *families* of long similar sequences. We present a novel method for repeat family detection and characterization in cases where the target genome sequence is not completely known. Therefore we first establish the sequence graph, a compacted version of sparse de Bruijn graphs. Using appropriate analysis of the structure of this graph and its connected components after local modifications, we are able to devise two algorithms for repeat family detection. The applicability of the methods is shown for both simulated and real genomic data sets.

1   Introduction

In the later 1980's, scientists had the first contact with genome sequences of higher-order organisms. At that time, they were amazed by the amount of "junk" in these sequences. Examining this junk in the following decades, they discovered that these portions of the genome were less useless than they first suspected. In fact, there is a myriad of active elements between coding sequences, some of them being able to replicate themselves, acting like virus DNA, termed insertion sequences in bacterial genomes, and mobile elements in eukaryotes. They are in fact believed to be the vestiges of virus infections in ancestral species.

Although repetitive elements may not be active parts of the genome, since they encode only proteins which are related to their own replication, they are able to change the genome in many ways. It is known that pairs of insertion sequences act sometimes together and duplicate not only themselves, but the whole sequence between them [11]. Also when mobile elements work alone, the position where the new copy is inserted may belong to important regions in the genome, like active genes. In fact, insertions of mobile elements are observed in several genetic disorders, like Duchenne muscular dystrophy, type 2 retinitis pigmentosa, β-thalassemia, or chronic granulomatous disease [15].

A maybe less noble, but really important motivation to study repetitive elements is the waste of time and money they cause in genomic research. Finishing

K.A. Crandall and J. Lagergren (Eds.): WABI 2008, LNBI 5251, pp. 342–353, 2008.
© Springer-Verlag Berlin Heidelberg 2008

a whole eukaryotic genome sequencing project is neither cheap nor fast, and the study of specific regions in these huge genomes still depends on specific primer design. But even when using very specific primers, PCR experiments may result in garbage, if the sequence to which the primer was designed appears thousands of times in the whole genome. In many cases, however, there may be enough sequenced information available to give an overview of the repetitive elements in the genome. Finding these elements in an incompletely sequenced, unfinished genome is the aim of this work.

Strategies for *de novo* repeat identification usually assume that two similar sequences in a given collection cannot be different copies of the same locus of the genome. Therefore they may assume that alignments with quality above a certain threshold provide evidence of a repeat family. We overcome this limitation by accepting any kind of sequence sets as input, including sets with several copies of the same locus. For doing this, we do not align the sequences, like the traditional approaches [1], but partially assemble them using a de Bruijn graph.

The first use of de Bruijn graphs in Bioinformatics was probably the Eulerian path approach to sequence assembly proposed by Idury and Waterman [9] and extended by Pevzner, Tang and Waterman [14]. Despite the success achieved by Pevzner and colleagues' Euler assembler in assembling bacterial genomes, the use of de Bruijn graphs for other biological applications does not seem to be further explored. We find many extensions of the de Bruijn graph based assembly approach in the recent literature [2,3,4,5,19], but they usually focus either on improvements in error correction methods or in adapting the original method to new sequencing data. Other works present graphs that slightly remind of de Bruijn graphs, but miss the main feature of them, namely, the unique representation of tuples of a given size [13,17].

The main problem with de Bruijn graphs becomes clear as soon as one starts working with them. As Myers [12] points out, de Bruijn graphs are simply space inefficient. And we believe Myers is right when he says that in the context of sequence assembly the whole process of cutting reads into small pieces to finally build the de Bruijn graph may not be necessary. To circumvent this, we propose an efficient implementation of sparse de Bruijn subgraphs, detailed in [16], and two strategies for using them as repeat family detection tools in incompletely sequenced genomes.

2   Sparse de Bruijn Graphs

A d-dimensional *de Bruijn graph* $G = (V, A)$ on an alphabet Σ is the graph defined as follows:

$$V = \Sigma^d$$
$$A = \{(u, v) \mid u, v \in V \text{ and } u_{i+1} = v_i, \text{ for all } i, 1 \leq i < d\}$$

where u_i denotes the ith character of string u. Strings of length at least d over the same alphabet describe walks on the d-dimensional de Bruijn graph.

Let s be a string over Σ. The d-dimensional *spectrum* of s, spectrum(s, d), is the set of all substrings of s with length d. The spectrum of a set of sequences is the union of the individual spectra. Given a set $\mathcal{S}$ of strings, the associated d-dimensional *de Bruijn subgraph* is the graph $G_{\mathcal{S}} = \{V_{\mathcal{S}}, A_{\mathcal{S}}\}$, where:

$$V_{\mathcal{S}} = \text{spectrum}(\mathcal{S}, d)$$
$$A_{\mathcal{S}} = \{(u, v) \mid u, v \in V_{\mathcal{S}} \text{ and } u_1 \ldots u_d v_d \in \text{spectrum}(\mathcal{S}, d+1)\}.$$

A vertex in an associated de Bruijn subgraph is called a *junction* when it has in-degree greater than 1. A vertex with out-degree greater than 1 is called *bifurcation*.

Sequence associated de Bruijn subgraphs have a very nice asymptotic behavior. Their maximum number of nodes increases linearly with the size of the input, and even decreases with the dimension of the graph. The main problem is that, although these graphs scale very well with the sequence set size, the graphs corresponding to genomes as small as bacterial genomes are already huge.

2.1   Sequence Graph

De Bruijn graphs are by definition sparse [6, Chapter 7]. Even in applications where smaller dimensions are required [8], their number of edges in the DNA world is not greater than four times the number of nodes. Their subgraphs are surely sparser. In a typical sequence analysis application [14,20], the probability of having a node with maximum in- or outdegree is very low. Therefore the graph construction in such applications is usually followed by a step where long branch-free paths are collapsed to single nodes. In [16], we present a way to directly construct the compact representation of a sparse de Bruijn graph, called a *d-dimensional sequence graph*, or simply *sequence graph*.

An example of a sequence graph is shown in Figure 1. Like a d-dimensional de Bruijn subgraph, every d-tuple over the given alphabet may be represented by at most one vertex in the sequence graph. Furthermore, a sequence graph may contain an arc (u, v) only if the suffix of length $d-1$ of u matches perfectly the prefix of v. The main difference between a sequence graph and a de Bruijn graph is that vertices in a sequence graph are not limited to the size d, but may have any size between d and $|\Sigma|^d + d - 1$. This allows the representation of non-branching paths in a single node. The compression, however, depends on the way the structure is built.

There is an index mapping every d-tuple represented by the sequence graph to the node in which it is found. The tuple position in the node is also stored, so that it may be directly accessed, after the index search. We also extend the natural concept of neighborhood from nodes to tuples. In a sequence graph, two d-tuples a and b are called *neighbors* if either there is an arc (u, v) such that the suffix of length d of u is a, and the prefix of length d of v is b; or both a and b are in the same node and the occurrence of a precedes the occurrence of b by one position.

Apart from the inclusion of nodes, there are two operations that can be applied on the set of nodes.

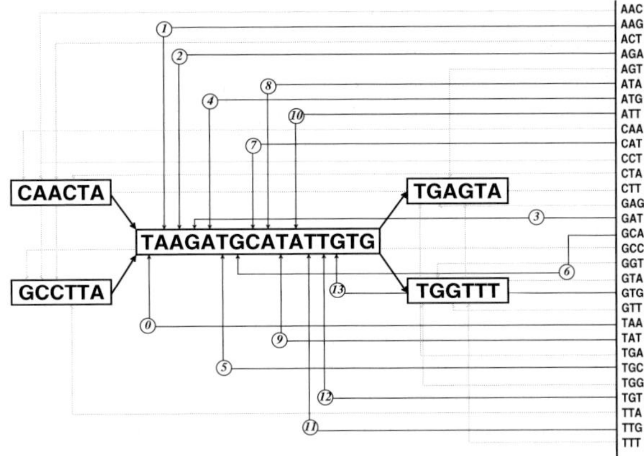

Fig. 1. Sequence graph corresponding to a 3-dimensional de Bruijn subgraph on the alphabet $\Sigma = \{A, C, G, T\}$. Links connecting the index to the node TAAGATGCATATTGTG are shown as black arrows with offsets, all other such connectors are shown in gray.

Cut: Transforms a single node in two neighbor nodes. A cut does not change the set of sequences represented by the graph, since no new tuple of size d or greater is created, and the new edge binds two tuples that were neighbors before.

Merge: Is the inverse operation of cut. Notice that this operation can only be applied on neighbor nodes u and v such that u has outdegree 1 and v has indegree 1. The operation removes the edge (u, v) by merging its nodes into a single node.

2.2   Repeat Families in Sequence Graphs

The length of repetitive elements may vary from the few bases of short tandem repeats to the thousands of bases of long transposons. We know that only exact repeats with length greater than the underlying graph dimension can be identified in such a graph, since they are represented by single nodes both in the sequence graphs, and in the original form of de Bruijn graphs.

Although they may be much smaller than the graph dimension, the exhaustive, uninterrupted succession of almost perfect copies in tandem repeats is able to create tangled patterns in the graph. In these cases, the large number and perfection of copies is responsible for the rising of larger perfect matches.

In the case of interspersed repeats, their replication mechanism allows the appearance of copies which are physically far away from each other in the DNA molecule. On the other hand, the content of a single copy is usually unique. Apart from the usual reverse short repeats in their extremities, the sequence inside mobile elements often lacks exact repeats. Therefore the portion of a sequence graph corresponding to a repeat family is much better organized than the tangled tandem repeats regions. Often the sequence graphs of repeat families are directed acyclic graphs. This can be used as a starting point for repeat identification.

3   Repeat Family Detection

Although the real challenge we want to address with this work is the detection of repeat families in eukaryotic genomes, our first subject of study are the simpler, easier to understand bacterial genomes. A typical bacterial genome is not bigger than six or seven million base pairs, and roughly the same number of l-tuples, when l is relatively small. The number of possible strings of length l for an alphabet of size 4, by contrast, is already huge for very small values of l. For typical sequence analysis applications, like approximate string matching, the value of l is chosen large enough to allow the assumption that very few tuples appear twice in the genome just by chance.

The number of repeat families in a single genome is quite small. The average number of different families in a single genome found in [11] is 2.79. The copy number of a family in a single genome is also not big. Although the number of copies can be as big as 14, like the number of copies belonging to the family IS1 in Mycoplasma, the average number of distinct elements of the same family is 2.27. Therefore, assuming that copies are uniformly spread along the genome sequence, we may expect repeats to be separated by quite long non-repetitive sequences. This may be also true for some eukaryotes, like *Arabidopsis thaliana*, which has 10% of its genome composed by mobile elements [7], while other eukaryotes have a much more complicated genome structure.

3.1   Connected Components

Nodes corresponding to repetitive sequences may be discovered and marked during the sequence graph construction. Nodes corresponding to unique sequences either represent larger sequences from the unique parts of the genome, or are the result of small dissimilarities between elements of the same repeat family. In the second case, unique nodes are not larger than a repeat family element, since entities of the same repeat family are similar enough to share perfect matches. On the other hand, unique sequences between repeat copies can be much longer.

The sequence graph for a genome must therefore be composed of clusters of small repetitive nodes, interconnected by longer single ones. As a result, the deletion of long unique nodes may decompose the graph into a few connected components, containing one or more repeat families. Based on this simple principle, we devised a method for separating repeat families in a genome. The procedure is described in Algorithm 1. The input is a set S of reads of some genome and a length threshold value l. We start building the sequence graph for this set of sequences. Originally, nodes with different sequence sets cannot be merged. As a result, every read end coincides with a node end, which leads in many cases to branch free paths in the sequence graph. Therefore we ignore this restriction and merge nodes in branch free paths, as long as they are either both marked as repeats, or both unmarked, even if their sequence sets are not identical. The resulting graph may contain long single nodes, exceeding the length threshold l. We delete them, and merge the repeated nodes that were not merged only because of the now deleted long nodes. Notice that, at this point, no other node can be merged or deleted.

The resulting graph is already a collection of separated connected components. However, some of them may be the result of unrelated small perfect matches. These repeats created by chance are easy to identify. They are in components with few nodes (not more than 5), with a single, short repeated node in the center. We call these components *small components*. The small components are removed from the graph as well, leaving only components corresponding to larger families.

We implemented this approach in the Java programming language and tested it with artificially created chromosomes, simulating situations from simple bacterial chromosomes to chromosomes which are more than 50% composed of repetitive elements. Details about simulation and results are given in Section 4.1.

Algorithm 1. Connected Components

1: **function** ISOLATECOMPONENTS($\mathcal{S}, l$)
2:     Build the sequence graph for $\mathcal{S}$
3:     Merge all possible pairs of nodes
4:     Remove all single nodes of length $\geq l$
5:     Merge all possible pairs of nodes
6:     Remove all small components
7:     **return** the resulting connected components
8: **end function**

3.2   Combining Nodes

In cases when elements of the same repeat family differ in many close bases, the connected component based detection method may miss some less represented families. This happens because the few sequences do not share any l-tuple in a certain region.

Algorithm 2. Combine

1: **procedure** COMBINE(n_1, n_2, t)
2:     Let n_1 be the longer of the two nodes
3:     Align the sequences of n_1 and n_2, creating a semi-global alignment of length l
4:     **if** the alignment score is smaller than $t \times \left\lceil \frac{l-d+1}{d} \right\rceil$ **then**
5:         Cut the node n_1 at the end of the aligned prefix
6:         Let $\overline{n_1}$ be the left portion of the cut node n_1
7:         Create a new node n with the consensus of $\overline{n_1}$ and n_2
8:         Bind the nodes in the neighborhood of $\overline{n_1}$ and n_2 to n
9:         Remove $\overline{n_1}$ and n_2
10:     **end if**
11: **end procedure**

As a result, although two (or more) long closely related single nodes can be found in the graph, they are discarded as long unique nodes, and the remaining part is either detected as a partial family, or is discarded as a small component. On the other hand, repeats of the same family often share at least one l-tuple somewhere along their sequence. And in the cases where sequences share only a few tuples, the nodes representing the common ones can be extended by combining the closely related unique nodes around them.

By *combining* two nodes we mean replacing both nodes by a single node whose sequence is the consensus between them. The procedure is shown in Algorithm 2 and represented in Figure 2. In the most general case, the two nodes to be combined, n_1 and n_2, are of different length. We assume w.l.o.g. that n_1 is longer than n_2.

In the first step, the node prefixes are aligned. We use a semi-global alignment algorithm [18, Section 3.2.3] for that. The score matrix considers matches between any two possible symbols found in the IUPAC standard code for nucleotides. For symbols that represent a single nucleotide (A, C, G, T), the score is simply 0 for a match and 1 for a mismatch. For matches involving at least one symbol representing a set of nucleotides, like W = {A, T}, the score is 0 if one set contains the other; otherwise it is the minimum number of replacements and deletions needed to transform one set into the other. For instance, the score for aligning W with G is 2, since we need to replace one of the elements of W by G, and delete the remaining one; on the other hand, the score for aligning W with T is 0, since W contains T.

Fig. 2. The *combine* operation. The two shaded nodes on the left are combined, and result in the shaded node on the right. Two nodes are only combined when the edit distance between their prefixes is below a certain threshold. The new node label contains the consensus sequence.

Alignments with a score below a certain threshold allow the combination. The threshold is defined by the minimum number of mismatches needed to separate the nodes, m, rescaled by a user defined scale factor t. The minimum number of mismatches is given by

$$m = \left\lceil \frac{l - d + 1}{d} \right\rceil,$$

where l is the alignment length and d is the underlying de Bruijn graph dimension.

When the alignment score allows a combination, the longer of the two nodes, n_1, is cut at the point where the aligned prefix ends. The prefix and the second node n_2 are then replaced by a node n representing the alignment consensus. This new node is finally connected to the neighborhood of the replaced nodes.

In the general case, combining two nodes does not reduce the graph size, but its complexity. In practice, a series of combinations may reduce tangled subgraphs to simple paths, which may be finally merged into a single longer node. Especially for more complex datasets sequenced at low genome coverage, this procedure gives a considerable advantage over the simple connected component approach, as shown by the results in Section 4.2.

4 Results

We applied our methods to both real and artificially created data. Comparing our methods to already published *de novo* repeat identification methods is not possible because they differ both in the input and in the output. We have as input not continous portions of an incompletely sequenced genome, but a low coverage set of reads. And we output connected components, which may be interpreted as collections of reads which belong to the same repeat family. Therefore, when measuring the success of our method in separating repeat families in a genome, the most successful scenario is clearly the situation where, we get each family in a different connected component, and components containing only members of a single family.

4.1 Connected Components

For a proof of concept, we applied the connected component strategy to artificially created chromosomes with different numbers of repeat families. Each simulated chromosome has a total length of 1 million base pairs and is composed by two kinds of sequences:

Background Sequence: The background sequence corresponds to the non-repetitive genome sequence. In our tests we used 19-dimensional de Bruijn subsequences as background, which means that the background sequences do not contain any duplicated substring of length 19.

Repeat Families: The repeat families are collections of similar sequences, called the *family members*. They originate from a 19-dimensional de Bruijn subsequence, called the family's *base sequence*, which is then used to create the other family members. Families are created in an incremental tree-like fashion: for creating a new member, we randomly take a pre-existing one and imperfectly duplicate it by simulating insertions, deletions and replacements. Each newly created sequence differs from its original in 6% of the nucleotides on average. This agrees with real cases, like the Alu family in the human genome, where the sequences diverge by up to 12% from other elements in the family [10]. The number of members in a family is called the *family size*.

The inserted repeat families were of size 2, 4, 16, and 256. In our tests, an artificial chromosome can have either 0 or 2 families of each size. All possible combinations were used, giving a total of 15 chromosome configurations. For each configuration we created 15 different chromosomes and read sets with 0.25, 0.5, 0.75, and 1 time coverage, simulating partially finished sequencing projects. The artificially created reads have average length of 250 base pairs.

Table 1. Summary of the experiments with artificial data described in Section 4.1. Each row corresponds to one of our 15 data sets. On the left we see the average number of different components containing sequences of the same family. In the middle are the average numbers of different families found in a single component. On the right we see the percentage of inserted families which could be found in the graph after eliminating long nodes.

Components per Family				Families per Component				Discovered Families (%)			
25	50	75	100	25	50	75	100	25	50	75	100
6.33	5.57	4.41	5.68	1.00	1.00	1.00	1.00	20	40	70	73
4.50	3.50	4.33	4.70	1.00	1.00	1.00	1.00	63	97	97	100
5.26	4.78	4.82	4.32	1.00	1.00	1.00	1.00	43	70	85	92
4.27	4.37	4.18	4.42	1.00	1.00	1.00	1.00	100	100	100	100
3.93	4.48	4.31	4.12	1.00	1.00	1.00	1.02	58	78	87	93
4.41	4.01	4.31	4.47	1.00	1.01	1.00	1.02	83	95	100	100
4.31	4.83	4.25	4.20	1.01	1.01	1.02	1.05	64	82	91	94
4.08	3.80	3.89	4.44	1.93	1.93	1.93	1.93	100	100	100	100
4.68	4.56	3.98	4.51	1.78	1.93	1.61	1.51	65	70	80	90
4.38	4.93	4.45	4.72	1.78	1.50	1.58	1.91	85	100	100	100
4.88	4.36	4.28	4.14	1.64	1.55	1.40	1.48	60	77	89	96
4.83	4.69	5.08	4.23	2.18	2.89	2.98	3.31	100	100	100	100
4.48	4.89	4.81	4.50	2.29	2.19	2.36	2.28	72	89	90	98
4.28	4.62	4.63	4.24	1.66	1.82	2.35	2.50	89	98	99	100
4.26	4.51	4.91	4.98	1.60	1.66	1.78	1.87	68	93	97	98

The sets of reads were given as input to the connected component based repeat family detector, resulting in a collection of connected components. Because we know which sequences correspond to family members, we were able to associate each resulting graph component to the families contained in it. The result of this association is shown in Table 1.

In the ideal case, we would find each family contained in a single connected component. Table 1 shows a different reality. In the left column ("Components per Family"), we see that families are usually split into more than three components. However, each component usually contains sequences of a single family, which is shown by the column "Families per Component". This shows that although the families are split, they are at least not so mixed up that their separation is impossible.

In the rightmost column ("Discovered Families (%)") we see how much of the inserted families could be detected by the method. The fact that we were never able to identify all the families in the odd rows is expected. These are cases where the chromosomes have families of size two. In such cases, depending on the underlying sequence graph dimension used, it can happen that the two family sequences do not share any tuple, or the number of shared tuples is so small that they end up being discarded as small components. In these cases, the combine operation plays an important role, as the next section shows.

4.2   Combine

In order to evaluate our more advanced algorithm, we created a dataset with real bacterial genome sequences and their known insertion sequences. The bacterial

Table 2. Percent of known insertion sequence families found in incompletely sequenced bacterial genomes at different coverages. Bacteria marked with a '*' symbol do *not* have any family with a single member. Numbers marked in **bold** indicate for which of the two combine factors more families were identified, on average.

Combine Factor (t)	1.0				3.0			
Sequencing Coverage (%)	25	50	75	100	25	50	75	100
*Bacillus anthracis (plasmid PX01)**	0	7	0	40	**17**	**58**	**92**	**100**
Bifidobacterium longum	14	54	63	64	**39**	**68**	**81**	**83**
Burkholderia xenovorans	44	**67**	73	78	**50**	67	**75**	**81**
*Colwellia psychrerythraea**	40	**100**	**100**	**100**	**83**	**100**	**100**	**100**
*Desulfitobacterium hafniense**	67	80	97	**100**	**71**	**96**	**100**	**100**
*Desulfovibrio desulfuricans**	**33**	47	**93**	100	33	**75**	92	100
Escherichia coli	17	50	62	70	**44**	**65**	**85**	**92**
Geobacter uraniumreducens	32	62	67	70	**46**	**68**	**75**	**80**
Gloeobacter violaceus	30	70	60	83	**54**	**75**	**88**	**100**
*Granulibacter bethesdensis**	7	7	40	53	**25**	**33**	**42**	**75**
Haloarcula marismortui	3	12	22	28	**13**	**25**	**50**	**63**
Halobacterium sp-plasmid pNRC100	**37**	**57**	**56**	**61**	37	42	53	57
Legionella pneumophila-Paris	0	**13**	**20**	7	**8**	0	0	**17**
Legionella pneumophila-Philadelphia 1	27	**63**	**93**	**93**	54	63	63	92
Methanosarcina acetivorans	88	98	98	**100**	**93**	**100**	**99**	**100**
Methylococcus capsulatus	22	65	77	83	**44**	**71**	**90**	**96**
*Nitrosospira multiformis**	53	93	100	100	**92**	**100**	**100**	**100**
Photobacterium profundum	87	100	100	100	**100**	100	100	100
Pseudomonas syringae	92	99	100	100	**97**	**100**	**100**	**100**
Pyrococcus furiosus	47	58	71	73	**50**	**72**	**81**	**89**
Ralstonia solanacearum	38	60	75	89	**53**	**78**	**93**	**95**
Rhodopirellula baltica	82	98	100	100	**97**	**100**	**100**	**100**
*Roseobacter denitrificans**	40	80	87	100	**42**	**92**	**100**	92
*Salinibacter ruber**	100	100	100	100	100	100	100	100
Shewanella oneidensis	10	26	18	23	**18**	**38**	15	**41**
Sulfolobus solfataricus	**94**	99	**100**	99	94	**100**	100	**100**

chromosomes were obtained from the NCBI Website[1], while the corresponding insertion sequences were obtained from the insertion sequence database *IS Finder*[2]. We created 15 read sets covering 25, 75, 50, and 100 percent of the genome on average. Each of the read sets was used twice as input for the *combine* method: once with combine scale factor $t = 1.0$, and a second time with scale factor $t = 3.0$. Again we associated the resulting connected components to the repeat families found in each of them.

In Table 2 we see the percentages of the known insertion sequence families which could be detected by the method at different coverages. The biggest percentage is shown in bold. We see that, by allowing more divergent nodes to combine, we are not only able to identify more families, but also to identify them at lower coverage.

5   Conclusion

We presented two methods for detecting repeat families in incompletely sequenced genomes. The methods are based on operations on the set of nodes and edges of a sequence graph, a compacted variant of a sparse de Bruijn graph.

---

[1] NCBI: http://www.ncbi.nlm.nih.gov
[2] IS Finder: http://www-is.biotoul.fr/is.html

The first method was based on the deletion of long nodes corresponding to non-repetitive genome portions. Based on experiments involving artificially created data, we showed that the use of this method for family detection is possible, although the families may be split in a few components in the resulting graph. We also noticed that families appearing in small copies are in many cases not detected by the method.

Another node operation was presented in order to combine similar, but not identical, nodes of different members in a family. This operation was applied before the deletion of long nodes, in order to avoid the splitting of family components. Although the combination of nodes leads to a reduction in the number of components per families (data not shown), the main advantage of this operation is better observed in experiments involving real bacterial genomes, where the node combination leads to the detection of only weakly represented families.

The main obstacle for using these methods in practical applications is the splitting of repeat families in separated components. This is for us the main problem to be tackled before applying the method in genomes of higher complexity, like eukaryotes.

References

1. Bao, Z., Eddy, S.R.: Automated de novo identification of repeat sequence families in sequenced genomes. Genome Res. 12, 1269–1276 (2002)
2. Bokhari, S.H., Sauer, J.R.: A parallel graph decomposition algorithm for DNA sequencing with nanopores. Bioinformatics 21(7), 889–896 (2005)
3. Butler, J., MacCallum, I., Kleber, M., Shlyakhter, I.A., Belmonte, M.K., Lander, E.S., Nusbaum, C., Jaffe, D.B.: ALLPATHS: De novo assembly of whole-genome shotgun microreads. Genome Res. 18, 810–820 (2008)
4. Chaisson, M., Pevzner, P., Tang, H.: Fragment assembly with short reads. Bioinformatics 20(13), 2067–2074 (2004)
5. Chaisson, M.J., Pevzner, P.A.: Short read fragment assembly of bacterial genomes. Genome Res. 18, 324–330 (2008)
6. Diestel, R.: Graph Theory, 3rd edn. Graduate Texts in Mathematics, vol. 173. Springer, Heidelberg (2005)
7. Hall, A.E., Fiebig, A., Preuss, D.: Beyond the arabidopsis genome: Opportunities for comparative genomics. Plant Physiol. 129, 1439–1447 (2002)
8. Heath, L.S., Pati, A.: Genomic signatures in de Bruijn chains. In: Giancarlo, R., Hannenhalli, S. (eds.) WABI 2007. LNCS (LNBI), vol. 4645, pp. 216–227. Springer, Heidelberg (2007)
9. Idury, R.M., Waterman, M.S.: A new algorithm for DNA sequence assembly. J. Comput. Biol. 2(2), 291–306 (1995)
10. Jelinek, W.R., Toomey, T.P., Leinwald, L., Duncan, C.H., Biro, P.A., Choudary, P.V., Weissman, S.M., Rubin, C.M., Houck, C.M., Deininger, P.L., Schmid, C.W.: Ubiquitous, interspersed repeated sequences in mammalian genomes. Proc. Natl. Acad. Sci. USA 77(3), 1398–1402 (1980)
11. Mahillon, J., Chandler, M.: Insertion sequences. Microbiol. Mol. Biol. Rev. 62(3), 725–774 (1998)
12. Myers, E.W.: The fragment assembly string graphs. Bioinformatics 21, ii79–ii85 (2005)

13. Pevzner, P.A., Tang, H., Tesler, G.: De novo repeat classification and fragment assembly. In: Proceedings of RECOMB 2004, pp. 213–222 (March 2004)
14. Pevzner, P.A., Tang, H., Waterman, M.S.: An Eulerian path approach to DNA fragment assembly. Proc. Natl. Acad. Sci. USA 98(17), 9748–9753 (2001)
15. Luning Prak, E.T., Kazazian Jr., H.H.: Mobile elements and the human genome. Nature Rev. 1, 134–144 (2000)
16. Amgarten Quitzau, J.A., Stoye, J.: A space efficient representation for sparse de Bruijn subgraphs. Report, Technische Fakultät der Universität Bielefeld, Abteilung Informationstechnik (2008),
    http://bieson.ub.uni-bielefeld.de/frontdoor.php?source_opus=1308
17. Raphael, B., Zhi, D., Tang, H., Pevzner, P.: A novel method for multiple alignment of sequences with repeated and shuffled elements. Genome Res. 14, 2336–2346 (2004)
18. Setubal, J.C., Meidanis, J.: Introduction to Computational Molecular Biology. PWS Publishing (1997)
19. Zerbino, D.R., Birney, E.: Velvet: Algorithms for de novo short read assembly using de Bruijn graphs. Genome Res. 18, 821–829 (2008)
20. Zhang, Y., Waterman, M.S.: An Eulerian path approach to local multiple alignment for DNA sequences. Proc. Natl. Acad. Sci. USA 102(5), 1285–1290 (2005)

Novel Phylogenetic Network Inference by Combining Maximum Likelihood and Hidden Markov Models

(Extended Abstract)

Sagi Snir[1] and Tamir Tuller[2]

[1] Dept. of Mathematics and Computer Science
Netanya Academic College, Netanya
`ssagi@Math.Berkeley.EDU`
[2] School of Computer Science
Tel Aviv University
`tamirtul@post.tau.ac.il`

Abstract. Horizontal Gene Transfer (HGT) is the event of transferring genetic material from one lineage in the evolutionary tree to a different lineage. HGT plays a major role in bacterial genome diversification and is a significant mechanism by which bacteria develop resistance to antibiotics. Although the prevailing assumption is of complete HGT, cases of partial HGT (which are also named *chimeric* HGT) where only part of a gene is horizontally transferred, have also been reported, albeit less frequently.

In this work we suggest a new probabilistic model for analyzing and modeling phylogenetic networks, the NET-HMM. This new model captures the biologically realistic assumption that neighboring sites of DNA or amino acid sequences are *not* independent, which increases the accuracy of the inference. The model describes the phylogenetic network as a Hidden Markov Model (HMM), where each hidden state is related to one of the network's trees. One of the advantages of the NET-HMM is its ability to infer partial HGT as well as complete HGT. We describe the properties of the NET-HMM, devise efficient algorithms for solving a set of problems related to it, and implement them in software. We also provide a novel complementary significance test for evaluating the fitness of a model (NET-HMM) to a given data set.

Using NET-HMM we are able to answer interesting biological questions, such as inferring the length of partial HGT's and the affected nucleotides in the genomic sequences, as well as inferring the exact location of HGT events along the tree branches. These advantages are demonstrated through the analysis of synthetical inputs and two different biological inputs.

1 Introduction

Eukaryotes evolve largely through vertical lineal descent in a tree-like manner. However, in the presence of HGT the right model of evolution is not a tree but is rather a *phylogenetic network*, which is a directed acyclic graph obtained by

K.A. Crandall and J. Lagergren (Eds.): WABI 2008, LNBI 5251, pp. 354–368, 2008.
© Springer-Verlag Berlin Heidelberg 2008

positing a set of edges between pairs of the branches of an organismal tree to model the horizontal transfer of genetic material [16]. In the case of a complete HGT the assumption is that a single tree (one of the trees induced by the network) describes the evolution of a gene, while in the case of a chimeric HGT more than one tree is needed (*i.e.* different parts of a gene evolve according to different trees).

HGT (partial or complete) is very common among *Bacteria* and *Archaea* [6,8,5], but evidence of HGT (partial or complete) between Eukaryotes are also accumulating [2,26]. A large body of work has been introduced in recent years to address phylogenetic network reconstruction and evaluation. Methods for dealing with this problem include a variety of approaches: *Splits Networks* (see e.g. [14]) which are graphical models that capture incompatibilities in the data due to various factors, not necessarily HGT or hybrid speciation; Maximum Parsimony (MP) [12,16,17] that are based on Occam's Razor approach; distance methods, that try to fit a distance matrix to a network [4] or use the minimum evolution criterion [3], and graph theoretical approaches that try to fit a gene tree to a species tree [1,10,23,11].

One of the most accurate and commonly used criteria for reconstructing phylogenetic trees is *Maximum Likelihood* (ML) [9]. Roughly speaking, this criterion considers a phylogenetic tree from a probabilistic perspective as a generative model, and seeks the model (i.e., tree) that maximizes the likelihood of observing the given input set of sequences at the leaves of the tree. Likelihood in the general network setting has been investigated in the past in various studies. Von Haeseler and Churchill [29] provided a framework for evaluating likelihood on networks and subsequently [28] provided an approach to assess this likelihood. These works consider a network as an arbitrary set of splits and do not correspond to a specific biological process. Likelihood on networks has also been considered in the setting of recombination networks (see e.g. [13]). These methods are tailored to identify breakpoints along the given sequences. However, their underlying model, the biological questions they investigate, and the algorithmic approaches they pursue are different from ours as they model a different biological process.

Recently, Jin *et al.* performed an initial step toward developing an HGT-oriented likelihood based model for evolutionary networks [16]. This work demonstrated the potential of using ML for inferring evolutionary networks. The main advantage of that work is its simplistic underlying model, that enables efficient implementation. Another related work is the study of Siepel and Haussler [27] who suggested a model that combines a phylogenetic tree along with a HMM and used it for aligning full genomes. Our work was inspired by these two works. However, while the main goal in [16] was to infer complete HGT events, here we focus on analyzing chimeric HGT events by adopting a more biologically relevant model for this task, the *NET-HMM* model.

The NET-HMM models a phylogenetic network by a Hidden Markov Model (HMM), where each of the network's trees corresponds to a state of the HMM (see figure 1); *i.e.* the emission probability in each state is according to its corresponding tree.

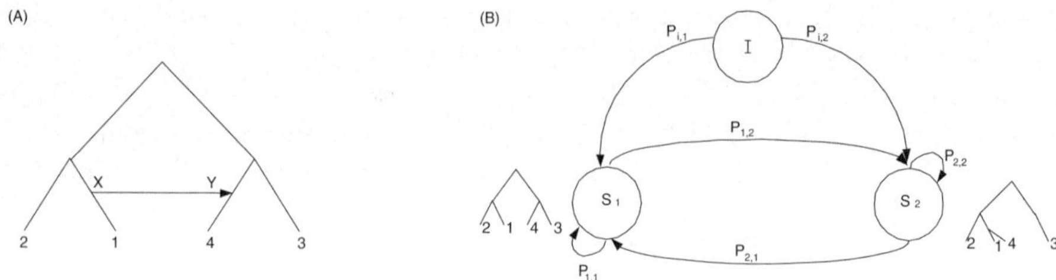

Fig. 1. Simple example of the NET-HMM model. (*A*) A phylogenetic network, with a single HGT from X to Y. (*B*) The HMM for the network: I denotes the initial state, all the other states are related to the networks' trees. The state S_1 is related to the underlining organismal tree. The state S_2 is related to the tree of realizing the horizontal transfer edge. We assume $P_{i,i} > P_{i,j}$, *i. e.* in each position the probability to stay in the same state/tree is higher than the probability of transition to other tree.

The model is supported by biological facts that (*i*) adjacent sites are not independent with respect to HGT, (*ii*) events of HGT are relatively rare. In the view of the NET-HMM model, events such as chimeric HGT [22] are described as a transition between states (trees) of the HMM. NET-HMM also reconstructs the exact HGT location on the tree edges, a task that methods such as MP or distance based methods [17,4] can not accomplish.

By applying the NET-HMM on synthetical data, we show very significant improvements over the i.i.d. model [16] in the ability to locate chimeric events and locations along tree edges. Results on biological data point out significant biological phenomena such as amelioration [15,25], chimeric HGT events, and occurrence of very recent HGT events. Applying the NET-HMM to data that were analyzed in the past suggest an excess of HGT inference. These findings suggest a further scrutiny.

In the statistical-algorithmic realm, we propose a novel algorithm, *EM NET-HMM*, that interweaves into the conventional EM algorithms a step of hill climbing to maximize emission probabilities, rendering a non trivial algorithmic approach. We also devised a novel permutation test to measure the fitness of a NET-HMM to a given dataset.

2   Methods

2.1   Preliminaries and Definitions

Let $T = (V, E)$ be a tree, where V and E are the *tree nodes* and *tree edges*, respectively, and let $F(T)$ denote its leaf set and $I(T)$ its internal nodes. Further, let χ be a set of taxa (species). Then, T is a phylogenetic tree over χ if there is a bijection between χ and $F(T)$. A tree T is said to be *rooted* if the set of edges E is directed and there is a single distinguished internal vertex r with in-degree 0. Let Σ denote the set of *states* (*e.g.* for DNA, $|\Sigma| = 4$). Then with each edge $e \in E$ we associate a *substitution probability* p_e indicating the probability of observing different states at the endpoints of e. For a wide variety of evolutionary models there is an invertible transformation from the p_e to *edge length* q_e s.t. q_e's are *additive* - applying the inverse transformation on the sum $q_\pi = \sum_{e \in \pi} q_e$ of edge lengths along a path π yields the probability p_π of observing different states at the endpoints of π. Therefore, under the mapping $\mathbf{q} : E \to \mathbb{R}$ of lengths, $T = (V, E, \mathbf{q})$ is a weighted tree (we omit $\mathbf{q}$ when it is clear from the context). The edge length and additivity are crucial for our formulation as is explained in the sequel.

In this work we consider the Jukes-Cantor (JC) model of sequence evolution [21]. However, all the results here can be generalized to other models of sequence evolution. Under JC, the length-substitution relationship is as follows: $p_e = \frac{3}{4}(1 - e^{-4/3q_e})$ and $q_e = -\frac{3}{4}\ln(1 - \frac{4}{3}p_e)$.

For a given set of input sequences S, the i-th site, S_i, is the set of states at the i-th position for every sequence in S [1]. Under the ML criterion, a phylogenetic tree is viewed as a probabilistic model from which input sites are assumed to be sampled. The probability of obtaining a site S_i given a tree T, $L(S_i|T)$, is defined as [9]:

$$L(S_i|T) = \sum_{\mathbf{a} \in \Sigma^{|I(T)|}} \prod_{e \in E(T)} m(p_e, S_i, a) \, , \tag{1}$$

where a ranges over all combinations of assigning states to the $I(T)$ internal nodes of T. Each term $m(p_e, S_i, a)$ is either $p_e/(|\Sigma| - 1)$ or $(1 - p_e)$, depending on whether, under a, the two endpoints of e are assigned different or the same states respectively.

A phylogenetic network $N = N(T) = (V', E')$ over the taxa set χ is derived from a rooted weighted tree $T = (V, E, \mathbf{q})$ by adding a set R of reticulation edges to T, where each edge $r \in R$ is added as follows: (1) split an edge $e \in E$ by adding a new node, v_e, s.t. the lengths of the newly created edges sum to the length of e; (2) split an edge $e' \in E$ by adding another new node, $v_{e'}$ (again by preserving lengths); (3) finally, add a directed *reticulation edge* r from v_e to $v_{e'}$. We add that the substitution probability (and hence the length) of r is zero as these events are instantaneous in time. The mathematical implication of the above is that it extends the partial order induced by T on V (see [19,18]). This results in having no cycles in which tree edges are traversed along their directionality and reticulation edges in either direction (see Figure 3 A.). A tree T' is induced by N by removing all but one incoming edges to the newly added nodes, and contracting degree-2 nodes (while summing the edge lengths). We denote by $T(N)$ the set of trees induced by N (see Figure 3 A.).

2.2   From a Network to a HMM

The i.i.d Model. Under the i.i.d model, each reticulation edge r has a probability denoting the corresponding event's probability. Under this formulation, every edge (including tree edges) is assigned an *occurrence* probability s.t. the

---

[1] Can be viewed as the i-th column when the sequences are aligned.

sum of occurrence probabilities of edges entering a node is 1. For a tree T, let E_T denote the set of (reticulation + tree) edges realized by (or giving rise to) T, and let $P(E_T)$ be the probability of observing E_T (It can be seen that this is the product of their individual probabilities).

Therefore, the likelihood of obtaining a site, S_i, given a phylogenetic network, N, is [16]: $L(S_i|N) = \sum_{t \in T(N)} P(E_t) \cdot L(S_i|t)$. The likelihood of obtaining the input sequences is $L(S|N) = \prod_i L(S_i|N)$. As different sites are modeled independently, a weak signal at a certain site will cause the inference of an erroneous tree at that site (a phenomenon we give more details on in the simulation section). To overcome this shortcoming, we treat the sites as a Markovian process, as we describe in the next subsection.

The NET-HMM. The NET-HMM is a tuple $M = \{N, H\}$ where $N = (V', E', \mathbf{q})$ is a phylogenetic network, and H is a Hidden Markov Model (HMM). We do not know the evolutionary history (a tree in $T(N)$) of every site in S, thus we assign a hidden state for each site in S, and an initial state, I. The hidden states correspond to the states of H and let Σ_H denote this set. Let $\Upsilon(h)$ denote the tree related to the hidden state h (the initial state is not related to a tree, so $h \neq I$). The meaning of relating the state h of the i-th site to a state of the HMM is that this site evolves on the tree $\Upsilon(h) \in T(N)$ (*i. e.* the i-th column was emitted by the tree T).

Let $p(h_{i-1} \to h_i)$ denote the transition probability between state h_{i-1} and state h_i in the HMM. The likelihood of a NET-HMM model M when observing a set S of n-long sequences, is defined as the probability of observing S evolving on M which is the sum of probabilities of all length-n paths of states from Σ_H. Thus $L(S|M)$ equals

$$\sum_{h_1^n \in (\Sigma_H)^n} p(I \to h_1) \prod_{i=2}^n p(h_{i-1} \to h_i) \cdot L(S_i|\Upsilon(h_i), \mathbf{q}) \tag{2}$$

where h_1^n is a sequence of n states (and $\forall_i h_i \in \Sigma_H$).

A different variant of the likelihood function scores a network by the probability of the most likely path, $\hat{h}_1^n$, in M. The latter is achieved by replacing the sum by a maximum relation.

Our goal is to find the model (network topology, edge length, and transition probabilities of the HMM) that maximizes the likelihood of the input sequences (equation 2). By using an HMM we gain two important advantages: 1) We gain dependencies among close sites, as in reality. 2) We indirectly infer the probabilities of reticulation events, while avoiding the use of arbitrary parameters for reticulation probability (as was assumed in [16]).

We emphasize three important constraints on the NET-HMM model; these constraints are biologically motivated but also decrease the parameter space (and thus reduce the running time), while improving the quality of the results:

1. The network induces both the topology and edge lengths of its trees (see Figure 2).
2. Temporal constraints on the reticulation edges decrease the number of valid networks (see Figure 3A.).

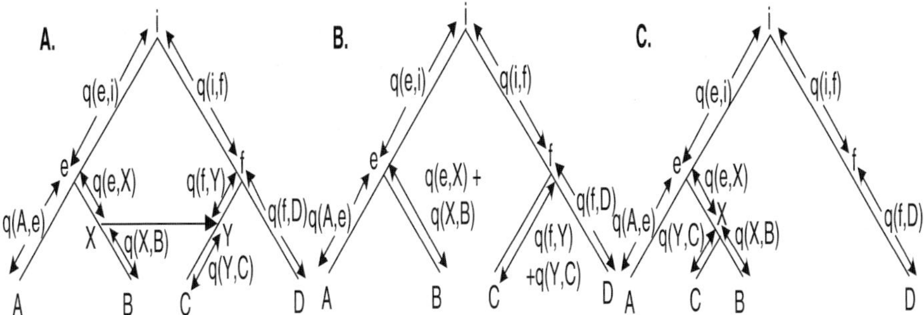

Fig. 2. A simple example of a phylogenetic network under the likelihood setting, and the set of induced trees. *A*. A phylogenetic network with one reticulation edge; each edges length denotes the expected number of subsections along the edge. *B*. One of the network's trees that does not include the reticulation edge. *C*. One of the network's trees that includes the reticulation edge.

3. By imposing constraints on the transition probabilities of H, we can drastically reduce the exponential number of paths. We add that there is a positive transition probability from the initial state to each of the other states (see Figure 3 *B*.).

Given a set of sequences associated with the tree leaves, we distinguish between three major versions of the problem, tiny, small, and large that are defined as follows:

1. **The tiny version:** Input: the network topology along with its edge lengths and transition probabilities between states in the induced HMM. Output: The most likely path in the HMM.
2. **The small version:** Input: The network topology. Output: The ML network edge probabilities and ML transition probabilities of the induced HMM.
3. **The large version:** Input: The initial organismal tree. Output: The ML NET-HMM (complete network+HMM).

The specific version of the problem should be employed depending on the given input. When only an organismal tree and a set of sequences, suspected of having undergone HGT, are given, the large problem is chosen. When some prior information about the topology of the network (i.e. a set of reticulation edges corresponding to the organismal tree) is given and we want to know the exact location of the reticulation edges (the "split" points) and the positions of the events along the sequences, we should solve the small problem. The tiny problem finds the positions of the events along the input sequences (a partial task of the small problem) by inferring the evolutionary history of each site (position) of the input sequences.

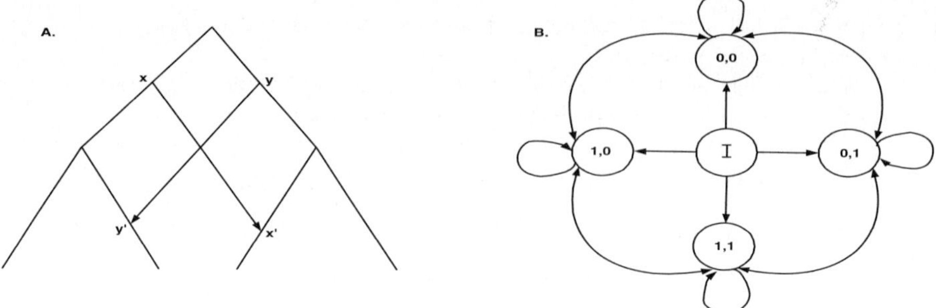

Fig. 3. Constraints on the structure of the NET-HMM. A. Example of a network that does not satisfy the temporal constraints. By the tree topology, x occurs before y' and y occurs before x'. By the reticulation edges, x and x' occur at the same time and y and y' occurs in the same time. Thus, y' occur before x' and x' occurs before y' - a contradiction. B. A simple example of the implementation of the assumption that HGTs are rare events. Suppose the phylogenetic network includes 2 edges. We name each of the network's four states/trees by its code (I denotes the initial state). The figure shows only the transitions with non-zero probabilities, *i. e.* transitions between trees (states) with hamming distance ≤ 1.

2.3   Algorithms

The Tiny Problem - Finding the Most Likely Path. Let $L(S_1^i|h', M)$ denote the likelihood of observing the first i positions of the sequences under M restricted that the i'th state is $h' \in H$. First we observe that $L(S|M) = \sum_{h' \in H} L(S_1^n|h', M)$. Moreover, $L(S_1^i|h', M)$ equals

$$\sum_{h'' \in \Sigma_H} L(S_1^{i-1}|h'', M) \cdot p(h'' \to h') \cdot L(S_i|\Upsilon(h'), \mathbf{q}). \tag{3}$$

Equation (3) is solved by the dynamic programming forward and backward algorithms [7]. The maximization variant is solved similarly by the Viterbi dynamic programming algorithm.

A related question is to infer the most likely state (tree), $\hat{h}_i$, at a certain site, i. For this problem we used the forward and backward algorithms and calculate: $p(S, h_i = k) = p(S_1^i, h_i = k) \cdot p(S_{i+1}^n|h_i = k)$. By the forward and backward algorithms we can calculate $p(S)$. Thus we get the results $p(h_i = k|S) = \frac{p(S, h_i = k)}{p(S)}$.

The Small Problem - Learning a Given Model. Given the network's topology (and hence the induced HMM) we use our extended EM algorithm, EM NET-HMM (depicted in Figure 4), for estimating the edge lengths and edge probabilities of the network as well as the state transition probabilities in the HMM.

EM NET-HMM:
1. Start with initial random edge lengths and transition probabilities.
2. Perform until convergence:
a. Given the edge lengths (that induce an emitting probability for all states); optimize the transitions probabilities by BW algorithm.
b. Given the transition probabilities of the HMM, optimize the edge probabilities (*i. e.* find edges that maximize the cost function that appear in equation 2) by hill climbing.

Fig. 4. The main algorithm, EM NET-HMM

Observation 1. *The algorithm EM NET-HMM terminates and converges to a local ML point.*

The Large Problem - Expanding a Given Model. Recall that the task in the large problem is to add reticulation edges s.t. the likelihood of the model is maximized. Literally, solving this problem boils down to iteratively adding edges to a given tree/network. By the fact that additional edges never decrease the likelihood of the model, some stopping criterion is required. One possibility is to observe the improvement in the likelihood score and stop when the improvement is insignificant (this approach was used in [16,17]). We however used a more rigorous approach as described in the next sub section.

2.4 Network Significance

Recall that the improvement of the NET-HMM model over the iid (see Section 3 for experimental results) was achieved by the coupling of neighboring sites. This is reflected by the transition probabilities of the HMM. Therefore in order to evaluate the significance of a given model (network+HMM) M' WRT the data order within the given input S', we devised the following test that resembles the conventional permutation test in statistics: For a given random sampling S_0 of the sites of the input, let $L(S_0|M')$ be the likelihood of M' WRT S_0 (i.e. the solution of the tiny problem). For a given big enough set of such samples, we obtain a probability distribution of the likelihood of M'. Given that distribution we can compute the *empirical p-value* of M' WRT S'. In our setting, we run this test after each time a model is built, that is after the application of the small problem. We note that another significance test could be obtained by randomizing over the space of networks, however, due to the small size of this space and the computational complexity of calculating this distribution, we used only the first test.

   We used this approach to determine stopping criterion. We stopped the algorithm of iteratively adding HGT edges when either the likelihood did not increase or the p-value increased.

3   Experimental Results

3.1   Synthetical Inputs

We first implemented our algorithms on synthetical data. In order to test the NET-HMM's accuracy we tried to solve the small problem on simulated data where we know the "true" model. We generated 20 synthetical phylogenetic networks, sampled them, and used these samples as inputs to our methods.

In the first test our aim was to check how well the NET-HMM reconstructs the set of events (correct positions along the correct sequences) occurred in the network. This is equivalent to finding the correct path between the HMM states (trees). We sampled a segment (a consecutive set of sites) from each of the trees in the network and concatenated them together into a basic concatenation block. We defined an error rate in the reconstruction of the most likely path as the fraction of sites where the NET-HMM inferred a wrong tree. The experiments were conducted for various HGT segment lengths and for various number of replications of the basic concatenation block. The results are shown in Figure 5.

In Figure 5.A. it can be appreciated that the NET-HMM was able to reconstruct each of the segments fairly accurately and significantly better than the i.i.d model. It demonstrates that segments of length 50 are enough for generating a surprisingly good reconstruction of the true path (error rate between 0.15 and 0.19), and segments of length 500 will give a very good reconstruction (error rate less than 0.01). The results also show that the method is independent of the number of replications of the basic concatenation block. On the other hand, as expected the error rate of the i.i.d model was around 0.8 irrespective of the number of replications and segment lengths. Figures 5.B and 5.C show the output of two typical runs of EM NET-HMM and the i.i.d. model compared with the true path, for segment lengths 50 and 500 respectively. The advantage of EM NET-HMM over the i.i.d. model is very clear and outstanding.

In our second test, we checked the ability to infer the correct HGTs by the NET-HMM (i.e. the big problem). When checking synthetic datasets of moderate sequence length (above 500 sites) and tree edges of significant length (so the correct HGT edge location is crucial) usually the improvement in the likelihood of the true HGT edges was much larger than arbitrary ones, and the NET-HMM successfully identified the correct HGTs.

3.2   Biological Inputs

As indicated above, the main strength of our method is its accuracy, making it appropriate for further analysis of results obtained or hypothesized by other methods. We believe this is the most practical way of using our model since it substantially decreases the number of learned parameters and hence the statistical and the computational complexity of the problem.

Hence, we solved a restricted version of the small problem where the task is to infer the exact position of the reticulation edges along the tree branches and the most likely tree at each site. The organismal tree (topology and edge lengths) was inferred using ML on various sets of genes and was taken as constant.

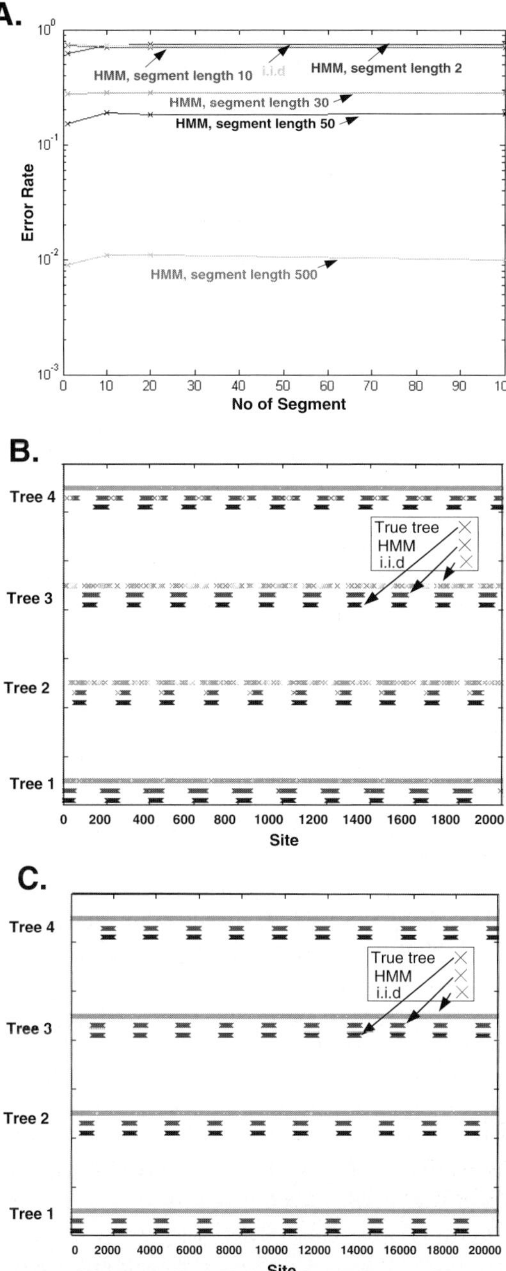

Fig. 5. The performances of the NET-HMM on synthetic inputs (a network with 8 leaves and two reticulation edges). *A*. Reconstructing the tree in each site: Error rate for different segment lengths. *B*. Segment length 50: The inferred tree in each site by NET-HMM vs the i.i.d model. *C*. Segment length 500: The inferred tree in each site by NET-HMM vs the i.i.d model.

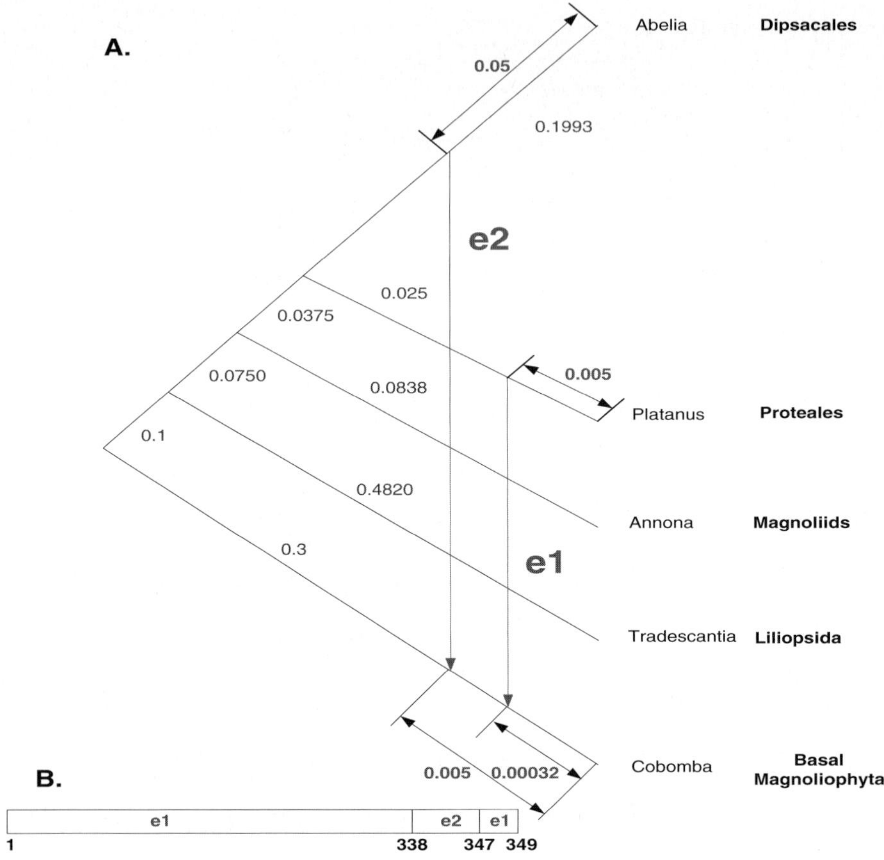

Fig. 6. Phylogenetic network of the ribosomal protein gene *rps11* of a group of 5 flowering plants. *A.* The inferred positions of the HGTs. *B* The inferred most likely path (the most likely reticulation edge in each position of the sequence).

In order to adjust the tree's branch lengths to the evolutionary rate of our set of genes/proteins, we used one scaling factor for all the branches that was estimated together with the other network parameters (similar idea to the proportional branch lengths approach [24]). The set of the reticulation edges (without their positions along the tree branches) was inferred by the fast algorithm that is based on the MP criterion [17].

The Ribosomal Protein Gene *rps11* of Flowering Plants. We analyzed the ribosomal protein gene *rps11* of a group of 5 flowering plants which was first analyzed by Bergthorsson *et al.* [2] who suggested that this dataset underwent chimeric HGTs. This dataset consists of 5 DNA sequences. The species tree was reconstructed based on various sources, including the work of [20], the edge lengths were computed by the gene *atp1* (1, 254 nucleotides) (see figure 6).

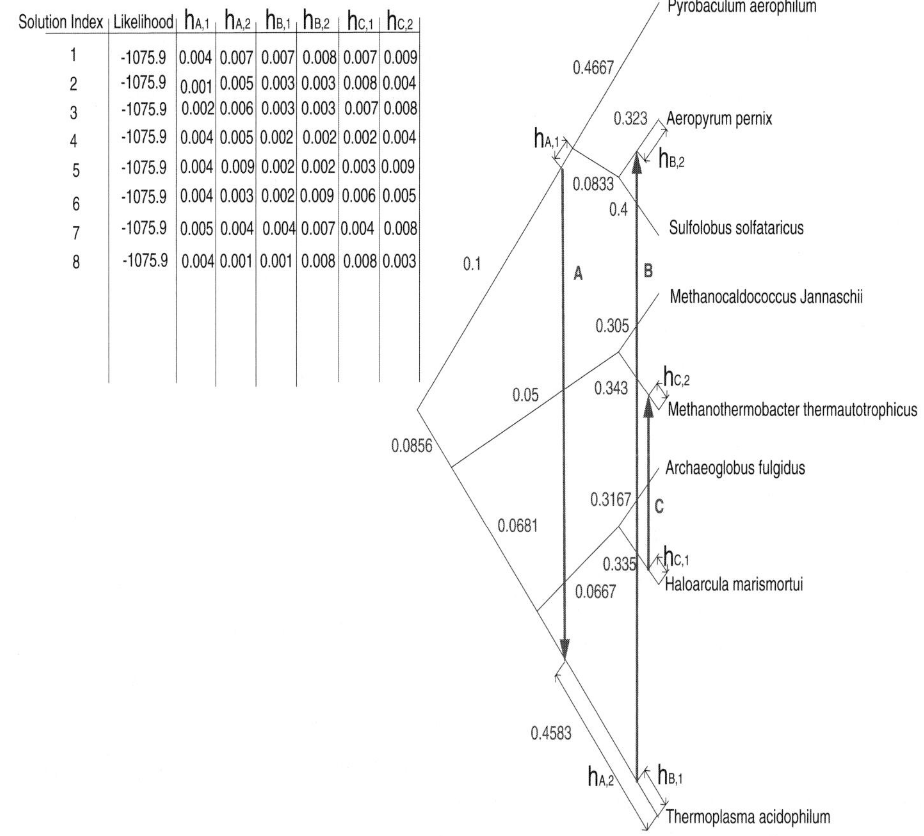

Fig. 7. Phylogenetic network of the ribosomal protein *rpl12e* of a group of 8 *Archaeal* organisms. The position of the HGTs ($h_{A,1}$, $h_{A,2}$, $h_{B,1}$, $h_{B,2}$, $h_{C,1}$, $h_{C,2}$) for each of the eight most likely solutions are plotted in the table beside.

The permutation p-value of the results is 0.02, and the plot of our EM NET-HMM supports the hypothesis of Bergthorsson *et al.* as the resulted ML paths consist of two different non-organismal trees (see figure 6). The tree that consists of one of the reticulation edges appear in most of the sites of the path but towards the end of the path the second tree (consisting of the second reticulation edge) appears (see figure 6).

Ribosomal Protein *rpl12e* of a Group of *Archaeal* Organisms. Next we analyzed the ribosomal protein *rpl12e* of a group of 8 *Archaeal* organisms, which was analyzed by Matte-Tailliez *et al.* [22]. This dataset consists of 14 aligned amino acid sequences, each of length 89 sites. We used the same organismal tree used by the authors [22]. In their work, they suggest that ribosomal protein *rpl12e* has different evolution history from the organismal evolutionary tree (probably due to HGT events). By using MP Jin *et al.* [17] indeed found three HGTs (see figure 7), that can explain the difference between the two trees.

The inferred positions of the HGTs are depicted in figure 7. All the ML solutions have permutation p-values around 0.02 (due to almost identical log likelihoods). The distances of events B and C from the leaves are very small (see $h_{B,1}$, $h_{B,2}$, $h_{C,1}$, $h_{C,2}$ in figure 7), suggesting that these two HGT events occurred fairly recently (in terms of the evolutionary time scale).

Amelioration [15,25] is a process by which a gene that was transferred horizontally acquires features (*e.g.* GC content, the percentage of nitrogenous bases on a DNA molecule which are either guanine or cytosine) similar to its new environment. This is particularly true in recent events as this process diminishes in time. The lengths along the branches of a phylogenetic network are related to the rate of mutation and time span [21]. The existence of two paths along a tree with the same time span but different lengths suggests a variation in mutation rate. By definition, the time span between the two ends of a reticulation edge and their corresponding leaves of the phylognetic network is identical. Interestingly, our results show $h_{B,1} < h_{B,2}$ and $h_{C,1} < h_{C,2}$ [2]. This fact suggests that after horizontal transfer events, genes have undergone an accelerated evolution or adaptation on the amino acid (protein) levels and not only on the nucleotides (gene) levels. This reasonable idea is new in this context.

4   Conclusions

In this work we have described a novel model for analyzing phylogenetic networks. We show that our model, along with its implementation, has advantages over other methods and it is complementary to the other methods in the field, by utilizing and extracting more information encompassed in the data. We also devised a novel test for the statistical significance of a hypothesis (network) under that model. The main strength of this model is its accuracy, however at the cost of increased complexity. The latter can be overcome by incorporating prior knowledge obtained by simple, less accurate models. We have devised an inference method for that model and have shown its performance on simulated data and subsequently applied it to real biological data that was analyzed previously by other methods. Using it, we are able to answer real biological questions such as existence of partial HGT, differences in the rate of mutation among various lineages, distribution of HGT over time and alike. On the computational side, we devised a novel algorithm, EM NET-HMM that employs an EM algorithm in the transition probability space combined with a hill climbing step in the network parameter space.

As a future work, it would be desirable to develop more efficient heuristics for optimizing our computations, and to implement our method on larger sets of organisms. By implementing the NET-HMM on large datasets we intend to

---

[2] Precisely, averaging over all ML solutions, $h_{B,1} = 0.003, h_{B,2} = 0.0053, h_{C,1} = 0.0056, h_{C,2} = 0.0063$.

answer the basic question of determining the length distribution of HGTs or partial HGTs.

Acknowledgment

T.T. was supported by the Edmond J. Safra Bioinformatics program at Tel Aviv University.

References

1. Addario-Berry, L., Hallett, M., Lagergren, J.: Towards identifying lateral gene transfer events. In: PSB 2003, pp. 279–290 (2003)
2. Bergthorsson, U., Adams, K., Thomason, B., Palmer, J.: Widespread horizontal transfer of mitochondrial genes in flowering plants. Nature 424, 197–201 (2003)
3. Birin, H., Gal-Or, Z., Elias, I., Tuller, T.: Inferring horizontal transfers in the presence of rearrangements by the minimum evolution criterion. Bioinformatics 24(6), 826–832 (2008)
4. Boc, A., Makarenkov, V.: New efficient algorithm for detection of horizontal gene transfer events. In: Benson, G., Page, R.D.M. (eds.) WABI 2003. LNCS (LNBI), vol. 2812, pp. 190–201. Springer, Heidelberg (2003)
5. Delwiche, C., Palmer, J.: Rampant horizontal transfer and duplication of rubisco genes in eubacteria and plastids. Mol. Biol. Evol. 13(6) (1996)
6. Doolittle, W.F., Boucher, Y., Nesbo, C.L., Douady, C.J., Andersson, J.O., Roger, A.J.: How big is the iceberg of which organellar genes in nuclear genomes are but the tip? Phil. Trans. R. Soc. Lond. B. Biol. Sci. 358, 39–57 (2003)
7. Durbin, R., Eddy, S.R., Krogh, A., Mitchison, G.: Biological Sequence Analysis: Probabilistic Models of Proteins and Nucleic Acids. Cambridge University Press, Cambridge (1999)
8. Paulsen, I.T., et al.: Role of mobile DNA in the evolution of Vacomycin-resistant Enterococcus faecalis. Science 299(5615), 2071–2074 (2003)
9. Felsenstein, J.: Evolutionary trees from DNA sequences: A maximum likelihood approach. J. Mol. Evol. 17, 368–376 (1981)
10. Hallett, M., Lagergren, J.: Efficient algorithms for lateral gene transfer problems. In: RECOMB 2001, pp. 149–156. ACM Press, New York (2001)
11. Hallett, M., Lagergren, J., Tofigh, A.: Simultaneous identification of duplications and lateral transfers. In: Proceedings of the eighth annual international conference on Research in computational molecular biology, pp. 347–356 (2004)
12. Hein, J.: Reconstructing evolution of sequences subject to recombination using parsimony. Math. Biosci. 98, 185–200 (1990)
13. Husmeier, D., McGuire, G.: Detecting recombination with MCMC. Bioinformatics 18, 345–353 (2002)
14. Huson, D.H., Bryant, D.: Application of phylogenetic networks in evolutionary studies. Mol. Biol. Evol. 23(2), 254–267 (2006)
15. Lawrence, J.G., Ochman, H.: Amelioration of bacterial genomes: rates of change and exchange. J. Mol. Evol. 44(4), 383–397 (1997)
16. Jin, G., Nakhleh, L., Snir, S., Tuller, T.: Maximum likelihood of phylogenetic networks. Bioinformatics 22(21), 2604–2611 (2006)

17. Jin, G., Nakhleh, L., Snir, S., Tuller, T.: Inferring phylogenetic networks by the maximum parsimony criterion: A case study. Mol. Biol. Evol. 24(1), 324–337 (2007)
18. Jin, G., Nakhleh, L., Snir, S., Tuller, T.: A new linear-time heuristic algorithm for computing the parsimony score of phylogenetic networks: Theoretical bounds and empirical performance. In: Măndoiu, I.I., Zelikovsky, A. (eds.) ISBRA 2007. LNCS (LNBI), vol. 4463, pp. 61–72. Springer, Heidelberg (2007)
19. Jin, G., Nakhleh, L., Snir, S., Tuller, T.: Parsimony score of phylogenetic networks: Hardness results and a linear-time heuristic (submitted, 2008)
20. Judd, W.S., Olmstead, R.G.: A survey of tricolpate (eudicot) phylogenetic relationships. Am. J. Bot. 91, 1627–1644 (2004)
21. Jukes, T., Cantor, C.: Evolution of protein molecules. In: Munro, H.N. (ed.) Mammalian protein metabolism, pp. 21–132 (1969)
22. Matte-Tailliez, O., Brochier, C., Forterre, P., Philippe, H.: Archaeal phylogeny based on ribosomal proteins. Mol. Biol. Evol. 19(5), 631–639 (2002)
23. Nakhleh, L., Ruths, D., Wang, L.S.: RIATA-HGT: A Fast and Accurate Heuristic for Reconstructing Horizontal Gene Transfer. In: Wang, L. (ed.) COCOON 2005. LNCS, vol. 3595, pp. 84–93. Springer, Heidelberg (2005)
24. Pupko, T., Huchon, D., Cao, Y., Okada, N., Hasegawa, M.: Combining multiple datasets in a likelihood analysis: which models are best. Mol. Biol. Evol. 19(12), 2294–2307 (2002)
25. Ragan, M.A.: On surrogate methods for detecting lateral gene transfer. FEMS Microbiol. Lett. 201(2), 187–191 (2001)
26. Richardson, A.O., Palmer, J.D.: Horizontal gene transfer in plants. J. Exp. Bot. 58(1), 1–9 (2007)
27. Siepel, A., Haussler, D.: Combining phylogenetic and hidden markov models in biosequence analysis. In: RECOMB 2003, pp. 277–286 (2003)
28. Strimmer, K., Moulton, V.: Likelihood analysis of phylogenetic networks using directed graphical models. Mol. Biol. Evol. 17(6), 875–881 (2000)
29. von Haeseler, A., Churchill, G.A.: Network models for sequence evolution. J. Mol. Evol. 37, 77–85 (1993)

A Local Move Set for Protein Folding in Triangular Lattice Models*

H.-J. Böckenhauer[1], A. Dayem Ullah[2], L. Kapsokalivas[2], and K. Steinhöfel[2]

[1] ETH Zürich, Department of Computer Science, CH-8092 Zurich, Switzerland
[2] King's College London, Department of Computer Science, London WC2R 2LS, UK

Abstract. The HP model is one of the most popular discretized models for the protein folding problem, i.e., for computationally predicting the three-dimensional structure of a protein from its amino acid sequence. This model considers the interactions between hydrophobic amino acids to be the driving force in the folding process. Thus, it distinguishes between polar and hydrophobic amino acids only and asks for an embedding of the amino acid sequence into a rectangular grid lattice which maximizes the number of neighboring pairs (contacts) of hydrophobic amino acids in the lattice.

In this paper, we consider an HP-like model which uses a more appropriate grid structure, namely the 2D triangular grid and the face-centered cubic lattice in 3D. We consider a local-search approach for finding an optimal embedding. For defining the local-search neighborhood, we design a move set, the so-called pull moves, and prove its reversibility and completeness. We then use these moves for a tabu search algorithm which is experimentally shown to lead into optimum energy configurations and improve the current best results for several sequences in 2D and 3D.

1 Introduction

One of the core problems in today's bioinformatics is the question of how proteins fold and whether we can efficiently predict their structure. Proteins regulate almost all cellular functions in an organism. They are built as linear chains of amino acids. These chains fold in three-dimensional space. This 3D structure, also referred to as tertiary structure, plays a key role in their function. According to Anfinsen's thermodynamic hypothesis, proteins fold into states of minimum free energy and their tertiary structure can be predicted from the linear sequence of amino acids [2]. In nature, proteins fold very rapidly, despite the enormous number of possible configurations. This fact motivated Levinthal [12] to suggest that the native protein state might have a higher energy value than the theoretical free-energy minimum, if the latter is not kinetically accessible; a phenomenon that is referred to as Levinthal's paradox and implies that a protein might be able to avoid kinetic traps of local minima or quickly escape them. Therefore, an almost greedy path to the native protein state exists [8]. The mechanism behind

---

* Research partially supported by EPSRC Grant No. EP/D062012/1.

K.A. Crandall and J. Lagergren (Eds.): WABI 2008, LNBI 5251, pp. 369–381, 2008.

© Springer-Verlag Berlin Heidelberg 2008

this paradox remains an open problem, which motivates many scientists to study protein folding simulations.

In order to speed up simulations, a variety of coarse-grained models have been introduced. Even in one of the most simplified models, namely the HP model [6], protein folding has been shown to be NP-complete [5]. In this original HP model, the amino acid chain of a protein is embedded into a two- or three-dimensional rectangular grid lattice. The quality of an embedding is measured by the number of neighboring pairs (contacts) of hydrophobic amino acids in the lattice. The problem is shown NP-hard for HP-like models into generalized lattices [10]. These hardness results motivated a study of approximation algorithms for the HP model [9]. Nevertheless, the approximation ratio is far from giving exact prediction and consequently a variety of heuristic methods have been proposed, such as genetic algorithms [18], Tabu search [13], Simulated Annealing [17].

Most approaches employ the HP model in rectangular lattices, which captures the computational properties of the problem but has the drawback of allowing interactions only between amino acids of opposite parity in the chain. One possibility for overcoming this problem is by using triangular grid lattices. The 3D equivalent of the 2D triangular lattice is the *face-centered cubic lattice* (FCC). Another strong motivation behind studying the FCC is the fact that its geometry has been rigorously shown to be the closest packing geometry for identical spheres [16], which concludes an old conjecture by Kepler. Although some other geometries might also exhibit the property of closest packing, the authors in [4] observed that in native foldings, the neighboring positions of amino acids closely approximate those of a distorted FCC lattice. Linear-time approximation algorithms [1] and various heuristic approaches have been developed for the triangular models as well. The latter include genetic algorithms [11] and various Monte Carlo search methods [19].

Especially in the context of local search methods, it is important to employ an efficient move set (neighbourhood function), since it determines the overall performance. The *pull move* set, originally introduced by Lesh et al. [13] for rectangular lattices, has been proven to be very efficient in the HP model under a variety of local search methods [13,17]. In [13], also the completeness and reversibility of the pull move set were shown for rectangular grid lattices; these are essential properties to guarantee reachability of the global minimum.

In this paper, we first formally introduce *pull moves* in the 2D triangular lattice. Secondly, we extend them to the FCC lattice based on the fact that the FCC lattice is isomorphic to a 3D extension of the triangular lattice. We also formally prove the *reversibility* and *completeness* of pull moves in both the 2D triangular and the FCC lattices. We then employ the set of pull moves in a tabu-search strategy for protein folding simulation. Experimental results on several benchmark problem instances show that the pull move set combined with the tabu-search algorithm is able to lead into known optimum energy configurations. Also tabu search gave improved results for several 2D and 3D benchmarks from [14,15].

2   The 2D Triangular Lattice Model

2.1   The Local Move Set in the 2D Case

To define the neighbourhood for our local search algorithms, we introduce a set of moves which allow us to transform a valid conformation into another. The main idea can be described as follows. We choose a vertex from the chain such that there exists a free position in the grid adjacent to both this vertex and either its predecessor or successor in the chain and move it to this free position. This might break the chain, so repairing the chain again also has to be part of a move. This repairing is done via pulling the chain, i.e., the now free old position of the moved vertex will be occupied by its successor (or predecessor), again leaving a free position where the next vertex of the chain is moved, and so on, until a valid conformation is reached. To describe these pull moves more precisely, we need a few formal definitions, where our notation will closely follow [3]. We start with a formal definition of the lattice that will serve as a spatial model for the folding of the protein.

Definition 1. *The* two-dimensional triangular grid lattice *is the infinite graph* $\mathcal{L}_{2D} = (V, E)$ *with vertex set* $V = (\sqrt{3} \cdot \mathbb{Z} \times \mathbb{Z}) \cup ((\sqrt{3} \cdot \mathbb{Z} + \frac{\sqrt{3}}{2}) \times (\mathbb{Z} + \frac{1}{2}))$ *and edge set* $E = \{\{x, x'\} \mid x, x' \in V, |x - x'|_2 = 1\}$, *where* $|\cdot|_2$ *denotes the Euclidean norm.*

As seen in Figure 1, a single vertex in the lattice can have exactly 6 neighbours. Note that, for simplicity, we defined the lattice as an infinite graph although we will only need a finite subgraph of it for each embedding of a protein.

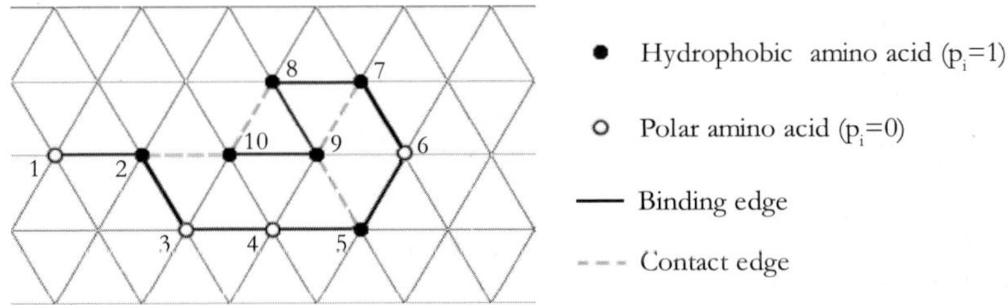

Fig. 1. An example of a conformation for string $p = p_1, \ldots, p_{10} = 0100101111$

We will now define the valid embeddings of a protein into this lattice according to our HP-like model. As usual in the HP model, we describe a protein as a string over the alphabet $\{0, 1\}$, where 0 represents a polar (hydrophilic) amino acid and 1 describes a hydrophobic amino acid. Intuitively, the embedding maps the symbols of the string injectively onto a subset of the vertices of the lattice in such a way that adjacent symbols in the string are also adjacent in the embedding. In other words, the embedding can be described as a self-avoiding walk along the grid edges. This embedding, called a *conformation* of the string, can be defined formally as follows.

Definition 2. *Let $p = p_1 \ldots p_m$ be a string of length m over the alphabet $\{0, 1\}$. A conformation of p in $\mathcal{L}_{2D}$ is an injective function $\phi : \{1, \ldots, m\} \to V$ from the positions of the string to the vertices of the lattice that assigns adjacent positions in p to adjacent vertices in $\mathcal{L}_{2D}$, i.e., $\{\phi(i), \phi(i+1)\} \in E$ for all $1 \le i \le m-1$. These edges $\{\phi(i), \phi(i+1)\} \in E$ for $1 \le i \le m-1$ are called* binding edges.

An edge $\{x, x'\}$ of $\mathcal{L}_{2D}$ is called a contact edge, *if it is not a binding edge and there exist $i, j \in \{1, \ldots, m\}$ such that $\phi(i) = x$, $\phi(j) = x'$, and $p_i = p_j = 1$ (see Figure 1).*

The protein folding problem in the HP model on the two-dimensional triangular grid lattice $\mathcal{L}_{2D}$ is now to find a conformation of a given string in the lattice $\mathcal{L}_{2D}$ with a maximum number of contact edges. Each contact edge contributes -1 to the energy. A special kind of local structure in a conformation are *loops* which consist of two consecutive binding edges forming a $60°$ angle (e.g., in Figure 1 the binding edges between vertices 8, 9 and 10 form a loop). We are now ready to formally define moves on conformations.

Definition 3. *Let $\mathcal{C}(p)$ denote the set of all conformations of a string p. A **move** on $\mathcal{C}(p)$ is a pair $\mu = (\phi, \psi)$ of conformations. We say that μ **transforms** the conformation ϕ into the conformation ψ.*

This notion of a move is very general. For our local search algorithms we want to consider only a special class of moves, so-called *pull moves*, defined as follows.

Definition 4. *Let $p = p_1 \ldots p_n$ be a string over $\{0, 1\}$, let ϕ and ψ be two conformations of p in $\mathcal{L}_{2D}$. We say that ϕ is transformed into ψ by a **forward pull move** if the following conditions hold:*

(a_1) There exists an index $i \in \{1, \ldots, n-1\}$ and a vertex $v \in V(\mathcal{L}_{2D})$ which is empty according to ϕ and adjacent to both $\phi(i)$ and $\phi(i+1)$.

(b_1) $\psi(i) = v$.

(c_1) If $\phi(i-1)$ is also adjacent to v, then $\psi(j) = \phi(j)$ for all $j \neq i$. Otherwise, let k be the maximum index from $\{1, \ldots, i-2\}$ such that edges $\{\phi(k), \phi(k+1)\}$ and $\{\phi(k+1), \phi(k+2)\}$ form a loop. If no such index exists, let $k = 0$. Then $\psi(j) = \phi(j)$ for all $j \in \{1, \ldots, k\} \cup \{i+1, \ldots, n\}$ and $\psi(j) = \phi(j+1)$ for $j \in \{k+1, \ldots, i-1\}$.

*We say that ϕ is transformed into ψ by a **front-end pull move** if the following conditions hold:*

(a_2) There exists a vertex $v \in V(\mathcal{L}_{2D})$ which is empty according to ϕ and adjacent to $\phi(1)$, but not to $\phi(2)$.

(b_2) $\psi(1) = v$.

(c_2) Let k be the minimum index from $\{3, \ldots, n\}$ such that the edges $\{\phi(k-2), \phi(k-1)\}$ and $\{\phi(k-1), \phi(k)\}$ form a loop. If no such index exists, let $k = n+1$. Then $\psi(j) = \phi(j)$ for all $j \in \{k, \ldots, n\}$ and $\psi(j) = \phi(j-1)$ for $j \in \{2, \ldots, k-1\}$.

Backward *and **back-end pull moves** can be defined in a similar fashion. A move is called a **pull move** if it is either a **forward pull move**, a **backward pull move**, a **front-end pull move** or a **back-end pull move**.*

An example of a pull move is shown in Figure 2. In local search, we want to employ a neighbourhood function that is *complete*. In other words, we want to ensure that every possible conformation is reachable from any other by a finite sequence of moves. Thus, our next goal is to show that our set of pull moves is complete. We start with a formal definition of completeness.

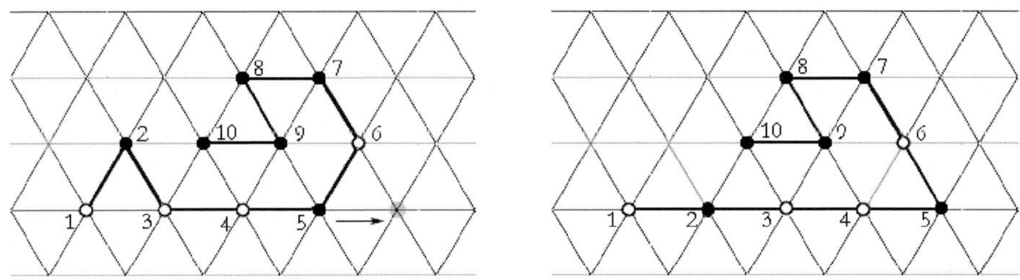

Fig. 2. An example of a forward pull move: Case 2a with $n = 10$, $i = 5$

Definition 5. *Let p be a string over $\{0,1\}$ and let $\mathcal{C}(p)$ be the set of conformations of p on $\mathcal{L}_{2D}$. A set $\mathcal{M}$ of moves on $\mathcal{C}(p)$ is called **complete** if, for every two conformations ϕ and ψ from $\mathcal{C}(p)$, there exists a sequence $\mu_1 = (\phi_1, \psi_1), \ldots, \mu_k = (\phi_k, \psi_k)$ of moves from $\mathcal{M}$ such that $\phi = \phi_1$, $\psi = \psi_k$ and $\psi_j = \phi_{j+1}$ for all $1 \le j \le k - 1$.*

To prove the completeness of the set of pull moves, we will first show reversibility, i.e., for each pull move transforming ϕ into ψ, we will exhibit a sequence of pull moves transforming ψ into ϕ. In a second step, we will show that an arbitrary conformation can be transformed into a straight line in $\mathcal{L}_{2D}$ using a sequence of pull moves. These two results together immediately imply the completeness of the pull move set. Let us first give a formal definition of reversibility.

Definition 6. *Let p be a string over $\{0,1\}$, $\mathcal{C}(p)$ the set of conformations of p on $\mathcal{L}_{2D}$ and $\mathcal{M}$ a set of moves on $\mathcal{C}(p)$. A move $\mu = (\phi, \psi) \in \mathcal{M}$ is called **reversible** with respect to $\mathcal{M}$, if there exists a sequence $\mu_1 = (\phi_1, \psi_1), \ldots, \mu_k = (\phi_k, \psi_k)$ of moves from $\mathcal{M}$ such that $\psi = \phi_1$, $\phi = \psi_k$ and $\psi_j = \phi_{j+1}$ for all $1 \le j \le k - 1$. The move set $\mathcal{M}$ is called **reversible** if every move $\mu \in \mathcal{M}$ is reversible.*

Lemma 1. *The pull move set as described in Definition 4 is reversible.*

Proof. Let $\mathcal{M}_{pull}(\mathcal{C}(p))$ denote the set of pull moves on $\mathcal{C}(p)$. In order to prove reversibility, it is sufficient to show that any single pull move $\mu = (\phi, \psi) \in \mathcal{M}_{pull}(\mathcal{C}(p))$ is reversible. We will proceed by examining all possible cases with respect to the number of vertices being relocated. Let μ be a forward pull move which starts by positioning $\phi(i)$ at vertex $v \in \mathcal{L}_{2D}$ adjacent to both $\phi(i)$ and $\phi(i+1)$, with $i \in \{1,\ldots,n-1\}$. Backward moves are symmetric, with the only difference being that vertices are moved in opposite direction. Possible cases then are **Case 1**, where only one vertex is relocated; **Case 2**, where two or more vertices are relocated (see Figure 2) until the first loop is found (**Case 2a**); or if no such loop exists, μ continues to the end of the chain $\phi(1)$ (**Case 2b**).

Case 1: Move μ relocates only one vertex if $\phi(i-1)$ is also adjacent to vertex v, i.e., μ consists of setting $\psi(i) = v$ only. Therefore, vertex q, previously occupied by $\phi(i)$, becomes free and is adjacent to both $\psi(i-1)$ and $\psi(i)$. This allows us to perform a backward move $\mu' = (\psi, \psi')$ that positions $\psi'(i) = q$ (see Figure 3). Note that vertex q is adjacent to $\psi(i+1) = \phi(i+1)$ and therefore no other vertices are relocated by μ'. Since $q = \phi(i)$ and no other vertices were relocated by μ, it follows that $\psi'(i) = \phi(i)$, where $i \in \{1,\ldots,n\}$ and move μ is reversed.

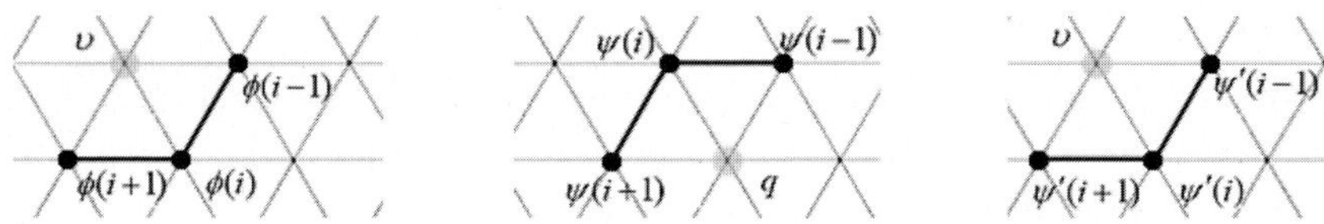

Fig. 3. The case of move μ, relocating vertex $\phi(i)$

Case 2: In **Case 2a**, the relocation of vertices stops because the loop formed by edges $\{\phi(k), \phi(k+1)\}$ and $\{\phi(k+1), \phi(k+2)\}$, with k the maximum index from $\{1,\ldots,i-2\}$, is encountered. We will try to reverse move μ by recreating this particular loop. We will show that the last vertex to be relocated by a reverse move is $\psi(i)$, resulting in deletion of the loop formed by the edges $\{\psi(i-1), \psi(i)\}$ and $\{\psi(i), \psi(i+1)\}$. Note that the move μ will always create this loop. Furthermore, in the new conformation ψ, there is a vertex q, previously occupied by $\phi(k+1)$, but now empty and adjacent to vertices $\psi(k)$ and $\psi(k+1)$. Thus, a backward move $\mu' = (\psi, \psi')$ can take place, starting by relocating $\psi'(k+1) = q$. Let r be the minimum index from $\{k+3,\ldots,n\}$, such that edges $\{\psi(r-2), \psi(r-1)\}$ and $\{\psi(r-1), \psi(r)\}$ form a loop. This gives the following:

$$r \in \{k+3,\ldots,n\}. \tag{1}$$

We should note that such an index exists, because there is a loop between edges $\{\psi(i-1), \psi(i)\}$ and $\{\psi(i), \psi(i+1)\}$ where $i \in \{1,\ldots,n-1\}$. Therefore index r can be at most equal to $i+1$ and we get $r \leq i+1$. By contradiction, we will show that also $r \geq i+1$. Suppose that $r < i+1$. This assumption along with (1), gives

$$r \in \{k+3,\ldots,i+2\}. \tag{2}$$

From (2) though, we get that a loop already existed in conformation ϕ between k and $i+2$, because no other loop was created as a result of move μ, apart from the one formed by edges $(\psi(i-1),\psi(i))$ and $(\psi(i),\psi(i+1))$. We also know that k is the maximum index, where move μ terminates. The existance of a loop between edges $\{\psi(r-2), \psi(r-1)\}$ and $\{\psi(r-1), \psi(r)\}$, though, means that $k \geq r$, otherwise k would not be the maximum index. This is a contradiction, because from (2) we have $k \leq r - 3 \rightarrow k < r$. Thus, $r = i + 1$ or, in other words, the last vertex to be moved by the backward move μ' is $\psi(i)$ and the move can only take us back to conformation ϕ. Finally, in **Case 2b** one can show in a similar way to **Case 2a** that the reverse move is a front-end pull move.   □

Before proceeding to the proof of completeness, we will introduce the notion of boundaries of a conformation ϕ in $\mathcal{C}(p)$. Without loss of generality, we can fix the orientation of our triangular grid $\mathcal{L}_{2D}$ by picking two adjacent grid points and then define a horizontal line as being any line of the grid parallel to the line crossing the two points we chose. Moreover, these horizontal lines are parallel to the x-axis, supposing that the grid is embedded on a cartesian coordinate system. We refer to all other lines of the grid as non-horizontal and the orientated version of $\mathcal{L}_{2D}$ is denoted with $\mathcal{L}'_{2D}$.

Definition 7. *We define the left boundary of ϕ as the set of vertices $L = \{r \times (\sqrt{3} \cdot \mathbb{Z})\} \cup \{(r + \frac{1}{2}) \times (\sqrt{3}\mathbb{Z} + \frac{\sqrt{3}}{2})\}$, where $L \subset V(\mathcal{L}'_{2D})$ and r is the minimum integer such that at least one edge of ϕ belongs to L.*

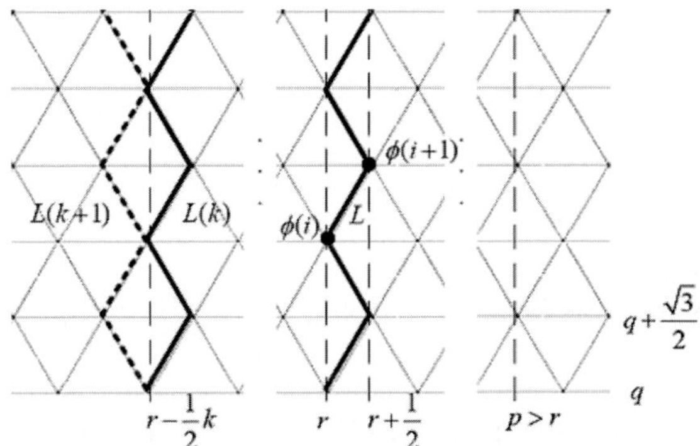

Fig. 4. The left boundary L and the k times shift to the left

The boundary defined above can be shifted k times to the left, as a result of a series of *pull moves*. We denote such a displacement as the new boundary $L(k) = \{(r - \frac{1}{2}k) \times (\sqrt{3} \cdot \mathbb{Z} - \frac{\sqrt{3}}{2}(-1)^k)\} \cup \{(r + \frac{1}{2} - \frac{1}{2}k) \times (\sqrt{3}\mathbb{Z} + \frac{\sqrt{3}}{2} + \frac{\sqrt{3}}{2}(-1)^k)\}$, where $k = 1, 2, \ldots$ (see Figure 4). The right boundary R and its shift $R(k)$ to the right can be defined symmetrically. Also we make use of the subscript t to denote a boundary at time step t, e.g., $L_t(k)$.

Theorem 1. *The pull move set as described in Definition 4 is complete.*

Proof. In order to show completeness, it is sufficient to show that there is a sequence of moves $\mu_1 = (\phi_1, \psi_1), \ldots, \mu_k = (\phi_k, \psi_k)$ from $\mathcal{M}_{pull}(\mathcal{C}(p))$, such that $\phi = \phi_1$, $\lambda = \psi_k$ and $\psi_j = \phi_{j+1}$ for all $1 \leq j \leq k - 1$, with ϕ being an arbitrary conformation and λ being a horizontal line conformation. Since the pull move set is reversible according to Theorem 1, the sequence of moves transforming ϕ to a horizontal line conformation can be reversed and another arbitrary conformation $\psi \neq \phi$ can be reached. In other words, there is a sequence of moves $\mu_1 = (\phi_1, \psi_1), \ldots, \mu_r = (\phi_r, \psi_r)$ from $\mathcal{M}_{pull}(\mathcal{C}(p))$, such that $\phi = \phi_1$, $\psi = \psi_r$ and $\psi_j = \phi_{j+1}$ for all $1 \leq j \leq r - 1$. According to Definition 5 this implies that $\mathcal{M}_{pull}(\mathcal{C}(p))$ is complete. Let us now show that λ can be reached from any conformation ϕ.

If the conformation ϕ is already horizontal, i.e., if none of its binding edges lies on a non-horizontal line of the lattice, then $\phi = \lambda$ and we are done. Otherwise, at least one of the binding edges in ϕ lies on the boundary. Let us consider the region between L and R, including the boundaries themselves, we call this region the *core* of the conformation. We will now distinguish between **Case a**, when either of the ends, namely vertices $\phi(1)$ or $\phi(n)$, lie outside the core and **Case b1** and **b2**, when both of the ends are located inside the *core*.

Case a: Any part of ϕ lying outside of the *core* must be connected via a horizontal chain with some vertex lying on the boundaries. Assume that there is a part of ϕ lying on the left of L, containing vertices $\{\phi(1), \ldots, \phi(j)\}$, with $\phi(j) \in L$ and $j \in \{2, \ldots, n - 1\}$. Since $\phi(1)$ lies outside the left boundary, there exists a vertex $v \in V(\mathcal{L}'_{2D})$, which is empty according to ϕ and adjacent to $\phi(1)$, but not to $\phi(2)$. Thus, a front-end pull move can be performed during time step t. In case edge $(\phi(j), \phi(j+1))$ happens to form a loop with edge $(\phi(j-1), \phi(j))$, the pull move erases this loop and shifts the boundary to the left. We get $L_t(1)$, but vertex $\phi(j)$ still lies on the boundary. Now an additional front-end pull move, can only bring vertex $\phi(j)$ outside the boundary, such that the horizontal part of the chain contains vertices $\{\phi(1), \ldots, \phi(j+1)\}$, with $\phi(j+1) \in L_{t+1}(2)$. In general, a sequence of either single front-end pull moves or pairs of front-end pull moves is needed to reach the horizontal line conformation.

Case b1: Let us consider the subcase when one of the end vertices lies exactly on one of the boundaries. Without loss of generality, we choose vertex $\phi(1)$ to lie on L, since the cases of the other boundary R or the other end $\phi(n)$ are symmetric. Then, there exists a vertex $v \in V(\mathcal{L}'_{2D})$, which is empty according to ϕ and adjacent to $\phi(1)$, but not to $\phi(2)$. This allows a front-end pull move to be executed during time step t. The result of this move is either to bring $\phi(1)$ outside boundary L, or to shift the boundary to the left. In the latter case, L becomes $L_t(1)$ and an additional front-end pull move can bring $\phi(1)$ outside $L_{t+1}(1)$. As already shown in Case a there is a sequence of moves to reach a horizontal line conformation, if either of the ends lies outside the core.

Case b2: Now we will consider the case of vertex $\phi(1)$ not lying on any of the boundaries, but rather being strictly inside the core. Initially, L contains the leftmost edge of ϕ. If more than one edges lie on L, let us pick edge $(\phi(i), \phi(i+1))$, where i is the minimum index from $\{2, \ldots, n - 1\}$. Then, there exists a vertex $v \in V(\mathcal{L}'_{2D})$, which is empty according to ϕ and adjacent to both $\phi(i)$ and $\phi(i+1)$.

This allows a forward pull move to be executed during time step t. Such a move always results in shifting the left boundary, i.e. L becomes $L_t(1)$. Depending on the initial orientation of edge $(\phi(i), \phi(i+1))$ on L, this edge might still lie on the new boundary $L_t(1)$. In that case, an additional forward pull move at time step $t+1$, is required to bring edge $(\phi(i-1), \phi(i))$ on $L_{t+1}(2)$, i.e. the boundary is shifted as well. Otherwise, one forward pull move at time step t is enough to bring edge $(\phi(i-1), \phi(i))$ directly on $L_t(1)$. In general, one forward pull move or a pair of forward pull moves always decreases by one the index of the leftmost vertex with minimum index. On the same time the leftmost boundary is shifted to the left, which ensures free space for future moves. Eventually, a sequence of such moves will bring $\phi(1)$ on the left boundary. For this case we have already described a sequence of moves to reach a horizontal line conformation. $\qquad\square$

3   The 3D Triangular Lattice Model

One of our aims is the extention of *pull moves* to the FCC lattice, using the isomorphism property between the 3D triangular grid and the FCC lattice. The 3D triangular grid can be described as a stack of 2D triangular grids, where every individual 2D grid is slightly offset with respect to the grids above and below it. The set $\mathcal{N}_{3D}$ of minimal vectors connecting its neighbors is given by

$$\mathcal{N}_{3D} = \{(0, \pm 1, 0), (\pm\tfrac{\sqrt{3}}{2}, \pm\tfrac{1}{2}, 0), (\tfrac{1}{\sqrt{3}}, 0, \sqrt{\tfrac{2}{3}}), (-\tfrac{1}{\sqrt{3}}, 0, -\sqrt{\tfrac{2}{3}}), (-\tfrac{1}{2\sqrt{3}}, \pm\tfrac{1}{2}, \sqrt{\tfrac{2}{3}}),$$
$$(\tfrac{1}{2\sqrt{3}}, \pm\tfrac{1}{2}, -\sqrt{\tfrac{2}{3}})\}.$$

A simple rotation of the $\mathcal{N}_{3D}$ along the y-axis by $\cos^{-1}(\tfrac{1}{\sqrt{3}})$ will give us precisely traditional neighborhood vectors in FCC lattice, namely $\mathcal{N}_{3D} = \{(0, \pm 1, 0), (\pm 1, 0, 0), (\pm\tfrac{1}{2}, \pm\tfrac{1}{2}, \pm\tfrac{1}{\sqrt{2}})\}$, thus establishing the isomorphism between the two models. The notion of a loop remains the same, since any two consecutive binding edges having end points as neighbours create a $60°$ angle. Hence the *pull moves* can be extended to the FCC without changing the definition, so as the reversibility and completeness proofs.

4   Experiments with Tabu Search

We implemented the Tabu Search (TS) strategy for evaluating pull moves in triangular lattice HP models. For this purpose, we used the Bioinformatics Utilities (BIU) (http://www.bioinf.uni-freiburg.de/Software/) library developed by the bioinformatics group of Albert-Ludwigs-University Freiburg, Germany. We extended the library by implementing the 2D triangular lattice and the pull move set for triangular lattice models.

Tabu search, originally proposed by Glover [7], is a form of greedy local search that begins by marching to a local minimum from an arbitrary solution state. To avoid retracing the steps used, the method records recent moves in the form of a tabu list. Normally, the TS strategy chooses a new better solution if one exists and is not tabu, but if such a choice is impossible, then a worse solution can be chosen. If a move is on the tabu list, but leads to a new optimal solution, then such a move can be done assuming that an additional condition, the so-called *aspiration level*, is met. The calculations start with the generation of a starting solution, using some problem-specific knowledge. In the following section, the key elements of tabu search are defined with respect to pull moves in triangular lattice models.

4.1   Tabu Search for Triangular Lattice HP-Models

For a given amino-acid sequence of arbitrary length n, a conformation is represented as a sequence of directions of length $n - 1$ in the triangular lattices. Therefore in our implementation, a conformation is also a string consisting of elements from the alphabet of 6 directions in the 2D case or of 12 directions in the FCC case. Note that, due to the self-avoidingness property of the embedding, not every sequence of directions describes a proper conformation.

The neighbourhood $\mathcal{N}(\phi)$ of an arbitrary conformation ϕ is the set of conformations obtained by applying all the valid pull moves to ϕ. Given the length n of the sequence, there will be at most $4(n - 1) + 6$ and $8(n - 1) + 14$ possible pull moves for the 2D triangular and the FCC lattice, respectively, but not all of them can be applied. Normally, the more compact the conformation is, the less is the number of applicable pull moves, and thus the smaller is the size of the neighbourhood.

The starting state is generated from a state with straight line conformation in the lattice, by applying a given number of moves consecutively, selected from the set of *pull moves* uniformly randomly, irrespective of the quality of the move. The neighbourhood $\mathcal{N}(\phi)$ of a given conformation ϕ can consist of many solutions with equally good energies. In such a case, we choose one of them uniformly randomly. The tabu list is a short time memory, implemented as a circular queue containing a limited number of forbidden moves. If the list is full, then adding a new move removes the oldest one. The searching procedure can make a transition from conformation ϕ to ψ by applying a forbidden move, if the energy value E_ψ is lower than E_{min} of the current best conformation, or E_ψ is lower than E_ϕ and all other conformations in the neighbourhood $\mathcal{N}(\phi)$ have equal or higher energy value than ϕ, given that ψ lies in the first half of the tabu list (aspiration level). Each execution of the search procedure is called an *iteration*. The algorithm terminates if an expected optimum value of the energy E_{opt} has been achieved, or a given number l_{it} of iterations, has been reached.

4.2   Experimental Results

The goal of our experiments was to establish the effectiveness of pull move set in finding optimal solutions in the 2D triangular lattice HP model, irrespective of the time, embedded in Tabu search algorithm. The preliminary results of these experiments look very promising. Our first set of experiments deals with 2D triangular benchmark sequences with known optimal energy [11]. The number of iterations was fixed to 100 000. The size of the tabu list was fixed to half of

Table 1. Test sequence-1 for 2D triangular lattice HP model

Sequence	Length	E_{opt}	E_{TS}
11001001010100101011	20	-17	-17
11010101010010010011	20	-17	-17
110010010100101001011	21	-17	-17
110100100101010010011	21	-17	-17
1100100101010010010011	22	-17	-17
11101010101010101010111	23	-25	-25
110010010010010010010011	24	-17	-17
111010100101010101010111	24	-25	-25
111001001001001010010100100111	30	-25	-25
111001001001010010100100100111	30	-25	-25
1110010010010101001001001000001010111	37	-29	-29
11110001010010101010101010100111111111111111111000101 01010010101111111110010101010001011111111111000111111	102	-122	-116

E_{opt}: Optimum energy in the literature.
E_{TS}: Minimum energy found by Tabu Search.

Table 2. Test sequence-2 for triangular lattice HP models

Sequence	Length	2D FCC		3D FCC	
		E_{HGA+TR}	E_{TS}	E_{HGA+TR}	E_{TS}
10100110100101100101	20	-15	-15	-29	-23
110010010010010010010011	24	-13	-17	-28	-23
0010011000011000011000011	25	-10	-12	-25	-17
000110011000001111 111001100001100100	36	-19	-24	-51	-38
001011101110000011111111 110000001100110010011111	48	-32	-40	-69	-74
110101010111101000100010001 000010001000101111010101011	54	-23	-31	-59	-77
001110111111110001111111111010 001111111111110000111111011010	60	-46	-70	-117	-130
111111111110101001100110010 0110 011001001100110010101111111111111	64	-46	-50	-103	-132

E_{HGA+TR}: Minimum energy found by Hybrid Genetic Algorithm with Twin Removal.
E_{TS}: Minimum energy found by Tabu Search.

the maximum number of pull moves for a given sequence. As the parameters of the Tabu Search were not fine tuned, we avoid mentioning the number of iterations required. The summary is presented in Table 1. As shown there, we got the optimum result for most of the sequences while only managed to find a local minimum for the longest sequence before the program terminates.

The reason for this is simple: As the number of conformations grows exponentially with the length of the sequence, it becomes harder for TS to escape the conformational space around local minima. Therefore we need to fine tune few parameters as well as allowing more time to converge to optimum value for all sequences. Incorporation of an effective diversification procedure can alleviate the problem by exploring new region earlier. Performance can be further improved by dynamically changing the tabu list size and aspiration level.

Our second set of experiments deals with benchmark sequences, for which optimal energy is not yet known. Previous work has been done [14,15] in 2D and 3D FCC lattice. We compared our results with this prior work and present a summary in Table 2. We had exceptionally good results for all 2D cases and longer sequences in 3D cases. Again, it is expected that fine tuning the parameters will help us to escape local minima and improve the results.

5   Conclusions

The main contribution of our paper can be summarized as the formal introduction of *pull moves* for local search in triangular grid lattices. We should also stress the importance of finding new ground states for several benchmarks in the 2D and 3D case. This clearly demonstrates the efficiency of *pull moves* as a neighborhood function for local search algorithms and sets the ground for further study of heuristic algorithms for protein structure prediction. We strongly believe that fine tuning the parameters of the tabu search can yield faster and more accurate results. We leave this as a subject of future research along with testing more sophisticated heuristics. Future research will also focus on landscape analysis of protein folding in triangular lattices.

References

1. Agarwala, R., Batzoglou, S., Dancik, V., et al.: Local rules for protein folding on a triangular lattice and generalized hydrophobicity in the HP model. In: Proc. SODA 1997, pp. 390–399 (1997)
2. Anfinsen, C.B.: Principles that govern the folding of protein chains. Science 181, 223–230 (1973)
3. Böckenhauer, H.-J., Bongartz, D.: Protein folding in the HP model on grid lattices with diagonals. Discrete Applied Mathematics 155, 230–256 (2007)
4. Bagci, Z., Jernigan, R.L., Bahar, I.: Residue coordination in proteins conforms to the closest packing of spheres. Polymer 43, 451–459 (2002)
5. Crescenzi, P., Goldman, D., Papadimitriou, C., et al.: On the complexity of protein folding. Journal of Computational Biology 5, 423–465 (1998)
6. Dill, K.A., Bromberg, S., Yue, K., et al.: Principles of protein folding - A perspective from simple exact models. Protein Sci. 4, 561–602 (1995)
7. Glover, F.: Tabu search. ORSA J. Comput. 1, 190–206 (1989)
8. Govindarajan, S., Goldstein, R.A.: On the thermodynamic hypothesis of protein folding. Proc. Natl. Acad. Sci. USA 95, 5545–5549 (1998)
9. Hart, W.E., Istrail, S.: Fast protein folding in the hydrophobic-hydrophilic model within three-eights of optimal. Journal of Computational Biology 3, 53–96 (1996)
10. Hart, W.E., Istrail, S.: Robust proofs of NP-hardness for protein folding: General lattices and energy potentials. Journal of Computational Biology 4, 1–22 (1997)
11. Krasnogor, N.: Studies on the Theory and Design Space of Memetic Algorithms. Doctoral Dissertation, University of the West of England, UK (2002)
12. Levinthal, C.: Are there pathways for protein folding? J. de Chimie Physique et de Physico-Chimie Biologique 65, 44–45 (1968)
13. Lesh, N., Mitzenmacher, M., Whitesides, S.: A complete and effective move set for simplified protein folding. In: Proc. RECOMB 2003, pp. 188–195 (2003)
14. Hoque, M.T., Chetty, M., Dooley, L.S.: A Hybrid Genetic Algorithm for 2D FCC Hydrophobic-Hydrophilic Lattice Model to Predict Protein Folding. In: Sattar, A., Kang, B.-h. (eds.) AI 2006. LNCS (LNAI), vol. 4304, pp. 867–876. Springer, Heidelberg (2006)

15. Hoque, M.T., Chetty, M., Sattar, A.: Protein Folding Prediction in 3D FCC HP Lattice using Genetic Algorithm. In: Proc. IEEE Congress on Evolutionary Computation 2007, pp. 4138–4145 (2007)
16. Sloane, N.J.A.: Kepler's conjecture confirmed. Nature 395, 435–436 (1998)
17. Steinhöfel, K., Skaliotis, A., Albrecht., A.A.: Stochastic protein folding simulation in the d-Dimensional HP-Model. In: Hochreiter, S., Wagner, R. (eds.) BIRD 2007. LNCS (LNBI), vol. 4414, pp. 381–394. Springer, Heidelberg (2007)
18. Unger, R., Moult, J.: Genetic algorithms for protein folding simulations. J. Mol. Biol. 231, 75–81 (1993)
19. Zhang, Y., Skolnick, J.: Parallel-Hat Tempering: A Monte Carlo Search Scheme for the Identification of Low-Energy Structures. Journal of Chemical Physics 115, 5027–5032 (2000)

Protein Decoy Generation Using Branch and Bound with Efficient Bounding

Martin Paluszewski and Pawel Winter

Department of Computer Science, University of Copenhagen,
Universitetsparken 1, 2100 Copenhagen, Denmark
palu@diku.dk, pawel@diku.dk

Abstract. We propose a new discrete protein structure model (using a modified face-centered cubic lattice). A novel branch and bound algorithm for finding global minimum structures in this model is suggested. The objective energy function is very simple as it depends on the predicted half-sphere exposure numbers of C_α-atoms. Bounding and branching also exploit predicted secondary structures and expected radius of gyration. The algorithm is fast and is able to generate the decoy set in less than 48 hours on all proteins tested.

Despite the simplicity of the model and the energy function, many of the lowest energy structures, using exact measures, are near the native structures (in terms of RMSD). As expected, when using predicted measures, the fraction of good decoys decreases, but in all cases tested, we obtained structures within 6 Å RMSD in a set of low-energy decoys. To the best of our knowledge, this is the first *de novo* branch and bound algorithm for protein decoy generation that only depends on such one-dimensional predictable measures. Another important advantage of the branch and bound approach is that the algorithm searches through the entire conformational space. Contrary to search heuristics, like Monte Carlo simulation or tabu search, the problem of escaping local minima is indirectly solved by the branch and bound algorithm when good lower bounds can be obtained.

1 Background

The contact number (CN) is a very simple solvent exposure measure that only depends on the positions of C_α-atoms. Given a fixed backbone structure, the CN of a residue A_i is the number of other C_α-atoms in a sphere of radius r centered at the C_α-atom of A_i. The CN of all residues of a given structure is called the *CN-vector*. A more information rich measure is called the *half-sphere-exposure* (HSE) measure [5]. Here, the sphere is divided into an upper and a lower hemisphere as illustrated in Figure 1. The up and down numbers of a residue therefore refer to the number of other C_α-atoms in the upper and lower hemispheres respectively. For a given fixed structure, the up and down numbers for all residues is called the *HSE-vector*. CN- and HSE-vectors therefore only depend on the radius of the spheres and the coordinates of C_α-atoms, which is very convenient when working with simplified models.

K.A. Crandall and J. Lagergren (Eds.): WABI 2008, LNBI 5251, pp. 382–393, 2008.
© Springer-Verlag Berlin Heidelberg 2008

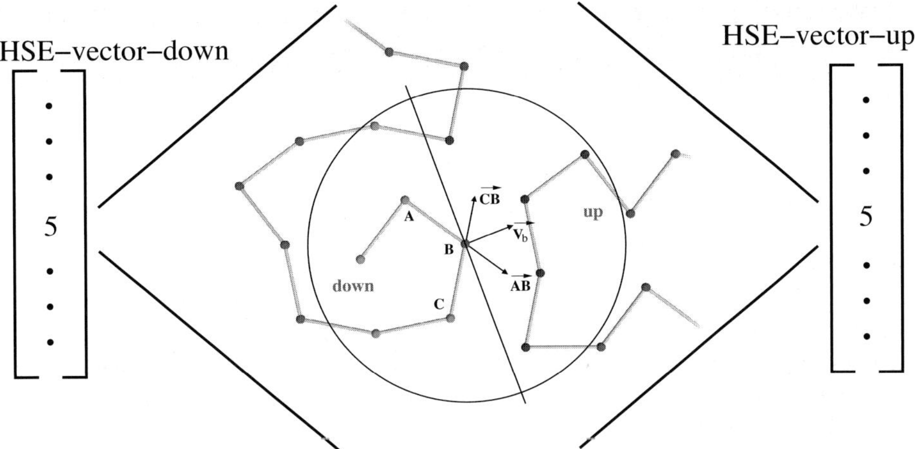

Fig. 1. Given the positions of 3 consecutive C_α-atoms (A, B, C), the approximate side-chain direction $\bar{V}_b$ can be computed as the sum of $\bar{AB}$ and $\bar{CB}$. The plane perpendicular to $\bar{V}_b$ cuts the sphere centered at B in an upper and a lower hemisphere.

Recently it was shown that it is possible to approximately reconstruct small protein structures from CN-vectors or HSE-vectors only [12]. These results showed that HSE-optimized structures tend to have better coordinate RMSD with the native structure and more accurate orientations of the side-chains compared to CN-optimized structures. This is very interesting in regards to *de novo* protein decoy generation, because CN- and HSE-vectors can be predicted with reasonable accuracy [19,14]. To use these results for *de novo* decoy generation, one could therefore first predict the HSE-vector from the amino acid sequence and then reconstruct the protein backbone from this vector. However, the results in [12] were only based on small proteins with up to 35 amino acids and it was conjectured that the reconstruction of larger proteins would require more information than what is contained in an HSE-vector [12]. Another difficulty is that HSE-based energy functions appear to have many local minima in the conformational space. This is often a problem for search heuristics like Monte Carlo simulation or tabu search, since they get trapped in these minima and must spend much time escaping them.

The problem of reconstructing protein structure from vectors of one-dimensional structural information has also been studied by Kinjo et al. [7]. They used exact vectors of secondary structure, CN and *residuewise contact order* (RWCO) together with refinement using the AMBER force field to reconstruct native-like structures. Their results indicated that secondary structure information and CN *without* the use of RWCO is *not* enough to reconstruct native-like structures. Unfortunately, RWCO is difficult to predict compared to CN, HSE and secondary structure [7] and it would therefore be difficult to use their method directly for *de novo* decoy generation.

Here we attack these problems by adding more predicted information to our model and use a thorough branch and bound algorithm for finding minimum energy structures. By adding more predicted information we expect to increase the probability of the energy function to have global minimum near the native structure. Furthermore, using a branch and bound approach we are able to implicitly search the whole conformational space and therefore avoid getting trapped in local minima. Besides using HSE-vectors, we also use *secondary structure* (SS) and *radius of gyration* (Rg). These three measures, (HSE, SS and Rg), can all be predicted from the amino acid sequence only [19,10,16], and can therefore be used for *de novo* protein decoy generation. The energy function is simple, and we show how a good lower bound of the energy for a subset of the conformational space can be computed in polynomial time. This lower bound enables the branch and bound algorithm to efficiently bound large conformational subspaces and to find global minimum energy structures in a reasonable amount of time. Throughout the text our branch and bound algorithm is referred to as EBBA (Efficient Branch and Bound Algorithm).

The idea of using secondary structure elements in a discrete model has been suggested by others, i.e., Fain et al. [4] and Levitt et al. [8]. However, their models have a relatively small conformational space and it is therefore possible to completely enumerate all structures allowed by the model. Branch and bound algorithms and other algorithms for determining global minimum structures have been used for protein structure prediction earlier. Some of these algorithms work on very simplified models like the HP-lattice model [1]. Even though these algorithms can solve most problems to optimality, the global minimum structures are often very far from the native structure. Another branch and bound algorithms, called αBB[9] uses more detailed potential energy functions which depend on several physical terms. In [9], the αBB is shown to be successful on small molecules. In [17], the αBB was improved and was used for prediction of real protein structures. Dal Palu et al.[11] use a constraint logic programming approach for protein structure prediction. They also use secondary structure segments in a simplified model. However, in their model, all C_α-atoms must be placed in a lattice (FCC). This differs from our approach, where we only demand lattice directions of the secondary structure segments. Dal Palu et al. use a standard solver (SICStus Prolog) which makes use of standard bounding techniques, while we have developed a much more efficient bounding algorithm specialized for this particular problem. Furthermore, the results published in [11,17] are not true *de novo* - the secondary structures are all derived from the native structure of the proteins. On the contrary, the results presented here are true *de novo*. All parts of the energy function are predicted from amino acid sequences only. EBBA is, to our knowledge, the first *de novo* branch and bound algorithm that only use one-dimensional predictable measures.

We use 6 benchmark proteins for evaluating EBBA. These benchmark proteins are chosen because they are used in similar studies before [15,6] and we are therefore able to compare our method with the state-of-the-art protein conformational sampler FB5-HMM [6]. Our results show that EBBA is able to find global minimum energy structures for most of these proteins in less than 48 hours. We have evaluated EBBA using both exact values and predicted values to estimate the importance of prediction quality. The results show that predicted structures having global minimum energy are *not always* native-like, however among the 10.000 lowest energy structures we typically find many good decoys (less than 6 Å RMSD).

2   Methods

A sequence of residues of the same secondary structure class is called a *segment*. Segments can be considered as rigid rods that describe the overall path of C_α-atoms belonging to the segment. Segments have a start coordinate and a direction, and for helices and sheets their end coordinate can also be determined because of their constrained geometry. A segment is therefore an abstract representation of a sequence of residues and it does not explicitly contain the coordinates of internal C_α-atoms. We define a *segment structure* to be the coordinates of all C_α-atoms of a segment. Note that a segment in principle allows for infinitely many different segment structures even though they are restricted to be of a specific secondary structure class. However, this model is discrete and therefore only a finite representative set of segment structures are generated. This is described in more detail in Section 2.1.

Any tertiary structure of a protein can be described in these terms; a list of segments and a segment structure for each segment. We call such a list of segments a *super structure* and a super structure with a fixed segment structure for each segment is called a *complete structure*.

To discretize and reduce the conformational space of this model, we reduce the degree of freedom for segments. Segments are therefore only allowed to have a discrete set of predefined directions between the first and last C_α-atoms. Ad-hoc experiments show that the 12 uniformly distributed directions acquired from the *face-centered cubic* (FCC) lattice is a good tradeoff between discretization and flexibility. The direction of a segment therefore has one of the following 12 direction vectors

```
[1,1,0], [1,0,1], [1,-1,0], [1,0,-1], [-1,1,0], [-1,0,1],
[-1,-1,0], [-1,0,-1], [0,1,1], [0,1,-1], [0,-1,1], [0,-1,-1]
```

To further discretize the model, we set an upper limit (u) on the number of possible segment structures allowed by a segment. Given an amino acid sequence with m segments and u possible segment structures for each segment, the total number of complete structures, N (disregarding symmetric structures), allowed by this model is

$$N = 4 \times 11^{m-2} \times u^m \tag{1}$$

2.1   Segment Structures

Here we briefly describe how the allowed segment structures of a given segment are computed. This computation depends on the secondary structure class of the segment.

Given a helix or sheet segment, we generate one segment structure having the angle properties of a right-handed helix or a beta strand. Then the other $u - 1$ segment structures are generated by rotating the first structure uniformly around the axis going through the first C_α-atom and ending at the beginning of the next segment.

There are no simple geometric constraints that describe coil structures. Experiments show that short sequences with similar amino acid sequences, so-called homologous sequences, often have similar tertiary structures [3]. Given a coil segment, we therefore query PDB Select (25) with protein sequences and their known structures and find the $\sqrt{u}$ best fragment matches in terms of amino acid similarity. Each of these structures is also rotated uniformly $\sqrt{u}$ times as for helices and sheets such that a total of u structures are obtained. The fragment database does of course not contain the proteins used in the experiments. Even though we are querying PDB Select (25) for coil fragments, we still consider our algorithm to be *de novo*, because we do not explicitly make use of templates. One of the most successful structure prediction algorithms (Rosetta[15]) also makes use of fragments from proteins in PDB and is also considered to be *de novo*.

2.2   Energy

The structures allowed by the model always have the desired secondary structure (from a prediction), however the HSE-vector and radius of gyration of the structures varies. Therefore, we want to identify those structures having correct radius of gyration and HSE-vectors similar to the predicted HSE-vectors. The radius of gyration can be predicted from the number of residues n of the protein [16]:

$$R_g = 2.2n^{0.38} \tag{2}$$

This prediction is often accurate for globular proteins. We therefore assign infinite energy to structures having radius of gyration more than 5% away from the predicted R_g. We assign infinite energy to structures if their subchain of amino acids from the first amino acid to the l'th ($l < n$) amino acid is more than 5% away from the predicted R_g. A structure is said to be *clashing* if the distance between two C_α-atoms is less than 3.5 Å. We also assign infinite energy to clashing structures and conformations where two succeeding segment structures have unlikely angle properties.

Let $\mathcal{P}$ denote the conformational space of a protein with n residues $A_1, A_2, ...,$ A_n. Let $P \in \mathcal{P}$. The total energy $Q(P)$ of P is defined as the sum of the residue energy contributions $Q_P(A_i)$, i.e.,

$$Q(P) = \sum_{i=1}^{n} Q_P(A_i) \tag{3}$$

with

$$Q_P(A_i) = \begin{cases} \Delta CN(A_i)^2 & \text{if } A_i \text{ is the first residue of a segment} \\ \Delta HD(A_i)^2 + \Delta HU(A_i)^2 & \text{otherwise} \end{cases}$$

$$(4)$$

where

- $\Delta CN(A_i)$ is the difference between the contact number of the i-th residue A_i in P and the desired (i.e., predicted) contact number of A_i.
- $\Delta HD(A_i)$ is the difference between the down half sphere exposure number of A_i in P and the desired down half sphere exposure number of A_i.
- $\Delta HU(A_i)$ is the similar difference for the up half sphere exposure.

The reason why CN instead of HSE is used for the first residue of a segment is that the HSE value depends on the position of the two neighbour residues as illustrated in Figure 1. On the other hand, HSE can be used for the last residue of a segment, because one of the neighbours are an interior residue and the other neighbour is the end position of the segment whose coordinates are always known. The radius of the contact sphere is set to 15 Å.

2.3   Branch and Bound

An explicit evaluation of all allowed structures is only feasible for proteins with very few segments and segment structures. A standard approach for overcoming such combinatorial explosion is to use the branch and bound technique [20].

Branching. The root of the branch and bound tree represents all complete structures allowed by the model. This is done by only fixing the direction of the first segment. Every other node s represents a smaller subset of complete structures P_s than its parent. This is done by either fixing a segment direction or by fixing a segment structure. Therefore, when branching on a node, either 11 children with fixed segment directions are created or u children with fixed segment structures are created. A node at level $2 \times m$ has all segment directions and segment structures fixed and therefore represents a complete structure. Nodes at level $2 \times m$ cannot be branched on further and are called leaves.

Bounding. A lower bound is a value that is less than, or equal to the lowest energy of any leaf in the subtree of the node. Such a value can be used to disregard, or *bound*, the subtree of a node if the lower bound is larger than some observed energy (*an upper bound*). An upper bound of the energy can be found using some advanced heuristic or a simple depth first search as described in section 2.4. Here we present a reasonable tight lower bound that can be computed fast. The use of this lower bound makes it possible to solve large problems to optimality as shown in the results section.

Let P_S denote the subset of the conformational space P at any node of the branch and bound tree where some segments might have fixed directions while others might have fixed segment structures (i.e., fixed coordinates of all C_α-atoms) as explained in the description of the branching strategy above. We are looking for a lower bound for $\min_{P \in \mathcal{P}_S} \{Q(P)\}$.

Consider the j-th segment S_j, $1 \leq j \leq m$, where m is the number of segments. Let

$$Q_P(S_j) = \sum_{A_i \in S_j} Q_P(A_i)$$

where $Q_P(A_i)$ is defined in Equation 4. Then the energy of a structure can be written as the sum of segment energies

$$Q(P) = \sum_{1 \leq j \leq m} Q_P(S_j)$$

Suppose that a lower bound for $\min_{P \in \mathcal{P}_S} \{Q_P(S_j)\}$ can be determined. Summing up these lower bounds for all m segments will therefore yield a lower bound for the energy of all conformations in $\mathcal{P}_S$. To compute such a lower bound for a segment S_j, the following problem is solved for all segment structures of S_j. For simplicity we only describe how a lower bound using CN-vectors can be computed, however it is straightforward to use a similar approach for HSE-vectors.

Given a segment structure for S_j, we determine for each of its C_α-atoms all possible values of CN when the super structure is fixed. This problem can clearly be solved in exponential time by complete enumeration of all possible segment structures. However, using the following dynamic programming approach, the problem can be solved much faster in polynomial time.

Let $c_{a,b}(i, r)$ where $(1 \leq i \leq m)$ and $(1 \leq r \leq u)$ be the number of contacts of residue a in segment b contributed by residues in segment i having segment structure r. Let (i, j) be an entry in the dynamic programming table and let $q_{a,b}(i, j) \in \{0, 1\}$ represent whether or not residue a in segment b can have a total of j, $(0 \leq j < n)$, contacts contributed by residues in segments S_l, $(l < i)$. Then the recursive equation of the dynamic programming algorithm is:

$$q_{a,b}(i,j) = \begin{cases} 1 \text{ if } i = 1 \text{ and } c_{a,b}(1, r) = j \text{ for some } r \\ 1 \text{ if } i > 1 \text{ and } q(i - 1, k) = 1 \text{ and } c_{a,b}(i, r) = j - k \text{ for some } r \\ 0 \text{ otherwise} \end{cases} \quad (5)$$

Each row can be computed in $\mathcal{O}(n \times u)$ time using the values from the previous row, so the total running time of the algorithm is $\mathcal{O}(m \times n \times u)$. The last row in the table represents all possible contact numbers for residue a in segment b. The last row can therefore easily be used to find the minimum difference between the desired CN and one of the possible CNs. The dynamic programming problem is solved for all residues of the segment and the sum of the minimum differences for each residue is the lower bound of the segment energy. For more details and examples of computing lower bounds, refer to [13].

In the above discussion, it was assumed that all C_α-atoms in S_j have their coordinates fixed in $\mathcal{P}_S$. Lower bounds can also be computed if the segment structure has not been fixed yet. The above lower bound computation is then merely repeated for each of the u possible segment structures, and the smallest one is selected as the overall lower bound of the segment.

Lower bounds can also be computed for nodes where a number of the last segment directions have not yet been fixed. Here, the input to the dynamic programming algorithm is only the first fixed segments. Then, the CN row for the last fixed segment is augmented by checking whether each C_α-atom on the free segments can possibly be in contact with the C_α-atom in question.

2.4   Searching

We search the branch and bound tree by keeping a set of nodes for which the lower bound has been computed but not bounded. Initially, the set contains only the root of the branch and bound tree. Iteratively the algorithm chooses the lowest cost node and replaces it with the children obtained by branching. When using this strategy, an optimal solution is found when the lowest cost node in the set is a leaf node. In practice the set of unbranched nodes might become very large and difficult to store in memory. We therefore combine it with a depth first search, such that when the node set contains more than 50.000 nodes we shift to depth first search until the set is less than 50.000 again.

3   Experiments

Here we predict the tertiary structures of 6 proteins. The tertiary structures of these proteins are known and we can therefore evaluate the quality of our results. These proteins have previously been used for benchmarks in the literature [15,6] and our results can therefore be directly compared with the state-of-the-art conformational sampler FB5-HMM[6].

The input to EBBA is a secondary structure assignment, HSE-vector and the radius of gyration. For each protein we obtain these values using prediction tools. Based on the amino acid sequence, we predict the secondary structure using PSIPRED [10] and we predict HSE-vectors using LAKI [19]. Note that PSIPRED and LAKI are neural networks trained on a selection of proteins from PDB. The 6 benchmark proteins used here also exist in PDB, so there is a slight chance that the training sets for PSIPRED and LAKI contain some of these proteins. However, the prediction quality of the 6 benchmark proteins is close to what should be expected from PSIPRED and LAKI. Here, the average Q_3 score of secondary structure prediction is 80.7% (compared to an average score of 80.6% on CASP targets). The average correlation of the HSE up and down values are respectively 0.74 and 0.66 (compared to the reported up and down correlations of 0.713 and 0.696 respectively). We do therefore not consider it to be a problem that the benchmark proteins exist in PDB. We predict the radius of gyration using Equation 2.

Branch and bound algorithms are typically used to find the global minimum solutions. However, we use EBBA for protein decoy generation and we therefore want to obtain a large number of structures. The 10.000 global best structures in terms of energy are therefore found and not just the global minimum. This can be done by maintaining a queue of 10.000 structures during the search. This number is still very small compared to the exponential size of the conformational space. For

Table 1. Column 2 shows the number of segments m and column 3 shows the number of segment structures u. Column 4 shows the order of helix, sheet and coil segments. Column 5 shows the size of the conformational space given by Equation 1 and column 6 shows the number of hours spent by the algorithm. Column 7 and 8 show the percentage of the 10.000 structures that fall below the given threshold. Column 9 shows the lowest RMSD of the 10.000 structures. Column 10 shows the energy of P^* which is the lowest energy structure. For each protein, there is an *exact* and a *predicted* row. Exact refers to HSE-vectors, radius of gyration and secondary structure obtained from the native structure. In the *predicted* rows, all input values are predicted from the amino acid sequence and the results can therefore be considered as *de novo*.

Type	m u SS	N	T	< 6 Å	< 5 Å	lowest	$Q(P^*)$
		Protein A (1FC2), 43 residues					
Exact	5 8 CHCHC	1.7×10^8	0.1	18.1	7.0	**2.8**	4.34
Predicted	7 8 CHCHCHC	1.4×10^{12}	6.9	33.0	13.8	**4.5**	5.26
		Homeodomain (1ENH), 54 residues					
Exact	6 8 CHCHCH	1.5×10^{10}	0.6	21.6	13.2	**3.1**	4.36
Predicted	7 8 CHCHCHC	1.4×10^{12}	6.1	4.1	0.8	**4.1**	5.70
		Protein G (2GB1), 56 residues					
Exact	9 8 SCSCHCSCS	1.0×10^{16}	18.2	60.8	36.6	**3.4**	4.22
Predicted	10 8 SCSCHCSCSC	9.2×10^{17}	4.7	73.1	0.0	**5.3**	6.22
		Cro repressor (2CRO), 65 residues					
Exact	11 4 CHCHCHCHCHC	4.0×10^{16}	24.1	5.7	1.4	**4.3**	6.49
Predicted	10 3 HCHCHCHCHC	5.1×10^{13}	7.4	1.5	0.0	**5.3**	5.89
		Protein L7/L12 (1CTF), 68 residues					
Exact	8 8 SCHCHCHC	1.2×10^{14}	5.6	5.1	1.9	**4.6**	7.19
Predicted	11 3 SCHSHCHCHCS	1.7×10^{15}	19.2	0.1	0.0	**5.4**	5.84
		Calbindin (4ICB), 76 residues					
Exact	11 2 CHCSHCHCHCH	1.9×10^{13}	3.56	4.5	0.7	**4.4**	6.18
Predicted	8 7 CHCHCHCH	4.1×10^{13}	31.4	0.5	0.0	**5.1**	6.79

comparison and evaluation of the model and prediction quality, all experiments are also done using the exact secondary structure and exact HSE-vectors obtained from the native structure of the proteins. All experiments were initially run with $u = 8$ (the number of segment structures). Some did not finish in 48 hours, and they were run with the highest value of u that could be solved in less than 48 hours. All computations were performed on a 2.4 GHz P4 with 512 RAM.

4   Results and Discussion

Table 1 shows the complexity of the models for different proteins and the running time of EBBA. The table also shows the results of running EBBA on the 6 benchmark proteins.

The maximum number of segment structures (u) that could be solved in less than 48 hours depends much on the number of segments of the protein. For the smallest proteins (1FC2 and 1ENH) the algorithm terminated in less than 48 hours using $u = 8$. Even though 2GB1 has relatively many segments the algorithm also

Table 2. Comparison between FB5-HMM and EBBA. Column 2 and column 4 show the percentage of good decoys for FB5-HMM and EBBA respectively. Column 3 and column 5 show the lowest RMSD of a structure found by FB5-HMM and EBBA respectively. Both algorithms uses predicted secondary structure information and predicted radius of gyration.

Protein	FB5-HMM		EBBA	
	< 6 Å	Min. RMSD	< 6 Å	Min. RMSD
Protein A (1FC2)	17.1	2.6	33.0	4.5
Homeodomain (1ENH)	12.2	3.8	4.1	4.1
Protein G (2GB1)	0.001	5.9	73.1	5.3
Cro repressor (2CRO)	1.0	4.1	1.5	5.3
Protein L7/L12 (1CTF)	0.3	4.1	0.1	5.4
Calbindin (4ICB)	0.4	4.5	0.5	5.1

terminated in less than 48 hours using $u = 8$. This is mainly because bounding occured early in the branch and bound tree.

The most difficult protein in terms of bounding efficiency is 4ICB (using predicted measures), where it turns out that significant bounding first occurs in level 5 of the branch and bound tree. In all instances, the conformational space is huge, and it is clear that finding global minimum structures could not have been done in reasonable time without efficient bounding.

Table 1 shows that the set of 10.000 low energy structures for all 6 proteins contains good decoys (RMSD less than 6 Å). Also, for all proteins the lowest RMSD is smallest when using exact values compared to the predicted values. This is expected since the energy landscape should have a global minimum closer to the native structure when using exact values. However, it is surprising that for two of the proteins (1FC2 and 2GB1) the fraction of good decoys (< 6 Å RMSD) is better when using predicted values compared to exact values.

The results have been compared directly with FB5-HMM [6] in Table 2. FB5-HMM is the state-of-the-art method for conformational sampling. The method is based on a Hidden Markov Model and generates a large set of structures which usually contains many good decoys when enforcing compactness. The major difference between FB5-HMM and EBBA is that FB5-HMM does not use an energy function. FB5-HMM can also benefit from the secondary structure prediction and radius of gyration prediction. The results we have shown for FB5-HMM are therefore obtained using predicted secondary structure and using a greedy collapse scheme. The results for FB5-HMM are from [6] where 100.000 structures are generated. The results show that EBBA finds a better percentage of good decoys for most of the proteins (1FC2, 2GB1, 2CRO and 4ICB). The high amount of good decoys for protein G is very interesting since protein G is known to be one of the more difficult structures in this benchmark set [15,6]. For all proteins, except 2GB1, FB5-HMM finds at least one structure with lower RMSD than EBBA. This is not surprising since FB5-HMM here generates 10 times as many decoys than EBBA and therefore has a much higher probability of hitting a low RMSD structure. Another advantage of structures generated by EBBA, is that the geometry of the secondary structure segments is perfect because they are constructed using the correct secondary structure geometry. The running time of FB5-HMM for producing the set of 100.000 decoys is comparable to the running time of EBBA.

5   Conclusions

We have presented a branch and bound algorithm for finding the lowest energy structures in a large conformational search space. The results show that the set of low-energy structures is a very good decoy set. The energy function is based on HSE which is a simple predictable measure. This algorithm is the first *de novo* branch and bound algorithm for protein decoy generation using only one-dimensional predictable information. We have shown experimentally that good decoys always exist among the 10.000 lowest energy structures for the proteins used here. We have also shown that the algorithm is comparable in performance with the state-of-the-art conformational sampler FB5-HMM. The energy function is not accurate enough to pinpoint the lowest RMSD structure in this set. An important future research direction is therefore to examine this set of low energy structures with a more detailed energy function and to identify the native-like structures. The largest protein considered have 76 residues. There is a problem using the branch and bound algorithm on larger proteins since then only a small fraction of the conformational space can be searched in reasonable time. However, we believe that exploiting how super secondary structures [18,2] arrange in nature, might be a way to solve this problem. Better search heuristics for finding upper bounds on the energy can also be relevant since a good upper bound on the energy also improves the performance of the branch and bound algorithm. Using a more probabilistic approach might also improve the quality of the results. It might also be possible to train a Bayesian network to predict the probability of a given HSE-vector given the amino acid sequence. This would be a more detailed usage of the HSE-vector compared to the simple energy function used here.

Acknowledgements

We would like to thank Thomas Hamelryck at the Bioinformatics Centre, University of Copenhagen for valuable contributions and insights. Martin Paluszewski and Pawel Winter are partially supported by a grant from the Danish Research Council (51-00-0336).

References

1. Backofen, R., Will, S.: A constraint-based approach to fast and exact structure prediction in three-dimensional protein models. Constraints 11(1), 5–30 (2006)
2. Boutonnet, N.S., Kajava, A.V., Rooman, M.J.: Structural classification of alphabetabeta and betabetaalpha supersecondary structure units in proteins. Proteins 30(2), 193–212 (1998)

3. Chothia, C., Lesk, A.M.: The relation between the divergence of sequence and structure in proteins. The EMBO Journal 5, 823–826 (1986)
4. Fain, B., Levitt, M.: A novel method for sampling alpha-helical protein backbones. Journal of Molecular Biology 305, 191–201 (2001)
5. Hamelryck, T.: An amino acid has two sides: a new 2D measure provides a different view of solvent exposure. Proteins 59(1), 38–48 (2005)
6. Hamelryck, T., Kent, J.T., Krogh, A.: Sampling realistic protein conformations using local structural bias. PLoS Computational Biology 2(9), 1121–1133 (2006)
7. Kinjo, A.R., Nishikawa, K.: Recoverable one-dimensional encoding of three-dimensional protein structures. Bioinformatics 21(10), 2167–2170 (2005)
8. Kolodny, R., Levitt, M.: Protein decoy assembly using short fragments under geometric constraints. Biopolymers 68(3), 278–285 (2003)
9. Maranas, C.D., Floudas, C.A.: A deterministic global optimization approach for molecular structure determination. J. Chem. Phys. 100, 1247–1261 (1994)
10. McGuffin, L.J., Bryson, K., Jones, D.T.: The PSIPRED protein structure prediction server. Bioinformatics 16, 404–405 (2000)
11. Palu, A.D., Dovier, A., Fogolari, F.: Constraint logic programming approach to protein structure prediction. BMC Bioinformatics 5(186) (2004)
12. Paluszewski, M., Hamelryck, T., Winter, P.: Reconstructing protein structure from solvent exposure using tabu search. Algorithms for Molecular Biology 1 (2006)
13. Paluszewski, M., Winter, P.: EBBA: Efficient branch and bound algorithm for protein decoy generation, Department of Computer Science, Univ. of Copenhagen, vol. 08(08) (2008)
14. Pollastri, G., Baldi, P., Fariselli, P., Casadio, R.: Prediction of coordination number and relative solvent accessibility in proteins. Proteins 47(2), 142–153 (2002)
15. Simons, K.T., Kooperberg, C., Huang, E., Baker, D.: Assembly of protein tertiary structures from fragments with similar local sequences using simulated annealing and Bayesian scoring functions. J. Mol. Biol. 268(1), 209–225 (1997)
16. Skolnick, J., Kolinski, A., Ortiz, A.R.: MONSSTER: A method for folding globular proteins with a small number of distance restraints. J. Mol. Biol. 265, 217–241 (1997)
17. Standley, D.M., Eyrich, V.A., Felts, A.K., Friesner, R.A., McDermott, A.E.: A branch and bound algorithm for protein structure refinement from sparse nmr data sets. J. Mol. Biol. 285, 1961–1710 (1999)
18. Sun, Z., Jiang, B.: Patterns and conformations of commonly occurring supersecondary structures (basic motifs) in protein data bank. J. Protein Chem. 15(7), 675–690 (1996)
19. Vilhjalmsson, B., Hamelryck, T.: Predicting a New Type of Solvent Exposure. In: ECCB, Computational Biology Madrid 2005, P-C35, Poster (2005)
20. Wolsey, L.A.: Integer Programming. Wiley-Interscience, Chichester (1998)

Author Index

Printing: Mercedes-Druck, Berlin
Binding: Stein+Lehmann, Berlin

Lecture Notes in Bioinformatics

Vol. 3745: J.L. Oliveira, V. Maojo, F. Martín-Sánchez, A.S. Pereira (Eds.), Biological and Medical Data Analysis. XII, 422 pages. 2005.

Vol. 3737: C. Priami, E. Merelli, P. Gonzalez, A. Omicini (Eds.), Transactions on Computational Systems Biology III. VII, 169 pages. 2005.

Vol. 3695: M. R. Berthold, R.C. Glen, K. Diederichs, O. Kohlbacher, I. Fischer (Eds.), Computational Life Sciences. XI, 277 pages. 2005.

Vol. 3692: R. Casadio, G. Myers (Eds.), Algorithms in Bioinformatics. X, 436 pages. 2005.

Vol. 3680: C. Priami, A. Zelikovsky (Eds.), Transactions on Computational Systems Biology II. IX, 153 pages. 2005.

Vol. 3678: A. McLysaght, D.H. Huson (Eds.), Comparative Genomics. VIII, 167 pages. 2005.

Vol. 3615: B. Ludäscher, L. Raschid (Eds.), Data Integration in the Life Sciences. XII, 344 pages. 2005.

Vol. 3594: J.C. Setubal, S. Verjovski-Almeida (Eds.), Advances in Bioinformatics and Computational Biology. XIV, 258 pages. 2005.

Vol. 3500: S. Miyano, J. Mesirov, S. Kasif, S. Istrail, P.A. Pevzner, M. Waterman (Eds.), Research in Computational Molecular Biology. XVII, 632 pages. 2005.

Vol. 3388: J. Lagergren (Ed.), Comparative Genomics. VII, 133 pages. 2005.

Vol. 3380: C. Priami (Ed.), Transactions on Computational Systems Biology I. IX, 111 pages. 2005.

Vol. 3370: A. Konagaya, K. Satou (Eds.), Grid Computing in Life Science. X, 188 pages. 2005.

Vol. 3318: E. Eskin, C. Workman (Eds.), Regulatory Genomics. VII, 115 pages. 2005.

Vol. 3240: I. Jonassen, J. Kim (Eds.), Algorithms in Bioinformatics. IX, 476 pages. 2004.

Vol. 3082: V. Danos, V. Schachter (Eds.), Computational Methods in Systems Biology. IX, 280 pages. 2005.

Vol. 2994: E. Rahm (Ed.), Data Integration in the Life Sciences. X, 221 pages. 2004.

Vol. 2983: S. Istrail, M.S. Waterman, A. Clark (Eds.), Computational Methods for SNPs and Haplotype Inference. IX, 153 pages. 2004.

Vol. 2812: G. Benson, R.D.M. Page (Eds.), Algorithms in Bioinformatics. X, 528 pages. 2003.

Vol. 2666: C. Guerra, S. Istrail (Eds.), Mathematical Methods for Protein Structure Analysis and Design. XI, 157 pages. 2003.